The Handbook of Environmental Chemistry

Founded by Otto Hutzinger

Editors-in-Chief: Damià Barceló • Andrey G. Kostianoy

Volume 35

More information about this series at
http://www.springer.com/series/698

The Handbook of Environmental Chemistry Also Available Electronically

The Handbook of Environmental Chemistry is included in Springer's eBook package *Earth and Environmental Science*. If a library does not opt for the whole package, the book series may be bought on a subscription basis.

For all customers who have a standing order to the print version of *The Handbook of Environmental Chemistry*, we offer free access to the electronic volumes of the Series published in the current year via SpringerLink. If you do not have access, you can still view the table of contents of each volume and the abstract of each article on SpringerLink (www.springerlink.com/content/110354/).

You will find information about the

- Editorial Board
- Aims and Scope
- Instructions for Authors
- Sample Contribution

at springer.com (www.springer.com/series/698).

All figures submitted in color are published in full color in the electronic version on SpringerLink.

Aims and Scope

Since 1980, *The Handbook of Environmental Chemistry* has provided sound and solid knowledge about environmental topics from a chemical perspective. Presenting a wide spectrum of viewpoints and approaches, the series now covers topics such as local and global changes of natural environment and climate; anthropogenic impact on the environment; water, air and soil pollution; remediation and waste characterization; environmental contaminants; biogeochemistry; geoecology; chemical reactions and processes; chemical and biological transformations as well as physical transport of chemicals in the environment; or environmental modeling. A particular focus of the series lies on methodological advances in environmental analytical chemistry.

Series Preface

With remarkable vision, Prof. Otto Hutzinger initiated *The Handbook of Environmental Chemistry* in 1980 and became the founding Editor-in-Chief. At that time, environmental chemistry was an emerging field, aiming at a complete description of the Earth's environment, encompassing the physical, chemical, biological, and geological transformations of chemical substances occurring on a local as well as a global scale. Environmental chemistry was intended to provide an account of the impact of man's activities on the natural environment by describing observed changes.

While a considerable amount of knowledge has been accumulated over the last three decades, as reflected in the more than 70 volumes of *The Handbook of Environmental Chemistry,* there are still many scientific and policy challenges ahead due to the complexity and interdisciplinary nature of the field. The series will therefore continue to provide compilations of current knowledge. Contributions are written by leading experts with practical experience in their fields. *The Handbook of Environmental Chemistry* grows with the increases in our scientific understanding, and provides a valuable source not only for scientists but also for environmental managers and decision-makers. Today, the series covers a broad range of environmental topics from a chemical perspective, including methodological advances in environmental analytical chemistry.

In recent years, there has been a growing tendency to include subject matter of societal relevance in the broad view of environmental chemistry. Topics include life cycle analysis, environmental management, sustainable development, and socio-economic, legal and even political problems, among others. While these topics are of great importance for the development and acceptance of *The Handbook of Environmental Chemistry,* the publisher and Editors-in-Chief have decided to keep the handbook essentially a source of information on "hard sciences" with a particular emphasis on chemistry, but also covering biology, geology, hydrology and engineering as applied to environmental sciences.

The volumes of the series are written at an advanced level, addressing the needs of both researchers and graduate students, as well as of people outside the field of "pure" chemistry, including those in industry, business, government, research establishments, and public interest groups. It would be very satisfying to see these volumes used as a basis for graduate courses in environmental chemistry. With its high standards of scientific quality and clarity, *The Handbook of Environmental Chemistry* provides a solid basis from which scientists can share their knowledge on the different aspects of environmental problems, presenting a wide spectrum of viewpoints and approaches.

The Handbook of Environmental Chemistry is available both in print and online via www.springerlink.com/content/110354/. Articles are published online as soon as they have been approved for publication. Authors, Volume Editors and Editors-in-Chief are rewarded by the broad acceptance of *The Handbook of Environmental Chemistry* by the scientific community, from whom suggestions for new topics to the Editors-in-Chief are always very welcome.

Damià Barceló
Andrey G. Kostianoy
Editors-in-Chief

Volume Preface

Since the publication of part 2 of Environmental Photochemistry in 2005, this topic has continued to be an area of extensive research activity. Consequently it was felt that there was an opportunity to produce a third volume of this series, which would provide a further platform to present some of the latest aspects of research in environmental photochemistry. As illustrated in this volume, photochemical processes continue to play a significant role in a variety of environmental applications. These range from energy conversion and storage through to environmental remediation and protection. The previous volumes considered a number of these topics already ranging from fundamental photochemical processes in the environment through to applications of photochemical technology for environmental protection. This volume follows in the same philosophy of parts 1 and 2 of Environmental Photochemistry considering both fundamental science and applications of photochemical technology.

The initial chapters consider some fundamental aspects of photochemical/catalytic processes and materials. The chapter by Minero presents a detailed consideration of surface modified photocatalytic materials. The modification of metal oxide photocatalysts is further developed in the chapter by Wang et al. on doping of layered transition metal oxides. Ismail and Bahnemann discuss developments, mechanistic studies and applications of mesoporous semiconductor photocatalytic materials for environmental remediation. The following chapters consider fundamental aspects of the overall photocatalytic process. A comprehensive consideration of the various kinetic processes involved in photocatalytic reactions is developed in the chapter by Rabani and Goldstein. The chapter by Howe demonstrates the application of infrared and electro paramagnetic resonance spectroscopy for probing reaction pathways and intermediates produced in photocatalytic reactions.

Engineering applications of photochemical processes are also presented. The chapter by Alfano et al. presents modelling photo-Fenton reactors for water treatment and is demonstrated for both lab and pilot scale reactor units. The use of LED light sources as alternative light sources to conventional UV lamps is described in the chapter by Tokode et al.

Applications of photochemistry for energy conversion, photofenton water treatment and chemical synthesis are described in the chapters by Skillen et al., Oliveros et al. and Kisch, respectively. The chapter on the use of photocatalysis for water splitting considers both one- and two-step photocatalyst systems together with an over view of the mechanistic reactions involved in the water splitting process. The development of photoreactors for water splitting is also discussed. The chapter by Oliveros et al. on photo-Fenton processes for water treatment covers the fundamental reactions involved in the Fenton reaction, through to the current state of process development and application. The uses of semiconductor photocatalysis for chemical synthesis, specifically for novel atom-economic organic reactions, are detailed in the chapter by Kisch.

In conclusion, environmental photochemistry remains a very active field of research from which a range of practical applications with vast commercial potential is emerging. Part 3 of Environmental Photochemistry further contributes to the knowledge of photochemical and photocatalytic processes for environmental applications. It will be fascinating to see how this topic further develops over the next decade.

Belfast, UK — Peter K.J. Robertson
Hannover, Germany — Detlef W. Bahnemann

Contents

Modelling of Photo-Fenton Solar Reactors for Environmental Applications

Orlando M. Alfano, Enrique D. Albizzati, and Leandro O. Conte

Abstract A proposal for modelling photo-Fenton reactors for water pollution remediation is presented. Reactor models, based on chemical reaction engineering principles and radiative energy transport fundamentals in homogeneous systems, are derived at both laboratory and solar pilot-plant scales. The proposed methodology is illustrated by presenting an example on the modelling and scaling up of a solar reactor for degradation of a model pollutant in aqueous solution: the herbicide 2,4-dichlorophenoxyacetic acid. Firstly, a kinetic model derived from a reaction sequence is proposed and its kinetic parameters estimated, using an isothermal, well-stirred tank laboratory photoreactor. Afterwards, the kinetic model is employed to predict the reacting species concentrations during the photo-Fenton degradation in a pilot-plant, nonisothermal solar reactor designed to capture the UV/Visible/IR solar radiation. This approach has proved to be appropriate to simulate the behaviour of the photo-Fenton reactor under different experimental conditions.

Keywords Photo-Fenton, Photon absorption, Pollutant degradation, Reactor modelling, Solar radiation

Contents

O.M. Alfano (✉) and L.O. Conte
Instituto de Desarrollo Tecnológico para la Industria Química (INTEC), Universidad Nacional del Litoral and CONICET, Paraje El Pozo. Colectora Ruta Nacional No 168, 3000, Santa Fe, Argentina
e-mail: alfano@intec.unl.edu.ar

E.D. Albizzati
Facultad de Ingeniería Química, Universidad Nacional del Litoral, Santiago del Estero 2654, 3000, Santa Fe, Argentina

D.W. Bahnemann and P.K.J. Robertson (eds.), *Environmental Photochemistry Part III*, Hdb Env Chem (2015) 35: 1–22, DOI 10.1007/698_2013_246,

1 Introduction

Water pollution is one of today's major environmental problems. It requires more effective and less expensive methods for the treatment of ground, surface, and wastewaters containing toxic and biologically nondegradable organic compounds. Advanced oxidation processes (AOPs) have received considerable attention as a nonconventional method to solve these problems. They are well known for their capacity for mineralising an important amount of organic pollutants and distinguished by the same chemical characteristic: the generation of hydroxyl radicals ($^{\cdot}OH$), one of the most powerful oxidising species [1]. Different AOPs can be identified, e.g. UV/ozone, UV/hydrogen peroxide, UV/titanium dioxide, and UV/Fenton's reagent, among others [2, 3].

The Fenton reaction is a complex chemical system that involves hydrogen peroxide and dissolved ferrous salts to generate the highly reactive hydroxyl radicals. It is also known that the degradation rate of the Fenton system is significantly enhanced when the mixture is irradiated with UV or UV/visible radiation [4, 5]. This photo-assisted Fenton process, or photo-Fenton process, is probably one of the most widely applied AOPs for the treatment of industrial wastewater in the homogeneous phase at ambient temperature [6]. Different operating conditions can affect the efficiency of this reaction: source and concentration of iron [5, 7–10], oxidant concentration [6], pH [5, 6], and reaction temperature [11, 12].

It has been demonstrated that relatively high temperatures increase the Fenton degradation rates [13–15]. Besides, to avoid the precipitation of iron compounds, the combined effects of temperature and iron concentration have also been investigated [9]. On the other hand, it has been reported that the generation of intermediates such as hydroquinone-like compounds can accelerate the regeneration of ferrous species, which is the slow step in the Fenton reaction mechanism [16–19].

The possible new application of renewable energy sources, namely, solar radiation, has transformed the photo-Fenton process into a non-expensive, competitive method for the chemical destruction of organic pollutants in contaminated wastewaters. Several attempts have been made to study the degradation of various organic compounds using the sun as an economic and ecological radiation source, e.g. persistent pharmaceutical compounds [20, 21], commercial pesticides [22–25], textile effluents [26], non-biodegradable dyes [27, 28], cellulose bleaching effluents [29], and real effluents from municipal treatment plants [30].

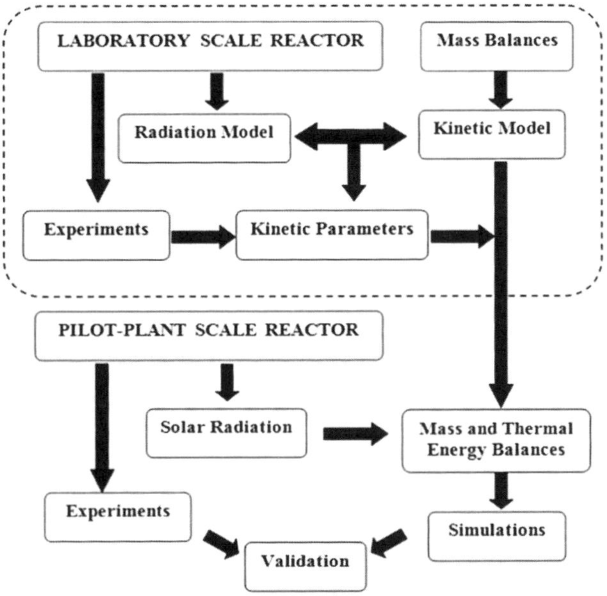

Fig. 1 Flow chart of the scaling up methodology

Both concentrating (several-suns) and non-concentrating (one-sun) collectors have been proposed in the design and construction of solar reactors for an efficient utilisation of the solar radiation at the Earth's surface [6, 31–33]. It has been claimed that one-sun reactors are more efficient than several-suns reactors since non-concentrating reactors effectively employ both direct beam and diffuse solar radiation fluxes that arrive at ground level [34–36]. In the ultraviolet wavelength range, according to the Solar Zenith angle, it is well known that the diffuse radiation can be equal to or greater than the direct radiation even for cloudless atmospheres [37].

This chapter presents a condensed description of a proposal for modelling photo-Fenton solar reactors to destroy a contaminant at both laboratory and pilot-plant scales. The design procedure is illustrated by presenting in detail the modelling and scaling up of a solar reactor for degradation of a model pollutant compound: herbicide 2,4-dichlorophenoxyacetic acid (2,4-D) in aqueous solution. This model, used in a predictive mode, is derived from chemical reaction engineering principles and radiative energy transport fundamentals. The proposed methodology is summarised in Fig. 1.

The first step involves the derivation of an intrinsic kinetic expression to represent the photo-Fenton degradation of the pollutant in a laboratory scale reactor. This kinetic expression must be independent of the actual values of the radiation source, reactor configuration, and experimental operating conditions. To achieve this goal, the proposed kinetic model should be based on a detailed reaction scheme and mainly incorporate the effects of (1) the ultraviolet radiation flux that arrives on the reactor window, (2) the pollutant, hydrogen peroxide, and iron concentrations, and (3) the reaction temperature.

In order to provide the theoretical evolution of the species involved in the laboratory reactor, the mass balances must include the pollutant and intermediate products of the reacting system. Besides, a precise evaluation of the radiation field inside this photoreactor is required to describe the effect of radiation absorption on the species degradation rates. To provide the values of the kinetic parameters of the Fenton and photo-Fenton reactions, the differences between model predictions and experimental data of the species concentrations are minimised applying a nonlinear optimisation algorithm; the Arrhenius parameters are also estimated with this numerical procedure.

The second step involves the simulation and experimental validation of the pilot-plant scale solar reactor for the photo-Fenton degradation of the pollutant in water. In this second part, mass and thermal energy balances are required and solved to predict the species concentrations and reaction temperatures in the reactor. The kinetic model derived in the first step is directly incorporated in the mass and thermal energy balances. At this point, values of direct and diffuse solar radiation fluxes arriving on the reactor window must be incorporated in the reactor model; they represent the boundary conditions for the resolution of the radiation transfer equation inside the reactor. Rigorous or simple computational codes can be employed with this purpose. Finally, in order to validate the proposed methodology, data obtained from experimental runs in the pilot-scale reactor are compared with the corresponding simulated results.

2 Mass Balances

The modelling of laboratory or pilot-plant scale reactors for photo-Fenton processes requires the solution of the multicomponent mass conservation equations. It is also worth noting that irradiated and nonirradiated zones inside the reactor should be considered for these reacting systems. Firstly, the local mass balance for a general component i in a well-mixed, batch photoreactor is written. Afterwards, one can integrate the mass balance over the total liquid volume of the system (V) and apply the divergence theorem to the molar flux term. The following expression is obtained:

$$\int_V \frac{\partial C_i(\underline{x},t)}{\partial t} dV + \int_A \underline{N}_i(\underline{x},t) \cdot \underline{n}\ dA = \int_V R_i(\underline{x},t)\ dV \tag{1}$$

Taking into account that a batch reactor does not have inlet or outlet streams (closed system), the second term on the left-hand side of Eq. (1) is null [38]. The integrals on the first term of the left-hand side and on the right-hand side of Eq. (1) can be divided into two terms: the irradiated liquid volume (V_{irr}) and the nonirradiated liquid volume (V_{nonirr}). Hence,

$$\frac{d}{dt}\int_{V_{\text{irr}}} C_i(\underline{x},t)\, dV + \frac{d}{dt}\int_{V_{\text{nonirr}}} C_i(\underline{x},t)\, dV = \int_{V_{\text{irr}}} R_i(\underline{x},t)\, dV + \int_{V_{\text{nonirr}}} R_i^T(\underline{x},t)\, dV \quad (2)$$

In the previous equation, R_i and R_i^T are the photochemical (photo-Fenton) and the thermal (Fenton) reaction rates for the i-component. The following averaged functions over the reactor volume can be defined:

$$\langle C_i(\underline{x},t)\rangle_{V_{\text{irr}}} = \frac{1}{V_{\text{irr}}}\int_{V_{\text{irr}}} C_i(\underline{x},t)\, dV \quad (3)$$

$$\langle C_i(\underline{x},t)\rangle_{V_{\text{nonirr}}} = \frac{1}{V_{\text{nonirr}}}\int_{V_{\text{nonirr}}} C_i(\underline{x},t)\, dV \quad (4)$$

$$\langle R_i(\underline{x},t)\rangle_{V_{\text{irr}}} = \frac{1}{V_{\text{irr}}}\int_{V_{\text{irr}}} R_i(\underline{x},t)\, dV \quad (5)$$

$$\langle R_i^T(\underline{x},t)\rangle_{V_{\text{nonirr}}} = \frac{1}{V_{\text{nonirr}}}\int_{V_{\text{nonirr}}} R_i^T(t)\, dV = R_i^T(t) \quad (6)$$

In the second and third members of Eq. (6), we have considered that the thermal reaction rate (R_i^T) is not a function of position.

Substituting Eqs. (3) to (6) into Eq. (2) and dividing by the total liquid volume of the system (V), one can write

$$\frac{d}{dt}\left[\frac{V_{\text{irr}}}{V}\langle C_i(\underline{x},t)\rangle_{V_{\text{irr}}} + \frac{V_{\text{nonirr}}}{V}\langle C_i(\underline{x},t)\rangle_{V_{\text{nonirr}}}\right] = \frac{V_{\text{irr}}}{V}\langle R_i(\underline{x},t)\rangle_{V_{\text{irr}}} + \frac{V_{\text{nonirr}}}{V} R_i^T(t) \quad (7)$$

In the second term of the left-hand side of Eq. (7), we can assume that (1) the molar concentration C_i is uniform and can be taken out of the averaged value and (2) the irradiated volume is much less than the total volume ($V_{\text{irr}} << V$). Thus,

$$\frac{d}{dt}\left[\frac{V_{\text{irr}}}{V}\langle C_i(\underline{x},t)\rangle_{V_{\text{irr}}} + \frac{V_{\text{nonirr}}}{V} C_i(t)|_{V_{\text{nonirr}}}\right] \cong \frac{d}{dt}\left[\frac{V_{\text{nonirr}}}{V} C_i(t)\right] \cong \frac{dC_i(t)}{dt} \quad (8)$$

Finally, substituting Eq. (8) into Eq. (7), the following expression for the mass balance and the initial condition for a general component i are obtained:

$$\frac{dC_i(t)}{dt} = \frac{V_{\text{irr}}}{V}\langle R_i(\underline{x},t)\rangle_{V_{\text{irr}}} + \frac{V_{\text{nonirr}}}{V} R_i^T(t) \quad (9)$$

$$C_i = C_i^0 \quad t = 0 \quad (10)$$

Fig. 2 View of the stirred tank laboratory photoreactor: *1* tank reactor, *2* cooling coils, *3* thermometer, *4* agitator, *5* liquid sample, and *6* emitting system

In Eq. (9), it should be noted that (1) the first term on the right-hand side corresponds to the *i*-component degradation by both the radiation activated and thermal reactions (photo-Fenton reaction) occurring inside the irradiated liquid volume and (2) the second term on the right-hand side represents the *i*-component degradation by the thermal reaction (Fenton reaction) taking place in the nonirradiated liquid volume.

Two different cases should be considered: (a) a reactor with a total irradiated volume and (b) a reactor with irradiated and nonirradiated volumes. In case (a), for example, a well-stirred tank reactor with a total irradiated volume (Fig. 2), a similar methodology may be applied for obtaining the mass balance. Then, for each reacting species *i*, the mass balance expression is

$$\frac{dC_i(t)}{dt} = \left\langle R_i\left(\underline{x}, t\right)\right\rangle_{V_{\text{irr}}} \tag{11}$$

In case (b), for example, a solar reactor placed inside a batch recycling system (Fig. 3), Eqs. (9) and (10) may be directly applied.

As explained in Sect. 6.1 below, it should be kept in mind that the reaction rate expression $R_i(\underline{x},t)$ included in Eqs. (9) and (11) is a function of (1) the pollutant, hydrogen peroxide and ferric (or ferrous) ion concentrations, (2) the primary quantum yield of the reaction, and (3) the Local Volumetric Rate of Photon Absorption (LVRPA) [39].

Fig. 3 View of a pilot-plant solar reactor inside a batch recycling system: *1* solar reactor, *2* valve, *3* pump, *4* tank, and *5* radiometer

3 Volumetric Rate of Photon Absorption

The evaluation of the radiation field is essential in the modelling and design of photoreactors. The radiation field expressed in terms of the LVRPA can be introduced into kinetic expressions, thus obtaining reaction rate equations independent of the experimental irradiation conditions.

Starting from the radiative transfer equation (RTE) [40–42], one can write

$$\frac{dI_\lambda(\underline{x},\underline{\Omega},t)}{ds} = -\underbrace{\kappa_\lambda\ I_\lambda(\underline{x},\underline{\Omega},t)}_{\text{ABSORPTION}} - \underbrace{\sigma_\lambda\ I_\lambda(\underline{x},\underline{\Omega},t)}_{\text{OUT-SCATTERING}} + \underbrace{\frac{\sigma_\lambda}{4\pi}\int_{\Omega'=4\pi} p(\underline{\Omega}' \to \underline{\Omega})\ I_\lambda(\underline{x},\underline{\Omega}',t)\ d\Omega'}_{\text{IN-SCATTERING}} \quad (12)$$

where I_λ is the spectral radiation intensity, s the linear coordinate along the direction $\underline{\Omega}$, Ω the solid angle, κ_λ the spectral volumetric absorption coefficient, σ_λ the spectral volumetric scattering coefficient, and p the phase function.

It is worth noting that two important assumptions have been considered in Eq. (12): (1) steady state conditions of the radiation field and (2) no radiation emission at the relatively low working temperatures used for the photo-Fenton process. Two sink terms (absorption and out-scattering) and one source term (in-scattering) are identified on the right-hand side of the previous equation.

In this chapter, we will consider the degradation of a specific organic pollutant in aqueous solution using the Fenton and photo-Fenton systems. From Eq. (12) for a homogeneous medium where there is only radiation absorption ($\sigma_\lambda = 0$), the resulting RTE is given by

$$\frac{dI_{\lambda}(\underline{x},\underline{\Omega},t)}{ds} = - \; \kappa_{\lambda} \; I_{\lambda}(\underline{x},\underline{\Omega},t) \tag{13}$$

Once the radiation intensity for any given wavelength is calculated solving Eq. (13), one can compute the LVRPA [or $e_{\lambda}^{a}(\underline{x},t)$]. To do this, the following equation has to be solved:

$$e_{\lambda}^{a}(\underline{x},t) = \kappa_{\lambda}\int_{\Omega} I_{\lambda}(\underline{x},\underline{\Omega},t)\, d\Omega \tag{14}$$

Afterwards, for polychromatic radiation, an integration over the useful wavelength range must be performed, accounting for the overlapping wavelength intervals of incident radiation, reactor wall transmission, and species absorption coefficient:

$$e^{a}(\underline{x},t) = \int_{\lambda}\kappa_{\lambda}\int_{\Omega} I_{\lambda}(\underline{x},\underline{\Omega},t)\, d\Omega\, d\lambda \tag{15}$$

4 Solar Radiation Incident at the Reactor Window

Knowledge of solar radiation at ground level is essential for the design and performance prediction of solar photo-Fenton reactors. The behaviour of the solar reactor can have a strong dependence with the radiation flux that arrives on the reactor window. It is known that this solar radiation flux varies with geographic location, day of the year, hour of day, and atmospheric conditions; thus, these radiation flux fluctuations with position and time must be considered in the evaluation of the solar photo-Fenton degradation of pollutants.

In order to solve Eqs. (13) to (15), we need the values of the direct and diffuse solar radiation flux on horizontal and tilted planes at the Earth's surface. These values are the boundary conditions for the resolution of the RTE. In general, two types of approaches can be used with this purpose: (1) sophisticated rigorous codes (also called RTE codes), for example, MODTRAN [43] or TUV [44], and (2) simple, parameterised computational codes, such as the Simple Solar Spectral Model (SPCTRAL2, [45]) or the Simple Model for the Atmospheric Radiative Transfer of Sunshine (SMARTS2, [46]).

In this chapter, the parameterised computational codes will be used. Basically, these models consider that the spectral global solar radiation received at ground level on a horizontal surface ($q_{G,\lambda}$) is given by

$$q_{G,\lambda} = q_{B,\lambda}\cos\theta_{Z} + q_{D,\lambda} \tag{16}$$

Here, $q_{B,\lambda}$ is the spectral direct beam radiation at ground level on a surface normal to the sun direction, θ_{Z} the cosine of the Zenith angle, and $q_{D,\lambda}$ the spectral diffuse radiation on a horizontal surface.

Under cloudless sky conditions, the spectral direct beam radiation on a surface normal to the sun direction is computed by the expression

$$q_{B,\lambda} = H_{o,\lambda} D T_{r,\lambda} T_{a,\lambda} T_{w,\lambda} T_{o,\lambda} T_{u,\lambda} \quad (17)$$

where $H_{o,\lambda}$ is the extraterrestrial solar radiation at the mean Earth–Sun distance, D the correction factor for the Earth–Sun distance, and $T_{i,\lambda}$ the transmittance functions of the atmosphere at the wavelength λ. The subindex i represents the following spectral extinction processes: molecular (Rayleigh) scattering ($i = r$), aerosol extinction ($i = a$), water vapour absorption ($i = w$), ozone absorption ($i = o$), and uniformly mixed gases ($i = u$). The SMARTS2 code has also included in Eq. (16) an additional transmittance function to take into account the nitrogen dioxide (NO_2) absorption ($T_{n,\lambda}$).

On the other hand, for cloudless atmospheres, the spectral diffuse radiation received at ground level on a horizontal surface can be calculated by the equation:

$$q_{D,\lambda} = q_{r,\lambda} + q_{a,\lambda} + q_{g,\lambda} \quad (18)$$

Note that three different components are considered in this equation: Rayleigh scattering ($q_{r,\lambda}$), aerosol scattering ($q_{a,\lambda}$), and multiple reflection between the ground and the air ($q_{g,\lambda}$).

In the quoted references on the parameterised computational codes, the authors have provided different specific expressions to compute each one of the functions and parameters to be introduced into Eqs. (17) and (18).

For the evaluation of spectral solar radiation on non-horizontal surfaces, additional calculations should be performed because the tilted surface receives [46]: (1) diffuse radiation from only a fraction of the sky vault and (2) reflected radiation from the foreground. Thus, the spectral solar radiation on an inclined plane is calculated by the expression:

$$q_{T,\lambda} = q_{B,\lambda} \cos\theta_i + R_d q_{D,\lambda} + \rho_{g,\lambda} R_r q_{G,\lambda} \quad (19)$$

where $\rho_{g,\lambda}$ is the local reflectance of the ground adjacent to the tilted surface and θ_i the angle of incidence of the sunrays on the inclined surface. The mathematical expressions to calculate the angle of incidence (θ_i), the local reflectance of the ground ($\rho_{g,\lambda}$), and the conversion factors (R_d, R_r) defined in Eq. (19) can be found in Gueymard [46].

5 Thermal Energy Balance

Solar photo-Fenton reactors are able to capture the UV/visible and near-infrared solar radiation, yielding important temperature increases during an experimental run. This fact can be beneficial from the reactor performance point of view, due to the positive effects on the thermal (Fenton) reaction rate. Consequently, a thermal energy balance must be written and solved to evaluate this temperature increase.

The thermal energy balance and the initial condition for a solar reactor placed inside the loop of a batch recycling system are given by the following equations:

$$C_{\mathrm{T}}\frac{dT}{dt} = \eta_{\mathrm{o}} A_{\mathrm{irr}}\, q_{\mathrm{T}}(t) - \mathrm{UA}\ \ [T - T_{\mathrm{a}}(t)] + Q_{\mathrm{P}} \tag{20}$$

$$T = T^0 \qquad t = 0 \tag{21}$$

where C_{T} is the effective heat capacity of the system (reactor plus storage tank), η_{o} the optical efficiency, A_{irr} the irradiated window area, q_{T} the total broadband solar radiation flux incident on the reactor wall, T the system temperature, T_{a} the ambient temperature, UA the heat loss coefficient of the total system, and Q_{P} the heat input from the pump.

On the right-hand side of Eq. (20), it should be noted that (1) the first term gives the solar radiation absorbed by the system, (2) the second term represents the thermal energy losses to the surroundings, and (3) the third term corresponds to the heat input from the circulation pump. Also notice that the heat loss coefficient of the total system can be split in the reactor [$(\mathrm{UA})_{\mathrm{R}}$] and tank [$(\mathrm{UA})_{\mathrm{Tk}}$] heat loss coefficients, by means of expression $\mathrm{UA} = (\mathrm{UA})_{\mathrm{R}} + (\mathrm{UA})_{\mathrm{Tk}}$.

6 Photo-Fenton Degradation of an Herbicide

The proposed methodology previously described in Sects. 1–5 is used here to study the photo-Fenton degradation of the active principle of herbicide 2,4-dichlorophenoxyacetic acid (2,4-D). First, the kinetic model is derived using an isothermal, well-stirred tank laboratory reactor irradiated from the bottom. Afterwards, the kinetic model is applied to study the behaviour of a pilot-plant solar reactor under different operating conditions.

6.1 Laboratory Scale Photoreactor

The kinetic model for the Fenton and photo-Fenton degradation of 2,4-D in aqueous solution is derived from a reaction sequence obtained from the specific literature [5, 47–49]. The complete reaction scheme comprises 16 reaction steps, involving initiation, propagation, and termination elementary reactions (Table 1). The 2,4-dichlorophenol (DCP) has been reported as the main intermediate compound in the degradation of 2,4-D.

Table 1 Reaction scheme for 2,4-D degradation

Number	Reaction step	Constant
0	$Fe(III) + H_2O \rightarrow Fe(II) + OH^{\bullet} + H^+$	$\Phi_{Fe(II),\lambda}$
1	$Fe(III) + H_2O_2 \rightarrow Fe(II) + H^+ + HO_2$	k_1
2	$Fe(II) + H_2O_2 \rightarrow Fe(III) + OH^- + HO^{\bullet}$	k_2
3	$H_2O_2 + HO^{\bullet} \rightarrow HO_2^{\cdot} + H_2O$	k_3
4	$Fe(II) + HO^{\bullet} \rightarrow Fe(III) + OH^-$	k_4
5	$H_2O_2 + HO_2^{\cdot} \rightarrow HO^{\bullet} + H_2O + O_2$	k_5
6	$2HO^{\bullet} \rightarrow H_2O_2$	k_6
7	$2HO_2^{\cdot} \rightarrow H_2O_2 + O_2$	k_7
8	$HO_2^{\cdot} + HO^{\bullet} \rightarrow H_2O + O_2$	k_8
9	$Fe(III) + HO_2^{\cdot} \rightarrow Fe(II) + H^+ + O_2$	k_9
10	$Fe(II) + HO_2^{\cdot} + H^+ \rightarrow Fe(III) + H_2O_2$	k_{10}
11	$2,4\text{-D} + HO^{\bullet} \rightarrow DCP$	k_{11}
12	$DCP + HO^{\bullet} \rightarrow QH_2$	k_{12}
13	$QH_2 + HO^{\bullet} \rightarrow$ products	k_{13}
14	$QH_2 + Fe(III) \rightarrow Fe(II) + QH^{\bullet} + H^+$	k_{14}
15	$QH^{\bullet} + Fe(III) \rightarrow Q + Fe(II) + H^+$	k_{15}

In order to obtain the degradation rate expressions for 2,4-D, DPC, and H_2O_2, Conte et al. [50] have recently adopted the following assumptions: (1) the steady state approximation may be applied for highly reactive radicals ($OH^{\cdot}$ and $QH^{\cdot}$), (2) radical–radical termination reactions are neglected when compared with the propagation reactions, (3) the oxygen concentration is in excess, (4) the reaction of hydroxyl radical with ferrous ion is neglected (very low ferric ion concentrations), (5) reaction step 5 is neglected (slow compared to others of $HO_2^{\cdot}$), (6) the reaction of hydroxyl radical with quinone intermediates (QH_2) is neglected as compared with ferric ion reactions, and (7) the ferrous ion and quinone intermediate concentrations are very low.

Based on these assumptions, the following reaction rate expressions can be derived:

$$R_{2,4-\mathrm{D}}(x,t) = R_{2,4-\mathrm{D}}^{\mathrm{T}}(t) - \frac{2}{\delta}\left(\overline{\Phi}_{\mathrm{Fe(II)}}\sum_{\lambda} e_{\lambda}^{\mathrm{a}}(x)\right) \tag{22}$$

$$R_{\mathrm{DCP}}(x,t) = R_{\mathrm{DCP}}^{\mathrm{T}}(t) + \frac{2}{\delta}\left(\overline{\Phi}_{\mathrm{Fe(II)}}\sum_{\lambda} e_{\lambda}^{\mathrm{a}}(x)\right)\left(1 - K_{\mathrm{DCP}}\frac{C_{\mathrm{DCP}}}{C_{2,4-\mathrm{D}}}\right) \tag{23}$$

$$R_{\mathrm{P}}(x,t) = R_{\mathrm{P}}^{\mathrm{T}}(t) - \frac{1}{\delta}\left(\overline{\Phi}_{\mathrm{Fe(II)}}\sum_{\lambda} e_{\lambda}^{\mathrm{a}}(x)\right)\left(1 + 3K_{\mathrm{DCP}}\frac{C_{\mathrm{DCP}}}{C_{2,4-\mathrm{D}}} + 3K_{\mathrm{III}}\frac{C_{\mathrm{P}}}{C_{2,4-\mathrm{D}}}\right) \tag{24}$$

To complete the kinetic model, the thermal reaction rate expressions are given by

$$R^{\mathrm{T}}_{2,4-\mathrm{D}}(t) = -\frac{1}{\delta}(K_1 C_{\mathrm{Fe}^{3+}} C_{\mathrm{P}}) \tag{25}$$

$$R^{\mathrm{T}}_{\mathrm{DCP}}(t) = \frac{1}{\delta}(K_1 C_{\mathrm{Fe}^{3+}} C_{\mathrm{P}})\left(1 - K_{\mathrm{DCP}}\frac{C_{\mathrm{DCP}}}{C_{2,4-\mathrm{D}}}\right) \tag{26}$$

$$R^{\mathrm{T}}_{\mathrm{P}}(t) = -(K_1 C_{\mathrm{Fe}^{3+}} C_{\mathrm{P}})\left(\frac{2 + 3K_{\mathrm{III}} C_{\mathrm{P}}/C_{2,4-\mathrm{D}}}{\delta}\right) \tag{27}$$

The following kinetic parameters have been defined in the previous equations:

$$\delta = 1 + K_{\mathrm{III}}\frac{C_{\mathrm{P}}}{C_{2,4-\mathrm{D}}} - K_{\mathrm{DCP}}\frac{C_{\mathrm{DCP}}}{C_{2,4-\mathrm{D}}} \quad K_{\mathrm{DCP}} = \frac{k_{12}}{k_{11}} \quad K_{\mathrm{III}} = \frac{k_3}{k_{11}} \quad K_1 = k_1 \tag{28}$$

To complete the theoretical description of the proposed kinetic model, it is necessary to introduce the expression of the spectral LVRPA on the right-hand side of Eqs. (22) to (24). As mentioned above, for the kinetic study of the photo-Fenton 2,4-D degradation, the employed apparatus was an isothermal, well-stirred tank laboratory reactor irradiated from the bottom with a low-pressure mercury-vapour fluorescent lamp (Philips TL 40W/09 N) (Fig. 2). Table 2 presents a summary of the main reactor, lamp, and reflector characteristics and dimensions.

For a similar laboratory scale photoreactor, a simple 1-D radiation field model was proposed to evaluate the spectral LVRPA as a function of the x-coordinate [51]:

$$e^{\mathrm{a}}_{\lambda}(x) = \kappa_{\lambda} q_{\mathrm{W}} f_{\lambda} \exp(-\kappa_{\mathrm{T},\lambda} x) \tag{29}$$

In Eq. (29) κ_{λ} is the volumetric absorption coefficient of the reacting species, q_{w} the net spectral flux at the reactor wall, f_{λ} the normalised spectral distribution of the lamp output power, and $\kappa_{\mathrm{T},\lambda}$ the volumetric absorption coefficient of the reacting medium.

Finally, the model equations for the laboratory scale photoreactor are given by the mass balances [Eqs. (11) and (10)]; the 2,4-D, DPC, and H_2O_2 reaction rates [Eqs. (22) to (28)]; and the LVRPA [Eq. (29)].

An experimental design method can be subsequently applied using the laboratory scale reactor previously described (Fig. 2 and Table 2). The operating variables of the experimental program are modified between minimum and maximum values of this reacting system. For example, the hydrogen peroxide to pollutant initial concentration ratios may be changed from 7 (corresponding to the stoichiometric ratio) to 50; the ferric iron concentrations, from 1 to 5 ppm; the reaction temperatures, from 20 to

Table 2 Reactor, lamp, and reflector characteristics and dimensions

	Value	Units
Reactor		
Total liquid volume	3.0	L
Diameter	14.2	cm
Length	18.9	cm
Lamp Philips TL 40 W/09 N		
Nominal power	40	W
Output power: 315 nm $\leq \lambda \leq$ 420 nm	6.6	W
Diameter	3.8	cm
Nominal arc length	61	cm
Reflector		
Parabola characteristic constant	2.1	cm
Distance vertex of parabolic reflector to reactor plate	8.4	cm
Length	15.8	cm

Adapted with permission from Farias et al. [12]. Copyright 2007 American Chemical Society

50°C; and the irradiation levels, from a dark (Fenton) to a totally irradiated (photo-Fenton) reactor. Note that in order to avoid the precipitation of ferric hydroxide and, consequently, the subsequent reduction of the photo-Fenton efficiency, low iron concentrations and reaction temperatures should be employed during the experimental work.

To provide the values of the kinetic parameters of the Fenton and photo-Fenton reactions, the differences between model predictions and experimental data should be minimised, employing an optimisation procedure. The Arrhenius parameters (the frequency factor A_1 and the activation energy E_1) and the remaining kinetic parameters (K_{III}, K_{DCP} and $\overline{\Phi}_{Fe(II)}$) for this reacting system are estimated applying a nonlinear, Newton–Gauss–Marquardt optimisation algorithm. The following results are obtained: $A_1 = 0.707$ L/(mol s), $E_1 = 83.66$ kJ/mol, $K_{III} = 0.036$, $K_{DCP} = 7.951$, $\overline{\Phi}_{Fe(II)} = 0.237$ mol/einstein. More details on the estimation of the kinetic parameters can be found in Conte et al. [50].

Figure 4 shows a 3-D plot with predicted and experimental 2,4-D conversions after 30 min of reaction time, as a function of the ferric iron concentration and the reaction temperature. The surfaces of lowest ($R = 7$) and highest ($R = 50$) ratios of the hydrogen peroxide to pollutant initial concentrations are also represented in this figure. When low and intermediate reaction temperatures are considered, it can be seen that the irradiated system (photo-Fenton reaction) always gives 2,4-D conversions higher than those reached with the nonirradiated system (Fenton reaction). Nonetheless, for relatively high temperatures ($T \cong 50$°C) and hydrogen peroxide to pollutant initial concentration ratios ($R \cong 50$), the Fenton and photo-Fenton reactions exhibit similar 2,4-D conversions after 30 min.

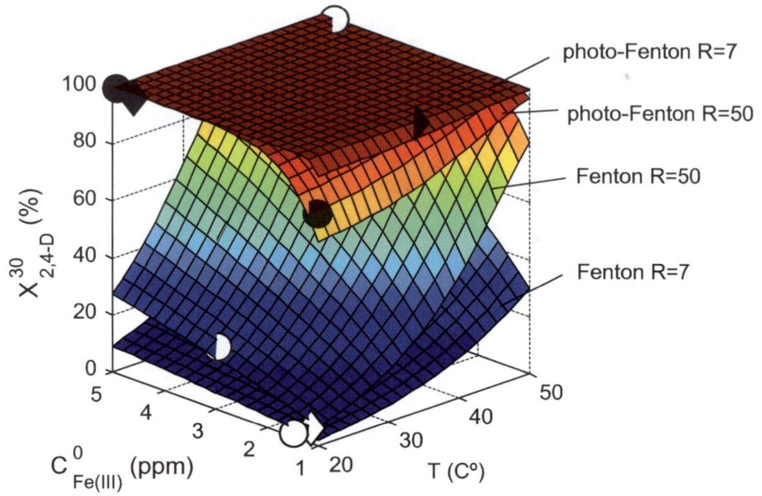

Fig. 4 Predicted (*surfaces*) and experimental (*symbols*) 2,4-D conversions after 30 min vs. reaction temperatures and ferric iron concentrations, for $R = 7$ and 50. Keys: Fenton (*open symbols*), photo-Fenton (*filled symbols*), $R = 50$ (*circles*), $R = 7$ (*diamonds*). Reprinted with permission from Conte et al. [50]. Copyright 2012 American Chemical Society

6.2 Pilot-Plant Scale Solar Reactor

The kinetic model developed in Sect. 6.1 can now be applied to simulate the reactant concentrations during the photo-Fenton degradation in a pilot-plant solar reactor designed and built to capture the UV/Visible and IR solar radiation [52, 53]. Figure 3 shows a schematic representation of the pilot-plant solar reactor placed inside the loop of a batch recycling system and Table 3 presents a summary of the main solar reactor dimensions and parameters.

The mass balances and initial conditions for this solar reactor are given by Eqs. (9) and (10) with i = 2,4-D, DCP, P, and the reaction rate expressions by Eqs. (22) to (28). The value of the LVRPA defined for monochromatic radiation can be obtained assuming that the reactor window is irradiated by direct beam ($e^{\mathrm{a}}_{B,\lambda}$) and diffuse ($e^{\mathrm{a}}_{D,\lambda}$) solar radiation. Thus, the spectral LVRPA for the total solar radiation is given by the following equations:

$$e^{\mathrm{a}}_{\lambda}(x,t) = e^{\mathrm{a}}_{\mathrm{B},\lambda}(x,t) + e^{\mathrm{a}}_{\mathrm{D},\lambda}(x,t) \tag{30}$$

$$e^{\mathrm{a}}_{\mathrm{B},\lambda}(x,t) = \kappa_{\lambda}(t)\, q_{\mathrm{B},\lambda}(t)\, \mathrm{Y}_{\mathrm{B},\lambda}(\mu_{\mathrm{i}})\exp[-\kappa_{\mathrm{T},\lambda}(t)\, x/\mu_{\mathrm{ref}}] \tag{31}$$

$$e^{a}_{D,\lambda}(x,t) = 2\kappa_{\lambda}(t)\, q_{D,\lambda}(t)\, \mathrm{Y}_{\mathrm{D},\lambda}(\mu_i)\, E\left[\kappa_{T,\lambda}(t), x\right] \tag{32}$$

Table 3 Pilot-plant solar reactor dimensions and parameters

	Value	Units
Reactor		
Total liquid volume (V)	35	dm^3
Irradiated volume (V_{irr})	6.1	dm^3
Reactor depth (L)	30	mm
Window area (A_c)	0.24	m^2
Plate thickness (e)	3.2	mm
Optical parameters		
Polycarbonate plate refractive index	1.49	
Transmittance of acrylic windows		
$\lambda = 300$ nm	0.60	
$\lambda = 350$ nm	0.87	
$\lambda = 360$ nm	0.89	
$\lambda = 373$ nm	0.90	
$\lambda = 400$ nm	0.90	
Water refractive index	1.33	
Air refractive index	1.00	
Solar radiation parameters		
UV/visible spectral range	[280:450]	nm
Total radiation spectral range	[305:2800]	nm
Tilt angle	30	deg
Azimuth angle (counted clockwise from north)	0	deg

where $q_{B,\lambda}$ and $\Upsilon_{B,\lambda}$ are the spectral direct beam radiation flux and transmittance, $q_{D,\lambda}$ and $\Upsilon_{D,\lambda}$ are the spectral diffuse radiation flux and transmittance, κ_λ is the spectral volumetric absorption coefficient of the absorbing species, and $\kappa_{T,\lambda}$ is the total absorption coefficient of the reacting medium. Besides, μ_{ref} is the cosine of refraction angle, μ_i is the cosine of incident angle, and $E(t,x)$ is the second-order exponential integral function [42]. Further details for the calculation of both the monochromatic direct beam ($\Upsilon_{B,\lambda}$) and diffuse ($\Upsilon_{D,\lambda}$) transmittances can be found elsewhere [50].

The thermal energy balance and the initial condition [Eqs. (20) and (21)] are employed here to predict the temperature evolution in the solar reactor. Equation (20) may also be written as

$$\frac{dT}{dt} = \Omega A_{irr} q_T(t) - \Gamma \ [T - T_a(t)] + K \quad (33)$$

where the following thermal energy parameters are defined:

$$\Omega = \frac{\eta_0}{C_T} \quad \Gamma = \frac{(UA)_R + (UA)_{Tk}}{C_T} \quad K = \frac{Q_P}{C_T} \quad (34)$$

In Eq. (33), note that q_T must be a function of time to take into account the variation of the Solar Zenith angle during the experimental run.

Values of the thermal energy parameters were previously estimated by Farias et al. [52] performing irradiated and nonirradiated experimental runs without chemical reaction. A nonlinear regression procedure was then applied to minimise the differences between measured and simulated temperature variations in the storage tank. The following thermal energy parameters were reported: $\Omega = 9.10 \times 10^{-5}$°C/J, $\Gamma = 5.39 \times 10^{-6}$ 1/s, and $K = 8.28 \times 10^{-4}$°C/s.

To solve the solar reactor model equations, four computational steps should be considered for each value of the sun position (Zenith angle). Firstly, the spectral direct and diffuse solar radiation fluxes on tilted planes at the Earth's surface are calculated using Eqs. (16) to (19). Then, direct beam and diffuse values of the spectral LVRPA are evaluated with Eqs. (30) to (32). The following step is the computation of the reaction rates for species 2,4-D, DCP, and H_2O_2 employing Eqs. (22) to (28). Finally, the reactant species concentrations and reaction temperatures are calculated solving the ordinary differential Eqs. (9), (10), (33), (34), and (21). It is noted that during the numerical solution of the model equations, the SMARTS2 code was called in every loop of the algorithm. More details on the numerical method can be found in Conte et al. [50].

Typical simulated results of the solar radiation absorption rate inside the pilot-plant reactor are shown in Fig. 5. A 3-D representation of the LVRPA for a constant value of the Zenith angle is illustrated as a function of the initial ferric ion concentration ($C^0_{Fe(III)}$) and the reactor spatial position (x). As expected, for a ferric ion concentration of 5.0 ppm, the radiation field presents a marked variation along the x-axis; for this case, more than 50% of the solar radiation entering the reactor is absorbed in the volume. On the other hand, when the initial iron concentration is low, the radiation profiles become much more uniform.

These simulation results of the LVRPA spatial distribution for a given initial ferric ion concentration provide valuable information to design the thickness of a solar reactor.

Finally, experimental results and theoretical predictions obtained with the complete model equations reported in this section are shown in Fig. 6a, b. In these figures, 2,4-D, DCP, and H_2O_2 concentrations as a function of time are presented. The experimental runs were performed with a similar initial 2,4-D concentration and different values of the initial ferric ion concentrations, hydrogen peroxide to pollutant initial molar ratios, and initial reaction temperatures.

By comparison of Fig. 6a, b, a more efficient 2,4-D degradation was observed in the first figure, as a result of higher ferric ion concentrations (2.8 ppm against 1.0 ppm) and hydrogen peroxide to pollutant initial molar ratios (31.5 against 24.6); for example, after a reaction time of 30 min, the following 2,4-D conversions are observed: 95% (Fig. 6a) and 60% (Fig. 6b). Total organic carbon (TOC) conversions after 210 min are also determined under the same operating conditions (results not shown here); as a result of the temperature increases, $\Delta T_{210\ \mathrm{min}} = 15.3$°C in Fig. 6a

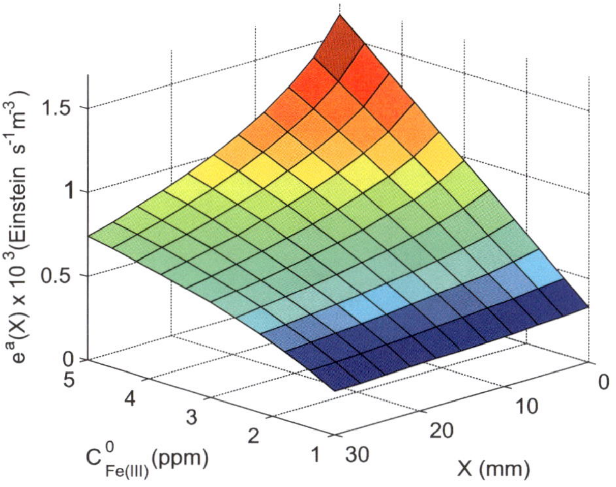

Fig. 5 Typical simulated results of LVRPA for $\theta z = 28°$, as a function of the ferric ion concentrations and the position x inside the solar reactor. Reprinted with permission from Conte et al. [50]. Copyright 2012 American Chemical Society

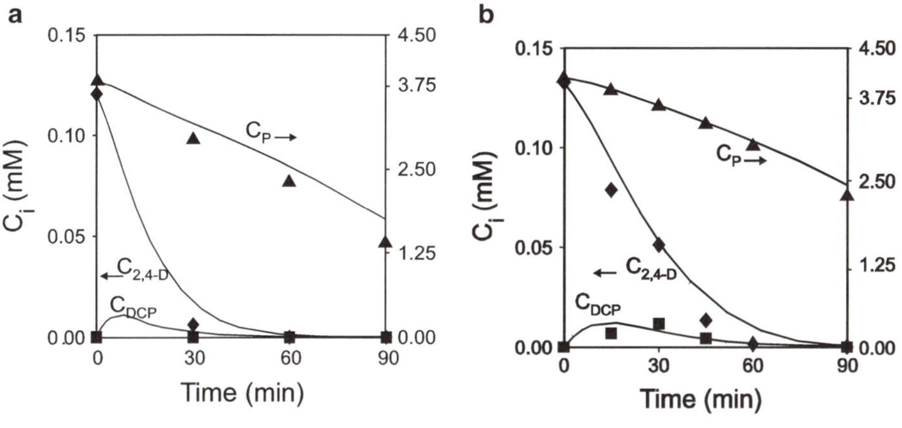

Fig. 6 Model predictions (*lines*) and experimental (*symbols*) concentrations vs. time for $C^0_{2,4\text{-}D}$ = 0.13 mM. (**a**) $C^0_{Fe^{3+}} = 2.8$ ppm, $R = 31.5$, $T^0 = 28°C$. (**b**) $C^0_{Fe^{3+}} = 1.0$ ppm, $R = 24.6$, $T^0 = 33°C$. Keys for photo-Fenton: 2,4-D (*filled diamond*), DCP (*filled square*), hydrogen peroxide (*filled triangle*). Reprinted with permission from Conte et al. [50]. Copyright 2012 American Chemical Society

and $\Delta T_{210\,\text{min}} = 21.0°C$ in Fig. 6b, the following TOC conversions are obtained: 91.6 and 98.6%, respectively.

For experiments carried out with solar radiation, it should be remarked that this pilot-plant reactor was able to reach an almost complete degradation of the herbicide 2,4-D and the main intermediate compound 2,4-dichlorophenol after a reaction time of 60 min. Moreover, under similar operating conditions and after 210 min of treatment, a very high pollutant mineralisation was reached.

7 Conclusions

The main results from this chapter can be summarised as follows:

- The modelling of a photo-Fenton reactor to describe the degradation of an organic pollutant requires the solution of the multicomponent mass conservation equations. The kinetic expressions for each reacting species should include the degradation rates for both nonirradiated (Fenton or dark reaction) and irradiated (photo-Fenton reaction) zones inside the reactor.
- The complete kinetic expressions for the photo-Fenton degradation of a specific pollutant are derived from a reaction sequence obtained from the specific literature and the corresponding kinetic parameters determined using an isothermal, well-stirred tank laboratory photoreactor. By applying a nonlinear regression procedure to the kinetic model equations and the experimental data at different reaction temperatures, the Arrhenius kinetic parameters can be estimated.
- The evaluation of the radiation field, expressed in terms of the Local Volumetric Rate of Photon Absorption, should be introduced into the kinetic expressions in order to obtain reaction rate equations independent of the experimental irradiation conditions. To do this, it is necessary to employ a mathematical modelling involving (1) the evaluation of radiation flux that arrives on the reactor window and (2) the solution of the radiative transfer equation inside the photoreactor.
- The temperature increase in the solar reactor during an experimental run can produce an advantageous result on the reactor performance, due to the positive effects that temperature produces on the thermal (Fenton) reaction rate. Thus, a thermal energy balance must be written and solved to take these effects into account.
- Then, the complete kinetic model expressions and parameters can be applied to investigate the behaviour of a pilot-plant solar reactor as a function of geographic location, day of the year, hour of day, and atmospheric conditions. From the reported results under different operating conditions, it is possible to conclude that the proposed methodology has proved to be appropriate to simulate the performance of a photo-Fenton solar reactor for water pollution remediation.

Acknowledgments The authors are grateful to Universidad Nacional del Litoral (UNL), Consejo Nacional de Investigaciones Científicas y Técnicas (CONICET), and Agencia Nacional de Promoción Científica y Tecnológica (ANPCyT) for the financial support. They also thank Antonio C. Negro for his valuable help during the experimental work and Claudia M. Romani for her technical assistance.

References

1. Malato S, Blanco J, Alarcón D et al (2007) Photocatalytic decontamination and disinfection of water with solar collectors. Catal Today 122:137–149
2. Legrini O, Oliveros E, Braun A (1993) Photochemical processes for water treatment. Chem Rev 93:671–698
3. Comninellis C, Kapalka A, Malato S et al (2008) Perspective advanced oxidation processes for water treatment: advances and trends for R&D. J Chem Technol Biotechnol 83:769–776
4. Ruppert G, Bauer R, Heisler G (1993) The photo-Fenton reaction – an effective photochemical wastewater treatment process. J Photochem Photobiol A 73:75–78
5. Pignatello J, Oliveros E, MacKay A (2006) Advanced oxidation processes for organic contaminant destruction based on the Fenton reaction and related chemistry. Crit Rev Environ Sci Technol 36:1–84
6. Malato S, Fernández-Ibáñez P, Maldonado M et al (2009) Decontamination and disinfection of water by solar photocatalysis: recent overview and trends. Catal Today 147:1–59
7. Sun Y, Pignatello J (1992) Chemical treatment of pesticide wastes. Evaluation of Fe(III) chelates for catalytic hydrogen peroxide oxidation of 2,4-D at circumneutral pH. J Agric Food Chem 40:322–327
8. Nogueira R, Silva M, Trovó A (2005) Influence of Iron source on the solar photo-Fenton degradation of different classes of organic compound. Sol Energy 79:384–392
9. Farias J, Albizzati E, Alfano O (2009) Kinetic study of the photo-Fenton degradation of formic acid. Effects of temperature and iron concentration. Catal Today 144:117–123
10. Monteagudo A, Durán J, Corral A et al (2012) Ferrioxalate-induced solar photo-Fenton system for the treatment of winery wastewaters. Chem Eng J 181:281–288
11. Lee Y, Lee C, Yoon J (2003) High temperature dependence of 2,4-diclorophenoxyacetic acid degradation by Fe^{3+}/H_2O_2 system. Chemosphere 51:963–971
12. Farias J, Rossetti G, Albizzati E et al (2007) Solar degradation of formic acid: temperature effects on the photo-Fenton reaction. Ind Eng Chem Res 46:7580–7586
13. Sagawe G, Lehnard A, Lübber M et al (2001) The insulated solar Fenton hybrid process: fundamental investigations. Helv Chim Acta 84:3742–3759
14. Lee C, Yoon J (2004) Temperature dependence of hydroxyl radical formation in the hv/Fe^{3+} / H_2O_2 and Fe^{3+}/H_2O_2 systems. Chemosphere 56:923–934
15. Gernjak W, Fuerhacker M, Fernández-Ibáñez P et al (2006) Solar photo-Fenton treatment-process parameters and process control. Appl Catal B Environ 64:121–130
16. Chen R, Pignatello J (1997) Role of quinone intermediates as electron shuttles in Fenton and photoassisted Fenton oxidations of aromatic compounds. Environ Sci Technol 31:2399–2406
17. Chen F, Ma W, He J et al (2002) Fenton degradation of malachite green catalyzed by aromatic additives. J Phys Chem A 106:9485–9490
18. Nichela D, Carlos L, García Einschlag F (2008) Autocatalytic oxidation of nitrobenzene using hydrogen peroxide and Fe(III). Appl Catal B Environ 82:11–18
19. Nichela D, Haddou M, Benoit-Marquié F et al (2010) Degradation kinetics of hydroxy and hydroxynitro derivatives of benzoic acid by Fenton-like and photo-Fenton techniques: a comparative study. Appl Catal B Environ 98:171–179
20. Pérez-Estrada L, Malato S, Gernjak W et al (2005) Photo-fenton degradation of diclofenac: identification of main intermediates and degradation pathway. Environ Sci Technol 39:8300–8306
21. Radjenovic J, Sirtori C, Petrovic M et al (2009) Solar photocatalytic degradation of persistent pharmaceuticals at pilot-scale: kinetics and characterization of major intermediate products. Appl Catal B Environ 89:255–264
22. Farré M, Maldonado M, Gernjak W et al (2008) Coupled solar photo-Fenton and biological treatment for the degradation of diuron and linuron herbicides at pilot scale. Chemosphere 72:622–629

23. Zapata A, Oller I, Bizani E et al (2009) Evaluation of operational parameters involved in solar photo-Fenton degradation of a commercial pesticide mixture. Catal Today 144:94–99
24. Mendoza-Marín C, Osorio P, Benítez N (2010) Decontamination of industrial wastewater from sugarcane crops by combining solar photo-Fenton and biological treatments. J Hard Mater 177:851–855
25. Jiménez M, Ollera I, Maldonado M et al (2011) Solar photo-Fenton degradation of herbicides partially dissolved in water. Catal Today 161:214–220
26. Pérez M, Torrades F, Domènech X et al (2002) Fenton and photo-Fenton oxidation of textile effluents. Water Res 36:2703–2710
27. García-Montaño J, Pérez-Estrada L, Oller I et al (2008) Pilot plant scale reactive dyes degradation by solar Photo-Fenton and biological processes. J Photochem Photobiol A 195:205–214
28. Monteagudo J, Durán M, San Martín I et al (2009) Effect of continuous addition of H_2O_2 and air injection on ferrioxalate-assisted solar photo-Fenton degradation of Orange II. Appl Catal B Environ 89:510–518
29. Torrades F, Pérez M, Mansilla H et al (2003) Experimental design of Fenton and photo-Fenton reactions for the treatment of cellulose bleaching effluents. Chemosphere 53:1211–1220
30. Klamerth N, Malato S, Maldonado M et al (2010) Application of photo-Fenton as a tertiary treatment of emerging contaminants in municipal wastewater. Environ Sci Technol 44:1792–1798
31. Alfano O, Bahnemann D, Cassano A et al (2000) Photocatalysis in water environments using artificial and solar light. Catal Today 58:199–230
32. Bahnemann D (2004) Photocatalytic water treatment: solar energy applications. Sol Energy 77:445–459
33. Blanco J, Malato S, Fernández-Ibañez P et al (2009) Review of feasible solar energy applications to water processes. Renew Sust Energ Rev 13:1437–1445
34. Ollis D (1991) Solar-assisted photocatalysis for water purification: issues, data, questions. In: Pelizzetti E (ed) Photochemical conversion and storage of solar energy. Kluwer Academic Publishers, The Netherlands
35. Bockelmann D, Goslich R, Weichgrebe D et al (1993) Solar detoxification of polluted water: comparing the efficiencies of a parabolic trough reactor and a novel thin-film-fixed-bed reactor. In: Ollis D (ed) Photocatalytic purification and treatment of water and air. Elsevier, Amsterdam
36. Bockelmann D, Weichgrebe D, Goslich R et al (1995) Concentrating versus non-concentrating reactors for solar water detoxification. Sol Energ Mat Sol C 38:441–451
37. Rossetti G, Albizzati E, Alfano O (1998) Modeling and experimental verification of a flat-plate solar photoreactor. Ind Eng Chem Res 37:3592–3601
38. Rossetti G, Albizzati E, Alfano O (2002) Decomposition of formic acid in a water solution employing the photo-fenton reaction. Ind Eng Chem Res 41:1436–1444
39. Rossetti G, Albizzati E, Alfano O (2004) Modeling of a flat.plate solar reactor. Degradation of formic acid by the photo-Fenton reaction. Sol Energy 5:461–470
40. Ozisik M (1973) Radiative transfer and interactions with conduction and convection. Wiley, New York
41. Cassano A, Martin C, Brandi R et al (1995) Photoreactor analysis and design: fundamentals and applications. Ind Eng Chem Res 34:2155–2201
42. Siegel R, Howell J (2002) Thermal radiation heat transfer, 3rd edn. Hemisphere, Washington
43. Anderson P, Chetwynd J, Theriault J et al (1993) MODTRAN2: suitable for remote sensing. In: Kohnle A, Miller W (eds) Proceeding conference atmospheric propagation and remote sensing II, Orlando vol 1968 pp 514–525
44. Madronich S, Flocke S (1997) Theoretical estimation of biologically effective UV radiation at the Earth's surface. In: Zerefos C (ed) solar ultraviolet radiation – modeling, measurements & effects. NATO ASI Series I: Global Environmental Change. Springer, Berlin

45. Bird R, Riordan C (1986) Simple spectral model for direct and diffuse irradiance on horizontal and tilted planes at the Earth's surface for cloudness atmospheres. J Clim Appl Meteorol 25:87–97
46. Gueymard C (1995) SMARTS2, a simple model of the atmospheric transfer of sunshine: algorithms and performance assessment. Report FSEC-PF-270-95, Florida Solar Energy Center, Florida
47. Sun Y, Pignatello J (1993) Organic intermediates in the degradation of 2,4-dochlorophenoxyacetic acid by Fe^{3+}/H_2O_2 and Fe^{3+}/H_2O_2/UV. J Agric Chem 41:1139–1142
48. Sun Y, Pignatello J (1993) Photochemical reactions in the total mineralization of 2,4-D by Fe^{3+}/H_2O_2/UV. Environ Sci Technol 27:304–310
49. Brillas E, Calpe J, Casado J (2000) Mineralization of 2,4-D by advanced electrochemical oxidation processes. Water Res 34:2253–2262
50. Conte L, Farias J, Albizzati E et al (2012) Photo-Fenton degradation of the herbicide 2,4-D in laboratory and solar pilot-plant reactors. Ind Eng Chem Res 51:4181–4191
51. Alfano O, Romero R, Cassano A (1985) A cylindrical photoreactor irradiated from the bottom. I. Radiation flux density generated by a tubular source and a parabolic reflector. Chem Eng Sci 40:2119–2127
52. Farias J, Albizzati E, Alfano O (2010) New pilot-plant photo-Fenton solar reactor for water decontamination. Ind Eng Chem Res 49:1265–1273
53. Farias J, Albizzati E, Alfano O (2010) Modelling and experimental verification of a solar reactor for the photo-fenton treatment. Water Sci Technol 61:1419–1426

Surface-Modified Photocatalysts

Claudio Minero

Abstract The surface properties of TiO_2 play a very important role in determining photocatalytic reaction efficiencies because heterogeneous photocatalytic reactions take place on the surface. Various parameters such as composition, phase structures, surface hydroxyl group, particle size, crystallinity, surface defects, and adsorbates or surface complexes play a key role. TiO_2 surfaces have been actively modified through manipulating the above parameters to enhance the photocatalytic performance. Here the main effects that influence the surface electron transfer are reported.

Keywords Adsorption of anions, Electron traps, Fluorides, Glycerol, Interfacial electron transfer

Contents

C. Minero (✉)
Dipartimento di Chimica, Università di Torino, via Pietro Giuria 5, 10125 Torino, Italy
e-mail: claudio.minero@unito.it

D.W. Bahnemann and P.K.J. Robertson (eds.), *Environmental Photochemistry Part III*, Hdb Env Chem (2015) 35: 23–44, DOI 10.1007/698_2013_250,

1 Introduction

After photon absorption, charge carriers are generated. The high dielectric constant of TiO_2 assists in their separation, as charges become screened from their counter charges by the lattice. Charge separation follows energetic restrictions, and electrons/holes spatially migrate where a negative change of their free energy is allowed. Hot electrons and deep holes are more likely to separate than charge carriers generated with near-bandgap energy light [1]. Conversely, higher temperatures can negate the charge separation effectiveness [2]. An imposed electric field can assist in charge separation, just as a space charge region can [3].

Surface modification by extraneous phases as noble metal (platinum, silver, gold, palladium, and rhodium), copper, nickel, and oxide-on-oxide heterojunctions and also with other semiconductor will not be treated. All of these modifications are intended to favor charge separation and in addition can offer the potential for new reaction sites at the interface. The focus will be instead on naked surfaces and on the effect that the composition of the second phase at the interface has on the first.

The movement of charge carriers in spatial regions where the system gains free energy (minimum of their electrochemical potential) is called trapping. The interface or a bulk charge trapping site can separate one carrier from the other. Trapping could be considered beneficial if it is localized at the surface or if it promotes charge carrier separation. Conversely, trapping could be considered detrimental if trap sites are far from the site of electron transfer or lead to recombination. Trapping energies, if too large, could diminish the oxidizing potential of holes or the reducing potential of electrons, or inhibit rapid transport. If the trapping energy is small, any benefits of trapping would disappear. Then charge carrier could recombine transforming photon absorption energy in other photons or heat, less stored energy is available to the redox photocatalysis or for generating chemical fuels, and the rate of the overall process is slowed down.

The energies of surface electron trap states can be affected by applied potential (in an electrochemical system) [4], by local structure, by the deep-stabilizing self-polarization potential well induced by the air- or water-semiconductor dielectric mismatch, which is a function of the pore size and the bulk parameters of the matrix material [5], or by the presence of adsorbates. The trapping event likely involves structural relaxation that contributes to localization.

2 Charge Carrier Trapping

Electron traps are believed to be localized in the TiO_2 lattice as Ti^{3+} sites. TiO_2 surfaces in some cases act as charge trapping agents [6], depending on the spatial energy-band distribution. The transient absorption of nanocrystalline TiO_2 films in the visible (450–800 nm) and IR (2,500 nm) wavelength regions showed that surface-trapped electrons and surface-trapped holes were generated within 200 fs.

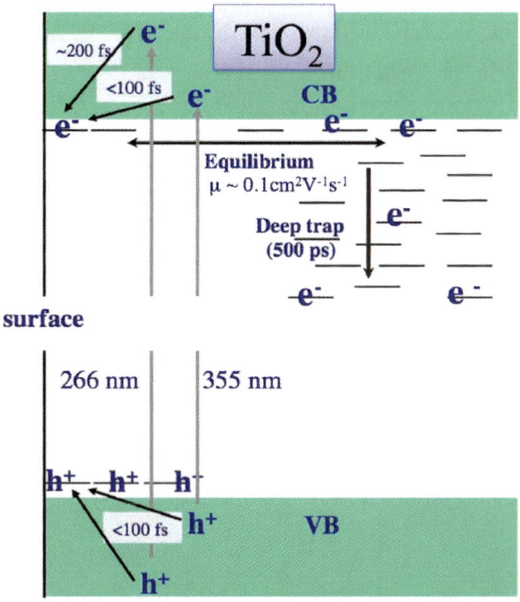

Fig. 1 The trapping and relaxation dynamics of electrons and holes in TiO_2 excited at 266 and 355 nm (from [9] with permission)

Surface-trapped electrons, which gave an absorption peak at around 800 nm, and bulk electrons, which absorbed in the IR wavelength region, decayed with a 500-ps time constant due to relaxation into deep bulk trapping sites. It is already known that, after this relaxation, electrons and holes survive for microseconds.

The electron trapping competes well with e−/h+ pair recombination even in the absence of a suitable hole scavenger [7]. The rate constant of electron-hole recombination in the bulk of a rutile TiO_2 single crystal was estimated to be $k_{rec} = 5 \times 10^{-13}$ cm^{-3} s^{-1} through sub-nanosecond time-resolved transient absorption spectroscopy. This value is more than four orders of magnitude smaller than the diffusion-limited rate constant ($k_{rec}^{diff} = 1.4 \times 10^{-8}$ cm^{-3} s^{-1}) for a rutile TiO_2 crystal, suggesting low recombination reactivity between electrons and holes in the crystal (Fig. 1) [8, 9].

There is a general agreement that electrons prefer trapping at the surfaces of TiO_2 [10–13] on defective sites just below the conduction band edge of TiO_2 in ranges of 0–0.35 V for anatase and 0–0.25 V for rutile [14]. On a chemical basis the most stable electron trap sites should be undercoordinated Ti cation sites located at surfaces. Based on Fourier transform infrared (FTIR) analysis [12, 15] photoexcited electrons are preferentially trapped at surface Ti-OH groups. A good examination of electron trapping is reported by the recent review of Henderson [16]. The electron trapping energy could range between 0.1 and 1 eV [14, 17], in agreement with photoemission results for electronic defects on the surface of TiO_2 single crystal surfaces [18].

While the trapping timescales are very short, the trapping lifetimes can be very long, particularly in the absence of electron scavengers such as O_2. Ethanol was used as a hole scavenger to study electron trapping in synthesized nanocrystalline TiO_2 films and in films of commercial TiO_2 with transient absorption spectroscopy [13]. In the absence of ethanol and O_2, the transient signal at 800 nm (associated with excitation of trapped electrons) had a long half-life (~25 μs), suggesting a slow e^-/h^+ pair recombination at trap sites. In the presence of ethanol and the absence of O_2, electron trapping lifetimes were even longer (~0.5 s). These observations highlight the ability of TiO_2 surfaces to rapidly stabilize CB electrons in trap sites.

Hole trapping at rutile and anatase surfaces was also deeply explored [6, 19–26] particularly by means of hole scavenging chemicals such as SCN^-, organics, and I^-. In the absence of any hole scavengers at least two different types of traps have to be considered: (1) deeply trapped holes, rather long-lived and unreactive toward DCA^- nor to SCN^- ions, and (2) shallowly trapped holes that are at a thermally activated equilibrium with free holes and exhibit a very high oxidation potential [10]. EPR results indicate that the most likely hole trap site is a surface Ti^{4+}–O– site, where the hole resides on an undercoordinated surface oxygen atom, for which it is still uncertain what is its coordination (e.g., bridged or oxo type).

If the excited electron is removed with an electron scavenger, lifetimes for holes trapped at TiO_2 surfaces in the absence of hole scavengers have been estimated on the order of milliseconds to minutes [6, 27]. Hole trapping on the surfaces of TiO_2 appears to occur rapidly (~50 fs), as does electron trapping.

In the absence of an electron scavenger, electron trapping competes well with surface recombination when the number of e^-/h^+ pairs per particle is ~1, but the recombination is greatly favored as their number increases [7, 20]. The presence of a scavenger decreases the amount of recombination, which usually follows a second-order kinetics as required also by early kinetic models developed to explain chemical kinetics [28] and later validated [29]. The recombination rate depends on the surface-to-volume ratio of particles [30] because the bulk density influences absorptivity and the surface influences recombination. Several studies have explored the relationship between surface-trapped charges and charge recombination. For example, it was observed using EPR under bandgap irradiation of P25 that the majority of recombination events in P25 occurred at the surface or at the interfaces between particles or phases [23]. The analysis of the cyclic voltammetry response on thin rutile films revealed a strong dependence of the trap state concentration on the morphological structure. On the basis of results concerning the surface modification of the electrodes, these bandgap states where located at grain boundaries [31].

The trapped holes (and electrons) are less reactive than their bulk photogenerated precursors, although reaction of free carriers with surface chemisorbed molecules can be very rapid and efficient [10]. If charge recombination is governed predominately by the catalyst surface and its inherent structural heterogeneity, the texture and composition of the photocatalyst surface is of paramount importance.

Different TiO_2 specimens have a great chance to have different surface texture and composition and thus different probability that recombination events occur at the surface. The published data on particles morphology suggest a good uniformity of the Merck TiO_2 particle surfaces and a more defective surface for the P25 particles. In a recent paper [32] we reported by a comparative FTIR analysis under various conditions the presence of a variety of surface OH on P25, that is, at least 3 types of linear hydroxyl groups (in which OH is bound to a surface Ti, let's say Ti-OH) and 3 types of bridged hydroxyl groups (Ti-OH-Ti). There is a large consensus on the action of surface hydroxyls as surface hole traps, so their chemical nature is relevant to photocatalysis.

2.1 Electron Transfer Across the Interface

The electron transfer across the interface is at the heart of redox photocatalysis. Electron transfer is an interfacial phenomenon between a surface-trapped charge carrier and a chemisorbed or physisorbed species. In this session the electron transfer across the interface refers to the reaction of trapped species with solution or physisorbed species only.

In aqueous solutions metal oxides have pH-dependent surface charge which can be explained by the existence of acid/base equilibria. From infrared studies of the TiO_2-gas interface, it is known that the surface is hydroxylated when exposed to water vapor, with several types of hydroxyl species present. It has been suggested [33] that there is one chemically active face (001 on anatase and 110 on rutile) which contains both terminal and bridging OH groups. The other major crystal faces contain metal ions which are believed to only weakly bind water molecules. As already pointed out, depending on the specimen, different types of terminal and bridging OH groups could be present on TiO_2 [32].

Detailed studies on surface oxygen species on TiO_2 in a pH range 2.3–11.7 by internal reflection FTIR spectroscopy reported that Ti-OH is present in a pH range from 4.3 to 10.7 (maximum =8), $Ti\text{-}OH_2^+$ exists in a pH range below 5, and $Ti\text{-}OH^+\text{-}Ti$ exists in a pH range below 4.3 [34]. There is also evidence for a surface water species which has an increasing population with increasing positive charge. At high pH the surface will be deprotonated with a negative charge. At some intermediate pH, the surface has net zero charge, and this pH is called the point of zero charge (PZC). A surface metal ion may have a residual charge which depends on the surface group and the crystal face. For a general review of the charge distribution, surface hydration, and the structure of the interface of metal hydroxides, see a recent review [35].

Application of the revised MUSIC model [36] to the vacuum-terminated rutile (110) surface was able to predict the formation and protonation of three unique types of surface oxygens upon surface hydration: (1) oxygen atoms in the Ti surface plane bonded to three Ti atoms that do not protonate in the accessible pH range (ca. 0–14); (2) bridging oxygens protruding above the Ti surface plane and bound to two surface Ti atoms that undergo a single protonation step in this pH range; (3) bare Ti atoms exposed at the surface that chemisorb a single water molecule, which may dissociate to form a hydroxyl group at sufficiently high pH. Bridging

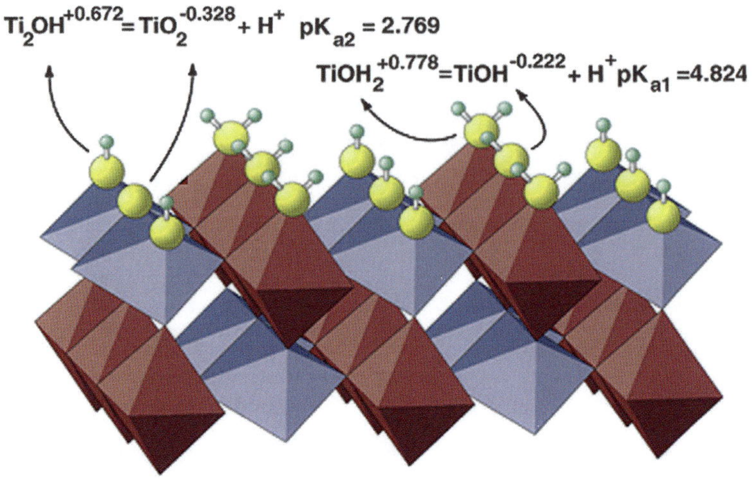

Fig. 2 Predicted protonation of hydrated (110) rutile surface by optimized MUSIC model. Protonation (*green spheres*) of surface oxygens (*yellow spheres*) reflects the four possible protonation states predicted by the revised MUSIC model: surface oxygens bridging two Ti octahedra (*blue* TiO_6 units) coordinate to one or zero protons, and surface oxygens at the apex of one Ti octahedral unit (*red* TiO_6 units) coordinate to one or two protons. The partial charges and equilibrium constants (pKa) were calculated with the MUSIC model using surface oxygen–Ti bond lengths predicted by ab initio calculations (from [37], with permission)

(Ti_2O and Ti_2OH) and terminal (Ti-OH and Ti-OH_2) protolytic surface oxygens are highlighted in Fig. 2 [37].

The effect of pH is also correlated to the anatase or rutile phase present and to their ratio [38]. In phenol decomposition the photocatalytic performance gradually and significantly increases with the anatase content. Photocatalysts containing only anatase as crystalline phase were up to three times more efficient than rutile ones. In salicylic acid decomposition, rutile-only catalysts were found to show no activity at all, but some of the prepared catalysts (both anatase-only and rutile-anatase mixtures) at pH 3 (but not at pH 7) displayed photocatalytic activity commensurable to that of Degussa P25.

In general pH variation changes the electrostatic interactions between the pH-dependent charge of the TiO_2 surface and the substrate. It is then apparent that the rate for the degradation of anionic dyes Methyl Orange, p-aminoazobenzene, Congo Red, and Brilliant Yellow was high at pH 5.6, while for cationic dyes like Rhodamine-B and Methylene Blue the highest rate was obtained in the alkaline pH 8.0 [39]. These differences can be accounted for by the adsorption capacity of the substrate on the catalyst surface under different pH conditions (see [40] for nitrophenols, where surface coverage can be manipulated with pH).

The TiO_2 CB edge exhibits Nernstian dependence with pH, shifting by −64 mV/pH unit in the range of −8 to +23 pH [18, 41].

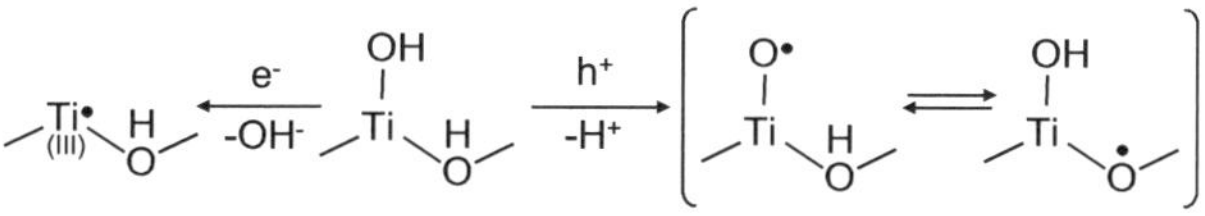

Fig. 3 Pictorial view of trapped electron and hole at the naked TiO_2 surface

From the pH dependence of transient absorption spectra in nanocrystalline TiO_2 films, two different trapped holes in the transient absorption spectrum have been identified. From their absorption spectra previously reported, spectra can be reproduced by their superposition [26]. In addition, they are significantly different from the expected shape of free carriers. This indicates that the holes in TiO_2 are trapped and that the spectral shape of holes reflects the condition of the trap states.

A very schematic and pictorial view of the trapped electron and hole at the naked TiO_2 surface is given in Fig. 3. As three types of linear hydroxyl groups and three types of bridged hydroxyl groups have been detected [32] on P25, the notation Ti(OH)(OH) refers to all of them. The species Ti(OH)(OH) can be protonated/deprotonated at various extent depending on solution pH.

The species in Fig. 3 at right are referred as (Ti(OH)(OH)) $\rightarrow$ (Ti(O•)(OH)) $\leftrightarrows$ (Ti(OH)(O•)), where the first (OH) refers to linear hydroxyl groups and the left (OH) to bridged hydroxyl groups. A shallow surface hole trap can be schematized as in the right of Fig. 3 where two limiting forms exist. The actual hole localization is formally regulated by their ratio $K = [\text{right}]/[\text{left}]$. The actual electron density distribution of the left and right forms in parenthesis are $1/(1{+}K)$ and $K/(1{+}K)$, respectively. The surface is considered hydrated to avoid to explicitly write charges that depend on the type of crystal face, the type of bridging oxygen, and the possible protonation and electron density [42]. The hole is stabilized at the surface by the surface hydration energy involved in the process $(TiO)_s + H_2O \rightarrow$ Ti(OH)(OH) and the pH-dependent release of protons that have a large hydration energy.

The schematic Ti(OH)(OH) for the (TiO_2) unit at the surface resumes many of the possible surface forms that a trapped hole (or electron) could have. In particular, in the framework of water oxidation, several other forms have been proposed, like $[Ti_2 = O\text{-}Ti]_s$ and $[Ti\text{-}O\text{-}Ti]_s$ that refer to a triply coordinated (normal) O atom and a bridging O atom at the surface, respectively [43]. For the trapped hole, the form $[Ti\text{-}O^{\bullet}\ HO\text{-}Ti]_s$ was also proposed, which is formed by a concerted hole and nucleophilic H_2O attack to $[Ti\text{-}O\text{-}Ti]_s$. This approach includes the possibility that surface oxygens are included in oxidation products and is consistent with surface reconstruction events and the phenomenon of superhydrophilicity [41]. The schematic of Fig. 3 loses this possibility and it is useful only for a general discussion of the electron transfer process. The resonance form of the trapped hole depicted inside parenthesis at left was supposed to be present only at $pH > 13$ [43]. According to the given picture, as Ti(O•)(OH) is not favored at acidic pH, and because the dissociation of the bridged hydroxyl occurring at $pH > 8$ with creation of a net negative charge on the bridged oxygen will further favor hole trapping on the oxygen, K will be >1 for naked titania and the most probable hole localization is Ti(OH)(O•), i.e., on bridged oxygen.

This also outlines the role of solution pH on the electron transfer rate. As the surface at acidic pH is positively charged, the electron abstraction from $Ti(OH_2^+)$ (OH) is more difficult, implying that the hole trap $Ti(OH_2^+)(O•)$ is shallower that Ti (OH)(O•). From this it follows a greater potential difference with the redox potential of the substrate to be oxidized and a greater reaction rate. At basic pH the $Ti(O^-)(O•)$ is stabilized, as an electron is easily abstracted from a negative surface. This implies a lower oxidizing potential and reduced trapped hole reactivity. Conversely the electron trapping at Fig. 3 (left) is favored by acidic pH, forming a trap with reduced reducing potential. This will limit reactivity toward physisorbed oxidants. This picture is in agreement with experimental results obtained by transient techniques. Tracking Methyl Orange photodecomposition on suspended P25 as a function of pH, a crossover at pH ~8 was detected between an oxidative and a reductive pathway for degradation [44]. Above a pH of 8, the rate-limiting step was oxidation and O_2 reduction was rate limiting at lower pH values. The reducing species becomes more reactive at higher pH, the converse being valid for the oxidizing carrier, as a result of the shift in electrical potential of TiO_2 particles. This conclusion accords also with recent results obtained using highly ordered TiO_2 nanotube prepared by an anodic oxidation method. It was found that 2,3-dichlophenol in alkaline solution was degraded and dechlorinated (a reductive process) faster than that in acidic solution, whereas dissolved organic carbon removal presented an opposite order in dependence of pH [45].

The pH can be varied to shift band positions and surface electrostatics that influence charge carrier dynamics as discussed above [44] and photoluminescence [43]. The shifts in band edges induced by changing pH can be used to control some electron transfer process or possibly shift conditions from oxidative to reductive [44]. The effect of pH is important for practical application. In the removal of cations from solution, the system pH controls the speciation in solution and the interfacial electron transfer. The photocatalytic removal of Hg(II) from aqueous solutions of $HgCl_2$ using TiO_2 as catalyst showed that the overall process strongly depended on pH, being enhanced as the pH was increased [46]. At pH 10, an efficient removal of Hg(II) was achieved even in the absence of organic additives, attaining final mercury concentrations in solution at trace levels ($\mu g\ L^{-1}$). In acidic conditions, the addition of sacrificial organic molecules significantly increased the rate and extent of aqueous Hg(II) removal. The nature and distribution of mercury products deposited on the catalyst were dependent on the reaction conditions. Reported evidences showed that it cannot be established a direct correlation between Hg(II) dark adsorption on the TiO_2 surface and the efficiency of Hg (II) photoreduction achieved.

Electron and hole transfer plausibly occurs at spatially separated surface traps. It has been demonstrated on both rutile and anatase microcrystals that the reductive and oxidative processes take place on different crystallographic faces [47]. Reductive facets are (110) and (101); oxidative facets are (011) and (001) for rutile and anatase, respectively. It is reasonable to assume that these two processes are not competing for the same surface sites. The presence of distinct sites for oxygen reduction supports the possibility that under complete coverage by ligands (see below) the scavenging of electrons will still be possible.

Recombination will occur by free electrons in CB with holes that are trapped on titania with both tautomeric forms of Fig. 3.

On the reductive side, electron transfer to O_2 forms an O_2 molecularly adsorbed state. O_2 can only physisorb to TiO_2 surfaces if reduced cation sites are not present [48]. Some OH groups associated with electron trap sites are altered by O_2 exposure, but others are not [12]. The interaction of O_2 with bridging OH groups (OH_{br}) results in the extraction of charge and a proton from the OH_{br} groups [49]. Berger et al. [2] using EPR results and simulations attributed signals to reaction of O_2 with a trapped electron to form O_2^-. Bahnemann et al. [10] have shown that the relative rate of O_2^- formation from the reaction of O_2 with trapped electrons was roughly 100 times slower than for the reaction of O_2 with solvated electrons in solution and that this represents a major bottleneck for photooxidation reactions. The O_2^- species appeared to be the intermediate through which a variety of potentially important reactive oxygen-containing species, such as O_2^{2-} and H_2O_2/HO_2, were photochemically formed. The primary step of photocatalytic O_2 reduction is the formation of the surface peroxo species, $Ti(O_2)$, giving the 943 cm^{-1} band, probably with the surface peroxo species, $TiOO^\bullet$, as a precursor, in neutral and acidic solutions. The surface peroxo species is then transformed to the surface hydroperoxo, TiOOH, giving the 838 and 1,250–1,120 cm^{-1} bands, by protonation in the dark [50]. Spectroscopic observations of $HO_2^\bullet$ and production of H_2O_2 [51] have been linked to reactions of O_2 with water-related species on or near the TiO_2 surface. It is extensively conjectured that O_2 chemistry in the reductive side of photocatalytic systems results in hydroxyl radicals that can participate in oxidative reactions.

On the oxidative side, the electron transfer event across the interface is usually quite fast. For example, for an aqueous I^-, hole transfer occurs on less than 10 ns [7]. The amount of I_2^- (product of oxidation) was observed relatively stable during the first 4 μs, in contrast to the decay of the electron population due to recombination which is only slightly different than in iodide-free solutions.

Reaction of holes with organic (reduced) molecules could be direct (direct electron transfer from the substrate to the valence band hole) or mediated by OH radicals either free or bound. Examples of remote oxidation [52, 53] through gas phase have been reported. In this phenomenon, oxidation events occur at regions of a TiO_2 sample not exposed to light or at locations that are apart from irradiated TiO_2 surfaces. Although oxidizing species could be formed in the gas phase from O_2 by subsequent reduction, evidence supports the surface generation of OH radicals [54].

Reports on the presence of free OH in solution are conflicting. The ESR detection of OH radicals on irradiated TiO_2 (anatase) at 77 K was reported [55], although the ESR signals showed no spectral change by H_2O/D_2O exchange. Ultraviolet photoelectron spectroscopic studies combined with scanning tunneling microscopy revealed that the O_{2p} levels for bridging hydroxyls groups (Ti-OH-Ti) at the (110) face and terminal hydroxyls groups (Ti-OH) at the (100) face of rutile are both far below the top of the valence band [56] to oxidize water. These results are confirmed by theoretical calculations [57]. Nosaka et al. pointed out that the water photooxidation reaction at TiO_2 produced no free OH radical, and spin-trapping agents reacted with surface-trapped holes (adsorbed OH radicals, see Fig. 3) [58]. Early ESR measurements showed that photogenerated holes were trapped at lattice O atoms (or Ti-O-Ti sites) at low temperatures of 4.2 or 77 K and did not produce OH radicals [59]. Micic et al. confirmed this conclusion and showed clearly that not OH radicals but surface-trapped radicals were involved in the methanol oxidation that proceeded by charge transfer from the oxygen lattice holes [60]. Serpone et al. reported a fast trapping in some adsorbed state of the OH radicals generated by pulse radiolysis onto the TiO_2 surface [61]. It is argued that the trapped hole and a surface-bound OH radical are indistinguishable species. Later studies showed that the reaction of TiO_2 with OH radicals probably does not form an adsorbed radical on the surface, but OH radicals inject holes into TiO_2 and subsequently form a charge transfer complexes between the holes and OH^- ions [26].

A direct convincing evidence that photogenerated species do not migrate far from the catalyst surface has been derived from the study of the degradation of decafluorobiphenyl (DFBP), a substrate that is strongly adsorbed (>99%) on alumina and TiO_2 [62]. When DFBP is adsorbed on alumina particles and mixed with titania particles, the amount exchanged is very low (<5%). Irradiation of DFBP adsorbed on alumina in the presence of H_2O_2 (generating OH in solution) or titania colloids (supposed to generate OH free or bound to the surface on mobile particles) leads to degradation, while DFBP is not degraded when larger titania particles (P25, TiO_2 beads) are present. Pentafluorophenol, which is easily exchanged between the two oxides, is photodegraded in all cases. This makes clear that active species formed upon irradiation do not migrate in solution, while the organic substrate may or not migrate to the catalyst surface. Only when both are present at the catalyst surface the degradation takes place.

For a poorly adsorbed substrate like phenol [63], the detailed kinetic analysis and the time evolution of the intermediates in the presence of different alcohols (tert-butyl alcohol, 2-propanol, and furfuryl alcohol used as hydroxyl radical scavengers) suggested that the oxidation of phenol proceeds for 90% through the reaction with surface-bound hydroxyl radical (Ti(OH)(O•), the remaining 10% via a direct interaction with the VB holes [64].

2.2 *Electron Transfer at the Interface*

If species are chemisorbed, it is trivial that the physical and electronic structures of the adsorbed state of the molecule have been changed. How a molecule binds on the TiO_2 surface influences its electronic structure and redox properties. There are many reports dealing with the thermodynamic of adsorption on the TiO_2 surface. The review by Thornton et al. [65] reports many of these cases, in particular the cases of HCOOH and C2–C8 alcohols.

The electronic structure and redox properties of undercoordinated Ti cations located at the surface, which likely act as more powerful adsorbing sites, will

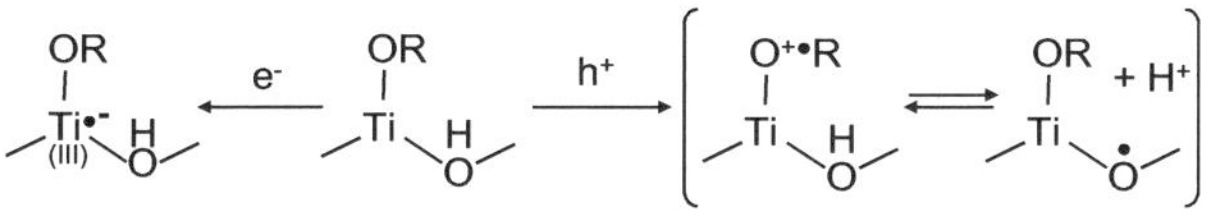

Fig. 4 Pictorial view of trapped electron and hole at an adsorption site

change with adsorption, *changing the energetic properties of electron traps* and the related charge carrier dynamics. The same holds for hole traps, as localization of the surface hole on bridged oxygen is influenced by near complexed titanium cations.

A very schematic and pictorial view of trapped electron and hole at the adsorption site on the TiO_2 surface, with limits already discussed for Fig. 3, is given in Fig. 4. The electron transfer at the interface implies the reaction of the charge carrier with adsorbed species giving a trapped charge carrier. Under this assumption an inner sphere electron transfer is implicated. The energy of the trap (trapped charge carrier) depends on adsorption (or complexing) free energy of the adsorbate. For the electrochemical half reaction depicted at right of Fig. 4, according to the Nernst law, the energy of the trap depends on the ratio of the formation constants β_{ox} of the substrate with the oxidized Ti(OH)(O•)) site to give (Ti(OR$^+$ •)(OH)) ⇆ (Ti(OR)(O•)) and the formation constants β_{rid} of the substrate with the site Ti(OH)(OH)) to give the reduced surface complex Ti(OR)(OH)). If $\beta_{rid}/\beta_{ox} < 1$, then the oxidant potential of the trap is reduced, forming an intra-bandgap surface-trapped hole. As Ti(OR)(O•) is more stable than Ti(OH)(O•), which is always true when −OR replaces −OH, the Ti(OR)O•) trap is deeper than Ti(OH)(O•).

In the cited work of Rabani et al. [7], alcohols (methanol and 2-propanol) at high concentrations unexpectedly reduce the initially observed electron population by up to fourfold, without affecting the shape of the nanosecond time profile. The alcohol effect was assigned to the formation of an alcoholic-positive ion radical (free) which is more reactive in recombination with conduction band electrons than the original hole. This is in our opinion the first spectroscopic evidence of a substrate-mediated charge recombination that was supposed in macroscopic kinetic model (called back reaction) [28] to explain the peaked dependence of the rate from the substrate concentration.

More recently by highly sensitive femtosecond and nanosecond spectroscopy under low-intensity excitation conditions to avoid fast electron-hole recombination, it was observed [27] that electron transfer from trapped holes in TiO_2 occurs over a wide timescale depending on the alcohol. The lifetimes of trapped holes in the films in methanol, ethanol, and 2-propanol are 0.3, 1.0, and 3.0 ns, respectively. The authors suggested that it is likely that the rate-limiting step in this experiment is hole transfer from trapping sites to alcohol molecules and that the oxidation reactions of alcohols can be regarded as electron transfer from alkoxy species to adjacent surface holes (unoccupied surface states just above the valence band), but concluded that the reason for the different oxidation times for different alcohols was unclear. According to the simplified picture given in Fig. 4, the hole transfer does not follow an outer sphere mechanism, but is an inner sphere electron transfer at the surface complex. The different alcohols would give different hole traps, and the rate-limiting step is the release of products from (Ti(OR^{+}•)(OH)).

By comparing the photocurrent yields versus the extent of organic photooxidation on P25-covered anodes under anaerobic conditions, it emerged that the efficiency of hole acceptors in four adsorbed organics followed the trend: oxalate > formate > acetate > methanol [66]. This trend did not correlate with the relative reaction rates of these molecules with free OH• in solution, but is consistent with the trap energy that depends as discussed above from the formation constant of the surface complex. Recently it was demonstrated that the oxidation of oxalate proceeds on the surface of rutile nanowires mainly via the bridging bidentate bioxalate (large formation constant), followed by a fast replenishment of photooxidized species by monodentate bioxalate [67].

A different reactivity from that of free OH• in solution has been also seen changing the substrate concentration and consequently the amount of substrate adsorbed on the TiO_2 surface. The main reaction products obtained from glycerol on irradiated TiO_2 P25 depend on glycerol concentration [68]. At low concentration glyceraldehyde and dihydroxyacetone (C3 compounds) are formed in a relative ratio about 2, together with formaldehyde and glycolaldehyde. As the glycerol concentration increases and glycerol rate is strongly depressed, the main products are formaldehyde and glycolaldehyde. For Merck TiO_2 the main products of phototransformation are the two C3 carbohydrates with relative ratio ranging from 1.3 to 1.8, according to an •OH-related chemistry. The evident experimental result is that mainly the two C3 carbohydrates are produced from glycerol in the presence of fluorides that competitively adsorb on the surface (see also below). On Merck TiO_2 almost no change is observed with addition of fluorides, while on P25 the production of formaldehyde and glycolaldehyde was strongly depressed. As fluorides impede the surface complexation on the surface Ti-OH site, the formation of formaldehyde and glycolaldehyde is strictly connected with glycerol chemisorption.

All these findings (except the oxalate case for which hole transfer is followed by electron injection in CB) lead to the conclusion that when a chemical species is strongly bound to the surface, its overall degradation rate is generally different from that in the physisorbed state and could be depressed (see the case of catechol [63]). Although the strong binding to the surface favors reactions with holes, this implies that the species is more subject to back reactions with electrons, resulting in a decreased overall rate of oxidation. Furthermore, the two mechanisms, the electron transfer across the interface and the electron transfer at the interface, depend at a large extent on the adsorption of substrates and thus depend on the concentration of the substrate in the free phase or on the concurrent equilibria (acid/base, concurrent complexation) that influence its adsorption. So the literature results could be somehow confusing because the concentration effect is mistreated. To add some complication, often the net result of a primary step of a chemical reaction could be the result of both oxidative and reductive processes. For example, it was recently showed through isotopic-labeling examinations combined with DRIFTS and electrochemical experiments that, contrary to the radical Kolbe decarboxylation, the decarboxylation of saturated carboxylic acids is possible by a concerted mechanism [69]. At first step, the pristine acid is oxidized to α-keto acid by holes or (adsorbed) OH radicals without loss of carbon atoms, and then, the intermediate α-keto acid is decarboxylated to shorter-chain acid through an e_{CB}/O_2 process, in which an oxygen atom of O_2 is selectively incorporated into product.

3 Change of TiO_2 Surface Speciation

A modification of the surface speciation would change the nature of surface traps and the overall charge carrier dynamics. As the pH effect and the complexing ability of the substrate on the hole trap energy have been already discussed, here two main additional categories of processes that change the state of the surface will be considered.

The anatase/water interface and, more generally, the metal/oxide water interface are characterized by a charge build-up whose entity and sign not only is pH dependent but also results from the specific adsorption of ions [70]. Various cations, anions, and neutral molecules were found to affect the rate of chloroform degradation. At pH 7 addition of 1 mM Co^{2+} decreased the reaction rate by more than 50%, 0.2 mM Al^{3+} reduced the rate by 70%, and 0.5 mM Zn^{2+} reduced the rate by 60%. The addition ClO_4^- did not affect the rate, F^- increased by 15%, and Cl^- and PO_4^{3-} decreased the rate depending on pH [71]. It is recognized that the flat-band potential of semiconductor oxides, and, consequently, the positions of the band edges, depends on the nature and composition of the electrolyte. Specific adsorption of ions can shift the flat-band position significantly. The adsorption of ions can thus change the driving force of electron transfer and introduce surface states that can act as carrier trapping sites and recombination centers and can inhibit the adsorption of other species.

3.1 Adsorption of Cations

Metal cations adsorbed on a TiO_2 surface can act as sites where electron transfer is enhanced/inhibited, as sites at which charge carriers are separated and/or trapped, and as sites where charges recombine. Metal cations could be incorporated in the TiO_2 lattice or segregated at the surface during a sol–gel or impregnation methods. Because such syntheses typically result in both surface and bulk modifications of TiO_2, it becomes difficult to distinguish between bulk and surface effects. Transition metal cations that have redox properties (Cu^+/Cu^{2+}, Fe^{3+}, V^{5+}, and Cr^{3+}) can act as electron acceptors from the TiO_2 CB, promoting charge carrier separation and efficient photooxidation [72]. Particularly in the case of gas/solid reactions, these deposited metals generate new catalytic sites. For example, isolated surface Cu^+ sites on TiO_2 promoted CO_2 photodissociation to CO [73], and surface Fe^{3+} cations enhanced maleic acid photooxidation [74]. Metal cations like Cu, V, or Cr loaded onto TiO_2 during the sol–gel procedure raised NO photooxidation because it enhanced adsorption [75]. In some cases, the ability of adsorbed metal cations to scavenge CB electrons had a competitive effect on O_2 photoreduction, particularly when the reaction products of O_2 photoreduction were needed to promote indirect oxidation processes [76].

The metal cation adsorption could enhance the adsorption by altering the electrostatics at the surface that promote or inhibit electron transfer processes. When TiO_2 plays a passive role as in DSSC, the oxidation of I^- and formation of I_2^- were facilitated with alkali promoters (Mg^{2+}, Li^+, Na^+ and K^+) that can reverse the surface charge from negative to positive, stabilize I^- at the surface, and enhance its rate of oxidation by a photoionized dye [77].

Few examples are reported for adsorption from solutions. The metal cation adsorption on the surface of TiO_2 could *block/impede the adsorption* of the substrate redox species. This could limit the application of photocatalysis to high ionic strength media, like seawater [78]. However, adsorbed Al^{3+} cations halted poisoning caused by adsorption of strongly bound surface intermediates on TiO_2 during salicylic acid photodegradation [79]. In anatase/water systems under bandgap irradiation, both the organic substrate (formate) oxidation initiated by photogenerated valence band holes and the formation of hydrogen peroxide from O_2 reduction (by conduction band electrons) are strongly influenced by the presence of Zn^{2+} cations [80]. Depending on the pH, the formate oxidation rate can be enhanced or nearly completely inhibited. The observed result can be rationalized by considering the fraction of Ti-OH surface sites blocked by inner sphere complexation of Zn^{2+} as a function of pH. When this fraction is low, the more positive surface charge favors formate oxidation, whereas when the fraction is high, the almost complete blockage of Ti-OH surface sites by Zn^{2+} stops almost entirely formate oxidation.

3.2 Adsorption of Anions

The interest on the effect of anions (carbonate, chloride, sulfate, and nitrate) started from application to wastewater treatment. The surface occupation by anions may be competitive with adsorption of organic molecules. This effect is directly related to their coverage fraction. At the surface anions are subjected to redox transformations after electron transfer with photogenerated charge carriers (as for ClO_2^-, ClO_3^-, NO_2^-, and NO_3^- [30]). This second effect could produce inhibition by competition of inorganic ions with the organics. In a recent paper, the inhibition of photocatalytic degradation of 2,3-dichlophenol on highly ordered TiO_2 nanotube arrays prepared on titanium sheets was larger for $SO_4^{2-} > Cl^- > H_2PO_4^- > NO_3^-$. The observed inhibition effect was attributed to the competitive adsorption and the formation of less reactive radicals during the photocatalytic reaction [45]. For example, phosphate binds strongly to TiO_2, with a Langmuir binding constant at pH $= 2.3$ $K_{ads} = (3.8 \pm 0.8) \times 10^4$ dm^3 mol^{-1}, which is similar to the binding constants onto TiO_2 for bidentate ligand species such as oxalate and catechol [81]. The adsorption strongly changes the water structure at titanium dioxide interface [82]. In the presence of Cl^- anions, the surface had an isoelectric point near pH 5.5 and showed the least degree of water organization near this pH. The phosphate ions shifted the isoelectric point of the interface to pH 2.0, and the intensity of the 3,400 cm^{-1} peak was significantly increased in comparison with the chlorides data at both neutral and acidic pH values. Flat-band potentials determined by Mott–Schottky analysis in the absence of phosphate were Nernstian only for pH 3–7. With the addition of phosphate, impedance spectroscopy results showed additional space charge capacitance, peaking at potentials 150 mV positive of the flat-band potential [70].

Then drastic changes are caused on reactivity by manipulation of the surface chemical composition via exchange of surface hydroxyl groups.

It is long known that fluoride adsorbs onto TiO_2 surfaces (see, e.g., [83]), and the adsorption of fluoride inhibits the adsorption of other ligands, e.g., catechol and hydrogen peroxide [51, 63]. Fluorination of P25 greatly simplifies the surface IR spectrum [32], leaving only the component at 3,674 cm^{-1} that was assigned to one type of bridged hydroxyl groups. The νOH components removed by fluorination can be ascribed to hydroxyls sitting on defective sites, which interact more strongly with ligands. The surface of TiO_2 P25 is characterized by the presence of at least two different hydroxyls, with different coordination strength toward fluorides (and presumably to other ligands). The confirmation of this picture comes from the evolution of νOH patterns for Merck TiO_2 and their comparison with P25. Pristine and fluorinated Merck TiO_2 show similar νOH pattern, with a dominant spectroscopic feature at 3,674 cm^{-1}. The effect of fluorination in this case is the decrease of the intensity at 3,674 cm^{-1}, but the pattern does not change. The spectra of pristine and fluorinated Merck TiO_2 are very similar to that of fluorinated P25. As demonstrated by Sun et al. [84], fluoride preferentially adsorb on the {001} facets. Surface Ti(VI) ions on these facets are more exposed, so more coordinatively unsaturated, with respect to the more stable {101} facets.

In general, the complexation between surface Ti(IV) sites and ligands should affect the surface charge density and consequently the zeta potentials. The zeta potentials of suspended TiO_2 particles in water as a function of pH and $[F^-]$ have been reported [85]. The PZC of TiO_2 is measured to be ca. pH 6.2, which is in agreement with the literature values. In the presence of F^-, the PZC is shifted to lower pH values, and the positive charge on TiO_2 surface at acidic pH region is much reduced since no more surface hydroxyls on Ti(F)(OH) are available to be protonated. As a result, the concentration of tetramethylammonium ion $(CH_3)_4N^+$) at the TiO_2/water interface at pH 3 was higher on F-TiO_2 than on naked TiO_2 film due to a reduced electrostatic repulsion between the cation and the surface [86].

Since the first reports in 2000 [63, 64], hundreds of papers have been published on this issue. Surface fluorination improves the photocatalytic degradation of a number of simple organic compounds, such as phenol [63], benzoic acid [87, 88], benzene [89], cyanide [90], and N-nitrosodimethylamine [91], and for a variety of organic dyes [92–96]. The positive effect on the photocatalytic degradation has been directly associated with the displacement of OH terminal groups from the

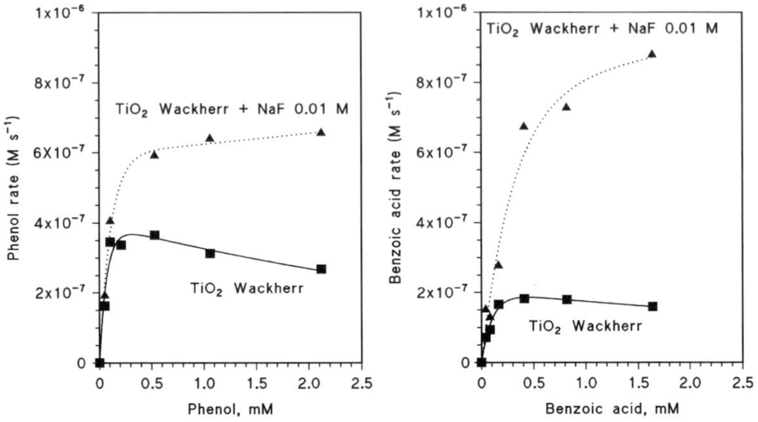

Fig. 5 Degradation rate of phenol and benzoic acid on Wackherr TiO_2 as a function of the substrate concentration in the absence and presence of fluoride ions at pH 3 (adapted from [87])

TiO_2 surface. The first hypothesis was that fluorination would enhance the generation of free OH radicals [63]. Acid Red 1 degradation mainly occurs via direct electron transfer because its rate is not depressed by 2-propanol addition on naked TiO_2. On the contrary, a strong decrease in the Acid Red 1 bleaching rate was observed upon 2-propanol addition on fluorinated TiO_2, indicating that oxidative paths through hydroxyl radicals play a major role under these conditions [97]. The convincing demonstration of the relevance of hydroxyl radicals on fluorinated TiO_2 was achieved by detecting the DMPO-OH adduct produced by irradiating unmodified and F-modified TiO2 suspensions [97]. As a consequence, the oxidation of the organic substrates would occur in solution, where the probability of back reaction (reduction) is reduced. Along the same line is the remote photocatalytic oxidation of stearic acids over the surface-fluorinated TiO_2 film monitored by Fourier transform infrared measurement and gas-chromatographic CO_2 production analysis, which was markedly faster with F-TiO_2 than with the pure TiO_2 film [98]. The production of CO_2 that evolved as a result of the remote oxidation of stearic acids was enhanced when H_2O_2 vapor was present but was strongly inhibited in the presence of ammonia gas that should scavenge OH radicals. These evidences suggested that the airborne oxidants in remote photocatalytic oxidation are most likely OH radicals and the surface fluorination of TiO_2 seems to facilitate desorption of OH radicals.

A comparison between the rate of degradation as a function of substrate concentration for phenol and benzoic acid shows that the effect of fluoride is more marked for benzoic acid than for phenol (see Fig. 5) [87]. Benzoic acid adsorbs on the surface of naked TiO_2 much more than phenol does. Interestingly the functional form of the rate is far from a Langmuir type and it is similar in the two cases, showing a decrement with increasing concentration. The rate of benzoic acid is lower than that of phenol and the reverse is seen in the presence of fluorides. This suggests that the adsorption is detrimental to the rate of degradation and that the

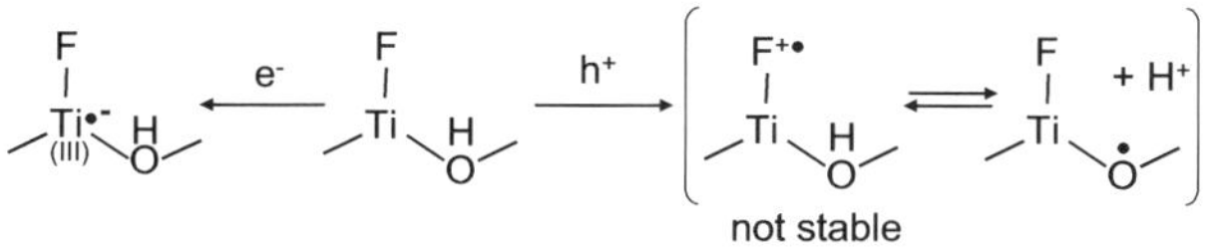

Fig. 6 Pictorial view of trapped electron and hole at the fluorinated TiO_2 surface

substrate surface complex can act as a recombination center (back reaction with CB electrons). This will be further discussed below using glycerol as substrate.

A decrease in photoactivity upon illumination was observed for formic acid [88] and dichloroacetate [85], species that are strongly bound to the TiO_2 surface. In the degradation of formic acid that adsorbs on the TiO_2 surface [51], it has been shown that no H_2O_2 is formed in the absence of (1) fluoride ions; (2) a hole scavenger; (3) oxygen, even in the presence of fluoride and Ag+ as electron scavenger. The surface fluorination of TiO_2 strongly promoted the photochemical production of hydrogen peroxide, the production rate of which was proportional to the surface Ti-F coverage. On the other hand, the degradation of H_2O_2 under photocatalytic conditions was inhibited by the presence of fluoride. Anions without surface complexing abilities (e.g., nitrate) did not lead to H_2O_2 formation. This suggests that the peroxides produced by O_2 reduction, when their adsorption is inhibited, are left in solution to be further reduced to H_2O_2, and when they are adsorbed, they act as recombination centers for holes. For all the above evidences [51, 85, 88], the explanation was that fluoride can displace peroxides from the Ti(IV) surface sites [99] hindering the direct hole transfer [100].

However, also the importance of fluorination on the electron transfer rate has been recognized [101]. By means of photopotential decay measurements, it was demonstrated that TiO_2 surface fluorination retards the reactivity of photogenerated electrons both for recombination with surface-trapped holes and for transfer to oxygen, upward shifts the electronic levels in the potential energy scale, changes the mechanism from direct to indirect for strong adsorbing species, and inhibits the adsorption of intermediates that could serve as recombination centers [102, 103].

A very schematic and pictorial view of trapped electron and hole at the adsorption site on the fluorinated TiO_2 surface, with limits already discussed for Fig. 3, is given in Fig. 6.

For fluorinated titania, some (but not all [32]) of the terminal hydroxyls are exchanged with fluoride, and the left tautomeric form is impeded by the high fluoride electronegativity. Thus $K/(1+K) \approx 1$ and $K >> 1$. In addition, as fluoride is more electronegative than −OH, the radical on bridging oxygen is less stable, i.e., the surface trap Ti(F)(O•) is more shallow than Ti(OH)(O•), and its energy level is more resonant with free holes in the valence band. This is consistent with the application to the half reaction at right of Fig. 6 of the Nernst law. The energy of the trap depends on the ratio of the formation constants β_{ox} of fluoride ion with the oxidized Ti(OH)(O•)) site to give (Ti(F)(O•)) and the formation constants β_{rid} of the fluoride ion with the site Ti(OH)(OH)) to give the reduced surface complex Ti(F)(OH)). As fluoride ions are able to displace from the surface most of organics, β_{rid} is very large. Conversely, as the left resonant form in parenthesis is not stable, $\beta_{rid}/\beta_{ox} > 1$, and the oxidant potential of the trap is increased. Its *value could then be very close to the oxidant potential of the VB free hole*. Since fluorination increases the rate for substrates that react with OH radicals and depresses the rate for substrates that react by direct electron transfer, the right form *Ti(F)(O•) performs as an OH radical*. In recent reconsideration of surface fluorination [104], the decrease of the recombination rate concurrent with the increase of the electron transfer rate with reduced dissolved species is invoked to explain the fluoride effect.

Due to the different electron affinity of the groups involved, also free electrons are more stabilized on −Ti(F)− than on −Ti(OH)− and the surface-assisted recombination is reduced. No experimental evidence is reported on the stability of the $Ti^{\bullet -}$(F)(OH) species depicted at left of Fig. 6. A release in solution of F− would change this species to $Ti^{\bullet}$()(OH) as in Fig. 3 (left). However, the reduced rate of O_2 reduction on fluorinated TiO_2 with respect to the naked one [102, 103] suggests that chemisorption is not allowed and the possible electron trap is that depicted in Fig. 6.

Alcohols, polyols, and carboxylic acids show good coordinative abilities toward Ti(IV) ions. At suitable concentration, these species are able to occupy surface sites. Surface complexation will form a surface deep trap for holes, as the oxidation potential of the surface complex Ti(OR)(OH) is lowered with respect to Ti(OH)(OH). Besides being an efficient recombination center, the oxidized surface complex (Ti($^{+}$•OR)(OH)) is an alkoxy radical-like species that has a chemical reactivity very different from the carbon-centered radical •OR generated via H-abstraction by the surface adsorbed OH•, namely, the Ti(OH)(O•) hole trap [105]. This was demonstrated using glycerol as substrate.

In this case the two produced carbon-centered radicals evolve to dihydroxyacetone (C3) or glycerolaldehyde (C3), while the surface complex (Ti($^{+}$•OR)(OH)) undergoes β-scission, giving formaldehyde (C1) and a second carbon-centered radical, which by reaction with molecular oxygen at diffusion controlled rates produces glycolaldehyde (C2) and hydroperoxydes. The reported case is quite interesting because, as β-scission is a slow process, an inhibition of the reaction rate is observed when all the surface hydroxyls are exchanged with glycerol. The deep-trapped hole (Ti($^{+}$•OR)(OH)) must recombine free electrons more easily than (Ti(OH)(O•)). In fact numerical simulation of the rate dependence on glycerol concentration showed that the kinetic profile is correct only if the deep-trapped hole (Ti($^{+}$•OR)(OH)) has a very high kinetic constant for electron recombination. It is also worth noting that the shift of the mechanism from an oxidative attack to not chemisorbed glycerol mediated by Ti(OH)(O•) shallow surface hole traps, to a direct hole transfer to the surface complex, leads to very different intermediates.

In the presence of fluorides that competitively adsorb on the surface, the evident experimental result is that mainly the two C3 carbohydrates are produced from glycerol. On Merck TiO_2 almost no change is observed with addition of fluorides, while on P25 the production of formaldehyde and glycolaldehyde was strongly depressed. As fluorides impede the surface complexation on the surface Ti-OH site, the formation of formaldehyde and glycolaldehyde is strictly connected with glycerol chemisorption.

When the surface is fluorinated, the surface complex (Ti(OR)(OH)) is no more allowed and only oxidative attack mediated by Ti(F)(O•) shallow surface hole traps is possible. In this case the two produced carbon-centered radicals evolve to dihydroxyacetone (C3) or glycerolaldehyde (C3) as in OH radical chemistry. The rates for glycerol degradation measured on fluorinated P25 [68, 105] (with most adsorption sites masked) are very similar to that on fluorinated and pristine Merck, except for a higher factor 4–5 that was ascribed to the different surface area of catalysts.

All the above results suggest that fluorination impedes surface complexation by substrates to be oxidized and renders the surface hole trap more shallows and resonant with valence band hole. The electron transfer will then follow the mechanism and energetics similar to that of free OH radicals.

4 Conclusions

The concepts of electron transfer *across* the interface and *at* the interface, together with the very nature of surface traps, are able to qualitatively (and in some case quantitatively [105]) explain the reactivity of electrons and holes produced by light absorption in semiconductors. The nature of surface traps depends on the crystalline phase and the specimen, as different surface hydroxyls are present with different complexing ability toward inorganic and organic species. The electron transfer *across* the interface is limited to few physisorbed species in some materials (e.g., P25) and is more common with others (e.g., anatase Merck). The reverse is true for electron transfer *at* the interface. The trap energy position, and consequently the trapping timescale and the overall process rate, depends on the photocatalyst, the substrate, and its concentration. Experiments with defined substrates under arbitrary chosen concentrations can be misleading. The functional dependence of the experimental response (IR, EPR, luminescence, overall rate... also time resolved) from the surface coverage of the substrate and other competitive complexing species is the suggested recipe to better understand the catalytic sites at the TiO_2 surface.

Acknowledgements The University of Torino support by Ricerca Locale 2012 is kindly acknowledged.

References

1. Watanabe M, Hayashi T (2005) J Lumin 112:88–91
2. Berger T, Sterrer M, Diwald O, Knözinger E, Panayotov D, Thompson TL, Yates JT Jr (2005) J Phys Chem B 109:6061–6068

3. Emeline AV, Frolov AV, Ryabchuk VK, Serpone N (2003) J Phys Chem B 107:7109–7119
4. Fu Y, Cao WH (2006) J Appl Phys 100:084324
5. Planelles J, Movilla JL (2006) Phys Rev B 73:235350
6. Tamaki Y, Furube A, Murai M, Hara K, Katoh R, Tachiya M (2007) Phys Chem Chem Phys 9:1453–1460
7. Rabani J, Yamashita K, Ushida K, Stark J, Kira A (1998) J Phys Chem B 102:1689–1695
8. Katoh R, Murai M, Furbe A (2008) Chem Phys Lett 461:238–241
9. Tamaki Y, Hara K, Katoh R, Tachiya M, Furube A (2009) J Phys Chem C 113:11741–11746
10. Bahnemann DW, Hilgendorff M, Memming R (1997) J Phys Chem B 101:4265–4275
11. Corrent S, Cosa G, Scaiano JC, Galletero MS, Alvaro M, Garcia H (2001) Chem Mater 13:715–722
12. Szczepankiewicz SH, Moss JA, Hoffmann MR (2002) J Phys Chem B 106:2922–2927
13. Peiro AM, Colombo C, Doyle G, Nelson J, Mills A, Durrant JR (2006) J Phys Chem B 110:23255–23263
14. Ikeda S, Sugiyama N, Muratami S, Kominami H, Kera Y, Noguchi H, Uosaki K, Torimoto T, Ohtani B (2003) Phys Chem Chem Phys 5:778–783
15. Szczepankiewicz SH, Colussi AJ, Hoffmann MR (2000) J Phys Chem B 104:9842–9850
16. Henderson MA (2011) Surf Sci Rep 66:185–297
17. Warren DS, Mcquillan AJ (2004) J Phys Chem B 108:19373–19379
18. Diebold U (2003) Surf Sci Rep 48:53–229
19. Micic OI, Zhang YN, Cromack KR, Trifunac AD, Thurnauer MC (1993) J Phys Chem 97:7277–7283
20. Serpone N, Lawless D, Khairutdinov R, Pelizzetti E (1995) J Phys Chem 99:16655–16661
21. Ke SC, Wang TC, Wong MS, Gopal NO (2006) J Phys Chem B 110:11628–11634
22. Deskins NA, Dupuis M (2009) J Phys Chem C 113:346–358
23. Hurum D C, Gray K A, Rajh T, Thurnauer M C (2005) J Phys Chem B 109:977–980, erratum 5388
24. Dimitrijevic NM, Saponjic ZV, Rabatic BM, Poluektov OG, Rajh T (2007) J Phys Chem C 111:14597–14601
25. Berger T, Sterrer M, Diwald O, Knozinger E, Panayotov D, Thompson TL, Yates JT Jr (2005) J Phys Chem B 109:6061–6068
26. Yoshihara T, Tamaki Y, Furube A, Murai M, Hara K, Katoh R (2007) Chem Phys Lett 438:268–273
27. Tamaki Y, Furube A, Murai M, Hara K, Katoh R, Tachiya M (2006) J Am Chem Soc 128:416–417
28. Minero C (1999) Catal Today 54:205–216
29. Minero C, Vione D (2006) Appl Catal B Environ 67:257–269
30. Gao R, Safrany A, Rabani J (2003) Radiat Phys Chem 67:25–39
31. Berger T, Lana-Villarreal T, Monllor-Satoca D, Gómez R (2007) J Phys Chem C 111:9936–9942
32. Minella M, Faga MG, Maurino V, Minero C, Pelizzetti E, Cosuccia S, Martra G (2010) Langmuir 26:2521–2527
33. Primet M, Pichat P, Mathieu M-V (1971) J Phys Chem 75:1216–1220; 75:1221–1226
34. Connor PA, Dobson KD, McQuillan AJ (1999) Langmuir 15:2402–2408
35. Hiemstra T, Van Riemsdijk WH (2006) J Colloid Interface Sci 301:1–18
36. Machesky ML, Wesolowski DJ, Palmer DA, Ridley MK (2001) J Colloid Interface Sci 239:314–327
37. Fitts JP, Machesky ML, Wesolowski DJ, Shang XM, Kubicki JD, Flynn GW, Heinz TF, Eisenthal KB (2005) Chem Phys Lett 411:399–403
38. Ambrus Z, Mogyorósi K, Szalai A, Alapi T, Demeter K, Dombi A, Sipos P (2008) Appl Catal A Gen 340:153–161
39. Gomathi DL, Narasimha MB, Girish KS (2009) Chemosphere 76:1163–1166
40. Vargas R, Núñez O (2009) J Mol Catal A Chem 300:65–71

41. Fujishima A, Zhang X, Tryk DA (2008) Surf Sci Rep 63:515–582
42. Kimmel GA, Petrik NG (2008) Phys Rev Lett 100:196102
43. Imanishi A, Okamura T, Ohashi N, Nakamura R, Nakato Y (2007) J Am Chem Soc 129:11569–11578
44. Cornu CJG, Colussi AJ, Hoffmann MR (2003) J Phys Chem B 107:3156–3160
45. Liang H-C, Li X-Z, Yang Y-H, Sze K-H, (2008) Chemosphere 73:805–812
46. López-Muñoz MJ, Aguado J, Arencibia A, Pascual R (2011) Appl Catal B Environ 104:220–228
47. Ohno T, Sarukawa K, Matsumura M (2002) J Chem 26:1167–1170
48. Petrik NG, Zhang Z, Du Y, Dohnálek Z, Lyubinetsky I, Kimmel GA (2009) J Phys Chem C 113:12407–12411
49. Henderson MA, Epling WS, Peden CHF, Perkins CL (2003) J Phys Chem B 107:534–545
50. Nakamura R, Imanishi A, Murakoshi K, Nakato Y (2003) J Am Chem Soc 125:7443–7450
51. Maurino V, Minero C, Mariella G, Pelizzetti E (2005) Chem Commun (20):2627–2629
52. Haick H, Paz Y (2003) ChemPhysChem 4:617–620
53. Tachikawa T, Majima T (2010) Chem Soc Rev 39:4802–4819
54. Thiebaud J, Thévenet F, Fittschen C (2010) J Phys Chem C 114:3082–3088
55. Anpo M, Shima T, Kubokawa Y (1985) Chem Lett (12):1799–1805
56. Henderson MA (2002) Surf Sci Rep 46:1–308
57. Di Valentin C, Pacchioni G, Selloni A (2006) Phys Rev Lett 97:166803
58. Nosaka Y, Komori S, Yawata K, Hirakawa T, Nosaka YA (2003) Phys Chem Chem Phys 5:4731–4739
59. Howe RF, Graetzel M (1987) J Phys Chem 91:3906–3911
60. Micic OI, Zhang Y, Cromack KR, Trifunac AD, Thurnauer MC (1993) J Phys Chem 97:13284–13289
61. Lawless D, Sermone N, Meisel D (1991) J Phys Chem 95:5166–5170
62. Minero C, Catozzo F, Pelizzetti E (1992) Langmuir 8:481–486
63. Minero C, Mariella G, Maurino V, Pelizzetti E (2000) Langmuir 16:2632–2641
64. Minero C, Mariella G, Maurino V, Pelizzetti E (2000) Langmuir 16:8964–8972
65. Pang CL, Lindsay R, Thornton G (2008) Chem Soc Rev 37:2328–2353
66. Byrne JA, Eggins BR, Linquette-Mailley S, Dunlop PSM (1998) Analyst 123:2007–2012
67. Berger T, Rodes A, Gómez R (2010) Phys Chem Chem Phys 12:10503–10511
68. Bedini A, Maurino V, Minella M, Minero C, Rubertelli F (2008) J Adv Oxid Technol 11:184–192
69. Wen B, Li Y, Chen C, Ma W, Zhao J (2010) Chem Eur J 16:11859–11866
70. Nelson BP, Candal R, Corn RM, Anderson MA (2000) Langmuir 16:6094–6101
71. Kormann C, Bahnemann DW, Hoffmann MR (1991) Environ Sci Technol 25:494–500
72. Murakami N, Chiyoya T, Tsubota T, Ohno T (2008) Appl Catal A General 348:148–152
73. Liu L, Zhao C, Li Y (2012) J Phys Chem C 116:7904–7912
74. Franch MI, Ayllon JA, Peral J, Domenech X (2005) Catal Today 101:245–252
75. Wu JCS, Cheng Y-T (2006) J Catal 237:393–404
76. Chen CC, Li XZ, Ma WH, Zhao JC, Hidaka H, Serpone N (2002) J Phys Chem B 106:318–324
77. Pelet S, Moser JE, Grätzel M (2000) J Phys Chem B 104:1791–1795
78. Minero C, Maurino V, Pelizzetti E (1997) Mar Chem 58:361–372
79. Franch MI, Peral J, Domenech X, Ayllon JA (2005) Chem Commun 14:1851–1853
80. Maurino V, Minero C, Pelizzetti E, Mariella G, Arbezzano A, Rubertelli F (2007) Res Chem Intermed 33:319–332
81. Connor PA, McQuillan AJ (1999) Langmuir 15:2916–2921
82. Kataoka S, Gurau MC, Albertorio F, Holden MA, Lim S-M, Lim S-M, Yang RD, Cremer PS (2004) Langmuir 20:1662–1666
83. Stone AT (1996) Environ Sci Technol 30:1604–1613

84. Yang HG, Sun CH, Qiao SZ, Zou J, Liu G, Campbell Smith S, Cheng HM, Lu GQ (2008) Nature 453:638–642
85. Park H, Choi WJ (2004) J Phys Chem B 108:4086–4093
86. Vohra MS, Kim S, Choi WJ (2003) J Photochem Photobiol A 160:55–60
87. Vione D, Minero C, Maurino V, Carlotti ME, Picatonotto T, Pelizzetti E (2005) Appl Catal B Environ 58:81–90
88. Mrowetz M, Selli E (2006) J Chem 30:108–114
89. Park H, Choi W (2005) Catal Today 101:291–297
90. Chiang K, Amal R, Tran TJ (2003) Mol Catal 193:285–297
91. Lee J, Choi W, Yoon J (2005) Environ Sci Technol 39:6800–6807
92. Yu J, Zhang J (2010) Dalton Trans 39:5860–5867
93. Dozzi MV, Schiavello GL, Selli E (2010) J Adv Oxid Technol 13:305–312
94. Chen Y, Chen F, Zhang J (2009) Appl Surf Sci 255:6290–6296
95. Chen KT, Lu CS, Chang TH, Lai YY, Chang TH, Wu CW, Chen CCJ (2010) Hazard Mater 174:598–609
96. Lv K, Li X, Deng K, Sun J, Li X, Li M (2010) Appl Catal B Environ 95:383–392
97. Mrowetz M, Selli E (2005) Phys Chem Chem Phys 7:1100–1102
98. Park JS, Choi W (2004) Langmuir 20:11523–11527
99. Lv K, Xu Y (2006) J Phys Chem B 110:6204–6212
100. Monllor-Satoca D, Gomez R, Gonzalez-Hidalgo M, Salvador P (2007) Catal Today 129:247–255
101. Xu Y, Lv K, Xiong Z, Leng W, Du W, Liu D, Xue X (2007) J Phys Chem C 111:19024–19032
102. Monllor-Satoca D, Gomez R (2008) J Phys Chem C 112:139–147
103. Monllor-Satoca D, Lana-Villarreal T, Gómez R (2011) Langmuir 27:15312–15321
104. Montoya JF, Salvador P (2010) Appl Catal B Environ 94:97–107
105. Minero C, Bedini A, Maurino V (2012) Appl Catal B Environ 128:135–143

Photocatalytic Splitting of Water

Nathan Skillen, Cathy McCullagh, and Morgan Adams

Abstract The use of photocatalysis for the photosplitting of water to generate hydrogen and oxygen has gained interest as a method for the conversion and storage of solar energy. The application of photocatalysis through catalyst engineering, mechanistic studies and photoreactor development has highlighted the potential of this technology, with the number of publications significantly increasing in the past few decades. In 1972 Fujishima and Honda described a photoelectrochemical system capable of generating H_2 and O_2 using thin-film TiO_2. Since this publication, a diverse range of catalysts and platforms have been deployed, along with a varying range of photoreactors coupled with photoelectrochemical and photovoltaic technology. This chapter aims to provide a comprehensive overview of photocatalytic technology applied to overall H_2O splitting. An insight into the electronic and geometric structure of catalysts is given based upon the one- and two-step photocatalyst systems. One-step photocatalysts are discussed based upon their d^0 and d^{10} electron configuration and core metal ion including transition metal oxides, typical metal oxides and metal nitrides. The two-step approach, referred to as the Z-scheme, is discussed as an alternative approach to the traditional one-step mechanism, and the potential of the system to utilise visible and solar irradiation. In addition to this the mechanistic procedure of H_2O splitting is reviewed to provide the reader with a detailed understanding of the process. Finally, the development of photoreactors and reactor properties are discussed with a view towards the photoelectrochemical splitting of H_2O.

Keywords Photocatalysis, Photoelectrochemistry, Photoreactors, Semiconductors, Water splitting

N. Skillen (✉), C. McCullagh, and M. Adams
IDEAS, Institute for Innovation, Design and Sustainability Research, CREE, Centre for Research in Energy and the Environment, School of Engineering, The Robert Gordon University, Aberdeen AB10 7QB, UK
e-mail: n.c.skillen1@rgu.ac.uk

D.W. Bahnemann and P.K.J. Robertson (eds.), *Environmental Photochemistry Part III*, Hdb Env Chem (2015) 35: 45–86, DOI 10.1007/698_2014_261,

Contents

1 Introduction

1.1 Historical Overview of Water Splitting

The production of renewable and non-polluting fuels via the direct conversion of solar energy into chemical energy remains a fascinating challenge for the end of this century. Among various interesting reactions, the splitting of water into molecular hydrogen and molecular oxygen by visible light is potentially one of the most promising ways for the photochemical conversion and storage of solar energy [1–5]. Since the first reported photosplitting of water by Fujishima and Honda [6] in 1972 many authors have published their efforts to split water using semiconductor photocatalysis [7–10].

Photocatalytic water splitting was reported by Fujishima and Honda in 1972 where they used TiO_2 thin film as the photocatalyst. Since then, TiO_2 has become a widely used photocatalyst in photocatalytic water splitting. Nevertheless, the biggest disadvantage of TiO_2 is its inability to harvest the visible light which accounts for a major portion of sunlight. To overcome this shortcoming, several techniques, such as metal doping, ion doping and dye sensitisation, have been studied extensively [11–13]. Despite the successful development of several visible-light-driven photocatalysts, only low H_2 or O_2 production yield can be obtained, which is attributed to the intrinsic band gap limitation of the photocatalyst.

This chapter aims to give an insight into photocatalytic technology applied to H_2O splitting for the production of H_2 and O_2 covering the areas of catalyst development, mechanistic pathways and reaction chambers deployed. A comprehensive, yet not exhaustive, review has been compiled to provide the reader with an understanding of the fundamentals of this field of research. The topics covered herein include a review of one- and two-step photocatalysts, the mechanism of H_2O splitting and a brief overview of photoreactors including photoelectrochemical cell typically used for the production of H_2 and O_2.

1.2 Important Parameters to Be Considered for Water Splitting

The following chapter is written with a view towards overall H_2O splitting and as such a number of parameters, put forward by Kudo and Miseki in their review in 2009, should be considered when reviewing published work in the field. There are additional parameters which are discussed in review papers [9, 14]; however, detailed below are the key parameters:

1. Stoichiometric production of H_2 and O_2
2. Experimental time frame
3. The turnover number (TON)
4. Quantum yield
5. Photoresponse

The stoichiometric production of H_2 and O_2 should follow the ratio of 2:1 respectively in the absence of a sacrificial reagent. Often reported is the evolution of H_2 in the presence of a sacrificial reagent with minimal O_2 recorded. It is ambiguous as to whether this is overall H_2O splitting and not a sacrificial reaction.

The evolution of H_2 and O_2 should be directly proportional to time and should increase with an increasing irradiation time. The production of H_2 and O_2 should also be stable over the time course with evolution occurring after the system has been evacuated.

The turnover number (TON) refers to the production of H_2 and O_2 in relation to the photocatalyst. In overall H_2O splitting the production of H_2/O_2 should be significantly greater than the amount of catalyst deployed. If the quantity of H_2/O_2 is less than the amount of catalyst it is not clear if the process has occurred photocatalytically. The TON is typically defined as the ratio of the number of reacted molecules to the number of active sites. As the number of active sites for a photocatalyst is difficult to establish the TON is often calculated by the ratio of number of reacted electrons to the number of atoms in a photocatalyst or number of atoms at the photocatalyst surface. The number of reacted electrons can be established from the volume of H_2 evolved.

Variation in experimental conditions presents a problem in the comparison and review of data. While the majority of data is presented as a unit such as $\mu mol\ h^{-1}$, photocatalytic activity is dependent on conditions such as irradiation source and reactor geometry. As such the quantum yield is an important parameter to evaluate. The determination of the number of absorbed photons by a photocatalyst is difficult to ascertain; therefore the calculated quantum yield is referred to as the apparent quantum yield (AQY). The AQY is calculated as the ratio of the number of reacted electrons to the number of incident photons.

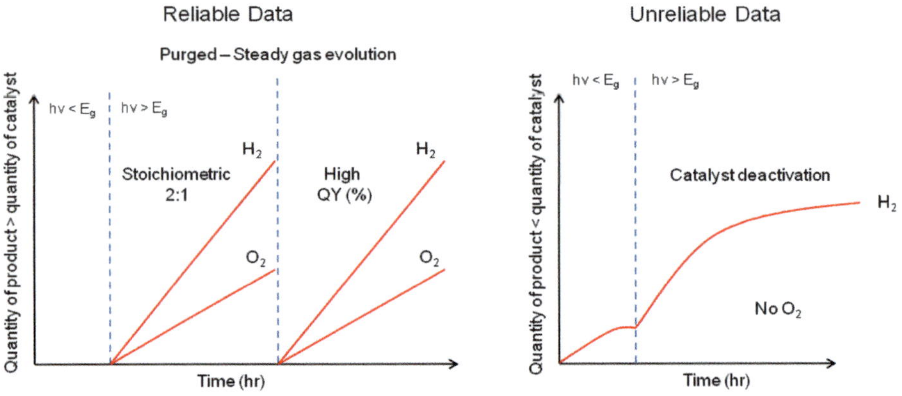

Fig. 1 Comparison of typical reliable and unreliable data for H_2O splitting

The response which is initiated by the absorption of light energy greater than the band gap of the catalyst should be evaluated especially when using a visible light catalyst. This can be achieved by using suitable control experiments which monitor H_2O splitting in the absence of a catalyst or illumination. Cut-off filters should also be used with visible emitting lamps to ensure activity being recorded is a result of the catalyst being excited by visible light photons.

Taking into consideration the above-mentioned parameters there is a general trend which indicates H_2O splitting results are reliable (Fig. 1). Reliable data should show steady stoichiometric evolution of both H_2 and O_2 with no activity prior to illumination and no significant deactivation of the catalyst over an increasing time frame. In contrast, unreliable data shows activity which could not be attributed to photocatalytic activity such as H_2 evolution under no illumination and lack of O_2 evolution.

2 Photocatalysts for Water Splitting

Water splitting over photocatalysts can be divided into two broad categories. The initial approach is the one-step photocatalyst stage, during which a catalyst is exposed to light to generate the production of H_2 and O_2. This approach requires the photocatalyst to have sufficient thermodynamic potential to allow the splitting of H_2O. Ideally it would also require a catalyst to have a narrow band gap to allow excitation by visible photons and have sufficient stability to prevent photocorrosion. The second approach uses a two-photocatalyst system, modelled on the photosynthesis process. The system is referred to as the Z-scheme.

When considering powdered photocatalysts for H_2O splitting a significant level of focus is given to transition and typical metal oxides and nitrides. Figure 2 details common elements which are used to compose photocatalysts for H_2O splitting. The red and green highlighted elements are transition and typical metals, respectively,

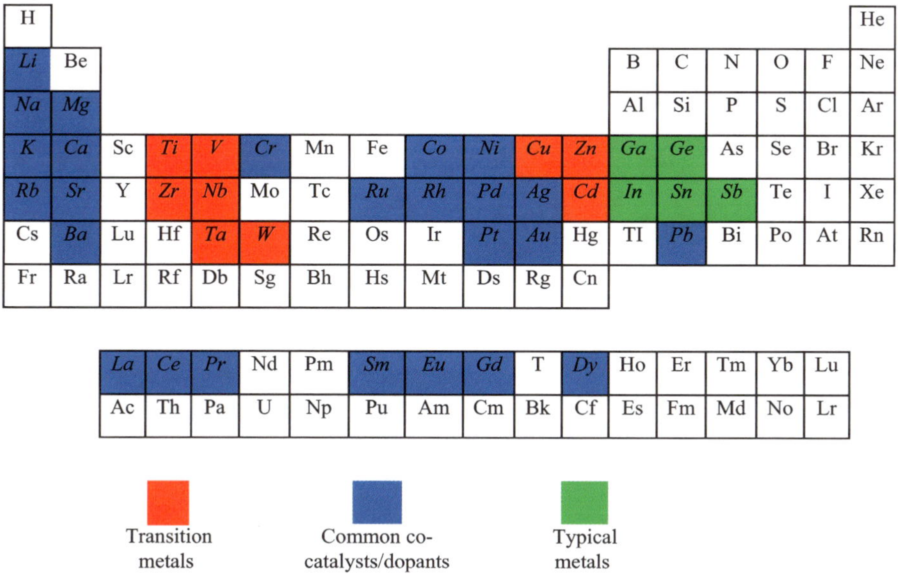

Fig. 2 Periodic table highlighting common elements used in H_2O splitting catalysts

while the blue highlighted elements are those that are frequently used as dopants and co-catalysts in photocatalytic complexes.

It is the electronic configuration of catalysts which makes them suitable for photocatalytic mechanisms. Transition metal oxides with a vacant d orbital (d^0 configuration) primarily include titanates, tantalates, niobates, vanadates and tungstanates. Typical metal oxides have shown photocatalytic H_2O splitting activity with an occupied d orbital (d^{10} configuration). Inoue [15] made the observation that d^0 and d^{10} configurations on a quantum chemistry level behave in a similar fashion. Catalysts with d^{10} configurations include gallates, germanates, stannates, antimonates and indates.

The catalysts discussed here are grouped firstly on their electronic configuration and secondly on the core metal ion of the compound. Within these groupings the geometric and electronic structure, distortion and performance of the catalysts are discussed. An overview of selected catalysts is also provided in Table 1; however, the reader is encouraged to refer to a number of excellent review papers which provide a detailed insight into catalyst development [9, 15, 34–38].

2.1 Water Splitting Over Powder Photocatalysts

The photocatalytic decomposition of H_2O over semiconductor powders has been studied for the conversion of photon energy into chemical energy. Thermodynamically, the photocatalytic splitting of H_2O into H_2 and O_2 Eq. (1) is described as an

Table 1 Overview of selected publications for overall water splitting

Electronic configuration	Core metal ion	Catalyst	Co-catalyst	E_g	Reaction solution	Activity (μmol h^{-1}) H$_2$	O$_2$	AQY (%)	References
d^0	Ti^{4+}	TiO_2	Pt	3.2	H_2O/2.2 M Na_2CO_3	568	287		Sayama and Arakawa [16]
		TiO_2	Rh	3.2	3 M NaOH	449	–	29	Yamaguti and Sato [17]
		$Sr_3Ti_2O_7$	NiO	3.2	Pure water	83	42	4.3	Jeong et al. [18]
		$K_2La_2Ti_3O_{10}$	NiO	3.5	H_20/0.1 M KOH	444	221		Takata et al. [19]
		$Na_2Ti_6O_{13}$	RuO_2	–	Pure water	~17	~8		Ogura et al. [20]
		$BaTi_4O_9$	RuO_2	–	Pure water	~18	~8		Inoue et al. [21]
	Zr^{4+}	ZrO_2	–	3.93	H_2O/Na_2CO_3	88.8	45		Reddy et al. [22]
	V^{5+}	$InVO_4$	NiO	1.8	Pure water	69.1	34.5		Lin et al. [23]
	Nb^{5+}	$K_4Nb_6O_{17}$	Pt	3.4	H_2O/2.2 M Na_2CO_3	451	217		Sayama et al. [24]
		$SbNbO_4$	RuO_2	3.1	Pure water	~2.4	–		Kim et al. [25]
		$H_{1.6}K_{0.2}La_{0.3}Bi_{0.1}Nb_2O_{6.5}$	–	3.49	Pure water	27.2	Trace		Chen et al. [26]
	Ta^{5+}	NaTaO3:La	NiO	4.0	Pure water	19.8	9.66	56	Kato et al. [27]
		$Sr_2Ta_2O_7$	NiO	4.55	Pure water	1,000	500		Kato and Kudo [28]
		$SbTaO_4$	RuO_2	3.7	Pure water	~5.8			Kim et al. [25]
		$K_2Ta_2O_6$	NiO	4.6	H_2O/NaOH	437	226		Ikeda et al. [29]
	W^{6+}	$PbWO_4$	RuO_2	3.24	Pure water	~18	~8		Kadowaki et al. [30]
d^{10}	Ga^{3+}	Ga_2O_3	NiO	4.6	Pure water	46	23		Inoue [15]
		Ga_2O_3:Zn	Ni	4.6	Pure water	4,100	2,200		Sakata et al. [31]
		GaN:ZnO	RuO_2	2.58	H_2O/H_2SO_4	0.98	0.48	0.14	Maeda et al. [1]
		$(Ga_{1-x}Zn)(N_{1-x}O_x)$	RuO_2	2.7	Pure water	3,835	1,988		Maeda et al. [32]
	In^{3+}	$NaInO_2$	RuO_2	3.9	Pure water	0.9	0.3		Kudo and Miseki [9]
		$SrIn_2O_4$	RuO_2	3.6	Pure water	7	3		Kudo and Miseki [9]
	Ge^{4+}	β-Ge_3N_4	RuO_2	3.85	Pure water	0.8	0.4	9	Maeda et al. [32]
	Sn^{4+}	Sr_2SnO_4	RuO_2	–	Pure water	5	2.5		Inoue [15]
	Sb^{5+}	$Ca_2Sb_2O_7$	RuO_2	3.6	Pure water	2.9	1		Sato et al. [33]
		$Sr_2Sb_2O_7$	RuO_2	4.0	Pure water	8	3		Sato et al. [33]
		$NaSbO_3$	RuO_2	3.6	Pure water	1.7	0.8		Sato et al. [33]

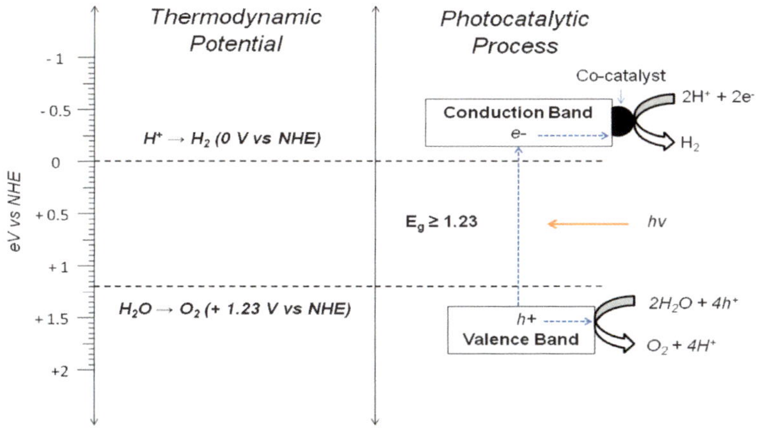

Fig. 3 Mechanism of photocatalytic H_2O splitting

"uphill reaction" where a large positive change in the Gibbs free energy is observed ($\Delta G_o = -238$ kJmol^{-1}) [35]. The properties of a catalyst including the positioning of the conduction band (CB) and valence band (VB) can dictate the level of activity. The CB of the catalyst must be more negative than the reduction potential of H^+ to H_2 (0 eV vs NHE at pH 0) (Eq. (2)). The VB must also be more positive than the oxidation potential of H_2O to O_2 (1.23 eV vs NHE) [35] (Eq. (3)):

$$2H_2O \rightarrow 2H_2 + O_2 \tag{1}$$

$$2H^+ + 2e^- \rightarrow H_2 \tag{2}$$

$$2H_2O + 4h^+ \rightarrow O_2 + 4H^+ \tag{3}$$

The mechanism of photocatalytic H_2O splitting along with the thermodynamic potential for the half reactions is shown in Fig. 3. The presence of a co-catalyst in the CB of the catalysts is typically required to allow the protonation of H^+ to H_2. Common co-catalysts are highlighted in Fig. 2 and include Pt [39], RuO_2 [15] and NiO [40].

The band positions of common catalysts are shown in Fig. 4 in relation to the required potential of H_2O splitting. The E_g of the catalysts are shown and as can be seen there are few catalysts which possess an E_g suitable for excitation by wavelengths in the visible region of the electromagnetic spectrum.

The band position of the catalyst is not the sole requirement for H_2O splitting to proceed as a number of additional requirements must be addressed. Suppressing the recombination of holes and electrons and efficient charge separation are key requirements. The creation and separation of active sites for H_2 and O_2 evolution are also important. It has been suggested that the separation of evolution sites should be on the nanometre scale to ensure the back reaction of O_2 and H_2 does not occur [27]. Suppressing the back reaction will also prevent the formation of any

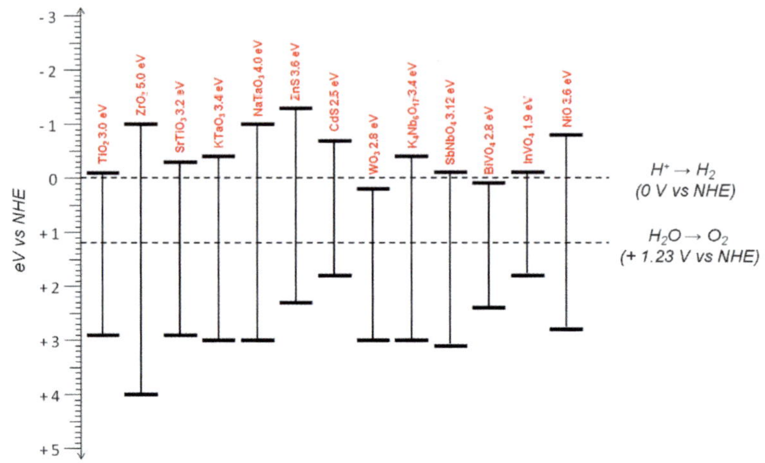

Fig. 4 Band position of selected catalysts and the potential for H_2O splitting

intermediates in the process. The catalyst structure has been investigated to improve the catalytic activity through charge separation and back reaction suppression. A number of catalyst structures have been reported including anatase, perovskite, layered structure, tunnel structure and cubic pyrochlore [9, 15, 19, 20, 26, 35, 41, 42]. While these structures may differ in chemical composition and crystal phase, MO_6/MO_4 (M=transition/typical metal) octahedral/tetrahedral units are often present and have been reported to play a key role in the activity exhibited by a catalyst. Furthermore, the distortion of this fundamental unit as a result of the M–O–M bond angle has also been found to impact activity.

2.2 *d^0 Configuration Catalysts*

Transition metals with a vacant d orbital have shown activity towards H_2O decomposition with a range of catalysts reported [9, 15]. The core metal ion of the catalysts reported include Ti^{4+}, Zr^{4+}, V^{5+}, Nb^{5+}, Ta^{5+} and W^{6+}. All of the metal ions have a d^0 electronic configuration. Discussed here are key examples of catalysts with a d^0 configuration based upon the core metal ion.

2.2.1 Titanates

Ti-based catalysts have been extensively used in the field of photocatalysis, specifically the use of TiO_2. Photocatalytic splitting of H_2O over TiO_2 was first reported by Fujishima and Honda using a photoelectrochemical system. The system comprised a TiO_2 electrode with Pt counter electrode, which used an external bias from a power supply. Upon UV illumination, electrons and holes were generated in the system which contributed to the reduction and oxidation of H_2O. On the Pt counter electrode the electrons reduced H_2O to H_2, while H_2O was oxidised to O_2 on the TiO_2 electrode. In contrast to this, powdered TiO_2 catalysts are not capable of splitting H_2O without modification to the catalyst such as the addition of a co-catalyst.

In the early stages of H_2O splitting research the capability of Pt-loaded TiO_2 was questionable due to low yields of products and no O_2 evolution. A number of publications, however, showed the evolution of both H_2 and O_2 over platinised TiO_2 using NaOH coatings [17, 39] and the addition of alkali carbonates [16]. Yamaguti and Sato found that a 10 wt% NaOH coating onto Pt–TiO_2 to be effective for H_2O splitting, producing a quantum yield of 17%. Following their study in 1984, Sayama and Arakawa [16] further increased the quantum yield to 29% when using a NaOH-coated Rh–TiO_2 catalyst. The enhanced activity of Pt–TiO_2 was due to the ability of Pt to restrict the recombination of electrons and holes and catalyse proton reduction [43–45]. Pt has improved electron acceptor properties as a result of a larger work function (5.65 eV), as described by Mizukoshi et al. [45], which prevents the recombination of photogenerated electrons and holes.

The addition of carbonate salts to a Pt–TiO_2 suspension was shown to increase the production of H_2 and O_2 from H_2O splitting. Sayama and Arakawa [16] found Pt–TiO_2 in suspension with 2.2 M Na_2CO_3 produced 568 and 287 μmol h^{-1} of H_2 and O_2, respectively, which conforms to the stoichiometric ratio of 2:1. The presence of carbonate species on the Pt–TiO_2 catalyst played a significant role in the reaction pathway. On the Pt, absorbed carbonates suppressed the H_2O splitting back reaction, while on TiO_2, the effective desorption of O_2 prevented photoadsorption.

The use of layered perovskite catalysts has been reported in literature using a range of titanates including $Rb_2La_2Ti_3O_{10}$ [19], $KaLaZr_{0.3}Ti_{0.7}O_4$ [22], $KTiNbO_5$ [46], $SrTiO_3$ [47], $K_2La_2Ti_3O_{10}$ [48], $La_2Ti_2O_7$ [49], $La_4CaTi_5O_{17}$ [49], $Gd_2Ti_2O_7$ [41] and $Y_2Ti_2O_7$ [41]. The activity exhibited by such layered structure catalysts is a result of the interlayer space formed resulting in the increased presence of evolution sites. Additionally, the evolution sites in a layered structure are sufficiently separated from one another to suppress the back reaction of H_2 and O_2. The catalyst $La_4CaTi_5O_{17}$, which had a 110-layered perovskite structure, was reported by Kim and colleagues in 1999 [49] to generate a high quantum yield of 20% under UV irradiation at <320 nm. Takata et al. [19] reported evolution rates of 444 μmol h^{-1} and 221 μmol h^{-1} for H_2 and O_2, respectively, over Ni–$K_2La_2Ti_3O_{10}$, while Ikeada et al. [29] later reported increased rates of 2,186 μmol h^{-1} and 1,131 μmol h^{-1} for H_2 and O_2, respectively, over the same catalyst synthesised by a polymerised complex method. The structure of $K_2La_2Ti_3O_{10}$ and $K_2La_2Ti_3O_{10}$ with a hydrated layer was described by Huang et al. in their publication in 2008 [49] and is shown in Fig. 5. The structure consists of negatively charged titanoanthanate perovskite layers and interlayer K^+ ions.

Abe et al. [41] also showed that catalysts with a perovskite structure were active for H_2O splitting and demonstrated for the first time that titanates with a pyrochlore structure were also active (Fig. 6). The cubic pyrochlore catalysts Y_2TiO_7 (3.5 eV) and Gd_2TiO_7 (3.5 eV) along with the monoclinic perovskite $La_2Ti_2O_7$ (3.8 eV) were capable of evolving both H_2 and O_2 from distilled water under irradiation from a 400 W Hg lamp. The catalysts were loaded with NiO to split pure H_2O into H_2 in

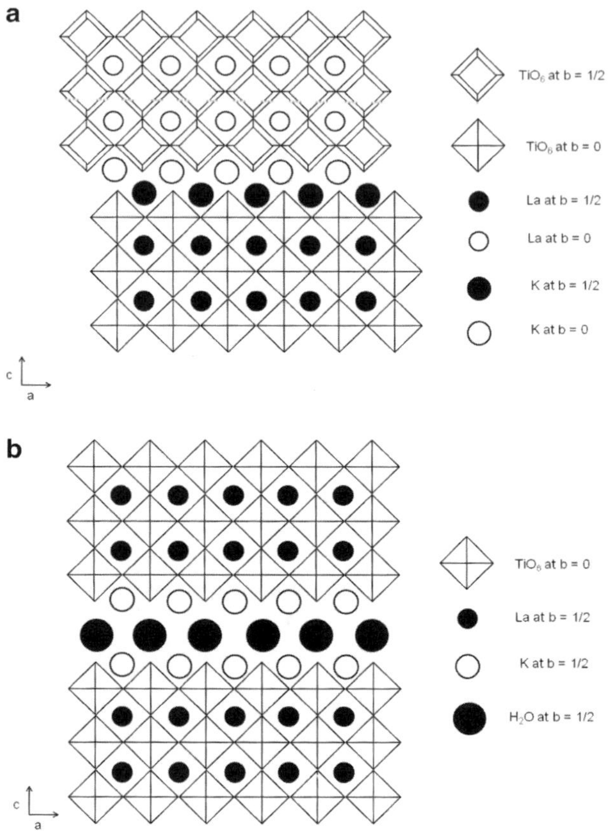

Fig. 5 Structure of (**a**) $K_2La_2Ti_3O_{10}$ and (**b**) $K_2La_2Ti_3O_{10}\cdot H_2O$ (reproduced from Huang et al. [48], with kind permission of Elsevier)

the range of 307–850 μmol h^{-1} and O_2 in the range of 152–420 μmol h^{-1}. Abe and colleagues found that the activity of $Y_2Ti_2O_2$, $Gd_2Ti_2O_7$ and $La_2Ti_2O_7$ was dictated by the crystal structure of the catalyst. The authors reported that the cubic pyrochlore structure of $Y_2Ti_2O_7$ and $Gd_2Ti_2O_7$ and the monoclinic perovskite structure of $La_2Ti_2O_7$ both possessed a network of corner-shared octahedral units of the metal cation TiO_6. The presence of the corner units allowed the formation of networks which were capable of increasing the mobility of charged particles. The study also showed that catalysts without this structure showed minimal activity, indicating the dependency of crystal structure on H_2O splitting.

The engineering of catalyst structures for H_2O splitting can play a key role in developing an efficient compound [44, 50–53]. The development of nanotubes arrays [44, 51, 52, 54] along with TiO_2 thin films [53] has been reported. Kitano and colleagues reported a visible light responsive system using a Pt–TiO_2 thin film to generate H_2 and O_2 from methanol and silver nitrate solutions respectively. Nanotube structures reported by Yang and colleagues used an Nb-doped TiO_2/$SrTiO_3$ catalyst in their photoelectrochemical studies, while Altomare et al. developed a system which consisted of TiO_2 nanotubes grown on a Ti support (Fig. 7). The

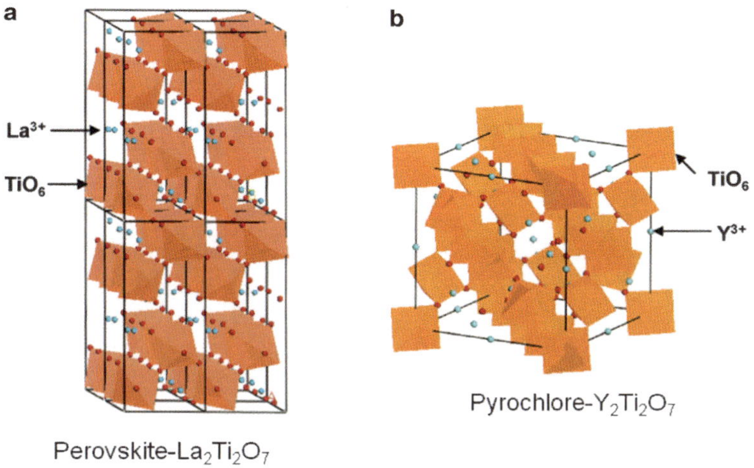

Fig. 6 Structure of (**a**) $La_2Ti_2O_7$ and (**b**) $Y_2Ti_2O_7$ (reproduced from Abe et al. [41], with kind permission of American Chemical Society)

Altomare et al. system yielded separate evolution streams of H_2 and O_2 without any electrical bias or sacrificial agent. An electrochemical anodisation method was used to synthesise self-organised nanotube arrays. The activity of the system was reported in terms of H_2 and O_2 evolution rates at 80 and 43 μmol h^{-1}, respectively. Based upon the production of H_2, a quantum yield of 15% was also reported under optimum conditions. This approach to H_2O splitting has numerous advantages over the use of more traditional catalysts for the decomposition of H_2O with a view towards clean H_2 production. There are a number of limitations which can restrict the large-scale applicability of powdered systems including fast rates of recombination, recovery from aqueous reaction solutions and downstream processing techniques associated with H_2 and O_2 separation from a mixture. Deploying a nanotube array system can suppress these limitations to construct a method of attaining high-purity H_2.

2.2.2 Tantalates

The use of tantalate-based catalysts has been used extensively in the field of photocatalytic H_2O splitting. Alkali and alkaline earth tantalates have exhibited high activity towards H_2O splitting to H_2 and O_2 in stoichiometric quantities. A number of Ta-based complexes have been studied including $InTaO_4$ [55], $K_3Ta_3Si_2O_{13}$, $ATaO_3$ (A=Li, Na, K), $Sr_2Ta_2O_7$, $RbLnTa_2O_7$, $K_2LnTa_5O_{15}$ and $A_2SrTa_2O_7{\cdot}nH_2O$ (A=H, K, Rb) [9, 15]. While these compounds vary in chemical composition and crystal structure, they all posses a TaO_6 octahedral interconnected in the 1-, 2- or 3-D manner. As discussed previously with Ti-based catalysts, the presence of a corner metal cation in the structure of catalysts has proven to be key for efficient H_2O splitting to proceed. In particular the distortion of this structure

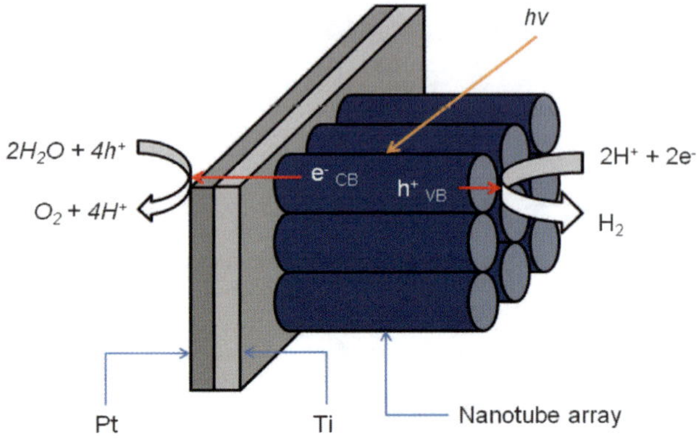

Fig. 7 Schematic of nanotube array on a titania support with a Pt electrode (reproduced from Altomare et al. [52], with kind permission of Elsevier)

has been found to further increase activity through increased charge separation [25]. The distortion of the catalyst structure is dictated by the angle of the M–O–M bond, where generally an angle of <180° results in distortion.

The presence of a TaO_6 corner unit has been reported to be present in catalysts with a pyrochlore (layered), perovskite and weberite structure. This fundamental unit has been reported to contribute to the formation of the CB, which was shown recently in work by Kim et al. [25] who developed niobate and tantalate layered crystal catalysts with ferroelectric properties. Catalysts $SbNbO_4$ (3.12 eV) and $SbTaO_4$ (3.7 eV) were synthesised using a conventional solid-state reaction method. The increased E_g of $SbTaO_4$ was found to be a result of the distortion of the crystal structure, which was dictated by the angle of the Ta–O–Ta bond (130°) in the TaO_6 octahedral. In contrast the 150° angle of Nb–O–Nb bonds in $SbNbO_4$ resulted in the compound being distortion free. Kato and Kudo also reported that a decrease in the Ta–O–Ta bond angle from 180° to 143° in their catalysts $KaTaO_3$ and $LiTaO_3$ resulted in distortion [56]. These observations by Kim et al. and Kato and Kudo were found to impact photocatalytic activity. Under optimum conditions 2.4 and 5.8 μmol h^{-1} H_2 was evolved over $SbNbO_4$ and $SbTaO_4$ respectively. The authors hypothesised that the 58.6% increase in H_2 production over $SbTaO_4$ was a result of the increased CB band edge of Ta 5d orbitals in the TaO_6 octahedral along with efficient charge separation due to the dielectric constant which was induced by the distorted crystal structure of the $SbTaO_4$ catalyst.

While the presence of a MO_6 metal octahedral structure has been shown to be significant, it is the metal at the M position that is essential for photocatalytic activity. In 2006 Abe et al. [41] showed that the level of distortion of the catalyst La_3TaO_7 and the presence of TaO_6 octahedral corner units increased H_2O splitting activity. The distortion of the crystal structure was a result of the ionic radius of rare earth metals in R_3TaO_7 (R=rare earth metal). The distortion caused a change in structure from cubic fluorite to cubic pyrochlore and then to orthorhombic weberite. Despite pyrochlore structure catalysts previously being reported active by Ikeada et al. [29], negligible activity was found when the cubic fluorite and pyrochlore catalysts were deployed. In contrast, the weberite structure catalyst showed significant H_2O splitting: 164 μmol h^{-1} H_2 and 80 μmol h^{-1} O_2.

The activity of La_3TaO_7 for H_2O splitting was found to increase with the shift in crystal structure from cubic pyrochlore to orthorhombic weberite. The weberite structure was composed of zigzag chains formed by corner-shared TaO_6 octahedrons, as shown in Fig. 8a. While it was found that the catalysts with a cubic pyrochlore (Gd_3TaO_7/La_3TaO_7) structure did contain TaO_6 units, the structure also revealed the presence of GdO_6 and LaO_6 corner units (Fig. 8b). The TaO_6 and Gd/LaO_6 units were in a ratio of 1:1. It was found that the network of octahedrons increases the mobility of the charged particles (electrons and holes) in the catalyst structure. Work conducted by Zheng and West [57] showed that the weberite structure of La_3TaO_7 exhibited high electrochemical conductivity. This conductivity can increase the mobility of charged particles, which can increase the probability of electrons and holes reaching an active site located on a catalyst surface.

The use of the catalyst $NaTaO_3$, with a distorted perovskite structure, has received an increased level of attention due to the high quantum yields reported [27, 56, 58]. As discussed earlier in the chapter, displaying catalyst activity as quantum yield is key to understanding the efficiency of the compound in relation to the quantity of photons absorbed and products evolved. The activity of this catalyst has been significantly improved by the addition of co-catalysts and doping of lanthanides. Kudo and Kato [58] investigated the production of H_2 and O_2 over $NaTaO_3$ catalysts doped with a range of lanthanides. The E_g of the catalysts synthesises ranged from 4.01 to 4.09 eV, which limited them to UV excitation by a 400 W high-pressure Hg lamp. All catalysts deployed displayed activity towards H_2 and O_2 evolution from distilled H_2O in the conventional ratio of 2:1. The highest activity recorded was over $NaTaO_3$:La using NiO as a co-catalyst at 5.9 and 2.6 μmol h^{-1} for H_2 and O_2 respectively. This rate of production produced an AQY of ~50% at 270 nm. Kato and Kudo further reported an AQY of 50% at 270 nm in 2003 along with a discussion into the crystal structure of the catalyst. Later in 2003, Kato et al. reported the highest apparent quantum yield for the catalyst at 56% at 270 nm. The optimum conditions were a La-doped loading of 2 mol% and NiO loading of 0.2 wt%, which produced 19.8 and 9.66 μmol h^{-1} H_2 and O_2, respectively. The activity of this catalyst was found to be due to the structure of the catalyst which consisted of a TaO_6 corner unit, as previously discussed. The authors, however, also discussed the importance of the catalyst nanostep structure in H_2O splitting. The authors used transmission electron microscope and extended X-ray absorption fine structure analysis to conclude that H_2 evolution occurred on the ultrafine NiO particles, which were loaded onto the nanostep structure of the catalyst while O_2 proceeded at the groove of the nanostep structure.

2.2.3 Niobates

As has previously been discussed the activity of niobate catalysts and the structure of the compound are closely linked. Common structures which have been reported in

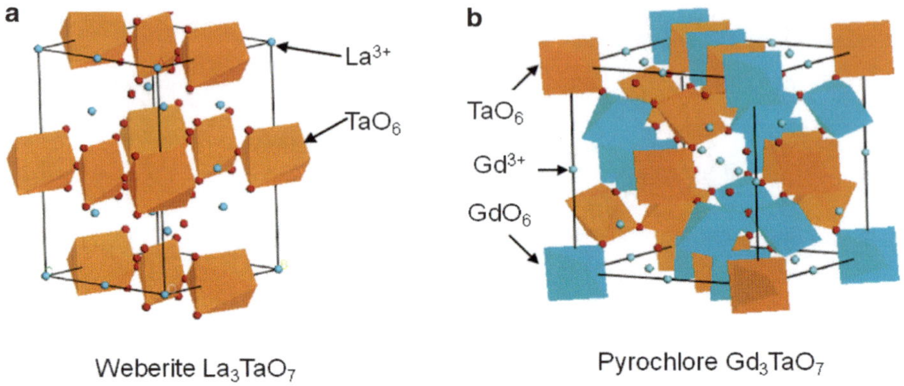

Fig. 8 Structure of orthorhombic (**a**) weberite La_3TaO_7 and (**b**) pyrochlore Gd_3TaO_7 (reproduced from Abe et al. [41], with kind permission of American Chemical Society)

Nb-based catalysts include layered perovskites, pyrochlores and wolframite [59]. The initial research into complexes with a layered structure was primarily conducted using Nb-based catalysts. A number of Nb catalysts, with a layered structure, have been reported active for H_2O splitting including $A_{2.23}Sr_{0.67}Nb_5O_{14.335}$ (A=K, H) [60], In–$H_2LaNb_2O_7$ [61], NiO–K_4NbO_{17} [23], Pt–K_4NbO_{17} [24], $K_{0.5}La_{0.5}Bi_2Nb_2O_9$ [26], $SbNbO_4$ [25], $Sr_2(Ta_{1-x}Nb_x)_2O_7$ [28] and Ni–$A_4Ta_xNb_{6-x}O_{17}$ (A=K, Rb) [62]. The structured catalysts have been shown to have two interlayer spaces (interlayers I and II) present, the advantages of which were discussed in Sect. 2.2.1. Co-catalysts were found to be present in the interlayers; Domen et al. [47] impregnated NiO into $K_4Nb_6O_{17}$ while Sayama et al. [24], Li et al. [60] and Wei et al. [61] all interlaced Pt into $K_4Nb_6O_{17}$, $A_{2.33}Sr_{0.67}Nb_{514.335}$ (A=K, H) and In–$H_2LaNb_2O_7$, respectively.

The publication by Domen and colleagues in 1986 was one of the earliest reports of the layered compound $K_4Nb_6O_{17}$ (3.3 eV) being used for H_2O splitting. Since the initial publication, $K_4Nb_6O_{17}$ and its variations have been reported frequently. While Domen et al. reported that stoichiometric evolution of H_2 and O_2 was achieved over bare $K_4Nb_6O_{17}$, an one order of magnitude increase was reported when using NiO as a co-catalyst. Under favourable conditions a maximum quantity of 80 μmol h^{-1} H_2 and 38 μmol h^{-1} O_2 was recorded.

Sayama and colleagues reported the synthesis of $K_4Nb_6O_{17}$ by photoreduction and H_2 reduction methods. Under photoreduction synthesis the Pt particles were found to be <6 Å, which allowed the fine particles to occupy interlayer I (Fig. 9). Under the H_2 reduction method particles were found to be >15 Å, which were too large to occupy the interlayer space and resulted in the destruction of the catalyst layers. The results of the research showed the photoreduction synthesised catalyst to have higher activity than the H_2 reduction catalyst. In a suspension of Na_2CO_3 451 and 217 μmol h^{-1}, H_2 and O_2 were produced respectively over Pt–$K_4Nb_6O_{17}$ by photoreduction, while over Pt–$K_4Nb_6O_{17}$ by H_2 reduction, 293 and 123 μmol h^{-1}

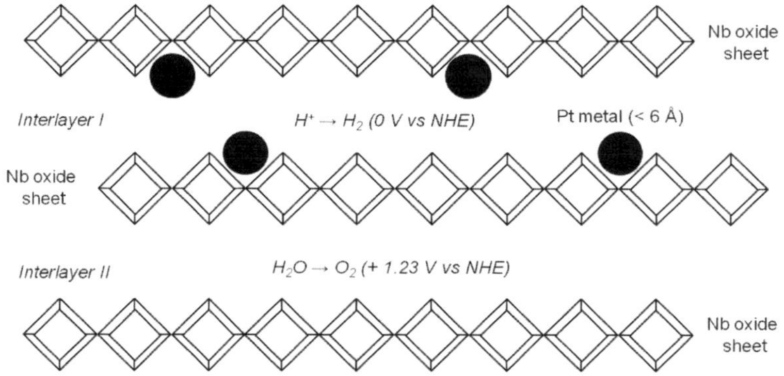

Fig. 9 Structure of interlayers in Pt-$K_4Nb_6O_{17}$ (reproduced from Sayama et al. [24], with kind permission of Elsevier)

H_2 and O_2 were produced respectively. The authors found that O_2 production occurred within interlayer II, while H_2 production proceeded in interlayer I with the addition of carbonate anions suppressing the back reaction of H_2 and O_2 over Pt.

In 1996, Sayama et al. found that $Rb_4Nb_6O_{17}$ also exhibited H_2O splitting activity. The research group observed an increase from 403 μmol h^{-1} H_2 and 197 μmol h^{-1} O_2 to 936 μmol h^{-1} H_2 and 451 μmol h^{-1} O_2 over Ni–$K_4Nb_6O_{17}$ and Ni–$Rb_4Nb_6O_{17}$, respectively. A series of catalysts were developed by Sayama and colleagues which comprised of Ni–$A_4Ta_xNb_{6-x}O_{17}$, where A=K or Rb. The highest activity was observed when $x = 0$ with a decreasing trend seen when the substitution of Nb for Ta increased. When $x = 6$, Ta was fully substituted for Nb and a significantly low level of H_2/O_2 was evolved: 11 μmol h^{-1} H_2 and 1 μmol h^{-1} O_2. The decrease in evolution was primarily a result of the difference in E_g of Ni–$Rb_4Ta_6O_{17}$ (4.1 eV) and Ni–$Rb_4Nb_6O_{17}$ (3.5 eV). Despite the difference in E_g, the quantum yields of the catalysts were similar, which was a result of $Rb_4Ta_6O_{17}$ absorbing less photons than $Rb_4Nb_6O_{17}$.

In 2003 a series of different structural Nb catalysts were investigated including an $A_2B_2O_7$ cubic pyrochlore, ABO_4 wolframite and ABO_4 stibiotantalite-type structures with a view towards visible light H_2O splitting [59]. As has been discussed in this chapter, despite the catalysts varying in crystal structure, they were all found to posses octahedral NbO_6 units. The layered wolframite structure was shown to contain both NbO_6 and InO_6 octahedral corner sharing units, which contributed to the formation of the structure. The NbO_6 units formed zigzag chains of [NbO_3], which were connected through the InO_6 units to form the network (Fig. 10). The E_g of $InNbO_4$ was calculated at 2.5 eV, which was suitable for absorption of visible wavelengths. Under irradiation from a 300 W Xe lamp with a 420 nm cut-off filter, the evolution of H_2 from pure H_2O was observed over NiO–$InNbO_4$. Under these experimental conditions, it should be noted that O_2 evolution was not recorded, with the authors suggesting any O_2 evolved was photoabsorbed onto the catalyst structure.

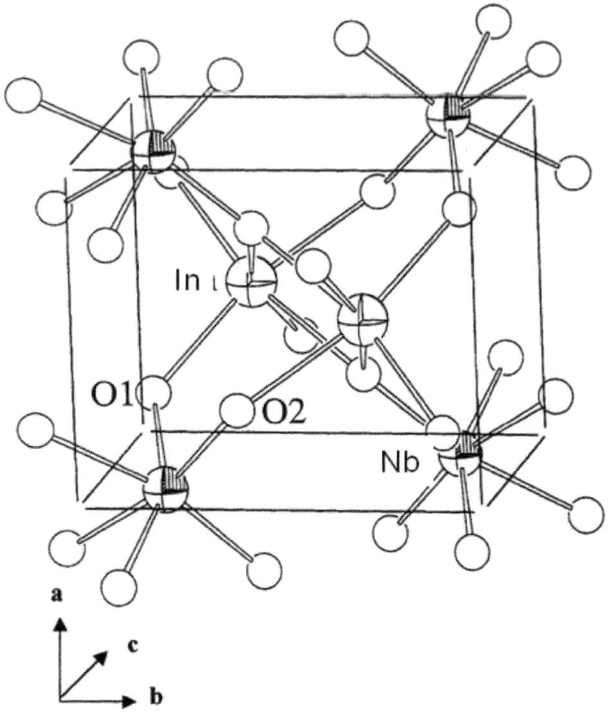

Fig. 10 Schematic structure of $InNbO_4$ (reproduced from Zou et al. [59], with kind permission of International Association of Hydrogen Energy)

2.2.4 Vanadates and Tungstanates

A number of additional transition metal oxides have been reported to be active for H_2O splitting including ZrO_2 [22], WO_3 [63–66], $PbWO_4$ [30], Bi_2WO_6 [67], $BiVO_4$ [68], Cu_2O [69], $InVO_4$ [70, 71] and Sr–CeO_2 [72].

The use of vanadium oxide (VO_4) catalysts for H_2O splitting is of particular interest due to the activity reported under visible light irradiation [68, 70]. Hau Ng and colleagues used $BiVO_4$ in conjunction with reduced graphene oxide (RGO) in a photoelectrochemical system to generate H_2 and O_2 (Fig. 11), while $InVO_4$ was deployed by Ye et al. Over RGO–$BiVO_4$, production rates of 0.75 and 0.21 μmol h^{-1} were achieved for H_2 and O_2, respectively, which was a 4.2% increase over pure $BiVO_4$. The improvement in activity was a result of photoexcited electrons injecting into the RGO, thus prolonging the lifetime and suppressing recombination. The rate of transition of the photoelectron to the Pt counter electrode was significantly faster using RGO over pure $BiVO_4$.

The activity of NiO–$InVO_4$ was reported in relation to H_2 production under irradiation from a 300 W Xe arc lamp using a 420 nm cut-off filter. The E_g of the catalyst was calculated at 1.9 eV which was formed from the V 4d orbital in VO_4. The structure of the catalyst was found to consist of both InO_6 octahedral units and VO_4 tetrahedral units, which was similar to that reported for $InNbO_6$ [59]. The production rate of H_2 from pure water was ~3.25 μmol h^{-1} over $InVO_4$ with a 1.0 wt % loading of NiO. There was no evolution of O_2, which the authors hypothesised

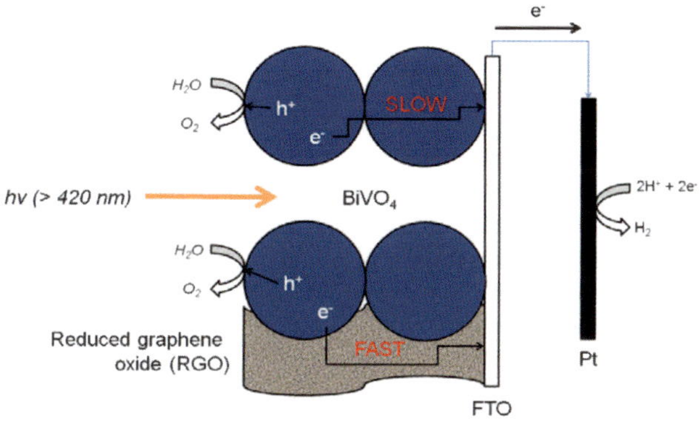

Fig. 11 Schematic representation of H_2O splitting mechanism on RGO-BiVO_4 (reproduced from Hau Ng et al. [68], with kind permission of American Chemical Society)

was due to O_2 photoabsorbing onto the catalyst surface in a physisorbed and/or chemisorbed molecular state [70].

2.3 *d^{10} Configuration Catalysts*

2.3.1 d^{10} Metal Oxides

The observation was made by Inoue in 2009 that since metal oxides with a vacant d orbital (d^0) are efficient photocatalysts, metal oxides with a filled d orbital will also be efficient. It has been reported that catalysts constructed using In^{3+}, Ga^{3+}, Ge^{4+}, Sn^{4+} and Sb^{5+} along with RuO_2 as a co-catalyst exhibit photocatalytic H_2O splitting activity. Work conducted primarily by Inoue's research group has highlighted the potential of d^{10} metal oxides for water splitting. A number of key catalysts were reviewed by Inoue [15] including layered catalysts such as $NaInO_2$ and $CaSb_2O_6$, weberite catalysts $Ca_2Sb_2O_7$ and $Sr_2Sb_2O_7$ and willemite catalyst Zn_2GeO_4. Inoue [15] and Kudo and Miseki [9] both discussed the importance of RuO_2 as a co-catalyst for d^{10} metal oxides. In the absence of RuO_2 the catalysts exhibited minimal activity. An exception was the catalyst Ga_2O_3, which previously displayed activity when loaded with NiO. The catalyst, with an E_g of 4.6 eV, produced 46 μmol h^{-1} H_2 and 23 μmol h^{-1} O_2. Using Ni as a co-catalyst, a Zn-doped Ga_2O_3 was found to be highly active [31]. Using irradiation from a 450 W Hg lamp, 4,100 μmol h^{-1} H_2 and 2,200 μmol h^{-1} O_2 were evolved.

The distortion of catalyst structures in a series of compounds researched by Inoue's group has been published. Catalysts investigated included alkaline earth metal MGa_2O_4 (M=Mg, Ca, Sr), $MSnO_4$ (M=Ca, Sr, Ba), $NaSbO_3$, $CaSb_2O_6$ and $M_2Sb_2O_7$ (M=Ca, Sr) [15, 33]. In 2002 Sato and colleagues found 1 wt% RuO_2-loaded $Sr_2Sb_2O_7$ to exhibit the highest level of activity over other antimonate catalysts due to its weberite structure consisting of two distorted SbO_6 units. The compression and elongation of Sb–O bonds gave rise to the distorted octahedral shapes shown in Fig. 12.

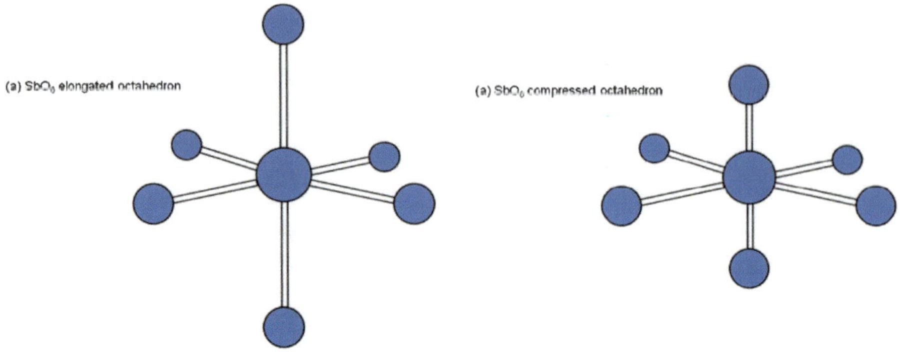

Fig. 12 Structure of (**a**) SbO_6 elongated octahedron and (**b**) SbO_6 compressed octahedron

The structures of MGa_2O_4 (M=Mg, Ba, Sr) were found to be cubic with a GaO_6 octahedron when M=Mg, monoclinic with a GaO_4 tetrahedron when M=Sr and hexagonal with a GaO_4 tetrahedron when M=Ba. The activity of these compounds was investigated in relation to their dipole moments, which were calculated to be 0 D for $MgGa_2O_4$, 0.80 and 1.2 D for $SrGa_2O_4$ and 1.70, 1.11 and 2.58 D for $BaGa_2O_4$. The dipole moment, which is a measure of difference in electronegativity, indicates the level of distortion in a compound. $MgGa_2O_4$ had a dipole moment of 0 which suggested the catalyst was distortion free, while $SrGa_2O_4$ and $BaGa_2O_4$ both had increased dipole moments suggesting they had a distorted structure. The authors found that the distorted structures of $SrGa_2O_4$ and $BaGa_2O_4$ were photoactive while the distortion-free structure of $MgGa_2O_4$ was not. A similar observation regarding distorted structures was made when using stannates doped with Ca, Sr and Ba. Ca_2SnO_4 and Sr_2SnO_4 were found to exhibit photocatalytic H_2O splitting activity, while Ba_2SnO_4 displayed negligible results. The fundamental unit, SnO_6, in the crystal structure of Ca_2SnO_4 and Sr_2SnO_4 was found to be distorted, while in Ba_2SnO_4 it remained distortion free.

The activity exhibited by catalysts with a distorted structure is interesting and a number of possible explanations have been reported. Inoue [15] suggested that the length of metal–oxygen bonds impacts photoexcitation. The poor symmetry formed as a result of distorted octahedral and tetrahedral structures results in isolated orbitals, which gives rise to different photoexcitation efficiencies. Inoue also suggested that the local internal fields of the compounds as result of the dipole moment could increase electron-hole separation and suppress recombination.

2.3.2 d^{10} Metal Nitrides

The investigation of metal nitrides (and oxynitrides) is an interesting area of H_2O splitting research with publications in this field increasing. The advantage of metal nitrides and oxynitrides is the potential of synthesising visible light responsive photocatalysts for overall H_2O splitting. The VB of metal nitrides is composed of N2p, which has a higher energy level than that of O2p in metal oxides. Therefore the E_g is narrower, presuming the CB remains the same, which allows for excitation by photons with a wavelength >400 nm. A number of d^0 transition metal oxynitrides have been investigated such as Ta_3N_5 [73], TaON [73], $MTaO_2N$ (M=Ca, Sr, Ba) [74] and $LaTiO_2N$ [75]; however, minimal activity was reported. Metal nitrides with a d^{10} configuration, however, have the advantage of forming large band dispersion at the CB, which yields highly mobile photoexcited electrons [15, 34].

Examples of metal nitrides being used include β-Ge_3N_4 [32], doped GaN and GaN in a solid solution with ZnO [1, 2, 32, 76, 77]. These catalysts have all shown activity towards H_2O splitting when used with RuO_2 as a co-catalyst. RuO_2-β-Ge_3N_4 was the first successful example of a non-metal oxide catalyst being deployed for overall H_2O splitting [15].

The activity of $(Ga_{1-x}Zn)(N_{1-x}O_x)$, synthesised from Ga_2O_3 and ZnO, has been reported by a number of research groups [1, 2, 32, 76–78]. The interest in this catalyst has grown due to its high level of activity and potential for visible light activation. Maeda et al. [1, 2] reported their samples of $(Ga_{1-x}Zn)(N_{1-x}O_x)$ to have an E_g in the range of 2.6–2.8 eV, which is lower than that of the individual catalysts ZnO (3.2 eV) and GaN (3.4 eV) and suitably low enough to permit excitation by absorbed visible photons. More recently, Lee et al. [77] have further minimised the E_g to 2.2 eV by increasing x to 0.87, which they concluded can significantly increase the absorption of solar photons. Using a 5 wt% of RuO_2, Maeda and colleagues [1, 2] found the catalyst was active under UV and visible illumination, producing stoichiometric quantities of H_2 and O_2. The highest activity was reported over $(Ga_{1-x}Zn)(N_{1-x}O_x)$ where $x = 0.12$ and the nitridation time was 15 h. The authors found that the longest wavelength available for water splitting was 460 nm, which corresponded to the absorption edges of the catalyst. Following this publication Maeda and colleagues published an additional paper in 2006 reporting an increased activity of $(Ga_{1-x}Zn)(N_{1-x}O_x)$ doped with Cr and a series of transition metals under UV irradiation. Using 1 wt% Rh and 1.5 wt% Cr, 3,835 μmol h^{-1} H_2 and 1,988 μmol h^{-1} O_2 were evolved over $(Ga_{1-x}Zn)(N_{1-x}O_x)$ from distilled water.

Recently it was reported that the activity of the Rh-doped $(Ga_{1-x}Zn)(N_{1-x}O_x)$ catalyst under visible irradiation was further increased using lanthanoid oxide layers [76]. Using a 300 W lamp with a cut-off filter (>400 nm), H_2 and O_2 was evolved over Rh–$(Ga_{1-x}Zn)(N_{1-x}O_x)$ catalysts doped with La, Pr, Sm, Gd and Dy, while un-doped Rh–$(Ga_{1-x}Zn)(N_{1-x}O_x)$ displayed no activity. It was found that the presence of lanthanoid oxide layers acts as Rh modifiers and suppresses the back reaction of H_2 and O_2 over Rh.

2.4 Z-Scheme Photocatalysts

The use of a single visible light responsive photocatalyst with a sufficient potential to achieve overall H_2O splitting is known as a one-step mechanism [1, 2]. As

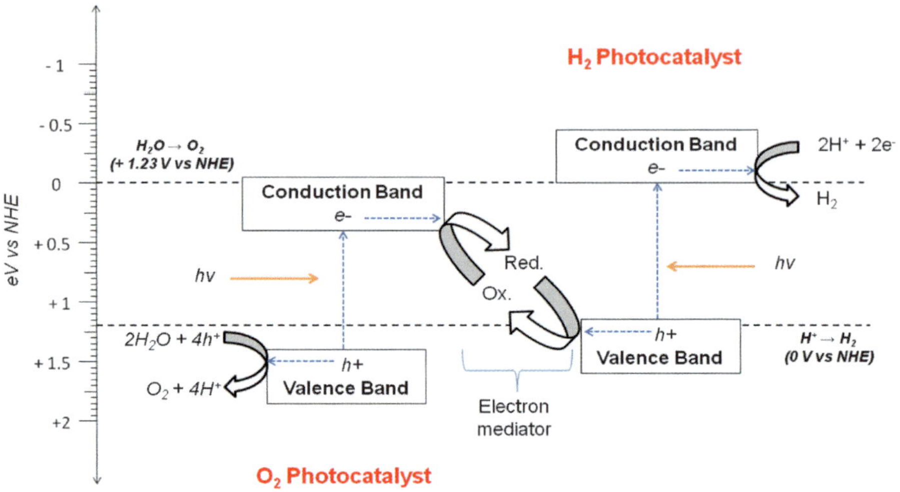

Fig. 13 Schematic of Z-scheme photocatalytic process

previously discussed, the photocatalyst employed should have a suitable thermodynamic potential for H_2O splitting, a sufficiently narrow band gap to harvest visible photons and stability against photocorrosion. These requirements are rather stringent and thereby limit the number of photocatalysts capable of photosplitting H_2O using the one-step mechanism [1–3]. Discussed here is the two-step Z-scheme mechanism for H_2O splitting (Fig. 13). This dual photocatalyst system for the photosplitting of H_2O was first suggested by Bard in 1979 [4]. A wider range of visible light is available because a change in Gibbs free energy required to drive each photocatalyst can be reduced when compared to the one-step system. It is also possible to separate evolved H_2 and O_2 which is a significant advantage with a view towards large-scale H_2 production. The problem of spontaneous backward reaction of redox products remains as in ordinary single-photocatalyst systems. Hence a useful system without any disadvantages has yet to be developed [5, 79–88].

Two-step Z-scheme H_2O splitting systems have a number of advantages over conventional one-step systems and currently appear to be the most promising way of achieving efficient H_2O splitting under visible light. Sayama et al. reported evidence of the photosplitting of H_2O into H_2 and O_2 using a Z-scheme photocatalytic system and visible light irradiation [5, 89, 90]. $SrTiO_3$ doped with Cr and Ta ($SrTiO_3$:Cr/Ta) for H_2 evolution, WO_3 for O_2 evolution and an iodate/iodide redox couple used as an electron mediator was capable of photosplitting H_2O under visible light. A version of the Z-scheme had been previously reported under UV irradiation using Pt-loaded anatase TiO_2 and bare rutile TiO_2 photocatalysts in the presence of iodate/iodide redox shuttle [80, 81]. This system was only capable of operation at wavelengths <400 nm. The large band gap of TiO_2 however meant that there was potential to extend this to the visible light region by appropriate doping of TiO_2 or inclusion of suitable visible light active catalysts.

The simultaneous evolution of H_2 and O_2 in a Z-scheme H_2O splitting system is difficult to achieve because the backward reactions of the redox mediator proceed readily over both photocatalysts and suppress the forward reactions (evolution of H_2 and O_2). Hence a redox shuttle is employed.

The most common redox shuttle employed within Z-scheme photocatalysis is the IO_3^-/I^- redox shuttle or the Fe^{3+}/Fe^{2+}. They facilitate this via the following reactions:

IO_3^-/I^-:

$$h\nu \rightarrow e_{CB^-} + h_{VB^+} \tag{4}$$

$$IO_3^- + 3H_2O + 6e_{CB^-} \rightarrow I^- + 6OH^- \tag{5}$$

$$2H_2O + 4h_{VB^+} \rightarrow O_2 + 4H^+ \tag{6}$$

Fe^{3+}/Fe^{2+}:

$$Fe^{3+} + e_{CB^-} \rightarrow Fe^{2+} \tag{7}$$

$$2H^+ + 2e_{CB^-} \rightarrow H_2 \tag{8}$$

$$Fe^{2+} + h_{VB^+} \rightarrow Fe^{3+} \tag{9}$$

Early reports of the use of the iodate/iodide redox couple demonstrated the advantage for the photosplitting of H_2O. Sayama et al. have reported O_2 evolution over four TiO_2 photocatalysts (A, anatase; R, rutile) in an aqueous solution containing 1 mM $NaIO_3$ under UV irradiation (>300 nm, 400 W high-pressure Hg lamp) [88]. Evolution of O_2 over rutile TiO_2 proceeded at a steady rate until the amount of O_2 produced reached 1,500 μmol in the absence of NaI; this was also observed in the presence of a significant amount of I^- anions. The amount of O_2 recorded agreed with the stoichiometric amounts expected based on the quantity of IO_3^- added to the solution before irradiation. The results when the photocatalyst used was Pt/WO_3 proceeded in a similar manner. For both catalysts when the amount of I^- was increased the O_2 evolution rate was reduced.

Also reported was the fact that the addition of excess I^- to the solution completely suppressed O_2 evolution over both anatase TiO_2 and $Pt/BiVO_4$ photocatalysts. The loading of a co-catalyst such as Pt or RuO_2 is necessary for the efficient evolution of O_2 over the WO_3 photocatalyst using IO_3^- anion as an electron acceptor. It has been reported that WO_3 alone is capable of O_2 generation from H_2O in the presence of other electron acceptors such as Ag^+ or Fe^{3+} [79, 91, 92]. The presence of the Pt co-catalyst mainly serves to provide reduction sites that enable the six-electron reduction of IO_3^- to I^-.

The use of a photocatalyst combination that favourably adsorbs either IO_3^- or I^- leads to the photosplitting of H_2O via a two-step mechanism. The first step involves the reduction of H_2O to H_2 and oxidation of I^- to IO_3^- over a Pt/anatase TiO_2

photocatalyst, while the second step involves reduction of IO_3^- to I^- and oxidation of H_2O to O_2 over a rutile TiO_2 photocatalyst.

The rapid reduction of IO^{3-} to I^- over rutile TiO_2 results in a very low IO_3^- concentration during the reaction. This effectively suppresses the undesirable backward reaction (IO_3^- reduction to I^-) over the Pt/anatase TiO_2 photocatalyst giving a higher H_2 evolution rate. The key for achieving H_2O splitting is to use different oxidation reactions; in other words, preferential oxidation of I^- to IO_3^- over the H_2 photocatalyst (e.g., Pt/anatase TiO_2) and preferential oxidation of water to O_2 over the O_2 photocatalyst (e.g., rutile TiO_2) must occur simultaneously in a single solution.

Another major redox couple used to facilitate the two-step process of the Z-scheme during photosplitting of H_2O is Fe^{3+}/Fe^{2+}. Kato et al. reported H_2O splitting under visible light using a Z-scheme photocatalytic system that consisted of Rh-doped $SrTiO_3$ ($SrTiO_3$:Rh) for H_2 evolution, $BiVO_4$ for O_2 evolution and a Fe^{3+}/Fe^{2+} redox couple as an electron mediator [93–95]. Previous studies had demonstrated the functionality of using Fe^{3+} ions as efficient electron acceptors over a RuO_2/WO_3 or rutile TiO_2 photocatalyst for O_2 evolution [79, 96]. However, the use of Fe^{2+} ions as electron donors for the efficient photocatalytic production of H_2 had not previously been achieved.

Doping with Rh cations [97] and co-doping with Cr^{3+}/Ta^{5+} [98], Cr^{3+}/Sb^{5+} [98] and Ni^{2+}/Ta^{5+} have been reported to sensitise $SrTiO_3$ to visible light. Kudo et al. [99] demonstrated that doped $SrTiO_3$ powders with a Pt co-catalyst exhibit photocatalytic activities for H_2 evolution under visible light. This proceeds in the presence of methanol as a sacrificial electron donor. Pt-loaded $SrTiO_3$:Rh photocatalyst demonstrated activity for H_2 evolution from H_2O in the presence of Fe^{2+} as a reversible electron donor. The reaction initiated in $FeCl_3$ aq. (2 mmol L^{-1}), predominantly produced O_2 during the initial stage of the first run. The rate of O_2 production subsequently decreased gradually. In a second run after evacuation of the gas phase, H_2 and O_2 were produced in the stoichiometric ratio (2:1) during the initial period. The reaction in $FeCl_2$ aq. (2 mmol L^{-1}) also exhibited stoichiometric evolution of H_2 and O_2 during the second run, while H_2 was predominantly evolved in the initial stage of the first run. In both cases, Fe^{3+} and Fe^{2+} ions in the steady state were found to have concentrations of ca. 1.5 and 0.5 mmol L^{-1}, respectively. H_2 and O_2 can be evolved in the (Pt/$SrTiO_3$:Rh)–($BiVO_4$) system even when the reductant (Fe^{2+}) and the oxidant (Fe^{3+}) coexist. This differs significantly when compared to the IO^{3-}/I^- redox couple, where H_2 production is suppressed by the coexistence of a small amount of oxidant (IO^{3-}) due to preferential re-reduction of IO^{3-} anions over water by the photoexcited electrons [5, 89, 100].

This demonstrates that the H_2 photocatalyst (Pt/$SrTiO_3$:Rh) and the O_2 photocatalyst ($BiVO_4$) have sufficiently high selectivities for the forward reactions. H_2 evolution on the Pt/$SrTiO_3$:Rh photocatalyst was enhanced rather than suppressed by the presence of Fe^{3+} [94]. The results of investigations of photocatalytic H_2 evolution and dark reactions in various aqueous solutions containing Fe^{3+} ions strongly suggest that the adsorption of $[Fe(SO_4)(H_2O)_5]^+$ and/or $[Fe(OH)(H_2O)_5]^{2+}$ on the Pt surface efficiently suppresses both the undesirable backward reactions on the Pt surface, H_2O formation from H_2 and O_2 and the reduction of Fe^{3+} with H_2. The increased H_2 production that accompanies an increase in the Fe^{3+} ion concentration is somewhat unusual in view of the electron acceptability of Fe^{3+}. The beneficial effect of Fe^{3+} suppressing the backward reactions should exceed the detrimental effect of Fe^{3+} trapping photoexcited electrons. In contrast to H_2 evolution, O_2 evolution over the $BiVO_4$ photocatalyst was remarkably inhibited by the presence of Fe^{2+} ions, indicating oxidation of Fe^{2+} to Fe^{3+} on the $BiVO_4$ photocatalyst [94]. $BiVO_4$ has a sufficiently high selectivity for the forward reaction, which enables O_2 production even when there is a high (5 mmol L^{-1}) Fe^{2+} ion concentration. The splitting of H_2O under visible light has also been achieved using WO_3 or Bi_2MoO_6 as an O_2 evolution photocatalyst in combination with a Pt/$SrTiO_3$:Rh photocatalyst in the presence of Fe^{3+}/Fe^{2+} electron mediator [93]. Despite recent developments, challenges remain in the promotion of electron transfer between two semiconductors and in the suppression of backward reactions involving shuttle redox mediators.

3 Water Splitting Mechanism

3.1 *Multimolecular Systems*

3.1.1 Ideal Functions

H_2 and/or O_2 production from H_2O by visible light requires one or several intermediates having ideally the following functions:

1. Visible light absorption
2. Conversion of the excitation energy to redox energy (charges)
3. Concerted transfer of several electrons to H_2O leading to the formation of H_2 as energy-storage compound and/or to the formation of O_2

Indeed, one of the main difficulties in achieving the splitting of H_2O by means of light-induced redox processes is that H_2 requires two electrons (Eq. (10)), while O_2 requires four electrons (Eq. (11)):

$$2H_2O + 2e^- \rightarrow H_2 + 2OH^-, E^{\circ}(\text{pH } 7) = -0.41 \text{ vs NHE} \quad (10)$$

$$2H_2O \rightarrow O_2 + 4H^+ + 4e^-, E^{\circ}(\text{pH } 7) = +0.82 \text{ vs NHE} \quad (11)$$

This number of charges corresponds to the most favourable thermodynamic conditions for Eq. (1). The reaction is a multi-electron transfer process which requires 1.23 eV per electron transferred. Hence, photons with $\lambda < 1{,}008$ nm corresponding to a minimum energy of 1.23 eV can induce the cleavage of H_2O.

As has been previously discussed, upon illumination with light of energy greater than the E_g, a semiconductor will form electrons and holes. The electron and holes formed are highly charged and initiate reduction and oxidation reactions. H_2O

molecules are reduced by the electrons to form H_2 and are oxidised by the holes to form O_2 for overall H_2O splitting.

3.1.2 General Schemes for H_2 and O_2 Production

Early research reported photochemical systems involving several compounds. A multimolecular system was designed with each compound fulfilling a particular role towards the photosplitting of H_2O. The first compound required was a photosensitiser (PS) capable of visible light absorption in order to generate the excited species PS* with useful redox properties (Eq. (12)):

$$PS \overset{h\nu_{vis}}{\rightarrow} PS^* \tag{12}$$

The presence of a second compound R is necessary, which can be reduced or oxidised by quenching of the excited species PS* in electron transfer reactions. This leads to the formation of charge pair, PS^+ and R^-, in the case of the oxidative quenching of PS (Eq. (13)):

$$PS^* + R \rightarrow PS^+ + R^- \tag{13}$$

Finally it requires a third compound capable of collecting several electrons to facilitate the exchange of two or four electrons with water. This multi-electron collection and transfer can be realised by a specific redox catalyst:

$$2R^- + 2H^+ \overset{Cat}{\rightarrow} 2R + H_2 \tag{14}$$

In such a system, the second compound R acts as an electron relay between the photosensitiser PS and the catalyst (Cat) mediating the electron collection. The redox potential of its reduced species R^- must be less than – 0.41 V (vs NHE, pH 7) to take part in Eq. (10).

The main problem associated with this process is the fast recombination of charge pairs (Eq. (15)):

$$PS^+ + R^- \rightarrow PS + R \tag{15}$$

The main challenge for these multimolecular systems is how to prevent the back electron transfer reaction in order to increase the charge separation lifetime.

In the case of multimolecular systems, the introduction of a fourth compound in the form of an electron donor should help to prevent this back reaction. The electron donor should scavenge the oxidised photosensitiser PS^+ in a competitive electron transfer reaction to give the initial PS and a donor oxidation product D^+ (Eq. (16)):

$$PS^+ + D \rightarrow PS + D^+ \tag{16}$$

$$D^{+} \rightarrow \text{Products} \tag{17}$$

The latter is a sacrificial system and rapidly decomposes irreversibly (Eq. (8)). D is the only compound, apart from H_2O (H^+), which is consumed. The other compounds PS, R and Cat follow catalytic cycles.

Two schemes for cyclic production of H_2 from H_2O were proposed [82]. The first called the "oxidative quenching mechanism" involved oxidation of the excited photosensitiser PS* to PS^+ by the electron relay R (Fig. 14a). It corresponds to reaction (Eqs. (12)–(17)).

The reduction of the excited state photosensitiser PS* by D is called reductive quenching (Fig. 14b):

$$PS^{*} + D \rightarrow PS^{-} + D^{+} \tag{18}$$

This primary reaction (Eq. (18)) yields the reduced photosensitiser PS^- and the oxidised donor D^+ which decompose irreversibly (Eq. (17)). In this way, PS^- can accumulate and react with an electron relay R to regenerate PS and to yield R^- (Eq. (19)):

$$PS^{-} + R \rightarrow PS + R^{-} \tag{19}$$

The inclusion of a suitable catalyst allows the formation of R^- which can lead to the production of H_2 as shown in Eq. (14).

PS^- is a more powerful reducing species than R^-. Therefore the reduction of H_2O to H_2 can be achieved directly by PS^- itself in the presence of a suitable catalyst. As a consequence, this scheme involves only three components (PS, D, Cat) and the mechanism becomes simplified (Fig. 15).

Similar three-component systems for O_2 production from H_2O have been proposed (Fig. 16). These systems require the formation, following visible light excitation of the photosensitiser PS, of a strong oxidising species PS^+, having a redox potential E° (PS^+/IPS) greater than 0.82 V (vs NHE, pH 7). This can be achieved by using an electron-acceptor, A, as a quencher which, once reduced to A^- (Eq. (20)), decomposes irreversibly (Eq. (21)):

$$PS^{+} + A \rightarrow PS^{+} + A^{-} \tag{20}$$

$$A^{-} \rightarrow \text{Decomposition products} \tag{21}$$

The oxidised PS^+ can accumulate and lead to O_2 evolution in the presence of a suitable catalyst capable of facilitating the exchange of four electrons with H_2O (Eq. (22)):

$$4PS^{+} + 2H_2O \overset{\text{Cat}}{\rightarrow} 4PS + 4H^{+} + O_2 \tag{22}$$

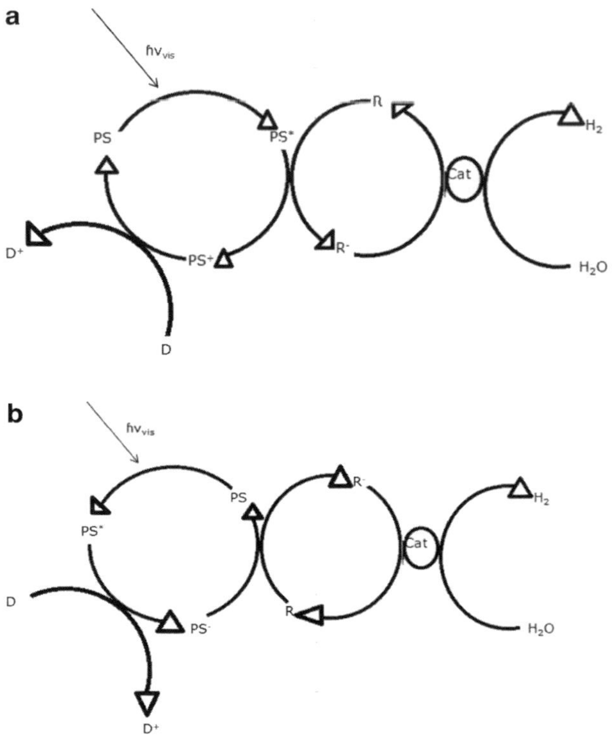

Fig. 14 Schematic representation of the redox catalytic cycles in the photoreduction of H_2O to H_2 by visible light irradiation of a four-component model system PS/R/D/Cat: (**a**) oxidative quenching mechanism and (**b**) reductive quenching mechanism (reproduced from Bolton [7], with kind permission of Elsevier)

3.2 Recent Developments

In recent years research into the development of materials capable of photosplitting H_2O have focused on two approaches:

1. One-step mechanism
2. Two-step mechanism

The one-step mechanism is shown in Fig. 3, while the two-step Z-scheme mechanism is represented in Fig. 13.

3.2.1 One-Step Mechanism

The one-step mechanism of H_2O photosplitting into H_2 and O_2 uses a single visible light active photocatalyst. However there are very few stable semiconductor materials available to achieve photosplitting of H_2O with a one-step mechanism. Band engineering of semiconductors is required to artificially develop new semiconductor materials that would satisfy the following criteria:

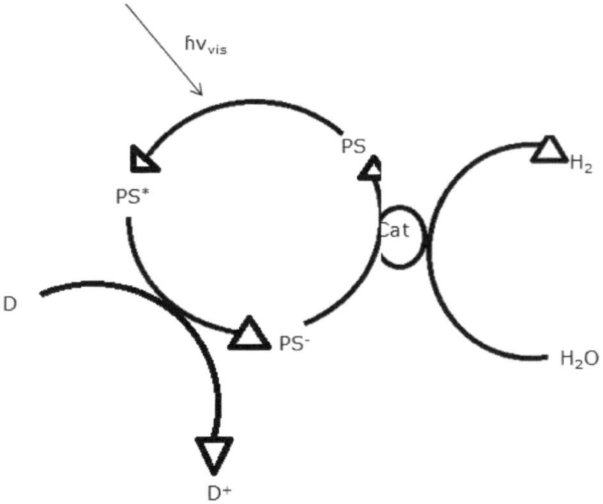

Fig. 15 Redox catalytic cycles in the photoreduction of H_2O to H_2, via reductive quenching mechanism for a three-component model system PS/D/Cat (reproduced from Bolton [7], with kind permission of Elsevier)

1. Have a narrow band gap
2. Stable under photo-irradiation
3. Suitable conduction and valence band levels for H_2 and O_2 production

It is impossible in principle to achieve separate production of H_2 and O_2 in a conventional one-step H_2O splitting system as H_2 and O_2 are evolved simultaneously on small semiconductor particles.

3.2.2 Two-Step Mechanism

A process inspired by natural photosynthesis in green plants is the two-step mechanism of photosplitting of H_2O, known as the Z-scheme. This mechanism uses two different photocatalysts, one tailored for H_2 production and the other for O_2 production, that splits H_2O into stoichiometric amounts of H_2 and O_2 in combination with a redox couple in the solution.

Within this system the photocatalyst for H_2 production and the photoexcited electrons reduce H_2O to H_2 and the holes in the valence band oxidise the reductant (Red) to an oxidant (O_x). The oxidant is reduced back to the reductant by photoexcited electrons generated over an O_2 evolving photocatalyst where the holes oxidise H_2O to O_2. This system allows a semiconductor to be used that has either a H_2O reduction or oxidation potential on one side of the system. As a result of this a variety of semiconductors can be used on the Z-scheme even if they do not satisfy all the stringent requirements for a one-step system.

Another advantage of Z-scheme systems is the ability to separate production of H_2 and O_2 by employing a separator, such as porous glass filter, that permits only redox mediators to be transferred. A disadvantage of Z-scheme systems, as this is a

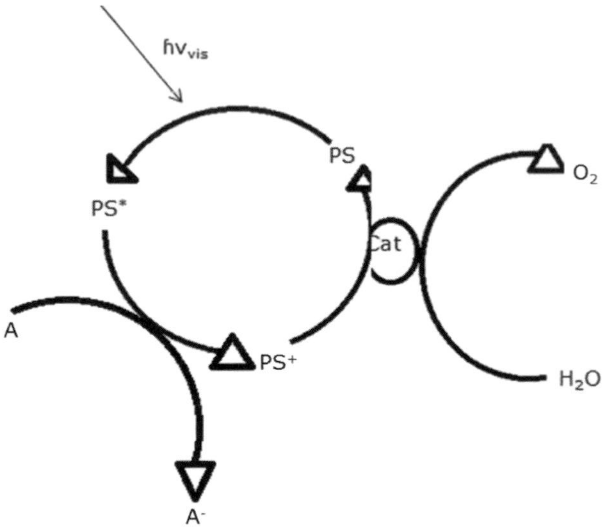

Fig. 16 Redox catalytic cycles in the photo-oxidation of H_2O to O_2 by visible light irradiation of a three-component model system PS/A/Cat (reproduced from Bolton [7], with kind permission of Elsevier)

two-step photoexcitation system, is that the number of photons needed to achieve H_2O splitting is twofold larger than with the one-step system.

4 Photoreactors for Water Splitting

The splitting of H_2O on illuminated semiconductors has long been studied as a clean and renewable means of converting solar energy into chemical energy in the form of H_2. The production of H_2 from H_2O through chemistry has been known for hundreds of years; the US military has been using a ferrosilicon reaction since the First World War [101]. NaOH, ferrosilicon and H_2O are added to a sealed vessel and as the hydroxide dissolves and heats to 200°F, H_2 and steam are produced. Many other methods are also used including steam reforming, CO_2 sequestration, partial oxidation, plasma reforming, coal, electrolysis, photobiological, sulphur-iodine, fermentative production, enzymatic production, renewable H_2 and lastly photocatalysis [102–110]. Photocatalytic H_2O splitting through two-step photoexcitation using two different semiconductor materials and a reversible donor–acceptor pair is one of the possible forms of "man-made photosynthesis" and is known as the Z-scheme system, which was described previously and can be seen in Fig. 13. The development of new semiconductor materials aids in the efficiency of Z-scheme H_2O splitting by reducing the number of back reactions. This section shall focus on reactors developed for H_2O splitting primarily with a view towards the production of H_2 via a photoelectrochemical system.

4.1 Photochemical Reactors

Photoreactors utilising a catalyst suspension have advantages over a photoelectrochemical cell including a higher surface area which results in more active sites for photocatalytic reactions to take place; also it is a far more simple process as no film deposition or coating is required.

Photocatalytic H_2O splitting can be divided into two types of reaction, photochemical and photoelectrochemical. The photochemical is more commonly referred to as taking place in a photoreactor or photocatalytic reactor, which consists of a powder catalyst in a suspension in an aqueous media, whereas a photoelectrochemical cell (PEC) consists of an immobilised photocatalyst on an anode.

Photoreactors for the splitting of H_2O take many forms, utilise numerous shapes and sizes of vessels and utilise many materials. These can range from the simplest of set-ups, single-chamber reactor (Fig. 17), to the more complex multichambered multilayer cell type. Whatever design the housing of the reactor takes, the fundamental operational principles remain the same, and it is those which shall be addressed here.

The most basic form of reactor is the beaker or cube-shaped vessel, the latter tends to be more acceptable as curved surfaces are not suitable for electrode-style set-ups. The biggest drawback of a single-chambered (or non-membrane) reactor is the dangerous mixture of O_2 and H_2. This is quite possibly the single biggest danger in H_2O splitting, the mixture of O_2/H_2 forming an explosive ratio. In the case of batch or stirred tank photoreactors, this proves to be the greatest danger but also the more complex problem, as separating H_2 and O_2 requires expensive membrane technology or cryo-processing [112, 113].

It can be seen in Fig. 18 that it is possible to use a batch-style stirred photoreactor to produce H_2 and O_2 in the same system, but the two gases are generated in isolation thanks to a proton membrane such as Nafion.

The same principles of format can be applied to multichambered PEC reactors (Fig. 19). The use of coated electrodes for the generation of H_2 and O_2 can be seen in a twin-chamber and triple-chamber arrangement.

In the triple-chamber reaction vessel, the H_2 and O_2 are evolved on the working electrode (WE) and counter electrode (CE) in ported chambers under illumination via a side window; this arrangement of porting prevents explosive mixing of the two gases. Similarly in the twin-chamber reactor, the evolution of gas occurs on the WE and CE electrodes, but this time the reactor contains a Nafion membrane which allows the H^+ to transfer to the CE side of the reactor.

There are many other formats of photochemical reactors which contain the same constituent components but vary in orientation and design [113, 117–127].

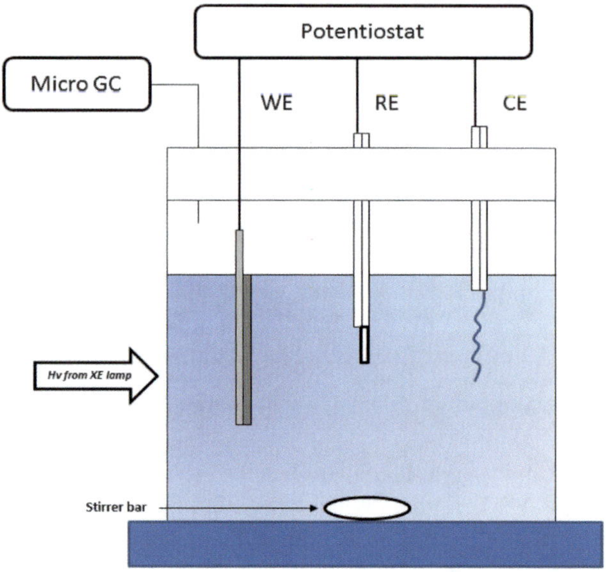

Fig. 17 Single-chambered triple electrode PEC reactor (reproduced from Xing et al. [111], with kind permission of Elsevier)

4.2 Photoelectrochemical Cell Reactors

In a PEC, the catalyst is prepared in a thin film on a substrate to form a photo-anode, with the application of an external circuit to direct the electrons from the photocatalyst to the cathode where H_2 is evolved (Fig. 20).

The mechanism involves fundamentally four steps:

1. Generation of electron hole pairs on the photo-anode
2. Oxidation of water by the holes to give O_2 and H^+
3. Transfer of photogenerated electrons that circulate on the cathode
4. Reduction of H^+ on cathode to give H_2

The first demonstration of this type of PEC was by Fujishima and Honda in 1972. The photoelectrochemical reactor is a bi-gas system where O_2 is generated, on the anode from the splitting of H_2O.

The major advantage of a PEC is the ability to create bias within the cell by altering the anode material configuration. The internal bias can be further complemented by applying an external bias between the electrodes. H_2O, under certain conditions, can be reversibly electrolysed at a potential of 1.23 V.

Semiconductors are the main photoactive material in a PEC and can be classified as either metal oxide or photovoltaic material. There are several types of PEC that exist: n-type, p-type, n-&p-type, hybrid, monolithic-bipolar and monolithic-electrical connection (Fig. 21). For example, an n-type semiconductor can be TiO_2 (a), p-type can be InP (b) and an n-&p-type can be n-GaAs/p-InP (c). A hybrid cell could consist of several n-type semiconductors with differing band gaps to cover more of the solar electromagnetic spectrum (d). It is possible to also have a

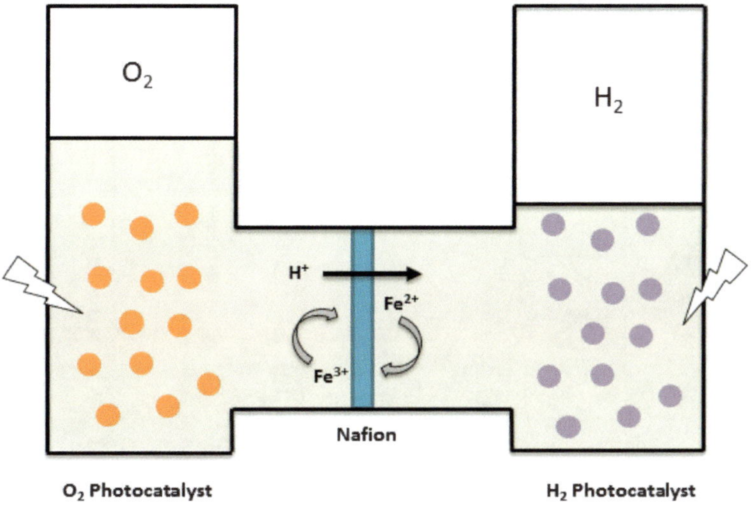

Fig. 18 Bi-gas sulphonated tetrafluoroethylene membrane cell (reproduced from Liao et al. [114, 115], with kind permission of International Association of Hydrogen Energy)

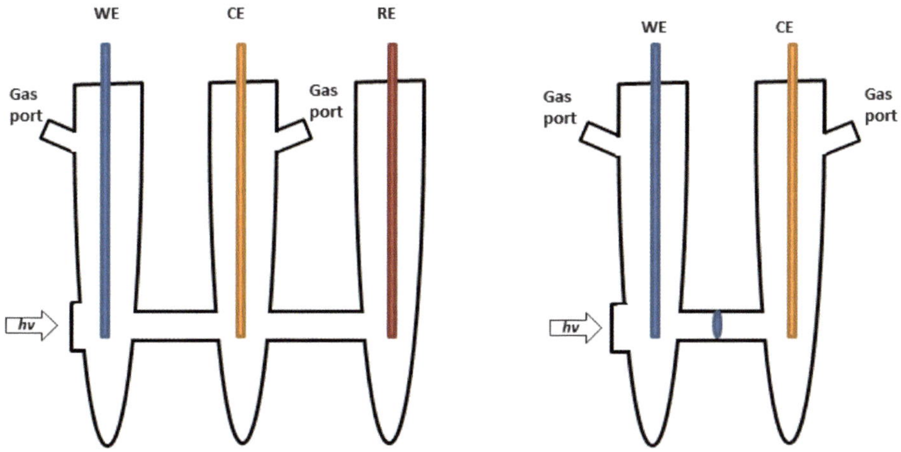

Fig. 19 H-type reaction vessels, triple and twin chamber (reproduced from Minggu et al. [116], with kind permission of International Association of Hydrogen Energy)

hybrid cell which combines both the anode and cathode in a monolithic structure (e) with a metal substrate coated on either side or to separate the anode and cathode with their own isolated substrate and connect the two with an electrical connection (f) ([119, 128–132].

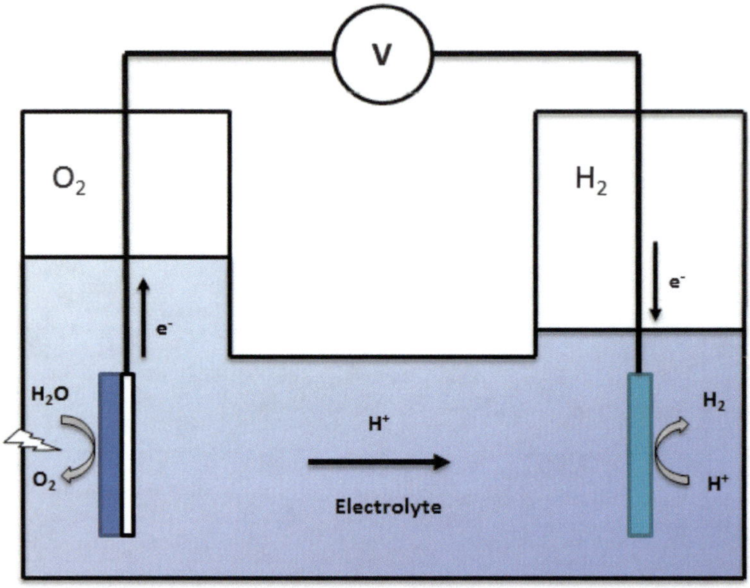

Fig. 20 Schematic of photoelectrochemical cell (reproduced from Liao et al. [114, 115], with kind permission of International Association of Hydrogen Energy)

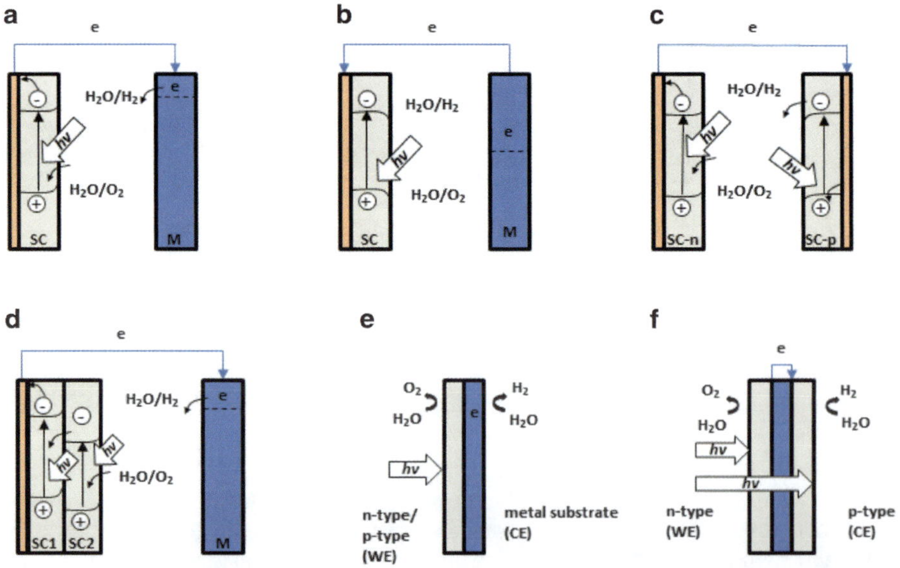

Fig. 21 Types of photoelectrochemical cells, (a) n-type, (b) p-type, (c) n-&p-type, (d) hybrid, (e) monolithic-bipolar and (f) monolithic-electrical connection (reproduced from Minggu et al. [116], with kind permission of International Association of Hydrogen Energy)

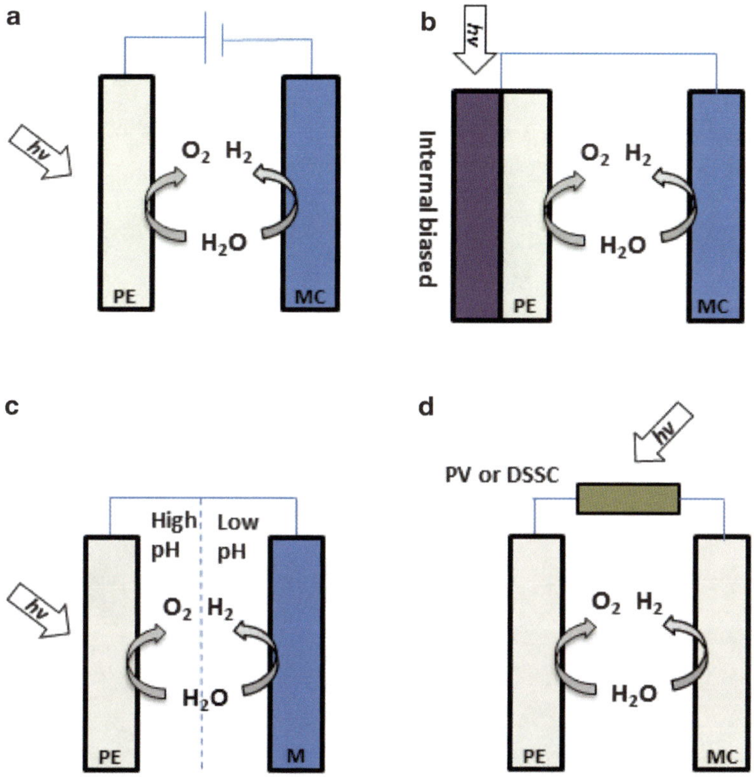

Fig. 22 Biasing methods for photoelectrochemical water splitting (reproduced from Minggu et al. [116], with kind permission of International Association of Hydrogen Energy)

4.2.1 Cell Biasing

The optimum performance for a PEC is the generation of H_2 without the application of an external current. This zero biasing only occurs when the band gap and band edges of the photocatalyst are correct to split H_2O. Currently there is no single semiconductor which can produce H_2 in a sufficient quantity under a zero-bias operation; the application of an external current is necessary to create viability [133, 134].

There are several types of external bias which can be applied to a PEC including an electrical bias, chemical bias, photovoltaic bias or internal bias (Fig. 22). In the case of electrical bias (a), the PEC is connected to an external power supply. In practice this is a reliable and effective method of biasing a PEC, but ultimately defeats the green energy credentials of photocatalytic systems, unless driven via renewable sources. For an internally biased PEC (b), it is the multilayer structure of the anode which creates the required bias. This can take the form of PV ~ a-SiGe,

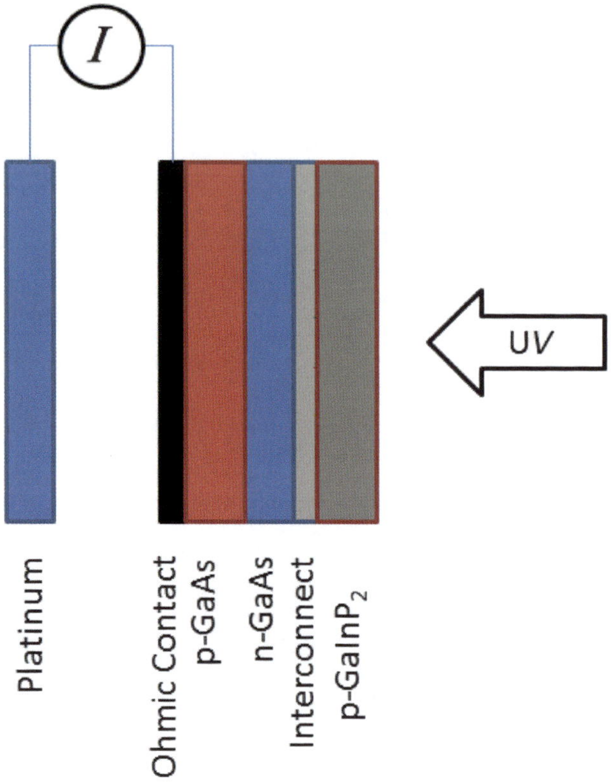

Fig. 23 Monolithic photoelectrochemical/ photovoltaic device (reproduced from Liao et al. [114, 115], with kind permission of International Association of Hydrogen Energy)

PEC ~ WO_3, PV_1 ~ GaInP, PV_2 ~ GaAs or PEC/PEC PEC_1 ~ DSSC PEC_2 ~ WO_3 whereby the structures achieve the correct band gap required. In the case of chemical bias (c), the pH of the electrolyte which the anode and cathode are submerged, separated by an ion exchange membrane, acid on one side and alkali on the other. This form of biasing proves un-cost-effective as a constant replenishing of the starting electrolytes is required as each move towards equilibrium as H^+ and OH^- are consumed. For photovoltaic bias (d), a solar photovoltaic cell is directly connected to the PEC; this could be a dye-sensitised cell, for example [36, 135–137].

4.2.2 Photovoltaic Photoelectrochemical Cell

As mentioned in the previous section, an increasingly common format for a PEC is the addition of a photovoltaic cell as shown in Fig. 23. For realistic low-impact H_2 production from H_2O spitting, the use of a photovoltaic cell allows a more energy-efficient process to occur [138–141]. Recent research has shown an increase in photovoltaic efficiency in the region of 12% [142].

4.3 Illumination

There are several laboratory-based illumination sources currently employed in H_2O splitting; ultraviolet, visible and solar simulator lamps are used with bench scale systems. In the case of larger scaleup systems, a move towards purely natural solar illumination is necessary as there is no economic viability in creating purely lamp-driven H_2. As the solar electromagnetic spectrum contains ~4% UV and almost 50% visible light, the move towards visible absorbing catalyst is needed to make photocatalytic H_2O splitting a viable technology. Current catalyst trends are directing formulation towards visible only activation, but efficiencies are still low for commercially viable catalysts.

References

1. Maeda K, Takata T, Hara M, Saito N, Inoue Y, Kobayashi H, Domen K (2005) GaN:ZnO solid solution as a photocatalyst for visible light driven overall water splitting. J Am Chem Soc 127:8286–8287
2. Maeda K, Teramura K, Takata T, Hara M, Saito N, Toda K, Inoue Y, Kobayashi H, Domen K (2005) Overall water splitting on $(Ga_{1-x}Zn_x)(N_{1-x}O_x)$ solid solution photocatalyst: relationship between physical properties and photocatalytic activity. J Phys Chem B 109(43):20504–20510
3. Lee Y, Terashima H, Shimodaira Y, Teramura K, Hara M, Kobayashi H, Domen K, Yashima M (2007) Zinc germanium oxynitride as a photocatalyst for overall water splitting under visible light. J Phys Chem C 111:1042–1048
4. Bard AJ (1979) Photoelectrochemistry and heterogenous photocatalysis at semiconductors. J Photochem Photobiol C 10:59–75
5. Sayama K, Mukasa K, Abe R, Abe Y, Arakawa H (2001) Stoichiometric water splitting into H_2 and O_2 using a mixture of two different photocatalysts and an IO_3-/I- shuttle redox mediator under visible light irradiation. Chem Commun 7(23):2416–2417
6. Fujishima A, Honda K (1972) Electrochemical photolysis of water at a semiconductor electrode. Nature 238:37–38
7. Bolton JR (1996) Solar photoproduction of hydrogen: a review. Sol Energy 57(1):37–50
8. Esswein A, Nocera D (2007) Hydrogen production by molecular photocatalysis. Chem Rev 107(10):4022–4047
9. Kudo A, Miseki Y (2009) Heterogeneous photocatalyst materials for water splitting. Roy Soc Ch 38:253–278
10. Amouyal E (1995) Photochemical production of hydrogen and oxygen from water: a review and state of the art. Sol Energy Mater Sol Cell 38(1–4):249–276
11. Dvoranova D, Brezova V, Mazur M, Malati M (2002) Investigations of metal-doped titanium dioxide photocatalysts. Appl Catal Environ 37(2):91–105
12. Ikuma Y, Bessho H (2007) Effect of Pt concentration on the production of hydrogen by a photocatalyst. Int J Hydrogen Energy 32(14):2689–2692
13. Jin Z, Zhang X, Li S, Lu G (2007) 5.1% apparent quantum efficiency for stable hydrogen generation over eosin-sensitized CuO/TiO_2 photocatalyst under visible light irradiation. Catal Commun 8(8):1267–1273
14. Ohtani B (2008) Preparing articles on photocatalysis—beyond the illusions, misconceptions, and speculation. Chem Lett 37(3):217–229
15. Inoue Y (2009) Photocatalytic water splitting by RuO_2-loaded metal oxides and nitrides with d0- and d10 -related electronic configurations. Energy Environ Sci 2:364–386

16. Sayama K, Arakawa H (1997) Effect of carbonate salt addition on the photocatalytic decomposition of liquid water over Pt–TiO_2 catalyst. J Chem Soc Faraday T 93(8):1647–1654
17. Yamaguti K, Sato S (1985) Photolysis of water over metallized powdered titanium dioxide. J Chem Soc Faraday T 81:1237–1246
18. Jeong H, Kim T, Kim D, Kim K (2006) Hydrogen production by the photocatalytic overall water splitting on: effect of preparation method. Int J Hydrogen Energy 31(9):1142–1146
19. Takata T, Shinohara K, Tanaka A, Hara M, Kondo JN, Domen K (1997) A highly active photocatalyst for overall water splitting with a hydrated layered perovskite structure. J Photoch Photobio A 106(1–3):45–49
20. Ogura S, Kohno M, Sato K, Inoue Y (1998) Photocatalytic properties of $M_2Ti_6O_{13}$ (M=Na, K, Rb, Cs) with rectangular tunnel and layer structures: behavior of a surface radical produced by UV irradiation and photocatalytic activity for water decomposition. Phys Chem Chem Phys 1:179–183
21. Inoue Y, Niiyama T, Asai Y, Sato K (1992) Stable photocatalytic activity of BaTi409 combined with ruthenium oxide for decomposition of water. J Chem Soc Chem commun 579–580
22. Reddy RV, Hwang DW, Lee JS (2003) Photocatalytic water splitting over ZrO_2 prepared by precipitation method. Korean J Chem Eng 20(6):1026–1029
23. Lin H, Lee T, Sie C (2008) Photocatalytic hydrogen production with nickel oxide intercalated $K_4Nb_6O_{17}$ under visible light irradiation. Int J Hydrogen Energy 33(15):4055–4063
24. Sayama K, Arakawa H, Asakura K, Tanaka A, Domen K, Onishi T (1998) Photocatalytic activity and reaction mechanism of Pt-interlaced $K_4Nb_6O_{17}$ catalyst on the water splitting in carbonate salt aqueous solution. J Photoch Photobio A 114:125–135
25. Kim SH, Park S, Lee CW, Han BS, Seo SW, Kim JS, Cho IS, Hong KS (2012) Photophysical and photocatalytic water splitting performance of stibiotantalite type-structure compounds, $SbMO_4$ (M=Nb, Ta). Int J Hydrogen Energy 37(22):16895–16902
26. Chen W, Li C, Gao H, Yuan J, Shangguan W, Su J, Sun Y (2012) Photocatalytic water splitting on protonated form of layered perovskites K0.5La0.5Bi2M2O9 (M=Ta; Nb) by ion-exchange. Int J Hydrogen Energy 37(17):12846–12851
27. Kato H, Asakura K, Kudo A (2003) Highly efficient water splitting into H_2 and O_2 over lanthanum-doped $NaTaO_3$ photocatalysts with high crystallinity and surface nanostructure. J Am Chem Soc 125:3082–3089
28. Kato H, Kudo A (2001) Energy structure and photocatalytic activity for water splitting of $Sr_2(Ta_1 - XNb_X)_2O_7$ solid solution. J Photoch Photobio A 145(1–2):129–133
29. Ikeda S, Fubuki M, Takahara YK, Matsumura M (2006) Photocatalytic activity of hydrothermally synthesized tantalate pyrochlores for overall water splitting. Appl Catal Gen 300 (2):186–190
30. Kadowaki H, Saito N, Nishiyama H, Kobayashi H, Shimodaira Y, Inoue Y (2007) Overall splitting of water by RuO_2-loaded $PbWO_4$ photocatalyst with d10s2-d0 configuration. J Phys Chem 111:439–444
31. Sakata Y, Matsuda Y, Yanagida T, Hirata K, Imamura H, Teramura K (2008) Effect of metal Ion addition in a Ni supported Ga_2O_3 photocatalyst on the photocatalytic overall splitting of H_2O. Catal Lett 125:22–26
32. Maeda K, Teramura K, Saito N, Inoue Y, Domen K (2006) Improvement of photocatalytic activity of $(Ga_{1-x}Zn_x)(N_{1-x}O_x)$ solid solution for overall water splitting by co-loading Cr and another transition metal. J Catal 243(2):303–308
33. Sato J, Saito N, Nishiyama H, Inoue Y (2002) Photocatalytic water decomposition by RuO_2-loaded antimonates, $M_2Sb_2O_7$ (M=Ca, Sr), $CaSb_2O_6$ and $NaSbO_3$, with d10 configuration. J Photoch Photobio A 148(1–3):85–89
34. Moriya Y, Takata T, Domen K (2013) Recent progress in the development of (oxy)nitride photocatalysts for water splitting under visible-light irradiation. Coord Chem Rev 257(13–14):1957–1969

35. Maeda K (2011) Photocatalytic water splitting using semiconductor particles: history and recent developments. J Photochem Photobiol C 12(4):237–268
36. Abe R (2010) Recent progress on photocatalytic and photoelectrochemical water splitting under visible light irradiation. J Photochem Photobiol C 11(4):179–209
37. Jang JS, Kim HG, Lee JS (2012) Heterojunction semiconductors: a strategy to develop efficient photocatalytic materials for visible light water splitting. Catal Today 185(1):270–277
38. Pai MR, Banerjee AM, Tripathi AK, Bharadwaj SR (2012) 14—Fundamentals and applications of the photocatalytic water splitting reaction. In: Banerjee S, Tyagi A (eds) Functional materials. Elsevier, London, pp 579–606
39. Yamaguti K, Sato S (1984) Photolysis of water over platinum/titanium dioxide catalyst. Nippon Kagaku Kaishi 2:258–263
40. Hu C, Teng H (2010) Structural features of p-type semiconducting NiO as a co-catalyst for photocatalytic water splitting. J Catal 272(1):1–8
41. Abe R, Higashi M, Sayama K, Abe Y, Sugihara H (2006) Photocatalytic activity of R_3MO_7 and $R_2Ti_2O_7$ (R=Y, Gd, La; M=Nb, Ta) for water splitting into H_2 and O_2. J Phys Chem B 110(5):2219–2226
42. Inoue Y, Asai Y, Sayama K (1994) Photocatalysts with tunnel structures for decomposition of water part 1.-BaTi, O, a pentagonal prism tunnel structure, and its combination with various promoters. J Chem Soc Faraday T 90(5):797–802
43. Zheng X, Wei L, Zhang Z, Jiang Q, Wei Y, Xie B, WEI M (2009) Research on photocatalytic H_2 production from acetic acid solution by Pt/TiO_2 nanoparticles under UV irradiation. Int J Hydrogen Energy 34(22):9033–9041
44. Yang J, Yan H, Wang X, Wen F, Wang Z, Fan D, Shi J, Li C (2012) Roles of cocatalysts in Pt–PdS/CdS with exceptionally high quantum efficiency for photocatalytic hydrogen production. J Catal 290:151–157
45. Mizukoshi Y, Makise Y, Shuto T, Hu J, Tominaga A, Shironita S, Tanabe S (2007) Immobilization of noble metal nanoparticles on the surface of TiO_2 by the sonochemical method: photocatalytic production of hydrogen from an aqueous solution of ethanol. Ultrason Sonochem 14(3):387–392
46. Takahashi T, Kakihana M, Yamashita K, Yoshida K, Ikeda S, Hara M, Domen K (1999) Synthesis of NiO-loaded $KTiNbO_5$ photocatalysts by a novel polymerizable complex method. J Alloys Compd 285:77–81
47. Domen K, Kudo A, Shinozaki A, Tanaka A, Maruya K, Onishi T (1986) Photodecomposition of water and hydrogen evolution from aqueous methanol solution over novel niobate photocatalysts. J Chem Soc Chem Commun (4):356–357
48. Huang Y, Wu J, Wei Y, Lin J, Huang M (2008) Hydrothermal synthesis of $K_2La_2Ti_3O_{10}$ and photocatalytic splitting of water. J Alloys Compd 456(1–2):364–367
49. Kim HG, Hwang DW, Kim J, Kim Y, Lee JS (1999) Highly donor-doped (110) layered perovskite materials as novel photocatalysts for overall water splitting. Chem Commun 1077–1078
50. Ikeda S, Hirao K, Ishino S, Matsumura M, Ohtani B (2006) Preparation of platinized strontium titanate covered with hollow silica and its activity for overall water splitting in a novel phase-boundary photocatalytic system. Catal Today 117(1–3):343–349
51. Yang Y, Lee K, Kado Y, Schmuki P (2012) Nb-doping of TiO_2/$SrTiO_3$ nanotubular heterostructures for enhanced photocatalytic water splitting. Electrochem Commun 17:56–59
52. Altomare M, Pozzi M, Allieta M, Bettini LG, Selli E (2013) H_2 and O_2 photocatalytic production on TiO_2 nanotube arrays: effect of the anodization time on structural features and photoactivity. Appl Catal Environ 136–137:81–88
53. Kitano M, Takeuchi M, Matsuoka M, Thomas JM, Anpo M (2007) Photocatalytic water splitting using Pt-loaded visible light-responsive TiO_2 thin film photocatalysts. Catal Today 120(2):133–138
54. Liu J, Liu J, Li Z (2013) Preparation and photocatalytic activity for water splitting of Pt–$Na_2Ta_2O_6$ nanotube arrays. J Solid State Chem 198:192–196

55. Chiou Y, Kumar U, Wu JCS (2009) Photocatalytic splitting of water on NiO/InTaO_4 catalysts prepared by an innovative sol–gel method. Appl Catal Gen 357(1):73–78
56. Kato H, Kudo A (2003) Photocatalytic water splitting into H_2 and O_2 over various tantalate photocatalysts. Catal Today 78:561–569
57. Zheng C, West A (1991) Compound and solid-solution formation, phase equilibria and Electrical properties in the ceramic system $Zr0_2$-La_2O_3-Ta_2O_5. J Mat Chem 1(2):163–167
58. Kudo A, Kato H (2000) Effect of lanthanide-doping into $NaTaO_3$ photocatalysts for efficient water splitting. Chem Phys Lett 331:373–377
59. Zou Z, Ye J, Arakawa H (2003) Photocatalytic water splitting into H_2 and/or O_2 under UV and visible light irradiation with a semiconductor photocatalyst. Int J Hydrogen Energy 28 (6):663–669
60. Li Y, Wu J, Huang Y, Huang M, Lin J (2009) Photocatalytic water splitting on new layered perovskite A2.33Sr0.67Nb5O14.335 (A=K, H). Int J Hydrogen Energy 34(19):7927–7933
61. Wei Y, Li J, Huang Y, Huang M, Lin J, Wu J (2009) Photocatalytic water splitting with In-doped $H_2LaNb_2O_7$ composite oxide semiconductors. Sol Energy Mater Sol Cell 93 (8):1176–1181
62. Sayama K, Arakawa H, Domen K (1996) Photocatalytic water splitting on nickel intercalated A4TaxNb6-xO17 (A=K, Rb). Catal Today 28(1–2):175–182
63. Hameed A, Gondal MA, Yamani ZH (2004) Effect of transition metal doping on photocatalytic activity of WO_3 for water splitting under laser illumination: role of 3d-orbitals. Catal Commun 5(11):715–719
64. He X, Boehm RF (2009) Direct solar water splitting cell using water, WO_3, Pt, and polymer electrolyte membrane. Energy 34(10):1454–1457
65. Lai CW, Sreekantan S (2013) Fabrication of WO_3 nanostructures by anodization method for visible-light driven water splitting and photodegradation of methyl orange. Mat Sci Semicon Proc 16(2):303–310
66. Rao PM, Cho IS, Zheng X (2013) Flame synthesis of WO_3 nanotubes and nanowires for efficient photoelectrochemical water-splitting. Proc Combust Inst 34(2):2187–2195
67. Lai K, Zhu Y, Lu J, Dai Y, Huang B (2013) N- and Mo-doping Bi_2WO_6 in photocatalytic water splitting. Comput Mat Sci 67:88–92
68. Ng YH, Iwase A, Kudo A, Amal R (2010) Reducing graphene oxide on a visible-light $BiVO_4$ photocatalyst for an enhanced photoelectrochemical water splitting. J Phys Chem Lett 1:2607–2612
69. Hara M, Kondo T, Komoda M, Ikeda S, Shinohara K, Tanaka A, Kondo JN, Domen K (1998) Cu_2O as a photocatalyst for overall water splitting under visible light irradiation. Chem Commun 357–358
70. Ye J, Zou Z, Arakawa H, Oshikiri M, Shimoda M, Matsushita A, Shishido T (2002) Correlation of crystal and electronic structures with photophysical properties of water splitting photocatalysts $InMO_4$ ($M=V^{5+}$, Nb^{5+}, Ta^{5+}). J Photoch Photobio A 148:79–83
71. Lin H, Chen Y, Chen Y (2007) Water splitting reaction on NiO/$InVO_4$ under visible light irradiation. Int J Hydrogen Energy 32(1):86–92
72. Kadowaki H, Saito N, Nishiyama H, Inoue Y (2007) RuO_2-Loaded Sr^{2+}-doped CeO_2 with d^0 electronic configuration as a new photocatalyst for overall water splitting. Chem Lett 36 (3):440–441
73. Hitoki G, Ishikawa A, Takata T, Kondo JN, Hara M, Domen K (2002) Ta_3N_5 as a novel visible light-driven photocatalyst $\lambda < 600$ nm). Chem Lett 31(7):736–737
74. Yamasita D, Takata T, Hara M, Kondo JN, Domen K (2004) Recent progress of visible-light-driven heterogeneous photocatalysts for overall water splitting. Solid State Ion 172(1–4):591–595
75. Kasahara A, Nukumizu K, Hitoki G, Takata T, Kondo JN, Hara M, Kobayashi H, Domen K (2002) Photoreactions on $LaTiO_2N$ under visible light irradiation. J Phys Chem A 106:6750–6753
76. Yoshida M, Maeda K, Lu D, Kubota J, Domen K (2013) Lanthanoid oxide layers on rhodium-loaded $(Ga_{1-x}Zn_x)(N_{1-x}O_x)$ photocatalyst as a modifier for overall water splitting under visible-light irradiation. J Phys Chem C 117:14000–14006

77. Lee K, Tienes B, Wilker M, Schnitzebaumer K, Dukovic G (2012) $(Ga_{1-x}Zn_x)(N_{1-x}O_x)$ nanocrystals: visible absorbers with tunable composition and absorption spectra. Am Chem Soc Nano Lett 12:3268–3272
78. Arai N, Saito N, Nishiyama H, Domen K, Kobayashi H, Sato K, Inoue Y (2007) Effects of divalent metal ion (Mg^{2+}, Zn^{2+} and Be^{2+}) doping on photocatalytic activity of ruthenium oxide-loaded gallium nitride for water splitting. Catal Today 129(3–4):407–413
79. Kato H, Hori M, Konta H, Shimodaira Y, Kudo A (2004) Construction of Z-scheme-type heterogeneous photocatalysis systems for water splitting into H_2 and O_2 under visible light irradiation. Chem Lett 33(10):1348–1349
80. Abe R, Takata T, Sugihara H, Domen K (2005) Photocatalytic overall water splitting under visible light by TaON andWO_3 with an IO_3-/I- shuttle redox mediator. Chem Commun 3829–3831
81. Abe R, Sayama K, Sugihara H (2005) Development of new photocatalytic water splitting into H_2 and O_2 using two different semiconductor photocatalysts and a shuttle redox mediator IO_3^-/I^-. J Phys Chem B 109:16052–16061
82. Higashi M, Abe R, Teramura K, Takata T, Ohtani B, Domen K (2008) Two step water splitting into H_2 and O_2 under visible light by $ATaO_2N$ (A=Ca, Sr, Ba) andWO_3 with IO_3-/I- shuttle redox mediator. Chem Phys Lett 452:120–123
83. Sasaki Y, Nemoto H, Saito K, Kudo A (2009) Solar water splitting using powdered photocatalysts driven by Z-schematic interparticle electron transfer without an electron mediator. J Phys Chem C 113:17536–17542
84. Abe R, Shinmei K, Hara K, Ohtani B (2009) Robust dye-sensitized overall water splitting system with two-step photoexcitation of coumarin dyes and metal oxide semiconductors. Chem Commun 3577–3579
85. Maeda K, Higashi M, Lu D, Abe R, Domen K (2010) Efficient nonsacrificial water splitting through two-step photoexcitation by visible light using a modified oxynitride as a hydrogen evolution photocatalyst. J Am Chem Soc 132:5858–5868
86. Tabata M, Maeda K, Higashi M, Lu D, Takata T, Abe R, Domen K (2010) Modified Ta_3N_5 powder as a photocatalyst for O_2 evolution in a two-step water splitting system with an iodate/iodide shuttle redox mediator under visible light. Langmuir 26:9161–9165
87. Moradour A, Amouyal E, Keller P, Kagan H (1978) Hydrogen production by visible light irradiation of aqueous solutions of Ru(bipy)32+. New J Chem 2(6):547–549
88. Navarro RM, del Valle F, Villoria de la Mano JA, Álvarez-Galván MC, Fierro JLG (2009) Photocatalytic water splitting under visible light: concept and catalysts development. Adv Chem Eng 36:111–143
89. Sayama K, Mukasa K, Abe R, Abe Y, Arakawa H (2002) A new photocatalytic water splitting system under visible light irradiation mimicking a Z-scheme mechanism in photosynthesis. J Photoch Photobio A 148(1–3):71–77
90. Abe R, Sayama K, Domen K, Arakawa H (2001) A new type of water splitting system composed of two different TiO_2 photocatalysts (anatase, rutile) and a IO_3^-/I^-shuttle redox mediator. Chem Phys Lett 344(3):339–344
91. Erbs W, Desilvestro J, Borgarello E, Gratzel M (1984) Visible-light-induced O_2 generation from aqueous dispersions of WO_3. J Phys Chem 88:4001
92. Sayama K, Yoshida M, Kusama H, Okabe K, Abe Y, Arakawa H (1997) Photocatalytic decomposition of water into H_2 and O_2 by a two-step photoexcitation reaction using a WO_3 suspension catalyst and an $Fe^{3+}Fe^{2+}$ redox system. Chem Phys Lett 277(4):387–391
93. Kato H, Sasaki Y, Iwase A, Kudo A (2007) Role of iron Ion electron mediator on photocatalytic overall water splitting under visible light irradiation using Z-scheme systems. Chem Soc Jpn 80(12):2457–2464
94. Sasaki Y, Iwase A, Kato H, Kudo A (2008) The effect of cocatalyst for Z-scheme photocatalysis systems with an Fe^{3+}/Fe^{2+} electron mediator on overall water splitting under visible light irradiation. J Catal 259:133–137

95. Ohno T, Haga D, Fujihara K, Kaizaki K, Matsumura M (1997) Unique effects of iron(III) ions on photocatalytic and photoelectrochemical properties of titanium dioxide. J Phys Chem B 101(33):6415–6419
96. Konta R, Ishii T, Kato H, Kudo A (2004) Photocatalytic activities of noble metal ion-doped $SrTiO_3$ under visible light irradiation. J Phys Chem B 108:8992–8995
97. Ishii T, Kato H, Kudo A (2004) H_2 evolution from an aqueous methanol solution on $SrTiO_3$ photocatalysts codoped with chromium and tantalum ions under visible light irradiation. J Photoch Photobio A 163(1–2):181–186
98. Kato H, Kudo A (2002) Visible-light-response and photocatalytic activities of TiO_2 and $SrTiO_3$ photocatalysts codoped with antimony and chromium. J Phys Chem B 106(19):5029–5034
99. Kudo A, Tanaka A, Domen K, Onishi T (1988) The effects of the calcination temperature of $SrTiO_3$ powder on photocatalytic activities. J Catal 111(2):296–301
100. Darwent JR, Mills A (1982) Photo-oxidation of water sensitized by WO_3 powder. J Chem Soc Faraday T 78:359–367
101. Weaver ER, Berry WM, Bohnson VL, Gordon BD (1920) The ferrosilicon process for the generation of hydrogen, Report No. 40. Annual Report National Advisory Committee for Aeronautics
102. Guo J, Chen X (eds) (2011) Solar hydrogen generation: transition metal oxides in water photoelectrolysis. McGraw Hill, New York
103. Musa A, Al-Saleh M, Loakeimidis ZC, Ouzounidou M, Yentekakis IV, Konsolakis M, Marnellos GE (2014) Hydrogen production by iso-octane steam reforming over Cu catalysts supported on rare earth oxides (REOs). Int J Hydrogen Energy 39(3):1350–1363
104. Kumar K, Roy S, Das D (2013) Continuous mode of carbon dioxide sequestration by *C. sorokiniana* and subsequent use of its biomass for hydrogen production by *E. cloacae* IIT-BT 08. Bioresour Technol 145:116–122
105. Sugai Y, Purwasena IA, Sasaki K, Fujiwara K, Hattori Y, Okatsu K (2012) Experimental studies on indigenous hydrocarbon-degrading and hydrogen-producing bacteria in an oilfield for microbial restoration of natural gas deposits with CO_2 sequestration. J Nat Gas Sci Eng 5:31–41
106. Huang L, Zhou J, Hsu A, Chen R (2013) Catalytic partial oxidation of *n*-butanol for hydrogen production over LDH-derived Ni-based catalysts. Int J Hydrogen Energy 38(34):14550–14558
107. Ge Z, Guo S, Guo L, Cao C, Su X, Jin H (2013) Hydrogen production by non-catalytic partial oxidation of coal in supercritical water: explore the way to complete gasification of lignite and bituminous coal. Int J Hydrogen Energy 38(29):12786–12794
108. Bundaleska N, Tsyganov D, Saavedra R, Tatarova E, Dias FM, Ferreira CM (2013) Hydrogen production from methanol reforming in microwave "tornado"-type plasma. Int J Hydrogen Energy 38(22):9145–9157
109. Kim HS, Kim YH, Ahn BT, Lee JG, Park CS, Bae KK (2014) Phase separation characteristics of the Bunsen reaction when using HI_x solution ($HI–I_2–H_2O$) in the sulfur–iodine hydrogen production process. Int J Hydrogen Energy 39(2):692–701
110. Liberatore R, Lanchi M, Caputo G, Felici C, Giaconia A, Sau S, Tarquini P (2012) Hydrogen production by flue gas through sulfur–iodine thermochemical process: economic and energy evaluation. Int J Hydrogen Energy 37(11):8939–8953
111. Xing Z, Zong X, Pan J, Wang L (2013) On the engineering part of solar hydrogen production from water splitting: photoreactor design. Chem Eng Soc 104:125–146
112. Zhang W, Li Y, Wang C, Wang P, Wang Q (2013) Energy recovery during advanced wastewater treatment: Simultaneous estrogenic activity removal and hydrogen production through solar photocatalysis. Water Res 47(3):1480–1490
113. Yu S, Huang C, Liao C, Wu JCS, Chang S, Chen K (2011) A novel membrane reactor for separating hydrogen and oxygen in photocatalytic water splitting. J Membr Sci 382 (1–2):291–299

114. Liao C, Huang C, Wu JCS (2012) Hydrogen production from semiconductor-based photocatalysis via water splitting. Catalysts 2:490–516
115. Liao C, Huang C, Wu JCS (2012) Novel dual-layer photoelectrode prepared by RF magnetron sputtering for photocatalytic water splitting. Int J Hydrogen Energy 37(16):11632–11639
116. Minggu LJ, Daud WRW, Kassim M (2010) An overview of photocells and photoreactors for photoelectrochemical water splitting. Int J Hydrogen Energy 35(11):5233–5244
117. Matsumoto H, Mashimo H, Kuroda C (2012) Process analysis of rotary-type solar reactor for hydrogen production systems. Comput Aided Chem Eng 30:1103–1107
118. Zhang C, Zhu X, Liao Q, Wang Y, Li J, Ding Y, Wang H (2010) Performance of a groove-type photobioreactor for hydrogen production by immobilized photosynthetic bacteria. Int J Hydrogen Energy 35(11):5284–5292
119. Zhang Z, Hossain MF, Takahashi T (2010) Photoelectrochemical water splitting on highly smooth and ordered TiO_2 nanotube arrays for hydrogen generation. Int J Hydrogen Energy 35 (16):8528–8535
120. Lo C, Huang C, Liao C, Wu JCS (2010) Novel twin reactor for separate evolution of hydrogen and oxygen in photocatalytic water splitting. Int J Hydrogen Energy 35(4):1523–1529
121. Jing D, Guo L, Zhao L, Zhang X, Liu H, Li M, Shen S, Liu G, Hu X, Zhang X, Zhang K, Ma L, Guo P (2010) Efficient solar hydrogen production by photocatalytic water splitting: from fundamental study to pilot demonstration. Int J Hydrogen Energy 35(13):7087–7097
122. Huang C, Liao C, Wu C, Wu JCS (2012) Photocatalytic water splitting to produce hydrogen using multi-junction solar cell with different deposited thin films. Sol Energy Mater Sol Cell 107:322–328
123. Yan W, Zheng CL, Liu YL, Guo LJ (2011) A novel dual-bed photocatalytic water splitting system for hydrogen production. Int J Hydrogen Energy 36(13):7405–7409
124. Babu VJ, Kumar MK, Nair AS, Kheng TL, Allakhverdiev SI, Ramakrishna S (2012) Visible light photocatalytic water splitting for hydrogen production from N-TiO_2 rice grain shaped electrospun nanostructures. Int J Hydrogen Energy 37(10):8897–8904
125. Ding L, Zhou H, Lou S, Ding J, Zhang D, Zhu H, Fan T (2013) Butterfly wing architecture assisted CdS/Au/TiO_2 Z-scheme type photocatalytic water splitting. Int J Hydrogen Energy 47(3):1480–1490
126. Ding L, Zhou H, Lou S, Ding J, Zhang D, Zhu H, Fan T (2013) Butterfly wing architecture assisted CdS/Au/TiO_2 Z-scheme type photocatalytic water splitting. Int J Hydrogen Energy 38(20):8244–8253
127. Bae SW, Ji SM, Hong SJ, Jang JW, Lee JS (2009) Photocatalytic overall water splitting with dual-bed system under visible light irradiation. Int J Hydrogen Energy 34(8):3243–3249
128. He Y, Yan F, Yu H, Yuan S, Tong Z, Sheng G (2014) Hydrogen production in a light-driven photoelectrochemical cell. Appl Energy 113:164–168
129. Li Y, Yu H, Zhang C, Song W, Li G, Shao Z, Yi B (2013) Effect of water and annealing temperature of anodized TiO_2 nanotubes on hydrogen production in photoelectrochemical cell. Electrachimica Acta 107:313–319
130. Zhu L, Qiang YH, Zhao YL, Gu XQ (2013) Double junction photoelectrochemical solar cells based on Cu_2ZnSnS_4/$Cu_2ZnSnSe_4$ thin film as composite photocathode. Appl Surf Sci 292:55–62
131. Danko DB, Sylenko PM, Shlapak AM, Khyzhun OY, Shcherbakova LG, Ershova OG, Solonin YM (2013) Photoelectrochemical cell for water decomposition with a hybrid photoanode and a metal-hydride cathode. Sol Energy Mater Sol Cell 114:172–178
132. Wang G, Ling Y, Wang H, Xihong L, Li Y (2014) Chemically modified nanostructures for photoelectrochemical water splitting. J Photochem Photobiol C 19:35–51
133. Li Y, Yu H, Song W, Li G, Yi B, Shao Z (2011) A novel photoelectrochemical cell with self-organized TiO_2 nanotubes as photoanodes for hydrogen generation. Int J Hydrogen Energy 36(22):14374–14380

134. Hsu C, Chen D (2011) Photoresponse and stability improvement of ZnO nanorod array thin film as a single layer of photoelectrode for photoelectrochemical water splitting. Int J Hydrogen Energy 36(24):15538–15547
135. Andrade L, Cruz R, Ribeiro HA, Mendes A (2010) Impedance characterization of dye-sensitized solar cells in a tandem arrangement for hydrogen production by water splitting. Int J Hydrogen Energy 35(17):8876–8883
136. Shin K, Yoo J, Hyeok JP (2013) Photoelectrochemical cell/dye-sensitized solar cell tandem water splitting systems with transparent and vertically aligned quantum dot sensitized TiO_2 nanorod arrays. J Power Sources 225:263–268
137. Lianos P (2011) Production of electricity and hydrogen by photocatalytic degradation of organic wastes in a photoelectrochemical cell: the concept of the photofuelcell: a review of a re-emerging research field. J Hazard Mater 185(2–3):575–590
138. Fujii K, Nakamura S, Sugiyama M, Watanabe K, Bagheri B, Nakano Y (2013) Characteristics of hydrogen generation from water splitting by polymer electrolyte electrochemical cell directly connected with concentrated photovoltaic cell. Int J Hydrogen Energy 38 (34):14424–14432
139. Zhang H, Huang S, Conibeer G (2012) Study of photo-cathode materials for tandem photoelectrochemical cell for direct water splitting. Energy Procedia 22:10–14
140. Avachat US, Jahagirdar AH, Dhere NG (2006) Multiple bandgap combination of thin film photovoltaic cells and a photoanode for efficient hydrogen and oxygen generation by water splitting. Sol Energy Mater Sol Cell 90(15):2464–2470
141. Mishra PR, Shukla PK, Srivastava ON (2007) Study of modular PEC solar cells for photoelectrochemical splitting of water employing nanostructured TiO_2 photoelectrodes. Int J Hydrogen Energy 32(12):1680–1685
142. Gibson STL, Kelly NA (2008) Optimization of solar powered hydrogen production using photovoltaic electrolysis devices. Int J Hydrogen Energy 33:5931–5940

Nonmetal Doping in TiO_2 Toward Visible-Light-Induced Photocatalysis

Xu Zong, Gaoqing (Max) Lu, and Lianzhou Wang

Abstract Over the past decade, the doping of nonmetal elements in wide band-gap semiconductors such as TiO_2 has been intensively investigated as an effective strategy of expanding the responsive solar spectrum of pristine semiconductors toward visible region. This chapter gives a review of this highly hot research topic. The fundamental principles involved and basic approaches are initially described. A range of nonmetal dopants are subsequently detailed with examples showing their effect on the photocatalytic performance such as pollutant degradation and water splitting under visible light. The problems simultaneously introduced by doping are also discussed.

Keywords Photocatalysis, Titanium dioxide, Visible light, Water splitting, Water treatment

Contents

X. Zong, G. Lu, and L. Wang (✉)
ARC Centre of Excellence for Functional Nanomaterials, School of Chemical Engineering and AIBN, The University of Queensland, St. Lucia, Brisbane, QLD 4072, Australia
e-mail: l.wang@uq.edu.au

D.W. Bahnemann and P.K.J. Robertson (eds.), *Environmental Photochemistry Part III*, Hdb Env Chem (2015) 35: 87–114, DOI 10.1007/698_2013_249,

1 Introduction

Heterogeneous photocatalysis has been the subject of intense studies as an attractive measure toward solving important environmental and energy problems [1–4]. Up to now, this technology has shown great potential in the applications of environmental remediation, water splitting to produce H_2, deactivation/killing of bacteria, and organic syntheses. As semiconductor photocatalyst is the principal component in this process, most efforts have been devoted to the development of novel semiconductor photocatalyst with desirable functionalities [5–12]. Up to now, various types of semiconductors such as oxide [13–15], sulfide [16–18], (oxy)nitride [19–23], and (oxy)sulfide [24, 25] have been developed and various strategies such as band-gap engineering, facet engineering, and surface modifications have been employed to further improve the performance of photocatalyst [26–28]. However, regardless of the effort, titania (TiO_2)-based materials still remain the focus of experimental as well as theoretical studies mainly due to its integrated fascinating properties such as nontoxicity, easy availability, high activity, high stability, and suitable electronic band structures for photocatalytic reactions. In fact, TiO_2 has been supposed to be the most promising photocatalyst in the practical applications instead of laboratory investigation [29]. However, a significant drawback of TiO_2 toward practical applications is its wide band gap that requires the use of UV light. Therefore, it is highly desirable to develop strategies that could extend the action spectra of TiO_2 toward utilizing an even larger fraction of the solar spectrum which consists of ~5% UV, ~43% visible light, and ~52% infrared light. In this regard, various strategies such as sensitization by organic dyes, coupling with a narrow band-gap semiconductor, and metal ion/nonmetal ion doping have been developed.

In this book chapter, we will focus on the nonmetal doping strategy (band-gap engineering) on modifying the electronic structures of TiO_2 toward enhanced light absorption in the visible region as well as improved visible-light-induced photocatalytic performance. The as-prepared TiO_2 photocatalysts are defined as second-generation titanium oxide photocatalysts as they surpass their ancestor in achieving visible-light response and enhanced photocatalytic activity in the visible region. The passion on investigating the second-generation TiO_2 photocatalyst was actually ignited by the work reported by Asahi et al. [30]. Up to now, the citation for this paper has been more than 3,500 times and will be continuously accumulating with the ever-increasing fever for the research in solar energy utilization. Because of the large volume of research papers available on the doping of TiO_2, it is not possible for this book chapter to be fully comprehensive, and hence only some breakthrough and representative work will be discussed. Readers interested in this field could refer to a series of excellent review papers [29, 31–44].

2 Doping Principles and Approaches

2.1 Doping Principles

To better understand the principles of doping strategy, we will first introduce the electronic structures of semiconductors and the basic principles for photocatalytic reactions. The unique electronic structures of semiconductors photocatalysts are characterized by a filled valence band (VB) and an empty conduction band (CB) separated by a forbidden band (Fig. 1). The energy difference between the VB and the CB is known as the "band gap," which is determined by the respective components for constructing the VB and CB. When the semiconductor absorbs incident light with energy greater or equal to the band-gap energy E_g of semiconductor, electrons have the possibility of being excited to the conduction band, while the holes are left in the valence band, thus creating negative-electron (e^-) and positive-hole (h^+) pairs. As electrons and holes have reductive and oxidative power, respectively, their subsequent migration to the surface of semiconductors will probably initiate redox reactions with the surrounding reaction media. Considering that the driving force for the redox reaction must be satisfied from the thermodynamic point, the band-edge positions of the conduction band and valence band of semiconductors should be correlated well with the potentials of the corresponding photocatalytic reactions. For example, in order to split water into hydrogen and oxygen both the reduction and oxidation potentials of water should locate within the forbidden band of the photocatalyst (Fig. 1). Therefore, the band-gap and the band-edge positions of the conduction and valence bands of semiconductors are two important parameters that will basically determine the photocatalytic performance of semiconductors under solar irradiation. Figure 2 lists a series of insulators, semiconductors, and metallic conductors. It's evident that semiconductors are characteristic of moderate band gap compared with insulators and conductors. Moreover, for different semiconductors, as the compositions and crystal structures are different, they have different band-gap and band-edge positions, therefore leading to distinct light absorption property and potential of photogenerated electrons and holes for photocatalytic reactions.

TiO_2 crystallizes in three structure forms anatase, rutile, and brookite with TiO_6 octahedra as the building unit. The VB of all the TiO_2 polymorphs is composed of O 2p orbitals and the CB is composed of Ti 3d orbitals. However, due to the variation of the crystal structures, the band gap is around 3.2 eV for anatase and brookite and 3.0 eV for rutile phase TiO_2. As for the band-edge positions of TiO_2, the conduction band edge of TiO_2 is slightly higher than the reduction potential of water and therefore can reduce protons to produce H_2 under UV light irradiation, and the

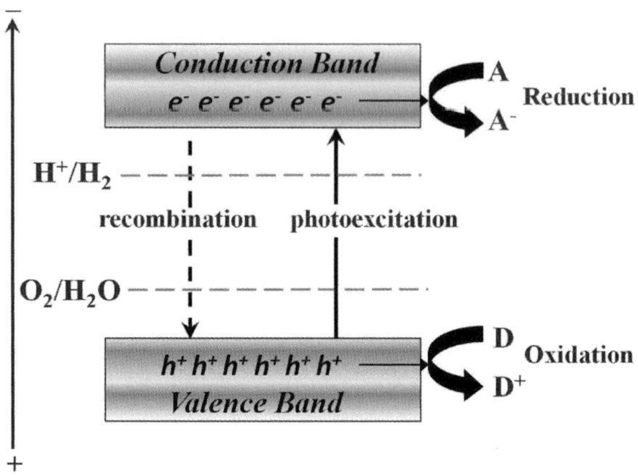

Fig. 1 Schematic illustration of the band structures and photocatalytic processes of semiconductor photocatalyst

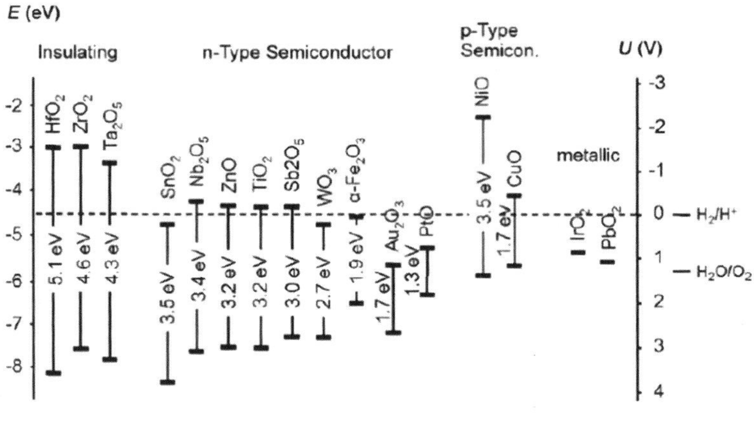

Fig. 2 Electronic band structure of different metal oxides and relative band-edge position to electrochemical scale (Reprinted with permission from [35]. Copyright 2010 Wiley-VCH)

valence band edge of TiO_2 is deep enough to oxidize water to produce O_2. In fact, from the thermodynamic point, the band-edge positions of TiO_2 are appropriate for different photocatalytic reactions such as photocatalytic and photoelectrochemical water splitting and pollutant degradation, therefore endowing TiO_2 with attractive electronic properties in different applications.

From the above analysis about the band structures of TiO_2, it is evident that the intrinsic electronic structure of TiO_2 is dominated by the O 2p and Ti 3d orbitals that construct the VB and CB. As the intrinsic band gap of TiO_2 is higher than 3.2 eV (anatase), in principle, five strategies could be used to induce visible-light response in TiO_2. The first strategy is to elevate the valence band maximum by forming a new band just above the valence band of TiO_2 by mixing the O 2p states

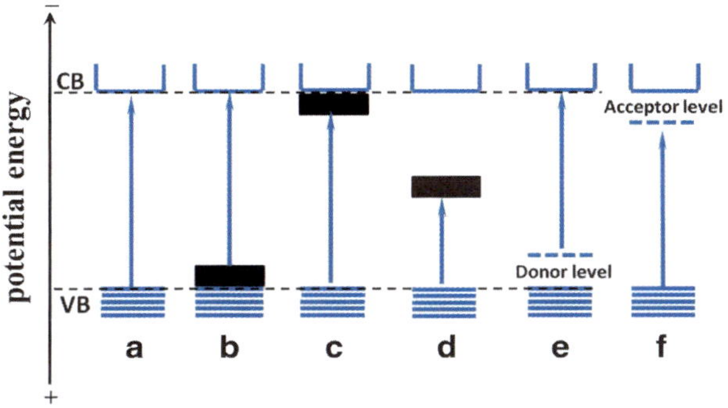

Fig. 3 Schematic illustration of the possible routes of modifying TiO_2 toward visible-light region through doping (*a*) pristine TiO_2, (*b*) form new band above the VB of TiO_2, (*c*) form new band below the CB of TiO_2, (*d*) form new band in the band gap, (*e*) form discrete donor level, and (*f*) form discrete acceptor level

with the dopant states (Fig. 3b). The second strategy is to lower the conduction band minimum by forming a new band just below the conduction band of TiO_2 by mixing the Ti 3d states with the dopant states (Fig. 3c). The third strategy is to form a new band in the band gap of TiO_2 by the dopant itself (Fig. 3d). The fourth strategy is to form a new donor level above the valence band of TiO_2 (Fig. 3e), and the last strategy is to form a new acceptor level below the conduction band of TiO_2 (Fig. 3f). In all the cases, the excitation route will not be directly from the intrinsic valence band to the conduction band of TiO_2 but from (or to) new formed band or levels as shown in Fig. 3. Therefore, the photon energy required to excite doped TiO_2 is greatly reduced; as a consequence, visible-light response in TiO_2 could be realized. Considering that the valence band of TiO_2 consists of O 2p orbitals, to induce visible-light activity by nonmetal doping, the electronegativity of the nonmetal dopant must be lower than that of oxygen, so that the dopant states can be involved in the formation of a new valence band by locating at its top. It is found that all nonmetals except F have lower electronegativity than O, and therefore, in principle these nonmetals have the potential to be used as dopants to introduce visible-light activity to TiO_2.

2.2 *Doping Approaches*

The doping approach will greatly influence the physiochemical properties such as crystallinity, surface area, surface properties, and chemical states of the dopants in the resulting materials and therefore play an important role in modifying the electronic as well as the crystal structures of the materials. Figure 4 shows the UV–vis absorption spectra of N-doped TiO_2 prepared by sputtering the TiO_2 target in a N_2 (40%)/Ar gas mixture (left) and radiofrequency magnetron sputtering (RF-MS) deposition method

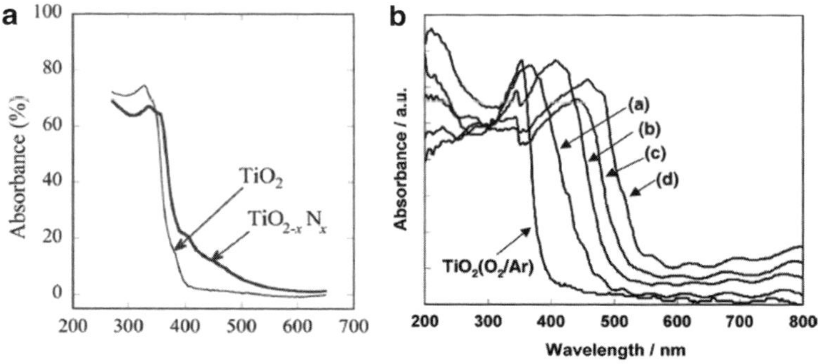

Fig. 4 (**A**) Optical-absorption spectra of $TiO_{2-x}N_x$ and TiO_2 films and (**B**) UV–vis absorption spectra of TiO_2 (O_2/Ar) and N–TiO_2– (X) thin films prepared on quartz substrates by a radiofrequency magnetron sputtering (RF-MS) method. X (%): (*a*) 2, (*b*) 4, (*c*) 10, and (*d*) 40 ((**A**) Reprinted with permission from [30]. Copyright 2001 AAAS. (**B**) Reprinted with permission from [45]. Copyright 2006 American Chemical Society)

using a N_2/Ar mixture sputtering gas (right) [45]. It is evident that the two kinds of films showed distinct visible-light absorption characteristics: taillike absorption for the former and band-to-band absorption for the latter. Therefore, by choosing appropriate doping strategies, desirable functionality could be achieved. Generally speaking, nonmetal dopants could be incorporated into TiO_2 with the following methods.

1. Solution-based approach

Solution-based method such as hydrothermal, sol–gel, and precipitation method is the most widely used method for doping nonmetal into TiO_2. In a typical synthesis, titanium species such as titanium isopropoxide, tetrabutyl orthotitanate, or titanium tetrachloride hydrolyze with reagents containing dopant species in a proper solvent to form a precursor. Further calcinations of the precursor at high temperatures in air or inert atmosphere or hydrothermal treatment will lead to the formation of the nonmetal doped TiO_2. The method of preparing the precursor as well as the following heat and hydrothermal treatment can greatly influence the photocatalytic properties of the resulting materials. Nonmetal can also be prepared with a one-step solution-based method. Materials containing both titanium and the doping species such as TiN or TiC directly hydrothermally treated in proper solvents can lead to the formation of corresponding nonmetal doped TiO_2. The solution-based method affords simplicity in controlling the dopant level and particle size by simple variations in the experimental conditions, such as hydrolysis rate, solution pH, and solvent systems.

2. Solid state treatment approach

Heat treatment of TiO_2 at high temperatures in atmosphere containing doping species can lead to the doping of the corresponding species in TiO_2. This method is generally used to prepare N- or C-doped TiO_2. For example, by heating TiO_2 in NH_3, N_2, or urea, N-doped TiO_2 nanoparticles or nanotubes can be prepared. By heating

TiO_2 in CO or acetylene, C-doped TiO_2 can be obtained. The reaction conditions such as the heat treatment temperature, time, and the atmosphere conditions have great influence on the photocatalytic properties of the final products. The other solid state reaction pathway is to directly calcine materials containing both titanium and dopant species such as TiS_2 and TiC in air. The heat treatment in air will lead to the partial oxidation of TiN, TiS_2, or TiC to TiO_2 and the remaining N, S, or C species act as the dopant in TiO_2.

3. Physical approach

There are several physical methods that are employed to prepare nonmetal doped TiO_2. Sputtering TiO_2 target in an atmosphere containing doping species such as N_2–Ar will incorporate N in the TiO_2 target [30]. Following heat treatment at high temperatures can improve the crystallinity of the N-doped TiO_2 target. Ion implantation of TiO_2 target with dopant species can incorporate the dopants into the TiO_2 lattice at low to medium doping levels [46]. The structural defects formed by ion implantation process, which lead to a decrease in the photoconversion efficiency, can be annealed out after heat treatment. Dopants can also be doped into TiO_2 by plasma-enhanced chemical vapor deposition (PECVD) method [47]. The titanium precursor together with the dopant species is carried into the reaction zone by the carrier gas and TiO_2 is deposited on the target substrate with the aid of plasma. Nevertheless, the physical method requires expensive apparatus for doping process, which makes this method not as attractive as solution-based as well as solid state reaction approach.

3 Nonmetal Doping Elements

The interest on the doping of nitrogen as well as other nonmetal species in TiO_2 toward decreased band gap and visible-light-induced photocatalytic activity was stimulated after the report by Asahi et al. in *Science* in 2001, even though there were some previous reports that didn't receive much attention [30]. In their work, Asahi et al. set the following requirements for inducing visible-light response and activity in TiO_2:

- Doping should produce states in the band gap of TiO_2 that absorb visible light.
- The CB minimum, including subsequent impurity states, should be as high as that of TiO_2 or higher than the H_2/H_2O level to ensure its photoreduction activity.
- The states in the gap should overlap sufficiently with the band states of TiO_2 to transfer photoexcited carriers to reactive sites at the catalyst surface within their lifetime.

Therefore, they chose nonmetal dopants for the doping instead of metal dopants because metal dopants often give quite localized d states deep in the band gap of TiO_2 and result in recombination centers of carriers. Moreover, to substitute lattice O atom, the dopant should have a radius comparable to that of the lattice O atoms to facilitate the substitution. In their work, the densities of states (DOSs) of the

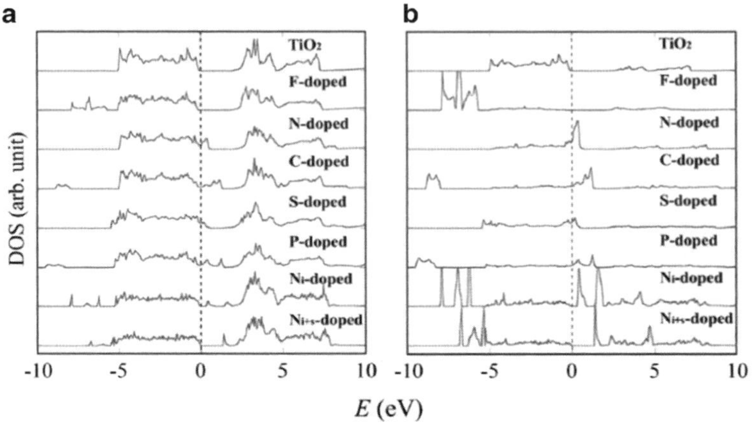

Fig. 5 (**a**) Total DOSs of doped TiO_2 and (**b**) the projected DOSs into the doped anion sites. The dopants F, N, C, S, and P were located at a substitutional site for an O atom in the anatase TiO_2 crystal. The results for N doping at an interstitial site (Ni-doped) and that at both substitutional and interstitial sites (Ni 1*s*-doped) are also shown (Reprinted with permission from [30]. Copyright 2001 AAAS)

substitutional doping of C, N, F, P, or S for O in the anatase TiO_2 crystal were calculated. Figure 5 shows the total DOSs and the projected DOSs of TiO_2 doped with different nonmetals in different sites. The substitutional doping of N was found to be the most effective because its p states contribute to the band-gap narrowing by mixing with O 2p states. Although doping with S shows a similar band-gap narrowing, the large ionic radius of S makes it difficult to be incorporated into the TiO_2 crystal. Moreover, the states introduced by C and P are too deep in the gap and therefore supposed hard to transfer photoexcited carriers to reactive sites at the catalyst surface within their lifetime. Therefore, N-doped TiO_2 was chosen as the first candidate in their investigation. In the following part, we will discuss N doping in TiO_2 followed by the introduction to other nonmetal dopants.

3.1 Nitrogen (N) Doping in TiO_2

Among all nonmetal doped TiO_2, nitrogen-doped TiO_2 has been most widely studied [30, 37, 48–60]. The doping of N in TiO_2 was actually first reported by Sato in 1986 [48]. In his work, it was found that the calcinations of titanium hydroxide at around 673 K can produce TiO_2 with pale yellow color and this yellow-colored TiO_2 showed enhanced photocatalytic activity in the visible-light region. It was concluded that the NO impurity which was formed from NH_4OH used for the preparation of titanium hydroxide led to the visible-light response and activity. However, this work didn't receive much attention until the work reported in *Science* in 2001 by Asahi et al. [30]. In their work, the substitutional doping of

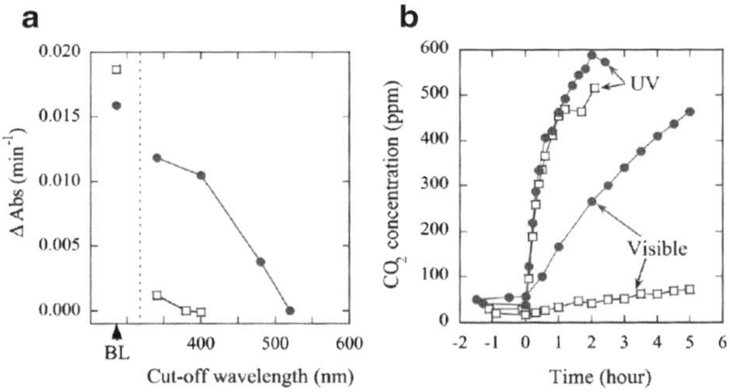

Fig. 6 Photocatalytic properties of $TiO_{2-x}N_x$ samples (*solid circles*) compared with TiO_2 samples (*open squares*). (**a**) Decomposition rates of methylene blue as a function of the cutoff wavelength of the optical high-path filters under fluorescent light. (**b**) CO_2 evolution as a function of irradiation time (light on at zero) during the photodegradation of acetaldehyde under UV irradiation and visible irradiation. (**c**) Contact angles of water as a function of time under interior lighting (Reprinted with permission from [30]. Copyright 2001 AAAS)

N was found to be the most effective because its p states contribute to the band-gap narrowing by mixing with O 2p states. Then in the experimental work, N-doped TiO_2 films were prepared by sputtering the TiO_2 target in a N_2 (40%)/Ar gas mixture and N-doped TiO_2 powders were prepared by heat treating anatase TiO_2 in the NH_3 (67%)/Ar atmosphere at 873 K for 3 h. The as-prepared films and powders revealed an improvement over pristine TiO_2 under visible light in optical absorption and photodegradation of methylene blue and gaseous acetaldehyde (Fig. 6). N 1*s* X-ray photoelectron spectroscopy (XPS) features at 396, 400, and 402 eV binding energies. The nitrogen species responsible for the overall band-gap narrowing exhibit the 396 eV N 1*s* binding energy. The nitrogen doped into substitutional sites of TiO_2 is supposed to be indispensable for band-gap narrowing and photocatalytic activity.

Following Asahi's work, an avalanche of reports on N-doped TiO_2 as well as other nonmetal doped TiO_2 emerged. Irie et al. prepared $TiO_{2-x}N_x$ powders by heating anatase TiO_2 powder under NH_3 flow at elevated temperatures and used the N-TiO_2 powder for the decomposition of gaseous 2-propanol (IPA). He proposed that the isolated N 2p states formed above the valence band in $TiO_{2-x}N_x$ are responsible for the visible-light response. Moreover, the increase of the nitrogen concentration lowered the quantum yield under UV illumination, indicating that excess doping sites could also work as recombination sites [49]. Ihara et al. synthesized nitrogen-doped TiO_2 with oxygen-deficient sites and they suggested that the visible-light absorption band of N-TiO_2 originates from the localized states of the oxygen deficiencies caused by nitrogen doping rather than from the nitrogen dopant itself [52]. The presence of nitrogen only improves the stabilization of these oxygen vacancies. Serpone reexamined the various claims and argued about the anion and cation doping of titanium dioxide with absorption edge red-shifted to lower energies. He proposed that the color centers formed in the band gap are the origin of the visible-light absorption band regardless of dopant species [59]. Meanwhile, by a combined experimental and theoretical approach, Livraghi et al. found that single-atom nitrogen impurity centers in the bulk of TiO_2 samples gave rise to localized states in the band gap of TiO_2. He proposes that these centers are responsible for visible-light absorption with promotion of electrons from the band-gap localized states to the conduction band or to surface-adsorbed electron scavengers [55]. Mitoraj and Kisch proposed a sensitization mechanism, in which higher melamine condensation products act as visible-light sensitizers [56].

Burda et al. reported the preparation of N-doped TiO_2 nanoparticles by employing the direct amination of 6–10-nm-sized TiO_2 particles [50]. The nanoscale synthesis route leads to increased nitrogen dopant concentration of up to 8% in titania and appreciable absorbance that extends to the visible region up to 600 nm. The synthesized N-doped TiO_2 nanoparticles are demonstrated to be photocatalytically active for the degradation of methylene blue under visible light, although the activity is quite low. Sathish et al. prepared N-doped TiO_2 nanocatalyst with a uniform size and spherical shape through a simple chemical method using $TiCl_3$ as precursor followed by calcination at different temperatures [58]. The light absorption of the as-prepared N-doped TiO_2 was shifted to the visible light, which is supposed to be due to the contribution of the N 2p states on the top of the valence band. N exists as N–Ti–O in the anatase TiO_2 lattice. N-TiO_2 demonstrated higher photocatalytic activity than the Degussa P25 TiO_2 photocatalyst in the visible region for the degradation of methylene blue. Chen investigated nitrogen-doped TiO_2 nanoparticles and Degussa P25 powder using XPS [54]. They concluded that O–Ti–N structure is the chemical structure formed during the substitutional doping of N into TiO_2, which is supposed to be responsible for the enhanced photocatalytic activity under visible light. Nakamura et al. prepared N-doped TiO_2 film electrodes and investigated the mechanism of the anodic photocurrents induced by visible-light irradiation [57]. They concluded that the visible-light responses for N-doped TiO_2 arise from an N-induced mid-gap level slightly above the top of the O 2p valence band. Therefore, the photocatalytic oxidation of organic compounds on N-doped TiO_2 under visible illumination mainly proceeds via reactions with surface intermediates of water oxidation or oxygen reduction, not by direct reactions with holes trapped at the N-induced mid-gap level. Lindgren et al. prepared nanocrystalline porous N-doped TiO_2 thin films with DC magnetron sputtering approach and investigated their photoelectrochemical properties as well as dye-sensitized performance [60]. The higher N content doped into TiO_2 was found to stabilize the anatase phase of TiO_2 and decrease surface roughness of the TiO_2 thin films. The states introduced by nitrogen doping were found to lie close to the valence band edge. The N-doped TiO_2 films showed visible-light absorption in the wavelength range from 400 to 535 nm and generated an incident photon-to-current efficiency response in good agreement with the optical spectra. Moreover, the electron-transfer properties in the conduction band of N-doped TiO_2 were found to be similar to those of undoped TiO_2. Under optimum preparation conditions, the N-doped TiO_2 electrodes demonstrated 200 times higher visible-light-induced photocurrent than that of undoped TiO_2 electrodes.

Up to now, there is a great deal of literature concerning on the topic of N doping in TiO_2 materials, most of which agrees that the doping of N to the lattice of TiO_2 leads to enhanced photocatalytic activity at lower photon energies. However, several distinct models have been proposed to elucidate the origin of the visible-light response as well as the visible-light-induced photocatalytic activity induced by nitrogen doping. A consistent conclusion on the origin of the visible-light response by nitrogen doping, the chemical nature, and the location of the species that allow to extend the absorption to the visible-light region has not yet been arrived at. For example, Asahi et al. assigned the peaks of N 1*s* core levels with binding energy of 396 and 400 eV to substitutional N (β–N, Ti–N bond) and chemisorbed N_2 (γ–N_2, N–N bond), respectively [30]. While Valentin et al. assigned the peak with binding energy of 400 eV to interstitial N rather than to γ–N_2 and Sato et al. correlated this peak to N in NO [37, 61, 62]. Debates on these issues will still continue.

3.2 Carbon (C) Doping in TiO_2

Following the work reported by Asahi et al., carbon doping in TiO_2 has received considerable attention [63–67]. Khan et al. investigated the substitutional doping of carbon in TiO_2 and employed the as-prepared electrodes for the photoelectrochemical water splitting for the first time [63]. The chemically modified n-type TiO_2 photoanode was prepared by controlled flame pyrolysis of Ti metal. Carbon was incorporated during the pyrolysis process that was carried out in the presence of natural gas. The resulting C-doped rutile TiO_2 has a much smaller band gap of 2.32 eV and can absorb light at wavelengths below 535 nm. Under light irradiation, the C-doped TiO_2 electrode can split water with a total conversion efficiency of 11% and a maximum photoconversion efficiency of 8.35% at an applied potential of 0.3 V, which is much higher that that obtained on n-type TiO_2 (1%) at an applied potential of 0.6 V (Fig. 7).

Irie et al. prepared C-doped TiO_2 anatase powders by oxidizing TiC powders with a two-step calcination approach in air [64]. The as-prepared C-doped TiO_2 powders demonstrated enhanced absorption in the visible region and could decompose IPA to acetone and CO_2 under visible-light (400–530 nm) irradiation. The doped carbon was found to be located at oxygen sites and supposed to be responsible for the visible-light responsibility. Sakthivel and Kisch prepared C-doped TiO_2 by heating precursor derived from titanium tetrachloride with tetrabutylammonium hydroxide at different temperatures in air [65]. The as-prepared C-doped TiO_2 demonstrated enhanced absorption in the visible region and photocatalytic activities for the mineralization of aqueous 4-chlorophenol and the azo dye remazol red and the oxidation of gaseous acetaldehyde, benzene, and carbon monoxide under direct artificial and diffuse natural light. Park et al. prepared C-doped TiO_2 nanotube arrays with a two-step procedure [66]. In the first step, TiO_2 nanotube arrays with high aspect ratios were prepared from a Ti substrate by anodization process. In the second step, the TiO_2 nanotube arrays were heated under flowing CO gas at different temperatures. The band-gap of the as-synthesized TiO_2 nanotube arrays was reduced from 3.2 to

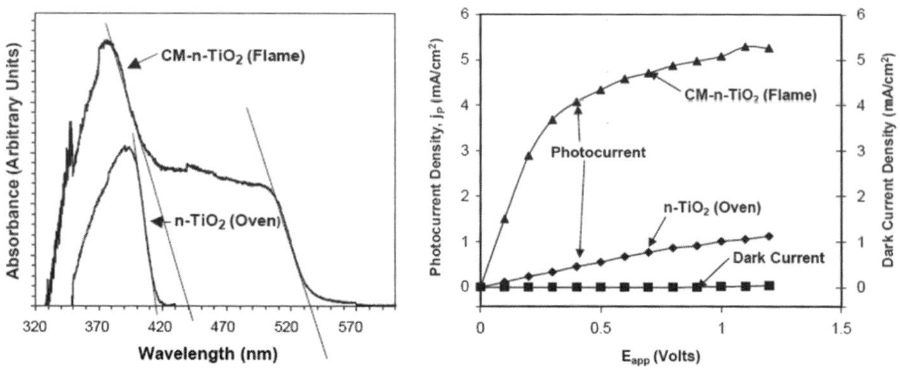

Fig. 7 (*Left*) The UV–visible spectra of CM-*n*-TiO_2 (flame-made) and reference *n*-TiO_2 (electric tube furnace- or oven-made). (*Right*) Photocurrent density j_p as a function of applied potential E_{app} at CM-*n*-TiO_2 (flame-made) and the reference *n*-TiO_2 (electric tube furnace- or oven-made) photoelectrodes under xenon lamp illumination at an intensity of 40 mW cm^{-2}. Dark current densities at CM-*n*-TiO_2 (flame-made) as a function of applied potential are also shown (Reprinted with permission from [63]. Copyright 2002 AAAS)

2.22 eV after doping with carbon. The $TiO_{2-x}C_x$ nanotube arrays demonstrated drastically enhanced photocurrent densities and more efficient water splitting under visible-light illumination (>420 nm) than pure TiO_2 nanotube arrays. Moreover, the total photocurrent was more than 20 times higher than that with a P25 nanoparticulate film under white-light illumination, implying the beneficial role of controlled morphology toward more efficient solar energy harvesting and utilization.

Valentin et al. investigated the role of substitutional and interstitial type carbon doping on the modification of the band structures of anatase as well as rutile polymorphs of TiO_2 using density functional theory (DFT) calculations [67]. At low carbon concentrations, substitutional C doping for oxygen atoms and oxygen vacancies formation is favored under oxygen-poor conditions, while interstitial and substitutional C doping is preferred under oxygen-rich conditions. Higher carbon concentrations undergo an unexpected stabilization caused by multidoping effects, interpreted as interspecies redox processes. For both anatase and rutile TiO_2, the two types of carbon doping induce the formation of several localized occupied states in the gap of TiO_2 depending upon the dopant type, the presence or absence of oxygen vacancies, and the partial pressure of oxygen. The presence of these states is supposed to lead to the decreased energy and observed red shift of the absorption edge toward the visible region.

3.3 *Sulfur (S) Doping in TiO_2*

Sulfur can be doped as an anion in TiO_2 by substituting O sites. However, due to the large ionic radius of S (1.8 Å for S^{2-}) compared to that of O (1.4 Å for O^{2-}), the

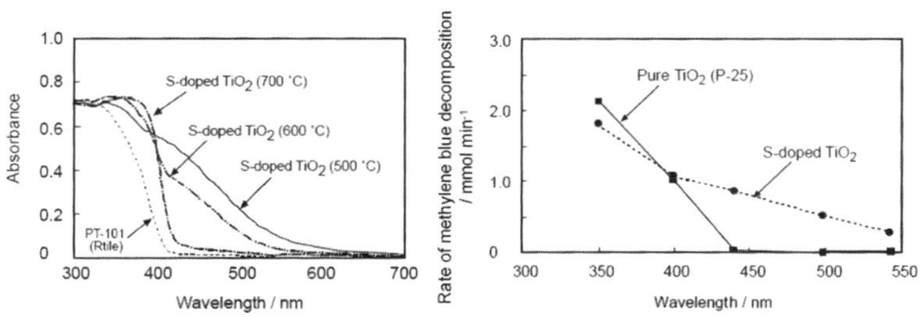

Fig. 8 (*Left*) Diffuse reflectance spectra of S-doped and pure TiO_2 powders (rutile) and (*right*) photocatalytic decomposition of methylene blue using S-doped TiO_2 or pure TiO_2 (P-25) as a function of the cutoff wavelength for irradiation from a 1,000 W Xe lamp (Reprinted with permission from [70]. Copyright 2004 Elsevier)

substitutional sulfur doping for oxygen sites will distort the crystal lattice of TiO_2. Sulfur dopant can also exist as cation (S^{4+}, S^{6+}) by replacing Ti^{4+} ions in bulk or at the surface. Umebayashi et al. prepared S-doped TiO_2 anatase polycrystalline powder by oxidation annealing of titanium disulfide (TiS_2) in air [68, 69]. During the calcination, most of the S atoms in TiS_2 were oxidized and the remaining S occupied O-atom sites in TiO_2. The substitution of S for O was found to cause a significant shift in the absorption edge to lower energy. DFT calculations showed that the band-gap narrowing due to the S doping originates from mixing the S 3p states with VB, leading to an increase in the VB width. In their following experimental work, they found that the as-prepared S-doped TiO_2 demonstrated activity for the degradation of methylene blue under the visible-light irradiation [69].

Ohno et al. prepared S-doped TiO_2 by calcining precursors derived from isopropoxide and thiourea at different temperatures under aerated conditions [70]. In contrast to the S-doped TiO_2 reported by Umebayashi et al., S atoms are supposed to be incorporated as cations and are expected to be replaced with Ti ions. The as-prepared S-doped TiO_2 showed strong absorption in the visible region and high activities for degradation of methylene blue in aqueous solution under irradiation at wavelengths longer than 440 nm (Fig. 8). In their following work, the chemical state of the S atoms doped into TiO_2 was determined to be mainly S^{4+}. Theoretical calculations indicated that the level above the VB of TiO_2 consisted of S 3s states and the transition between this level and CB of TiO_2 is supposed to be the origin of the visible light. The as-prepared S-doped TiO_2 demonstrated strong absorption in the visible region and high activities for the degradation of methylene blue and 2-propanol in aqueous solution and partial oxidation of adamantane under irradiation at wavelengths longer than 440 nm.

Yu et al. prepared S-doped TiO_2 with a solution-based method followed by calcination at 773 K in air [71]. XPS characterizations indicate that S atoms are in the state of S^{6+} in all S-doped TiO_2 samples. The S-doped TiO_2 exhibited strong absorption in the visible region and can effectively kill *Micrococcus lylae*. The formation of hydroxyl radicals on S-doped TiO_2 is supposed to be the origin of the considerable bactericidal activity under visible-light irradiation.

3.4 Fluorine (F) Doping in TiO_2

The F doping was initially found to improve the activity of TiO_2 under UV light irradiation [72, 73]. Hattori and Tada prepared F-doped TiO_2 film with a sol–gel (SG) method [72]. By adding a small amount of trifluoroacetic acid (TFA) as the F source into the precursor solution, F was successfully doped into TiO_2 film upon subsequent calcinations at 773 K. It was found that the F doping improved the absorption coefficient for ultraviolet light ($\lambda < 360$ nm) due to film densification and the improvement of crystallinity of TiO_2 film, as a result leading to significantly enhanced photoactivity, which can be confirmed by the two separate experiments of the methylsiloxane monolayer oxidation and the photocurrent measurements. However, only enhanced UV light activity was reported in this study. It was found that F doping could lead to the enhancement of surface acidity, formation of surface hydroxyl radicals, and creation of oxygen vacancies or Ti^{3+}, therefore changing the interfacial e^-/h^+ transfer, surface charge distribution, and substrate–surface interaction [74].

Yu et al. prepared F-doped TiO_2 with anatase and brookite phase by hydrolysis of titanium tetraisopropoxide in a mixed NH_4F-H_2O solution followed by calcinations at different temperatures in air [75]. During the calcinations, the thermal energy can trigger the substitution of F^- for O^{2-} in the lattice of TiO_2. The F-doped TiO_2 samples showed stronger absorption in the UV–visible range and a red shift in the band-gap transition. Moreover, the doping of F improved the crystallinity of TiO_2, suppressed the formation of brookite phase, and retarded the phase transition from anatase to rutile. The prepared F-doped TiO_2 powders demonstrated quite high activity for the photocatalytic oxidation of acetone and the activity of F-doped TiO_2 prepared under optimum preparation conditions even exceeded that of benchmark Degussa P25 by 39% (Fig. 9).

Yamaki et al. prepared F-doped rutile TiO_2 single crystals with an ion implantation technique followed by thermal annealing [76]. The thermal annealing treatment can recover the radiation damage and leads to the diffusion of F atoms to the outer surface. They suggested that visible-light absorption might be achieved on the as-obtained F-doped rutile TiO_2 due to the modification of density of states near the CB edge of rutile TiO_2 by fluorine doping.

Li et al. prepared F-doped TiO_2 by spray pyrolysis at different temperatures from an aqueous solution of H_2TiF_6 [77]. It was found that the doping of F did not change the absorption properties of as-prepared TiO_2 as the F 2p states were calculated to locate at a position positive than the VB maximum of TiO_2. However, F-doped TiO_2 demonstrated drastically enhanced photocatalytic activity for the degradation of gas-phase acetaldehyde and trichloroethylene under visible light. The F-doped TiO_2 prepared at optimum conditions showed much higher activity than P25. The high photocatalytic activity of F-doped TiO_2 was supposed to originate from the absorption induced by the oxygen vacancies instead of the absorption of bulk TiO_2. The absorption requires less energy to activate and therefore is supposed to induce surface charge carriers by visible light.

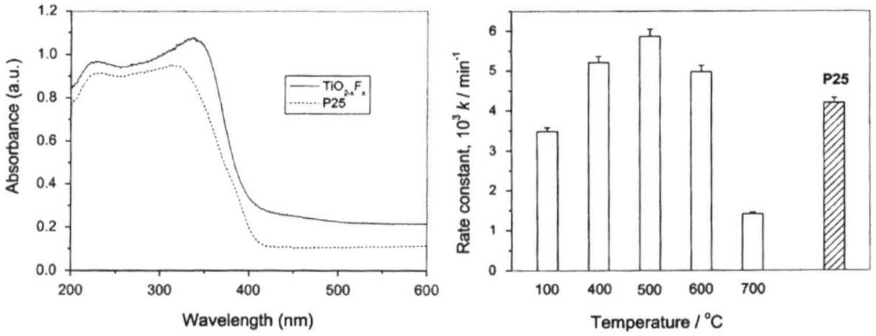

Fig. 9 (*Left*) UV–visible absorption spectra of Degussa P25 and the F-doped TiO_2 powders and (*right*) the dependence of the apparent rate constants on calcination temperature (Reprinted with permission from [75]. Copyright 2002 American Chemical Society)

3.5 Boron (B) Doping in TiO_2

Zhao et al. prepared B-doped TiO_2 loaded with Ni_2O_3 catalyst with a modified sol–gel method [78]. By calcining precursors containing appropriate amounts of Ti, B, and Ni salts in N_2 atmosphere, B is successfully doped into the lattice of TiO_2, while Ni_2O_3 remains as a separated phase from B-doped TiO_2. The doping of B into TiO_2 greatly extends the spectral response of TiO_2 to the visible region, which is explained by theoretical calculations. However, the photocatalytic activity of B-doped TiO_2 remains quite low under visible-light irradiation. After loading Ni_2O_3, the as-prepared $Ni_2O_3/TiO_{2-x}B_x$ demonstrated quite high activity for the degradation and mineralization of toxic organic pollutants such as trichlorophenol (TCP), 2,4-dichlorophenol (2,4-DCP), and sodium benzoate. The loaded Ni_2O_3 species are supposed to act as electron traps and thus suppress efficiently the recombination of photoproduced electron–hole. Therefore, by combining doping of B and loading Ni_2O_3 catalyst strategies, TiO_2 with greatly improved photocatalytic activity under visible light was achieved (Fig. 10).

In et al. prepared B-doped TiO_2 by using $TiCl_4$ and BH_3 as the Ti and B source, respectively [79]. The resulting B-doped TiO_2 exhibited red-shifted absorption spectra and high photocatalytic activity for the photocatalytic decomposition of methyl tertiary butyl ether (MTBE) under visible light. Based upon the XPS analysis, Boron was supposed to be incorporated within the TiO_2 lattice, thereby inducing the high visible-light photocatalytic activity of B-doped TiO_2. Chen et al. prepared B-doped TiO_2 with a sol–gel method by hydrolysis of titanium tetra-*n*-butyl oxide in H_3BO_3 aqueous solution [80]. The doped B is supposed to exist as B^{3+} in the doped TiO_2 and was likely to weave into the interstitial TiO_2 structure. The doping of B was found to efficiently inhibit the grain growth and facilitate the anatase-to-rutile transformation before the formation of B_2O_3. Moreover, all the B-doped TiO_2 samples showed increased photocatalytic activity over that of pure TiO_2 sample in the photocatalytic reaction of NADH regeneration under UV light irradiation. However, only UV light activity was observed on the

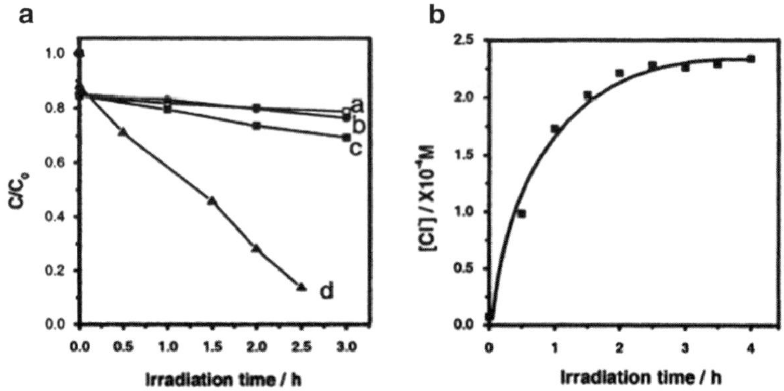

Fig. 10 (**A**) Temporal course of the photodegradation of TCP in aqueous dispersions containing 50 mg of catalysts under visible light irradiation: (*a*) pure TiO_2, (*b*) nickel-doped TiO_2, (*c*) boron doped TiO_2, and (*d*) boron- and nickel-doped TiO_2. (**B**) Formation of Cl during the degradation process in the boron- and nickel-doped TiO_2 system (Reprinted with permission from [78]. Copyright 2003 American Chemical Society)

B-doped TiO_2 prepared with this approach. Therefore, the preparation method has drastic influence on the chemical state of B and the corresponding photocatalytic capability.

3.6 *Iodine (I) Doping in TiO_2*

Similar to S, iodine (I) dopants could exist in TiO_2 matrix with multiple chemical states such as −1, 0, +5, and +7. The variation of the chemical states will lead to the different occupied sites by I dopants in TiO_2 and therefore will induce different band structures and optical properties of TiO_2. Hong et al. prepared I-doped TiO_2 with a solution-based method [81]. The precursor was first prepared by adding tetrabutyl titanate to a solution containing iodic acid under stirring conditions. After heat treatment of the precursor from 673 to 873 K in air, I-doped yellow crystals were obtained. I was found to exist as I^{5+} in the doped TiO_2 and I^{5+} was supposed to substituteTi^{4+} due to the equivalent ionic radius of 0.62 and 0.64 nm for I^{5+} and Ti^{4+}, respectively. The as-prepared I-doped TiO_2 nanoparticles show strong absorption in the visible region and much higher photocatalytic activity than Degussa P25 for the degradation of phenol under visible-light irradiation ($\lambda > 400$ nm). Moreover, the I-doped TiO_2 nanoparticles show similar activity with P25 under UV and visible-light irradiation (Fig. 11).

Liu et al. prepared I-doped mesoporous TiO_2 with a bicrystalline (anatase and rutile) framework by a two-step template hydrothermal synthesis route [82]. The as-prepared I-doped TiO_2 showed strong absorption in the visible region and much higher activity than P25 and undoped mesoporous TiO_2 for the photodegradation of

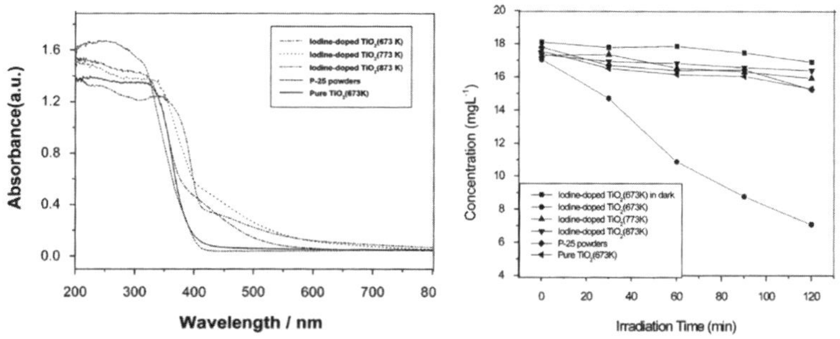

Fig. 11 (*Left*) UV–vis absorption spectra of P25 powders (Degussa), pure TiO_2 (673 K), and iodine-doped TiO_2 powders before reaction and (*right*) comparison of the photocatalytic degradation of phenol in the presence of I-doped TiO_2, P25 powders, and pure TiO_2 nanoparticles (673 K) under visible light irradiation and absorption of phenol on I-doped TiO_2 (673 K) in dark (Reprinted with permission from [81]. Copyright 2005 American Chemical Society)

methylene blue under visible light ($\lambda > 420$ nm) as well as UV–visible light. The high activity is supposed to be attributable to the bicrystalline framework, high crystallinity, large surface area, mesoporous structure, and high absorbance in the visible-light range by iodine doping [82]. Long et al. compared the photocatalytic activities of TiO_2, I-doped TiO_2, and N-doped TiO_2 for the degradation of phenol under visible-light irradiation ($k > 400$ nm) [83]. It was found that I-doped TiO_2 demonstrated higher activity than both N-doped TiO_2 and TiO_2 under visible light. DFT calculations indicated that I 5*p* orbitals contributed to the formation of the valence and conduction band of I-doped TiO_2 by mixing with O 2p and Ti 3d orbitals, respectively. Moreover, the band potentials of I-doped shifted downwards, leading to enhanced photoxidation ability. Compared with N-doped TiO_2, the TiO_6 octahedra distorted heavier in I-doped TiO_2, and the distortion will bring internal dipole moment and promote the charge separation. Tojo et al. prepared I-doped TiO_2 with a hydrothermal method followed by a heat treatment at different temperatures in air [84]. Compared with undoped TiO_2, the as-prepared I-doped TiO_2 shows strong visible-light absorption as well as high performance for the photodegradation of 4-chlorophenol (4-CP) in water under visible light. The photocatalytic processes of I-TiO_2 were investigated based on the steady-state and time-resolved spectral measurements. It is supposed that the recombination of electron and hole pairs is sufficiently inhibited because the doping I sites act as trapping site to capture the electrons during the photocatalytic reactions on I-doped TiO_2; therefore, the long-lived photogenerated holes were formed upon the laser excitation of I-TiO_2 powders, while no trapped electrons were observed. The I-induced continuous states mixed with the valence band of TiO_2 were supposed to facilitate the trapping processes of holes.

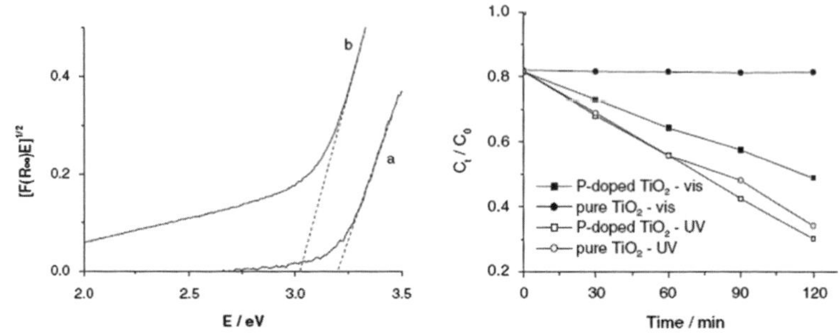

Fig. 12 (*Left*) Diffuse reflectance absorption spectra of (*a*) pure TiO_2 and (*b*) P-doped TiO_2 and (*right*) temporal course of the photodegradation of 4CP in aqueous dispersions containing 100 mg of catalysts under UV and visible light irradiation (Reprinted with permission from [86]. Copyright 2005 The Chemical Society of Japan)

3.7 Phosphor (P) Doping in TiO_2

Yu et al. prepared phosphated mesoporous TiO_2 with high surface area by incorporating phosphorus from phosphoric acid into the framework of TiO_2 with a surfactant-templated approach [85]. The incorporation of phosphorus was found to stabilize the TiO_2 framework, inhibit grain growth, and increase the surface area significantly. They suppose that the phosphated mesoporous TiO_2 is composed of amorphous titanium phosphate with embedded crystalline anatase, which is different from the framework of pure mesoporous TiO_2. The calcined phosphated mesoporous TiO_2 showed higher photocatalytic activity on the oxidation of *n*-pentane than both the calcined pure mesoporous TiO_2 and the commercial nonporous photocatalyst P25. The higher photocatalytic activity of phosphated mesoporous TiO_2 is supposed to be due to the extended band-gap energy, large surface area, and the existence of Ti ions in a tetrahedral coordination. However, this P-doped TiO_2 didn't show visible-light-induced activity.

Lin et al. prepared phosphor-doped anatase TiO_2 with a simple modified sol-gel method using hypophosphorous acid as precursor [86, 87]. By calcining the precursor containing both titanium and phosphor species in N_2 at 673 K, yellow-colored phosphor-doped TiO_2 was obtained. The doping of phosphor was found to significantly increase the surface area of the materials and lead to a higher content of surface hydroxyl groups. Moreover, the phosphor doping improved the thermal stability of titania and decreased the phase transformation of anatase to rutile to a certain extent. The as-prepared P-doped TiO_2 shows a narrower band gap than pure TiO_2 and an absorption tail in the visible range. Under visible-light irradiation (>420 nm), P-doped TiO_2 demonstrated high efficiency for the photocatalytic degradation of 4-chlorophenol (Fig. 12).

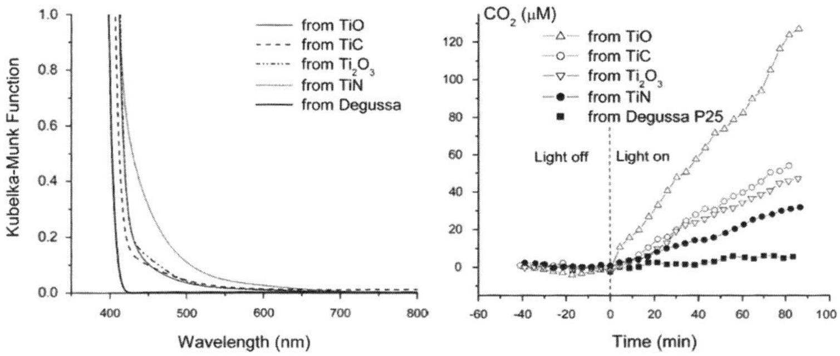

Fig. 13 (*Left*) Visible part of optical spectra of the titanium dioxides prepared from different precursors and (*right*) influence of the precursors on photoactivity of TiO_2 in the reaction of acetaldehyde oxidation in air under visible light (Reprinted with permission from [90]. Copyright 2004 The Royal Society of Chemistry)

3.8 *Self Doping in TiO_2*

During the doping process, some defects such as oxygen vacancies, titanium vacancies, and interstitial titanium may be simultaneously developed as a result of maintaining charge balance or stabilizing the crystal structure of TiO_2, which may change the electronic structure of doped TiO_2 [52, 61, 88]. It was reported that the doping of nitrogen can reduce the formation energy of oxygen vacancies in TiO_2 from 4.3 to 0.6 eV [61], therefore facilitating the generation of oxygen vacancies-related state. Oxygen vacancies and Ti interstitials can form donor levels at 0.75–1.18 and 1.23–1.56 eV below the CB, respectively, while Ti vacancies form acceptor levels above the VB. The presence of these defects can induce an additional shoulder absorption band in the visible-light range and/or a tail absorption band in the near-infrared and infrared ranges [89, 90].

Justicia et al. prepare TiO_2 film by metal-organic chemical vapor deposition method in the presence of oxygen-deficient atmosphere. The TiO_2 film is characteristic of strong oxygen sub-stoichiometry and showed enhanced photocatalytic activity in the visible region. The presence of oxygen vacancies is supposed to be the reason for the gap narrowing and the corresponding visible-light-induced photocatalytic activity [89]. Martyanov et al. prepared TiO_2 with structural defects by oxidation of TiO or Ti_2O_3 in air (Fig. 13). The as-prepared defective TiO_2 showed much higher visible-light-induced activity for the oxidation of acetaldehyde to form carbon dioxide than the TiO_2 samples obtained from TiN precursor, therefore pointing out the importance of oxygen defects/vacancies for extension of activity of TiO_2 into the visible region [90]. Kuznetsov et al. systematically analyze the absorption spectral features of various doped or undoped TiO_2 samples in the visible spectral domain and examine the origins of such bands. They concluded that the near-infrared and infrared absorption bands originate from Ti^{n+}-related (n = 3, 2) color centers, while the absorption bands in the visible region are associated with oxygen vacancies. Moreover, in most doped TiO_2, the absorption features of Ti-related centers are totally suppressed while oxygen vacancy-related absorption features are preferential [91]. After all, visible-light absorption observed in nonmetal doped TiO_2 could be partially due to the oxygen vacancy-related absorption. Therefore, it is very hard to differentiate the contribution of each factor to the visible-light-induced photocatalytic activity, and further studies are needed to clarify these points.

3.9 Co-doping of Nonmetal in TiO_2

In principle, co-doping with two and more suitable heteroatoms will induce visible-light response to TiO_2 with a mechanism similar to that of monodoping. However, co-doping may achieve substantial synergistic effects and lead to much enhanced doping concentration and absorption compared with that achieved with monodopant. As N doping TiO_2 is the most investigated candidate, we will discuss some reprehensive samples using TiO_2 co-doped with N and other nonmetal elements.

Liu et al. prepared B-co-doped TiO_2 with a solution-based approach and then co-dope N by heating B-TiO_2 in NH_3 atmosphere [92]. The co-doping of B and N was found to drastically improve visible-light absorption and the photocatalytic activity under visible light. Two factors induced by co-doping were supposed to be the main reason. First, the amount of doped N on the TiO_2 surface was increased due to the strong interaction between B and N. Secondly, specific O–Ti–B–N structure was formed on the photocatalyst surface, which could act as active sites (cocatalyst) for the surface separation and transfer of visible-light-induced carriers.

Wang et al. employed facet engineering technique to prepare N and F co-doped TiO_2 with dominant {001} facet. By nitriding $TiOF_2$ precursor in flowing NH_3, N and F are simultaneously doped into the anatase TiO_2 nanoparticles. The as-obtained N and F co-doped TiO_2 with dominant {001} facet demonstrated drastically enhanced absorption and excellent water oxidation performance in the visible region (Fig. 14) [93]. Domen et al. investigated the co-doping effects of F and N on the photocatalytic performance of TiO_2 [94, 95]. It was found that the band gap of TiO_2 was decreased from 3.3 to 2.2 eV after co-doping with N and F. Compared with monodoped TiO_2, N and F co-doped TiO_2 demonstrated a sharp absorption edge in the visible region, indicating that a band-to-band excitation will occur upon light irradiation. DFT calculations indicate that the doping level formed by N 2p orbitals in the forbidden band above the valence band consisting of O 2p orbitals contributes to the visible-light response of N/F co-doped TiO_2. The presence of F will help to maintain the charge balance and stabilize the structure of TiO_2. Under visible-light irradiation, N/F co-doped TiO_2 is quite active for photocatalytic O_2 production in the presence of $AgNO_3$ as the sacrificial reagent. N/F co-doped TiO_2 also demonstrated trace activity for H_2 evolution under visible light in aqueous methanol solution when loading Pt cocatalyst.

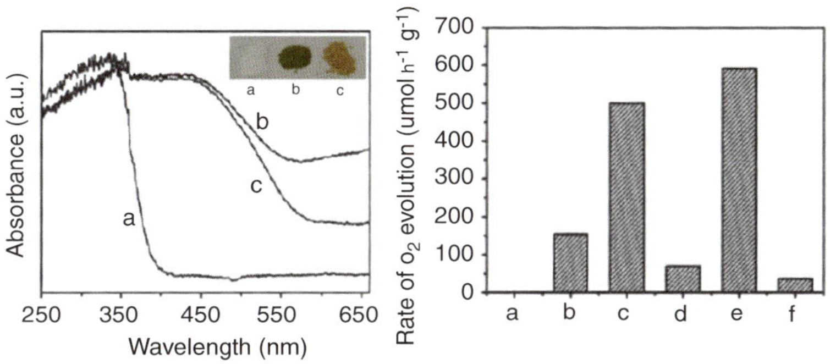

Fig. 14 (*Left*) Diffuse reflectance spectra of (*a*) $TiOF_2$ precursor, (*b*) TiOFN sample prepared by nitriding $TiOF_2$ in NH_3 gas flow at 773 K, and (*c*) TiOFN sample obtained by calcining sample (*b*) in air at 673 K. (*Right*) Rates of photocatalytic O_2 evolution on (*a*) $TiOF_2$ precursor, (*b–d*) TiOFN prepared by calcining $TiOF_2$ precursor in ammonia gas flow at 673, 773, and 873 K, and (*e* and *f*) TiOFN obtained by calcining sample (*c*) in air at 673 and 773 K (Reprinted with permission from [93]. Copyright 2011 The Royal Society of Chemistry)

Li et al. prepared F and N co-doped TiO_2 with a spray pyrolysis approach [96]. The as-prepared powders could absorb strong ultraviolet light and part of the visible light up to 550 nm. Four electronic energy states were found to exist between the valence band and conduction band of N/F co-doped TiO_2. The N/F co-doped TiO_2 powder demonstrated a higher photocatalytic activity than undoped TiO_2 as well as commercial P25 under both UV and visible irradiation. This high activity was ascribed to a synergetic effect of its unique surface characteristics, doped N atoms, and doped F atoms [96].

3.10 Homogeneous Doping

In most of the reported anion-doped TiO_2, only taillike weak absorption in the visible region can be realized due to the low amount of doping concentration and inhomogeneous distribution of the dopant in TiO_2. This will lead to inefficient light absorption and unsatisfactory photoactivity in the visible region. Although high-energy ion implantation techniques can achieve efficient homogeneous doping for transition-metal dopant atoms, it is inefficient for nonmetal ions with low atomic numbers. Therefore, only a thin surface layer of TiO_2 (9 nm with even 3 eV ion treatment) can be doped with physical method, which is typical for thin film fabrication. However, it is inappropriate for TiO_2 in the powder form, even TiO_2 nanopowder.

In order to realize homogeneous doping, there are two possible routes. One is to employ TiO_2 with extremely small particle size to decrease the diffusion length of the dopants from the surface to bulk. However, a suitable low-temperature technique is required to stabilize the TiO_2 nanoparticles with simultaneous doping.

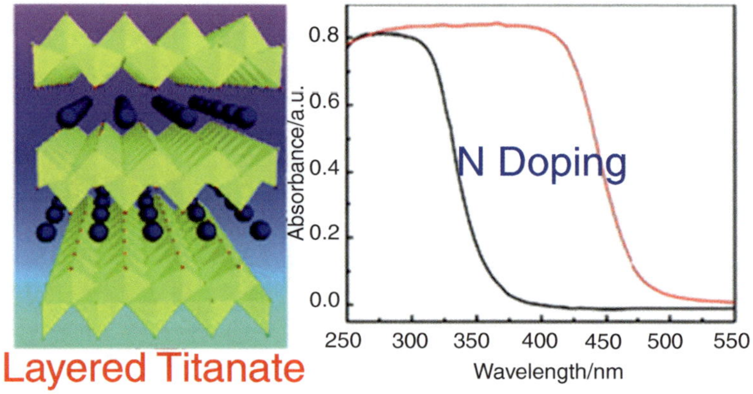

Fig. 15 (*Left*) The schematic illustration of the structure of $Cs_{0.68}Ti_{1.83}O_4$ and (*right*) the UV–vis spectra of $Cs_{0.68}Ti_{1.83}O_4$ (*black line*) and $Cs_{0.68}Ti_{1.83}O_{4-x}N_x$ (*red line*) (Reprinted with permission from [98]. Copyright 2009 American Chemical Society)

The other is to use layered-structured TiO_2 precursor consisting of interlay gallery, therefore facilitating the distribution and homogeneous doping throughout the TiO_2 particles [97–99]. Lu et al. developed this novel strategy to realize homogeneous doping in TiO_2-based material and strong band-to-band absorption in the visible region (Fig. 15). By nitriding layered-structured $Cs_{0.68}Ti_{1.83}O_4$ in flowing NH_3, bright yellow $Cs_{0.68}Ti_{1.83}O_{4-x}N_x$ was obtained. The subsequent protonation of $Cs_{0.68}Ti_{1.83}O_{4-x}N_x$ can lead to the formation of yellow color $H_{0.68}Ti_{1.83}O_{4-x}N_x$. The band gaps of $Cs_{0.68}Ti_{1.83}O_{4-x}N_x$ and $H_{0.68}Ti_{1.83}O_{4-x}N_x$ are around 2.73 and 2.85 eV, which are much smaller than those of undoped metals of around 3.62 and 3.47 eV. The contribution of the mixed N 2p states with O 2p states to the valence band is supposed to be responsible for the decreased band gap. XPS depth analysis and energy-filtered TEM images indicate that N is homogeneously doped among the whole range of the $Cs_{0.68}Ti_{1.83}O_{4-x}N_x$ and $H_{0.68}Ti_{1.83}O_{4-x}N_x$ nanoparticles. Under visible-light irradiation, the as-prepared $Cs_{0.68}Ti_{1.83}O_{4-x}N_x$ and $H_{0.68}Ti_{1.83}O_{4-x}N_x$ nanoparticles demonstrated high performance for the oxidation of OH^- into active $^{\bullet}OH$ radicals. This work gives very important implications that strong band-to-band absorption can be realized in TiO_2-based materials using layered-structured precursor, paving a new way for developing highly efficient anion-doped TiO_2-based materials in the visible region.

From $H_{0.68}Ti_{1.83}O_{4-x}N_x$ nanoparticles, with a simple delamination approach, single-layer N-$Ti_{0.91}O_2$ nanosheet with homogeneous nitrogen doping and extremely thin thickness can be obtained [97]. This TiO_2-based nanosheet possesses the unique structural feature of two-dimensional anisotropy and represents a new form of TiO_2 compared with traditional TiO_2 nanoparticles. Under visible-light irradiation, the photoanode fabricated from N-$Ti_{0.91}O_2$ nanosheet exhibited a remarkable enhancement of photocurrent compared with that of undoped $Ti_{0.91}O_2$ nanosheet.

4 Problems Involved in Doping

Despite the numerous works on the development of second-generation TiO_2 photocatalyst with visible-light response and activity, there are some problems simultaneously introduced by nonmetal doping.

Firstly, it is difficult to obtain TiO_2 with high doping concentration. Therefore, the modification of the band structure of TiO_2 can only attain to a small level. In most cases, the visible-light response was induced by isolated level and not by integrated band, which leads to inferior mobility of photogenerated charges and correspondingly inefficient migration of the carriers to the TiO_2 surface to react with surface-adsorbed reaction media. Secondly, the doping of foreign elements in the lattice of TiO_2 will lead to the formation of defects due to the incompatible atom size or unbalanced charge. These defects may act as recombination centers for photogenerated charges and therefore will decrease the photocatalytic performance. In many cases, although the introduction of dopants can lead to visible-light response, drastic activity loss in the UV range is observed. Thirdly, the visible-light response of nonmetal doped TiO_2 is induced by the new levels or band formed in the intrinsic band gap of TiO_2. Therefore, the redox potentials of the photogenerated electrons or holes will decrease as compensation to the new derived functionality. However, only when the redox potentials of the charge carriers induced by visible-light excitation are powerful enough for subsequent photocatalytic reactions, the visible-light absorption by doping does make sense. For example, as the conduction band of TiO_2 is only slightly higher than the reduction potential of protons to H_2, the small decrease of conduction band level will drastically influence the photocatalytic H_2 production activity of nonmetal doped TiO_2. If fact, most works reported the photocatalytic degradation and photoelectrochemical water splitting reaction while there are still few reports on the visible-light-induced H_2 production on nonmetal doped TiO_2. On the other hand, if the valence band level is increased, the oxidation of water and pollutant will be undoubtedly affected. For example, Mrowetz et al. investigated the oxidative powder of N-doped TiO_2 under visible-light irradiation [100]. N-doped TiO_2 was found to be able to photocatalyze the oxidation of $HCOO^-$ into $CO_2^{\bullet -}$ radicals under UV irradiation, while not under visible-light irradiation. Therefore, the holes generated on N-doped TiO_2 by visible photons are unable to oxidize $HCOO^-$ either by direct means or via intermediate species produced in the oxidation of water or the catalyst. Fourthly, the long-term stability of the nonmetal doped TiO_2 is still a big concern. For example, the photocatalytic performance of N-doped TiO_2 is found to become worse after photoelectrochemical or photocatalytic reactions under visible-light irradiation due to the self-oxidation of N by the photogenerated holes in the valence band. And lastly, as has been addressed in this chapter, the fundamental understanding on the origin of the visible-light response and the electronic structures of nonmetal doped TiO_2 as well as the chemical states of dopants needs further studies to draw a consistent conclusion.

5 Conclusion

This chapter gives an overview on the fundamental principles, fabrication, properties, as well as photocatalytic applications of the second-generation TiO_2 photocatalyst with nonmetal dopants. The nonmetal doping has proved to be an effective strategy in introducing visible-light response as well as photocatalytic functionalities such as pollutant degradation and photoelectrochemical and photocatalytic water splitting under visible-light irradiation. The tremendous effort regarding the synthesis, modifications, as well as fundamental studies has led to great development of this field in the past decade. Even though there exist problems associated with nonmetal doping in TiO_2, investigation on the second-generation TiO_2 photocatalysts still represents one of the hottest research topics in the field of photocatalysis as TiO_2 is supposed to be the most suitable candidate for practical applications and visible-light activation of TiO_2 will make this possibility more tangible. The total or part address of the problems involved in TiO_2 doping is the future effort that should be directed to and the related success will endow nonmetal doped TiO_2 more privileges toward the commercial applications as well as fundamental investigations.

References

1. Fujishima A, Honda K (1972) Nature 238(5358):37–38
2. Linsebigler AL, Lu GQ, Yates JT (1995) Chem Rev 95(3):735–758
3. Bard AJ, Fox MA (1995) Acc Chem Res 28(3):141–145
4. Walter MG, Warren EL, McKone JR, Boettcher SW, Mi QX, Santori EA, Lewis NS (2010) Chem Rev 110(11):6446–6473
5. Kudo A, Miseki Y (2009) Chem Soc Rev 38(1):253–278
6. Hernandez-Alonso MD, Fresno F, Suarez S, Coronado JM (2009) Energ Environ Sci 2(12):1231–1257
7. Inoue Y (2009) Energ Environ Sci 2(4):364–386
8. Kudo A (2003) Catal Surv Asia 7(1):31–38
9. Maeda K, Domen K (2007) J Phys Chem C 111(22):7851–7861
10. Maeda K, Teramura K, Domen K (2007) Catal Surv Asia 11(4):145–157
11. Chen XB, Shen SH, Guo LJ, Mao SS (2010) Chem Rev 110(11):6503–6570
12. Osterloh FE (2008) Chem Mater 20(1):35–54
13. Kato H, Asakura K, Kudo A (2003) J Am Chem Soc 125(10):3082–3089
14. Yi ZG, Ye JH, Kikugawa N, Kako T, Ouyang SX, Stuart-Williams H, Yang H, Cao JY, Luo WJ, Li ZS, Liu Y, Withers RL (2010) Nat Mater 9(7):559–564
15. Kudo A, Omori K, Kato H (1999) J Am Chem Soc 121(49):11459–11467
16. Yan HJ, Yang JH, Ma GJ, Wu GP, Zong X, Lei ZB, Shi JY, Li C (2009) J Catal 266(2):165–168
17. Tsuji I, Kato H, Kudo A (2005) Angew Chem Int Edit 44(23):3565–3568
18. Tsuji I, Kato H, Kobayashi H, Kudo A (2004) J Am Chem Soc 126(41):13406–13413
19. Maeda K, Takata T, Hara M, Saito N, Inoue Y, Kobayashi H, Domen K (2005) J Am Chem Soc 127(23):8286–8287
20. Maeda K, Teramura K, Domen K (2008) J Catal 254(2):198–204
21. Wang X, Maeda K, Thomas A, Takanabe K, Xin G, Carlsson JM, Domen K, Antonietti M (2009) Nat Mater 8(1):76–80

22. Hitoki G, Takata T, Kondo JN, Hara M, Kobayashi H, Domen K (2002) Chem Commun 16:1698–1699
23. Sato J, Saito N, Yamada Y, Maeda K, Takata T, Kondo JN, Hara M, Kobayashi H, Domen K, Inoue Y (2005) J Am Chem Soc 127(12):4150–4151
24. Ishikawa A, Takata T, Kondo JN, Hara M, Kobayashi H, Domen K (2002) J Am Chem Soc 124(45):13547–13553
25. Ishikawa A, Yamada Y, Takata T, Kondo JN, Hara M, Kobayashi H, Domen K (2003) Chem Mater 15(23):4442–4446
26. Yang HG, Sun CH, Qiao SZ, Zou J, Liu G, Smith SC, Cheng HM, Lu GQ (2008) Nature 453(7195):638–641
27. Han XG, Kuang Q, Jin MS, Xie ZX, Zheng LS (2009) J Am Chem Soc 131(9):3152–3153
28. Chen JS, Tan YL, Li CM, Cheah YL, Luan DY, Madhavi S, Boey FYC, Archer LA, Lou XW (2010) J Am Chem Soc 132(17):6124–6130
29. Hashimoto K, Irie H, Fujishima A (2005) Jpn J Appl Phys 44(12):8269–8285
30. Asahi R, Morikawa T, Ohwaki T, Aoki K, Taga Y (2001) Science 293(5528):269–271
31. Fujishima A, Zhang XT, Tryk DA (2008) Surf Sci Rep 63(12):515–582
32. Chen X, Mao SS (2007) Chem Rev 107(7):2891–2959
33. Thompson TL, Yates JT (2006) Chem Rev 106(10):4428–4453
34. Gaya UI, Abdullah AH (2008) J Photochem Photobiol C 9(1):1–12
35. Nah YC, Paramasivam I, Schmuki P (2010) ChemPhysChem 11(13):2698–2713
36. Chatterjee D, Dasgupta S (2005) J Photochem Photobiol C 6(2–3):186–205
37. Di Valentin C, Finazzi E, Pacchioni G, Selloni A, Livraghi S, Paganini MC, Giamello E (2007) Chem Phys 339(1–3):44–56
38. Rehman S, Ullah R, Butt AM, Gohar ND (2009) J Hazard Mater 170(2–3):560–569
39. Zhang JL, Wu YM, Xing MY, Leghari SAK, Sajjad S (2010) Energ Environ Sci 3 (6):715–726
40. Yates HM, Nolan MG, Sheel DW, Pemble ME (2006) J Photochem Photobiol A 179(1–2):213–223
41. Sun HQ, Wang SB, Ang HM, Tade MO, Li Q (2010) Chem Eng J 162(2):437–447
42. Ismail AA, Bahnemann DW (2011) J Mater Chem 21(32):11686–11707
43. Liu G, Wang LZ, Yang HG, Cheng HM, Lu GQ (2010) J Mater Chem 20(5):831–843
44. Navarro RM, Sanchez-Sanchez MC, Alvarez-Galvan MC, del Valle F, Fierro JLG (2009) Energ Environ Sci 2(1):35–54
45. Kitano M, Funatsu K, Matsuoka M, Ueshima M, Anpo M (2006) J Phys Chem B 110(50):25266–25272
46. Ghicov A, Macak JM, Tsuchiya H, Kunze J, Haeublein V, Frey L, Schmuki P (2006) Nano Lett 6(5):1080–1082
47. Maeda M, Watanabe T (2006) J Electrochem Soc 153(3):C186–C189
48. Sato S (1986) Chem Phys Lett 123(1–2):126–128
49. Irie H, Watanabe Y, Hashimoto K (2003) J Phys Chem B 107(23):5483–5486
50. Burda C, Lou YB, Chen XB, Samia ACS, Stout J, Gole JL (2003) Nano Lett 3(8):1049–1051
51. Gole JL, Stout JD, Burda C, Lou YB, Chen XB (2004) J Phys Chem B 108(4):1230–1240
52. Ihara T, Miyoshi M, Iriyama Y, Matsumoto O, Sugihara S (2003) Appl Catal A Environ 42(4):403–409
53. Sakthivel S, Kisch H (2003) ChemPhysChem 4(5):487–490
54. Chen XB, Burda C (2004) J Phys Chem B 108(40):15446–15449
55. Livraghi S, Paganini MC, Giamello E, Selloni A, Di Valentin C, Pacchioni G (2006) J Am Chem Soc 128(49):15666–15671
56. Mitoraj D, Kisch H (2008) Angew Chem Int Edit 47(51):9975–9978
57. Nakamura R, Tanaka T, Nakato Y (2004) J Phys Chem B 108(30):10617–10620
58. Sathish M, Viswanathan B, Viswanath RP, Gopinath CS (2005) Chem Mater 17(25):6349–6353
59. Serpone N (2006) J Phys Chem B 110(48):24287–24293

60. Lindgren T, Mwabora JM, Avendano E, Jonsson J, Hoel A, Granqvist CG, Lindquist SE (2003) J Phys Chem B 107(24):5709–5716
61. Di Valentin C, Pacchioni G, Selloni A, Livraghi S, Giamello E (2005) J Phys Chem B 109(23):11414–11419
62. Sato S, Nakamura R, Abe S (2005) J Photochem Photobiol A 284(1–2):131–137
63. Khan SUM, Al-Shahry M, Ingler WB (2002) Science 297(5590):2243–2245
64. Irie H, Watanabe Y, Hashimoto K (2003) Chem Lett 32(8):772–773
65. Sakthivel S, Kisch H (2003) Angew Chem Int Edit 42(40):4908–4911
66. Park JH, Kim S, Bard AJ (2006) Nano Lett 6(1):24–28
67. Di Valentin C, Pacchioni G, Selloni A (2005) Chem Mater 17(26):6656–6665
68. Umebayashi T, Yamaki T, Itoh H, Asai K (2002) Appl Phys Lett 81(3):454–456
69. Umebayashi T, Yamaki T, Tanaka S, Asai K (2003) Chem Lett 32(4):330–331
70. Ohno T, Akiyoshi M, Umebayashi T, Asai K, Mitsui T, Matsumura M (2004) Appl Catal A Gen 265(1):115–121
71. Yu JC, Ho WK, Yu JG, Yip H, Wong PK, Zhao JC (2005) Environ Sci Technol 39(4):1175–1179
72. Hattori A, Tada H (2001) J Sol-Gel Sci Technol 22(1–2):47–52
73. Hattori A, Shimoda K, Tada H, Ito S (1999) Langmuir 15(16):5422–5425
74. Vohra MS, Kim S, Choi W (2003) J Photochem Photobiol A 160(1–2):55–60
75. Yu JC, Yu JG, Ho WK, Jiang ZT, Zhang LZ (2002) Chem Mater 14(9):3808–3816
76. Yamaki T, Umebayashi T, Sumita T, Yamamoto S, Maekawa M, Kawasuso A, Itoh H (2003) Nucl Instrum Meth Phys Res B 206:254–258
77. Li D, Haneda H, Labhsetwar NK, Hishita S, Ohashi N (2005) Chem Phys Lett 401(4–6):579–584
78. Zhao W, Ma WH, Chen CC, Zhao JC, Shuai ZG (2004) J Am Chem Soc 126(15):4782–4783
79. In S, Orlov A, Berg R, Garcia F, Pedrosa-Jimenez S, Tikhov MS, Wright DS, Lambert RM (2007) J Am Chem Soc 129(45):13790–13791
80. Chen D, Yang D, Wang Q, Jiang Z (2006) Ind Eng Chem Res 45(12):4110–4116
81. Hong X, Wang Z, Cai W, Lu F, Zhang J, Yang Y, Ma N, Liu Y (2005) Chem Mater 17(6):1548–1552
82. Liu G, Chen Z, Dong C, Zhao Y, Li F, Lu GQ, Cheng H-M (2006) J Phys Chem B 110(42):20823–20828
83. Long MC, Cai WM, Wang ZP, Liu GZ (2006) Chem Phys Lett 420(1–3):71–76
84. Tojo S, Tachikawa T, Fujitsuka M, Majima T (2008) J Phys Chem C 112(38):14948–14954
85. Yu JC, Zhang LZ, Zheng Z, Zhao JC (2003) Chem Mater 15(11):2280–2286
86. Lin L, Lin W, Zhu YX, Zhao BY, Xie YC (2005) Chem Lett 34(3):284–285
87. Lin L, Lin W, Xie JL, Zhu YX, Zhao BY, Xie YC (2007) Appl Catal A Environ 75 (1–2):52–58
88. Nowotny MK, Sheppard LR, Bak T, Nowotny J (2008) J Phys Chem C 112(14):5275–5300
89. Justicia I, Ordejon P, Canto G, Mozos JL, Fraxedas J, Battiston GA, Gerbasi R, Figueras A (2002) Adv Mater 14(19):1399–1402
90. Martyanov IN, Uma S, Rodrigues S, Klabunde KJ (2004) Chem Commun 21:2476–2477
91. Kuznetsov VN, Serpone N (2009) J Phys Chem C 113(34):15110–15123
92. Liu G, Zhao YN, Sun CH, Li F, Lu GQ, Cheng HM (2008) Angew Chem Int Edit 47(24):4516–4520
93. Zong X, Xing Z, Yu H, Chen Z, Tang F, Zou J, Lu GQ, Wang L (2011) Chem Commun 47(42):11742–11744
94. Nukumizu K, Nunoshige J, Takata T, Kondo JN, Hara M, Kobayashi H, Domen K (2003) Chem Lett 32(2):196–197
95. Maeda K, Shimodaira Y, Lee B, Teramura K, Lu D, Kobayashi H, Domen K (2007) J Phys Chem C 111(49):18264–18270
96. Li D, Haneda H, Hishita S, Ohashi N (2005) Chem Mater 17(10):2596–2602

97. Liu G, Wang LZ, Sun CH, Chen ZG, Yan XX, Cheng L, Cheng HM, Lu GQ (2009) Chem Commun 11:1383–1385
98. Liu G, Wang LZ, Sun CH, Yan XX, Wang XW, Chen ZG, Smith SC, Cheng HM, Lu GQ (2009) Chem Mater 21(7):1266–1274
99. Liu G, Sun CH, Wang LZ, Smith SC, Lu GQ, Cheng HM (2011) J Mater Chem 21(38):14672–14679
100. Mrowetz M, Balcerski W, Colussi AJ, Hoffmann MR (2004) J Phys Chem B 108(45):17269–17273

Mechanisms of Reactions Induced by Photocatalysis of Titanium Dioxide Nanoparticles

Joseph Rabani and Sara Goldstein

Abstract Photochemical reactions induced by TiO_2 nanoparticles share common mechanistic features where electron and hole pairs are formed, migrate to the surface, and their recombination competes with their reaction with various substrates. The main interest in TiO_2 photocatalysis is related to its potential application for decontamination of water and air. However, the absorption of TiO_2, which is limited to UV light, does not enable the use of natural or cheap light sources, and therefore tremendous effort has been invested in inducing visible-light activity via modification of TiO_2 including doping with nonmetals and metals, surface coating, and bi- and multicomponent assembling. In addition, much research has been carried out to inhibit the electron–hole recombination and enhance the reactions of holes and electrons with substrates. The basic mechanism of bare and modified TiO_2 and the main principles of the photocatalytic processes remain similar, although the excitation energy is different and the energies of the electrons and holes and their reaction kinetic parameters may vary. These photocatalytic processes are reviewed and discussed.

Keywords Bilayers, Composites, Doping, Electron–hole recombination, Graphene, Surface modification, TiO_2

Contents

J. Rabani (✉) and S. Goldstein
Accelerator Laboratory, Institute of Chemistry, The Hebrew University of Jerusalem, Jerusalem 91904, Israel
e-mail: rabani@mail.huji.ac.il

D.W. Bahnemann and P.K.J. Robertson (eds.), *Environmental Photochemistry Part III*, Hdb Env Chem (2015) 35:115–158, DOI 10.1007/698_2013_248,

1 Introduction

Titanium dioxide, also known as titania, is a semiconductor, which occurs in nature as minerals rutile, anatase and brookite. The most common form is rutile, which is also the equilibrium phase at all temperatures. The metastable anatase and brookite phases are converted into rutile upon heating. Absorption of light by TiO_2 excites an electron from the valence band to the conduction band followed by a large number of reactions, which depend on the nature of the TiO_2 and on the surrounding conditions. Most TiO_2 photochemical studies involve anatase, rutile or their biphasic nanocrystals. Although differences in the photoactivities of these forms have been reported [1, 2], the fundamental mechanisms are similar. Furthermore, TiO_2 properties are considerably affected by surface defects, e.g., structural oxygen deficiency, and by the presence of dopants in the nanocrystals. Doped atoms have been supposed to change the band gap of TiO_2 and more commonly to introduce an intraband energy level. Absorption of visible light involves excitation of an electron from or to this new level. Inner level excitation of the doped atoms, which does not lead to charge separation, does not induce chemical reactions with additives and is not within the scope of this review. Besides the physicochemical properties of TiO_2 nanocrystals, its photochemical activity depends on the nature of the added substrate and its concentration, adsorption, and redox properties as well as on the environmental conditions such as pH and absorbed light density. Surface modification by chemisorbed and physisorbed metals adds a thermal catalytic feature, which may improve the photocatalytic performance. Composites of two or more components have been used in order to enhance charge separation. The present review concerns all forms of TiO_2 modifications, although photosensitization by organic dyes is not included. With the above exclusion, the general pattern of photocatalysis will be discussed together with selected examples, which contain quantitative information or comparative measurements under controlled conditions.

There are thousands of papers, which directly or indirectly involve reaction mechanisms. Therefore, we have addressed predominantly works related to basic mechanisms, which involve quantum yields or comparative tests under similar absorbed light densities and solutes that do not absorb the photocatalyst excitation light. Some qualitative works have been included to demonstrate the dispersed activity in branches of photocatalysis.

2 Fundamental Mechanisms

Absorption of UV light by TiO_2 produces conduction band electron (e_{CB}^-) and valence band hole (h_{VB}^+) pairs (Eq. 1).

$$TiO_2 \xrightarrow{hv} e_{CB}^- + h_{VB}^+ \quad (1)$$

Both e_{CB}^- and h_{VB}^+ become quickly localized at the nanocrystal surface as less mobile states, so-called "trapped" electrons and holes (reactions 2 and 3), and electron–hole recombination may take place via reactions 4–7.

$$e_{CB}^- \rightarrow e_T^- \quad (2)$$

$$h_{VB}^+ \rightarrow h_T^+ \quad (3)$$

$$h_{VB}^+ + e_{CB}^- \rightarrow TiO_2 \quad (4)$$

$$h_{VB}^+ + e_T^- \rightarrow TiO_2 \quad (5)$$

$$h_T^+ + e_{CB}^- \rightarrow TiO_2 \quad (6)$$

$$h_T^+ + e_T^- \rightarrow TiO_2 \quad (7)$$

The contribution of reaction 6 is insignificant since electron trapping is much faster than hole trapping [3, 4].

The electrons and holes can reduce and oxidize a large number of substrates at the TiO_2 surface. Hence, TiO_2 catalyzes the conversion of the photon energy into chemical reaction energy. In most cases, the photocatalysis enhances thermodynamically downhill reactions, although chemical energy storage by uphill reactions may also occur. Much research has been done on both downhill and uphill processes, which are closely associated with energy and environmental conservation. Photochemical energy storage has an indirect effect on the environment due to renewing of used energy while downhill reactions have direct applications for water and air purification as well as for self-cleaning surfaces coated with TiO_2. While h_{VB}^+ is a powerful oxidant, the ability of e_{CB}^- to drive redox reactions is limited by its relatively low redox potential, and the most common reaction is with adsorbed oxygen. In the absence of oxygen and any other additives, the absorbed light energy is converted into heat.

3 The Nature of Electrons and Holes

TiO_2 is a typical transition metal oxide with a band gap in the UV region of 3.0 eV for rutile and 3.2 eV for anatase [5]. The spectroscopic features of the electrons and holes in TiO_2 have been intensively studied. The main component of the conduction band is the Ti 3d orbital [6]. Both electrons and holes have very broad optical spectrum expanding from the UV to the IR region [3, 4, 7–18]. Transient absorption in this region has been assigned to trapped holes, trapped electrons, and bulk electrons as shown in Fig. 1.

The conversion of e_{CB}^- and h_{VB}^+ into their respective surface states (reactions 2 and 3) is a multistage process. First, a portion of the electrons is trapped near the surface at shallow sites establishing a quasi equilibrium with bulk electrons. These electrons become relaxed into deeper trapping sites in the bulk and eventually are trapped at the surface as Ti^{3+} [19] with a lifetime of about 500 ps [13, 18]. Evidence for migration of electrons to lower traps during the first 100 ns has been demonstrated in TiO_2 layers immersed in concentrated iodide solution where changes of the electron absorption at 600 nm take place while that of the oxidizing product $I_2^{\bullet-}$ at 390 nm remains nearly unchanged [20]. These results imply that the changes in the visible range do not involve electron reaction with $I_2^{\bullet-}$. Ultrafast kinetic measurements show that h_{VB}^+ is trapped near the surface of the nanoparticles within 170 fs [18] or 30 ps [4], although much slower hole trapping has been reported [3, 8, 21–25], e.g., $k_3 = 5 \times 10^5\ s^{-1}$ in TiO_2 colloid solution [3] and $k_3 > 4 \times 10^6\ s^{-1}$ in P-25 suspension [17].

Hole trapping involves ultrafast relaxation to states inside the nanoparticle followed by a slower migration to the surface [13]. The trapping rate depends on the nature of the nanoparticles such as size and number of trapping sites. The nature of h_T^+ has been a matter of controversy. The EPR spectrum obtained upon UV photolysis of TiO_2 particles has been assigned to h_T^+, to Ti^{IV}–O–Ti^{IV}–$O^\bullet$ [19], or to surface $^\bullet OH$ [26, 27], which desorbs and reacts with solute molecules in the bulk [28]. Evidence supporting one or another form of $^\bullet OH$ as the active species includes the detection of hydroxylated reaction intermediates and products [29–37] and kinetic isotope effects [38]. UV photolysis of TiO_2 suspension containing aromatic compounds produced the isomeric distribution found for $^\bullet OH$-radical attack [34, 39], suggesting that $^\bullet OH$ is formed in this system. On the other hand, the distribution of the hydroxylated products in the presence of 4-hydroxybenzyl alcohol was different from that obtained by $^\bullet OH$ formed via the Fenton reaction [37]. This was taken as evidence for direct hole oxidation as opposed to adsorbed or free $^\bullet OH$, although it is not clear whether the Fenton reaction produces $^\bullet OH$ or a higher valence state of iron [40, 41]. TiO_2-mediated photo-hydroxylation of relatively low concentrations of phenol demonstrated that the isomer distribution is similar to that obtained by radiolytically borne $^\bullet OH$ [39]. At high concentrations of phenol, the distribution was similar to that observed upon oxidation of phenol by $SO_4^{\bullet-}$, which reacts via an electron transfer as expected for a mobile hole. These results show distinct difference between mobile and trapped-hole reactions.

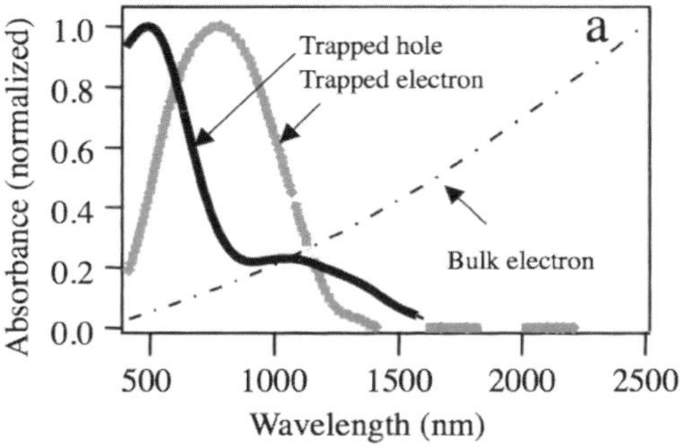

Fig. 1 Absorption spectra of trapped hole, trapped electron, and bulk electron in TiO_2 nanocrystalline film obtained upon excitation at 266 nm. Taken from [14]. Copyright 2004 American Chemical Society

The latter is assigned to $(Ti^{IV}–O–Ti^{IV}OH)^{\bullet}$, which apparently reacts with phenol via $^{\bullet}OH$ transfer to the benzene ring [39]. Pulse radiolysis has been applied to both acidic and alkaline colloid solutions to discriminate between trapped-hole and OH-adduct [23, 42]. It has been demonstrated that TiO_2 reaction with $^{\bullet}OH$ yields a completely different absorption spectrum compared to that produced via TiO_2 oxidation by $SO_4^{\bullet-}$, which have been ascribed to the formation of adsorbed $^{\bullet}OH$ and $–Ti^{IV}–O^{\bullet}$, respectively [42]. The absorption spectrum of the latter is continuously rising below 480 nm and steeply toward the UV region while that of adsorbed $^{\bullet}OH$ has an absorption maximum at 620 nm [23, 42]. Furthermore, the rate of $^{\bullet}OH$ reaction with 13 nm TiO_2 nanoparticles in acidic pH is near diffusion controlled [23]. The resultant surface-bound $^{\bullet}OH$ exhibits a broad absorption band centered around 350 nm and was presumed to be identical to h_T^+ [23]. The fast one-way reaction between free $^{\bullet}OH$ and TiO_2 rules out migration of adsorbed $^{\bullet}OH$ to the bulk of the solution. The identification of h_T^+ as lattice $O^{\bullet-}$ is further supported by the transient spectrum of photochemically produced h_T^+ with a maximum absorption at 430 nm, which differs from that of adsorbed $^{\bullet}OH$ [8, 22]. In a latter work the formation of deeply trapped and shallowly trapped holes in equilibrium with valence band holes has been suggested where the latter are responsible for photooxidation of dichloroacetate and SCN^- and the former are inert toward these substrates [24].

A pronounced advancement in understanding the nature of h_T^+ was achieved by EPR studies. The EPR signal obtained upon illumination of TiO_2 ice (6–200°K) was unaffected by replacing H_2O with D_2O for the colloid preparation. Thus, h_T^+ must be part of the TiO_2 structure and cannot be assigned to bound or free $^{\bullet}OH$. Results obtained with TiO_2 colloids doped with ^{17}O support the identification of h_T^+ as $–Ti^{IV}–O^{\bullet}$ [43]. Salvador [44] demonstrated that photooxidation of H_2O and OH^- adsorbed on surface Ti atoms is not feasible thermodynamically under bandgap illumination, which was based on the electronic structure and energetics of TiO_2 surface-bound water [45–50]. Consequently, trapping h_{VB}^+ involves oxidation of surface oxygen to $–Ti^{IV}–O^{\bullet-}$ or $–Ti^{IV}–^{\bullet}OH$ depending on the pH. These species are inherent surface states of the TiO_2 nanocrystal as opposed to adsorbed $^{\bullet}OH$ or $O^{\bullet-}$ and are unable to react in the bulk of the suspension [44]. A recent theoretical study has shown that the most favorable site for h_T^+ is a single lattice oxygen bridge atom

[51]. These calculations demonstrate, however, that electron transfer from free or physisorbed H_2O (adsorption energy 0.12 eV) to h_T^+ is thermodynamically feasible resulting in $^{\bullet}OH$ adsorbed on the Ti row. This reaction path has a kinetic barrier of ~0.46 eV. It is doubtful whether under common experimental conditions oxidation of H_2O can compete with electron–hole recombination and with oxidation of other additives.

3.1 Redox Properties

The electron trap site was found to be 0.8 V below the conduction band edge in single-crystal rutile electrodes [52]. On the other hand, the (0 0 1) surface of highly doped TiO_2 showed a trap depth of only 0.3 V [53]. Sintered anatase particles of 15 nm diameter, Degussa P-25, and films of small anatase particles showed trap depths of ~0.7 V [54], ~0.5–0.6 V [55], and ~0.5 V [56] below the conduction band edge, respectively. The redox (flat band) potential of the conduction band electron for TiO_2 nanoparticles of radius >3.5 nm is −0.60 V at pH 7.0 [57, 58].

3.2 Electron–Hole Recombination

It is generally accepted that reactions 5 and 7 represent the predominant electron–hole recombination at time longer than a few ps. Diffuse reflectance investigations (P-25 TiO_2) have shown, however, a dramatic increase in the population of trapped electrons within the first few ps upon the addition of the hole scavenger SCN^- implying that hole oxidation effectively competes with electron–hole recombination before hole trapping occurs [17]. More refined studies have demonstrated multicomponent recombination time profile in the ps and >ns time ranges, which has been attributed to shallow and deep-trapped states [13, 16, 18]. A first-order electron–hole recombination has been reported in the ns time range [4], which is expected when only one electron–hole pair is involved in the same nanoparticle. Electron–hole recombination is affected by the absorbed light density when more than one electron–hole pair per particle is produced. At high excitation light intensities a second-order time profile is observed, although the lifetime of the electron–hole recombination extends from 10 to 30 ps [16, 17, 21] to the ns [3, 15], μs [3, 15], and ms [59, 60] for different TiO_2 preparations.

3.3 Electron and Hole Reactions in the Presence of Oxygen

The interactions between TiO_2 surfaces and O_2 have been recently reviewed [61]. Adsorption of O_2 to TiO_2 involves surface defects such as oxygen atom deficiency [62–64]. O_2 is reduced to $O_2^{\bullet -}$ by e_{CB}^- and e_T^- (reactions 8 and 9).

$$e_{CB}^- + O_2 \rightarrow O_2^{\bullet -} \tag{8}$$

$$e_T^- + O_2 \rightarrow O_2^{\bullet -} \tag{9}$$

$O_2^{\bullet -}$ may react with electrons (reactions 10 and 11) and holes (reactions 12 and 14) or decompose via dismutation (reaction 14). The resulting H_2O_2 may react with electrons (reaction 15) and holes (reaction 16) forming $^{\bullet}OH$ and $O_2^{\bullet -}$, respectively.

$$e_{CB}^- + O_2^{\bullet -} + 2H^+ \rightarrow H_2O_2 \tag{10}$$

$$e_T^- + O_2^{\bullet -} + 2H^+ \rightarrow H_2O_2 \tag{11}$$

$$h_{VB}^+ + O_2^{\bullet -} \rightarrow O_2 \tag{12}$$

$$h_T^+ + O_2^{\bullet -} \rightarrow O_2 \tag{13}$$

$$O_2^{\bullet -} + HO_2^{\bullet} + H^+ \rightarrow H_2O_2 \tag{14}$$

$$H_2O_2 + e_T^- \rightarrow {}^{\bullet}OH + OH^- \tag{15}$$

$$H_2O_2 + h_T^+ \rightarrow O_2^{\bullet -} + H^+ \tag{16}$$

$^{\bullet}OH$ can be sacavenged by TiO_2 [23], $O_2^{\bullet -}$, H_2O_2, and e_T^- (reactions 17–20).

$$^{\bullet}OH + TiO_2 \rightarrow {}^{\bullet}OH\text{-}TiO_2 \tag{17}$$

$$^{\bullet}OH + O_2^{\bullet -} \rightarrow O_2 + H^+ \tag{18}$$

$$^{\bullet}OH + H_2O_2 \rightarrow O_2^{\bullet -} + H_2O + H^+ \tag{19}$$

$$^{\bullet}OH + e_T^- \rightarrow OH^- \tag{20}$$

This complex mechanism does not form any stable products, except very low steady-state concentrations of the above oxygen species. Small amounts of impurities may react with $^{\bullet}OH$ and simplify the reaction scheme while producing oxidation products.

3.4 Electron and Hole Reactions in the Presence of Solutes

The energy of both h_{VB}^+ and h_T^+ is sufficient to oxidize most organic and inorganic substrates. The reaction of h_{VB}^+ with adsorbed solutes may take place at highly reactive solute concentrations. Similarly, but to a smaller degree, reaction of h_T^+ with solutes competes with electron–hole recombination. When both h_{VB}^+ and h_T^+

produce the same product, it is usually difficult to distinguish between their reactions without the use of fast kinetics techniques. Once holes of either kind have reacted with the solute, the subsequent reactions involving radical intermediates are similar to radical reactions in TiO_2-free solutions, which have been studied mainly by pulse and steady-state radiolysis. The kinetic parameters, however, are strongly affected by the partial adsorption of intermediates and products. In the following section, the general behavior of organic solutes in photocatalytic systems will be discussed with particular emphasis on the effect of absorbed light density. Since most studies have been carried out in aerated systems, the electrons avoiding recombination with holes are converted into $O_2^{\bullet -}$, although solutes with high adsorption constant or at high concentrations may compete with O_2 for the electrons.

3.4.1 Organic Solutes

Almost all organic molecules (RH_2) can be oxidized by holes (reactions 21 and 22) and by $^{\bullet}OH$ (reaction 23).

$$h_{VB}{}^{+} + RH_2 \rightarrow {}^{\bullet}RH_2 + H^{+} \tag{21}$$

$$h_{T}{}^{+} + RH_2 \rightarrow {}^{\bullet}RH + H^{+} \tag{22}$$

$$^{\bullet}OH + RH_2 \rightarrow {}^{\bullet}RH + H_2O \tag{23}$$

Reactions 21 and 22 compete with electron–hole recombination, reaction 23 suppresses reactions 17–20, and the chemistry becomes that of $^{\bullet}RH$. Since $^{\bullet}OH$ is produced at the TiO_2 surface, reaction 17 is instantaneous. It is widely accepted that in steady-state photolysis the rate of product buildup (R_p) depends on the substrate concentration according to the Langmuir-Hinshelwood (LH) rate (Eq. 24), where K is the Langmuir adsorption constant and k is the limiting apparent rate constant at high substrate concentrations [65].

$$R_p = kK[\text{substrate}]/(1 + K[\text{substrate}]) \tag{24}$$

The LH kinetic law is based on fast adsorption equilibration followed by surface rate-controlling oxidation step. This reaction model has recently been challenged [66–71]. Thus, it is frequently found that K apparently depends on the absorbed light density implying that the equilibrated adsorption/desorption of reactants is not maintained under illumination [68]. An alternative "Direct–Indirect" model has been proposed, which considers the degree of electronic interaction of the semiconductor surface with dissolved reactant molecules [68]. However, the LH model still predominates because of its simplicity and its good agreement with the data as long as the absorbed photon density is constant (see for example [72–74]).

The quantum yield of RH_2 consumption may reach unity at sufficiently high $[RH_2]$. However, in many cases oxidation of RH_2 by $h_{VB}{}^{+}$ does not compete effectively with hole trapping and electron–hole recombination, and oxidation of RH_2 by h_T^+ is the predominant reaction. The oxidation yield increases with the solute concentration, and the highest oxidation yield is determined by the yield of hole trapping.

In aerated systems reactions 8 and 9 take place and $O_2^{\bullet-}$ is largely adsorbed at the surface [75] and is able to react with both electrons and holes, although the rate constants are not known. $^{\bullet}RH$ produced vis reactions 21–23 readily reacts with O_2 forming the respective peroxyl radical (reaction 25).

$$^{\bullet}RH + O_2 \rightarrow RHO_2^{\bullet} \tag{25}$$

The fate of the peroxyl radical depends on the nature of RH_2. When $RH^{\bullet}$ is derived from α-alcohol, $RHO_2^{\bullet}$ decomposes to R and superoxide according to reaction 26, which is base-catalyzed [76]. In other cases $RHO_2^{\bullet}$ decomposes bimolecularly via reactions 27 and 28 [77].

$$RHO_2^{\bullet} \rightarrow R + H^+ + O_2^{\bullet-} \tag{26}$$

$$RHO_2^{\bullet} + RHO_2^{\bullet} \rightarrow RHO\text{–}O\text{–}O\text{–}ORH \tag{27}$$

$$RHO\text{–}O\text{–}O\text{–}ORH \rightarrow O_2 + HROORH \tag{28}$$

Eventually the organic solutes are oxidized to CO_2 and water, which makes the photocatalytic processes useful for water decontamination.

4 Effect of Absorbed Light Density, I_d

The effect of the excitation light intensity has been intensively studied demonstrating a linear relationship between R_p and the square root of the photon flux [78–106]. A considerable numbers of works have been devoted to various kinetic models for different types of photocatalytic systems including suspensions, porous layers, and colloid solutions [4, 68–71, 90, 94, 107–122]. The square root law can be expressed by Eq. (29), where Φ and Φ_{max} are the quantum yields at absorbed light density I_d and $I_d \rightarrow 0$, respectively, and K_d is a constant typical to the TiO_2 preparation at fixed reaction conditions such as pH and oxygen level [91].

$$\Phi_{max}/\Phi = 1 + I_d^{1/2}K_d \tag{29}$$

Most published works, however, do not address the light density effect quantitatively and do not enable calculations of Φ nor use limiting conditions where Φ is independent of substrate concentration. Therefore, in most cases Eq. (29) is not applicable, and it is usually difficult or impossible to compare data from different TiO_2 preparations, particularly if made in different laboratories. In addition, the incident light intensity (einstein s^{-1} cm^{-2}) or radiant flux (einstein s^{-1}) is frequently reported instead of I_d (einstein s^{-1} g^{-1}). Application of a wide spectral range for excitation further complicates comparative analysis since I_d is a function of the illuminating wavelength. Furthermore, I_d is not uniform in layers. The square root law is expected in steady-state illumination when more than one electron–hole pairs are preset at the same time in the same TiO_2 nanocrystal [28]. In practice it is also observed in steady-state photolysis of powders, layers, and colloids when the time separation between two consecutive excitations of the same nanoparticle is in the order of seconds to minutes. Evidently, the lifetimes of all kinds of TiO_2 electrons and holes are too short to have more than one electron–hole pair at the same time. This apparent discrepancy will be discussed below. R_p can be derived numerically solving the simultaneous relevant kinetic equations for reactions 2–23, 25, and 26. This is a very complicated task because most rate constants, which depend on the specific TiO_2 preparation, are not known. However, under certain conditions a simplified calculation is possible as demonstrated below using methanol as a substrate.

4.1 The Methanol System

Methanol is one of the most popular substrate for studying the basic mechanisms of photooxidation mediated by TiO_2 [89, 104, 123–135]. TiO_2 photolysis in the presence of aqueous methanol produces formaldehyde. A limiting quantum yield $\Phi_{max} = 2$ has been reported when I_d was extrapolated to zero [91]. The constant K_d, determined experimentally from Eq. (29), has been suggested as a measure of the relative efficiency of hole trapping compared to electron–hole recombination, which defines the quality of TiO_2 preparation for photocatalysis. The system is simplified using 2–5 M CH_3OH where the yield of formaldehyde approaches a plateau corresponding to $\Phi_{HCHO} < 2$. Under such conditions, CH_3OH reacts with practically all h_T^+ while its reaction with h_{VB}^+ is not important. This is obvious since competition of CH_3OH for h_{VB}^+ is expected to show an increase of Φ_{HCHO} with [CH_3OH] until a value of 2 is reached. Thus, a simplified mechanism is obtained involving reactions 3 and 5.

In the absence of oxygen adsorbed $^{\bullet}CH_2OH$ radicals react with TiO_2 producing e_T^- via electron transfer to TiO_2. The photolysis builds up a steady-state concentration of e_T^- while h_T^+ is removed by CH_3OH. The steady-state concentration depends on the competition between electron–hole recombination and hole reaction with CH_3OH. If the average steady-state level corresponds to less than one electron per nanoparticle, the electron–hole recombination rate does not depend on the light flux since the lifetime of the transients is too short for interparticle reactions. On the other hand, if a number of electrons accumulate on the same nanoparticle, R_p is not proportional to I_d and Φ_{HCHO} responds to the light flux.

Simple computer simulations based on $I_d = 2 \times 10^{-7}$ einstein s^{-1} per gram TiO_2 and on the kinetic parameters $k_3 = 4 \times 10^6\ s^{-1}$ and $k_6 = 2 \times 10^{10}\ mol^{-1}\ L_{TiO_2}\ s^{-1}$ show that the buildup of steady-state electron concentration takes many hours.

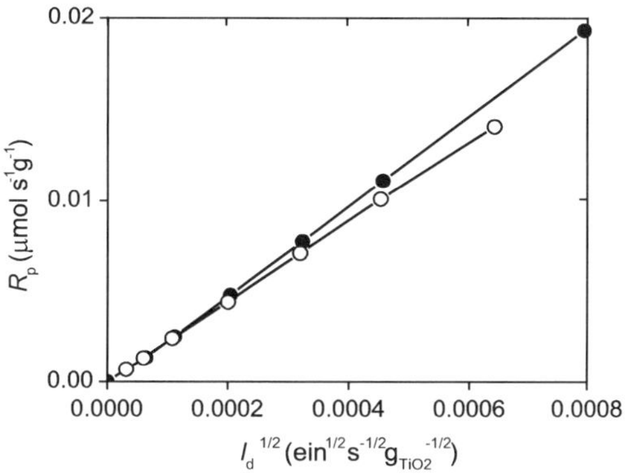

Fig. 2 Simulation of the square root dependency when electrons are slowly removed. Illumination time was taken as 10 min (*filled circles*) or 1,000 min (*open circles*). The kinetic constants taken from [3]. A pseudo-first-order rate constant $k = 0.01\ s^{-1}$ for the electrons removal was arbitrarily chosen

The square root dependency is observed only when the product concentration is measured at a given time irrespective of the light intensity. Thus, the square root law in deaerated photocatalytic systems is expected only under very specific conditions, which are rarely applied in practice. Usually, the illumination time is higher when the rate of the product formation is low. The simulations show that the presence of an electron scavenger has an important role in the observed square root behavior under a wide range of conditions.

Oxygen suppresses the buildup of e_T^- and therefore increases the rate of hole oxidation. Hence, a steady-state concentration of e_T^- is quickly obtained and remains constant during the illumination time. Under these conditions, the square root law is no longer restricted to constant illumination time. Figure 2 shows that computation results in a hypothetic system where e_T^- reacts with the solute forming unreactive products. A computed linear square root profile is obtained using $k = 0.01\ s^{-1}$ for the removal of the electrons. The two straight lines in Fig. 2 represent two selected illumination times for the light intensity effect, which differ by a factor of 100, demonstrating that R_p increases almost linearly with $I_d^{1/2}$. The deviation is within the usual experimental uncertainty when $k = 0.001\ s^{-1}$, but when $k = 0.1\ s^{-1}$ the steady-state number of e_T^- per particle is too low, and the square root dependency is not expected.

Adsorbed O_2 reacts with trapped and possibly also with conduction band electrons followed by the complicated multistage process involving reactions 10–12, 14, 15, 17–20, and 25–28. Reaction 11 is a multi-exponential process where the rate constants depend on the particle size [136]. The exponential lifetimes of the electrons range from sub ms to 0.5 s (4.7 nm average diameter, acid pH), which are several orders faster than those derived from $k = 0.01\ s^{-1}$ chosen for the simulations presented in Fig. 2. It is therefore suggested that $O_2^{\bullet -}$ formed via reaction 12 has the role usually assigned to e_T^- and is responsible for the square root dependency. Accumulation of a number of $O_2^{\bullet -}$ per particle is more reasonable than accumulation of e_T^- per se because $O_2^{\bullet -}$ has a longer lifetime under ordinary working conditions. The removal of $O_2^{\bullet -}$, which is essential for the square root low, takes place via diffusion of adsorbed $O_2^{\bullet -}$ to the bulk and its dismutation to H_2O_2 and O_2, which is catalyzed by metal impurities. It has been recently shown that very low concentrations of cupric ions induce a considerable increase of R_{HCHO}, which has been attributed to superoxide dismutation catalyzed by cupric ions [137]. Although there cannot be more than one h_{VB}^+ in a nanoparticle, accumulation of several $O_2^{•-}$ is sufficient to impose a second-order recombination rate law. It is concluded that the electron–hole recombination involves adsorbed $O_2^{•-}$ and h_{VB}^+.

5 Real-Time Kinetic Measurements of Electron and Hole Reactions with Solutes

Time-resolved techniques have been extensively applied to study the reaction of electrons and holes with selected solutes, and hole scavenging in the ps to ns time range has been observed [20, 24, 25, 138–157]. The interfacial electron transfer through the TiO_2–water boundary has been addressed in many studies [147–151, 156]. Oxidation of organic molecules produces free radicals, which can inject an electron to the conduction band of TiO_2 resulting in the conversion of holes into electrons. When holes are converted to less reactive species, time separation between the different fast reactions may exist and the kinetics is simplified. Otherwise, the differential equations of simultaneous reactions must be solved, which usually introduce some uncertainty in the reaction rate constants. The dynamics of interfacial charge transfer to selected organic molecules (formic acid, formaldehyde, and methanol) in TiO_2 aqueous colloid solution has been observed using transient absorption and EPR spectroscopy [158]. These techniques are useful tools for studying many processes including energy storage by CO_2 reduction to hydrocarbons [157, 159].

In some systems h_{VB}^+ and h_T^+ differ also with respect to reaction paths as demonstrated in the case of alcohols; alcohols at moderate concentrations react with h_T^+ via H-abstraction, although at 5 M CH_3OH or 100% $CH_3CH_2OHCH_3$ there is evidence for the formation of CH_3OH^+ and $CH_3CHOHCH_3^+$, respectively [20]. However, the kinetics is complex when electron–hole recombination cannot be ignored. Simplification of the system by complete removal of e_T^- or h_T^+ is difficult to achieve because of the relatively fast competing recombination. Furthermore, the reaction of a solute with electrons or holes often gives rise to a transient, which may also react with the remaining charge carrier. For example, h_T^+ and h_{VB}^+ are nearly completely removed by 5–7 M iodide forming $I_2^{•-}$, which reacts with the TiO_2 electrons [160].

Although fast reaction techniques are usually superior over steady-state illumination, the latter method may be successful for indirect mechanistic studies throughout product analysis as demonstrated above in the case of methanol.

6 Reactions of Electrons in the Absence of Paired Holes

Relatively stable electrons can be produced upon pulse- and γ-radiolysis of deaerated TiO_2 colloid solution in the presence of a hole scavenger, e.g., CH_3OH, $CH_3CHOHCH_3$ [7, 161]. In irradiated aqueous solutions, a mixture of reducing and oxidizing radicals is formed initially according to Eq. (30). The values in parentheses are the *G*-values, which represent the yields of the species in 10^{-7} M Gy^{-1}.

$$H_2O \rightsquigarrow e_{aq}^{-}(2.6),\ {}^{\bullet}OH(2.7),\ H^{\bullet}(0.6),\ H_3O^{+}(2.6),\ H_2O_2(0.72) \tag{30}$$

Irradiation of deaerated colloid TiO_2 containing high concentrations of RH_2 generates $^{\bullet}RH$ via efficient scavenging of $^{\bullet}OH$ by RH_2. $H^{\bullet}$, e_{aq}^{-} and $^{\bullet}RH$ are capable to inject electrons to the conduction band of TiO_2 (reactions 31 and 32), which are readily converted to e_T^{-}.

$$^{\bullet}RH + TiO_2 \rightarrow e_{CB}^{-} + R \tag{31}$$

$$e_{aq}^{-}(H^{\bullet}) + TiO_2 \rightarrow e_{CB}^{-}(+H^{+}) \tag{32}$$

When $RH_2 = CH_3OH$ or $CH_3CHOHCH_3$, R = HCHO or CH_3COCH_3, which do not react effectively with e_T^{-}, although two-electron back-reduction to the starting compounds is thermodynamically feasible. The overall result of the radiolysis of TiO_2 in the presence of RH_2 and absence of O_2 is the accumulation of relatively stable excess of electrons on the TiO_2 surface. The excess electrons on colloid TiO_2 particles possess the expected intense blue color, which is stable for at least several weeks and allows the study of both slow and fast reactions of these electrons with solutes. The kinetics of one-electron reduction of nitrate, nitrite, O_2, H_2O_2, copper ions, and noble metal catalytic deposits has been reported [7, 136, 162, 163]. Prolonged radiolysis in the presence of 2-propanol and absence of O_2 accumulates as much as 150 electrons per TiO_2 nanoparticle, which may be useful for the study of multi-electron transfer reactions. A simultaneous two-electron transfer has already been demonstrated for the reduction of H_2O_2 to OH^{-} [7].

Another approach to generate excess electrons on the TiO_2 surface is via band gap photolysis. In this case $^{\bullet}RH$ can inject the electron to TiO_2 and the accumulation of the electrons is limited by the efficiency of the electron–hole recombination. Nevertheless, the solution can be transferred under inert gas to a rapid mixing stopped-flow system, and the reduction of solutes by these electrons can be studied as recently demonstrated in the case of O_2 [134, 164], H_2O_2 [134, 164], NO_3^{-} [134, 164], Cu^{2+} [134, 164], Ag^{+} [165], Au(III) [166], and methyl viologen [166].

Scheme 1 Reduction of nitrate to ammonia via the formation of different nitrogen intermediates

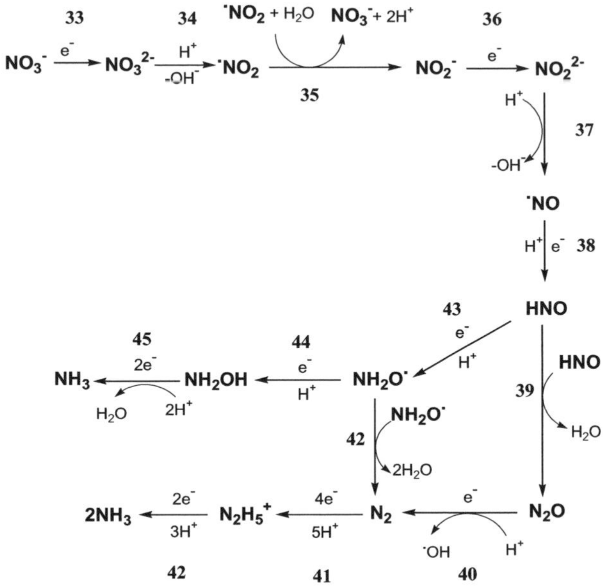

6.1 *Discrimination Between One- and Multi-Electron Transfer Reactions*

Reduction processes in solutions usually involve one- or consecutive one-electron transfer steps, although atom transfer, e.g., O, represents apparent multi-electron transfer. Real multi-electron transfer is well known in electrochemical processes. The probability to observe a real multi-electron reduction by excess electrons on TiO_2 nanoparticles increases with the number of electrons per particle. However, the distinction between consecutive one-electron and multi-electron transfer reactions is not simple as demonstrated below for nitrate reduction to ammonia.

6.1.1 Consecutive One-Electron and Multi-Electron Reduction of Nitrate

The reduction of nitrate to ammonia involves the formation of several nitrogen intermediates with known physical and chemical properties including NO_3^{2-}, $^{\bullet}NO_2$, NO_2^{2-}, NO_2^{-}, NO, HNO, $NH_2O^{\bullet}$, NH_2OH, N_2O, and N_2 (Scheme 1). Thermodynamically, the reduction of some of the valence states of nitrogen by two or more electrons is favorable compared to one-electron reduction.

Most if not all the reduction steps of NO_3^{-} by TiO_2 electrons are thermodynamically downhill processes. In view of the highly negative reduction potentials of organic radicals, the reaction of radicals such as $CH_3C^{\bullet}HOHCH_3$ and $^{\bullet}CH_2OH$ with the N-compounds are feasible and might compete with the electron injection to TiO_2.

It can be seen that reduction of the nitrogen species by consecutive one-electron transfer reactions leads to different products than reduction by one multi-electron transfer reaction, while in other cases the products are the same. For example, reduction of nitrate to nitrite can proceed via reactions 33–35 or via a two-electron transfer reaction (see Scheme 1). Recently [134, 164] UV photolysis of transparent small colloidal TiO_2 in deaerated aqueous solutions containing nitrate and methanol at pH 2.3 was studied. Under these conditions, a fraction of the holes is removed by the methanol as discussed above. Each hole reacting with methanol gives rise to two stored electrons, which have been shown to reduce nitrate ions to ammonia through the transfer of eight electrons suggesting N_2 as an intermediate. Although the reduction of $2NO_3^-$ to N_2 via ten electrons transfer is feasible ($E^o = 1.25$ V vs. NHE [167]), it is not clear whether other stable nitrogen intermediates are formed during this process as no attempt to analyze other nitrogen compounds such as nitrite, hydroxylamine, or N_2O has been reported. One-electron reduction of nitrate and nitrite has been observed under the conditions where there was on average less than one electron per TiO_2 nanoparticle [7, 136].

7 TiO_2 Modification

Bare TiO_2 has two major shortcomings: (a) The band gap is too high so that no visible light is absorbed, and therefore solar energy cannot be used for photocatalytic processes; (b) the fast electron–hole recombination competes with electron and hole trapping and/or substrate reactions. The major attempts to overcome these problems include doping and co-doping of metals and nonmetals for decreasing the band gap or introducing intra-gap states as well as bilayers, multilayers, and surface-modified TiO_2 to separate electrons and holes in different phases.

7.1 Doped TiO_2

The discovery that doping of TiO_2 extends the photoactive region from UV to visible light [168] has remarkably increased the interest in such materials for visible-light-driven solar conversion and photocatalytic oxidations. Several review articles have been published on semiconductor photocatalysts on TiO_2 composites including doped TiO_2 [169–173].

7.1.1 Doped TiO_2 with Nonmetals

Many elements have been applied for doping and surface modification of TiO_2 including N [168, 174–196], co-doping N with other elements [197–220], C [104, 221–233], S [221, 234–242], P [243, 244], and halogen [241, 245–251].

Analysis of the absorption spectra of visible-light-active TiO_2 photocatalysts has shown that the absorption spectrum of N-, S-, and C-doped titania can be described by the sum of two absorption bands at 427 and 486 nm. It has been taken as an indication that visible-light activation of TiO_2 is due to defects associated with oxygen vacancies rather than narrowing the original band gap of TiO_2 by mixing of dopant and oxygen states [178, 252, 253]. It has been observed that N-doped TiO_2 films catalyze the decomposition of gaseous 2-propanol upon UV illumination with $\Phi = 0.154$ compared to $\Phi = 0.184$ for undoped TiO_2. In the visible range $\Phi = 0.0041$ and 0.000 for N-doped TiO_2 and undoped TiO_2, respectively [254]. These results provide unequivocal evidence that the N states are not mixed N 2p with O 2p levels, namely, the N doping does not lower the band gap of TiO_2 but rather forms intra-gap states. Furthermore, Φ in the case of N-doped TiO_2 is lower than that of undoped TiO_2 under UV illumination, indicating that the N states act as centers for electron–hole recombination [254, 255]. This is supported by calculations based on density functional theory (DFT) assigning the visible-light photocatalytic activity to N 2p levels near the valence band. Oxygen vacancies and the associated Ti^{3+} species act as recombination centers for the photoinduced electrons and holes, which reduce the photocatalytic activity, although they contribute to the visible-light absorbance [256]. Similarly to the N-doped TiO_2, the C atoms act as efficient recombination centers in C-doped TiO_2 [104, 231]. This is based on measurements of CH_3OH photooxidation in C-doped TiO_2 suspensions, which show that the quantum yield of the formaldehyde product extrapolated to zero light intensity is about one order of magnitude higher under UV photolysis compared to visible illumination. Furthermore, undoped TiO_2 showed higher extrapolated yields than C-doped TiO_2 [104, 231]. On the other hand, F doping of TiO_2 enhanced the photocatalytic activity [248, 249, 257], which has been attributed to decreased terminal hydroxyl group content [257].

Oxidation of gaseous acetaldehyde by F-doped and N, F co-doped TiO_2 has shown enhanced activity compared to undoped TiO_2. The measured $\Phi(CO_2)$ in the doped TiO_2 was 3–6 times higher compared to undoped TiO_2, although $\Phi(CO_2)$ upon visible-light excitation was significantly lower compared to UV excitation [248, 249]. These results have been attributed to color centers produced by replacing O with F^- in the nanoparticle lattice, which apparently produce an intra-gap state without acting as a recombination center for electrons and holes [169].

N, S co-doping has shown higher photocatalytic activity compared to mono-doped TiO_2 apparently due to mixing of O 2p, N 2p, S 3p, and Ti 3d states [258]. Photocatalytic oxidation of CH_3OH, SCN^-, Br^-, I^-, and hydroquinone has been interpreted by mid-gap level introduced by the N doping formed slightly above the top of the O 2p valence band. In contrast, the mechanism of visible photocatalytic oxidation of organic compounds has been assigned to reactions with surface intermediates of water oxidation or oxygen reduction as opposed to direct reactions with the mid-gap trapped holes [49]. This is supported by EPR spin-trapping technique used to detect radical intermediates produced by UV or visible illumination of 4-chlorphenol in N-doped TiO_2 aqueous suspensions. The results suggest that surface $^{\bullet}OH$ and $O_2^{\bullet -}$ are responsible for the visible photodegradation [176].

7.1.2 Doped TiO_2 with Metals

With the exception of noble metals, doping with metal oxides or metal ions (salts) ends up with the metal incorporated in the TiO_2 nano-lattice. In the following we shall use the metal ion notation whenever the valence state has been reported.

Doping TiO_2 with metals as well as co-doping with metal and metal–nonmetal pairs have been reported to show visible absorption [259–271] although not always associated with enhanced photocatalytic activity. Many metals have been used for TiO_2 doping including Ag [263, 265, 270], Au [264], Pt [130], Fe^{3+} [130, 260, 272–283], Bi^{3+} [284], W [267, 277, 278, 285, 286], B [260, 262, 269, 270], Li [284], $Mn^{3+/4+}$ [271, 285, 287], Mo [277, 278, 288–290], Ni [291], La^{3+} [282, 292], Eu^{3+} [275, 293], Cr^{3+} [276–278, 294], Cu^{2+} [274, 277, 278, 295], V [274, 276–278, 290, 296], Co [263, 266, 277, 278], Th^{4+} [290], Zn^{2+} [296], Ce [297], and Si [263]. Several selected examples are given in more details: (1) Doping with metal ions such as Fe^{3+}, Mo^{5+}, Ru^{3+}, Os^{3+}, Re^{5+}, V^{4+}, and Rh^{3+} has shown increased photoactivity compared to undoped TiO_2 while Co^{3+} and Al^{3+} decreased the activity [289]. This has been correlated to the electron–hole lifetime. The efficiency of the dopant was shown to depend on whether it serves as a mediator of interfacial charge transfer or acts as a recombination center. The photoactivity depends on the ability of the dopant to function as an electron or hole trap, on the dopant concentration, on the energy level within the TiO_2 lattice, on the d-electronic configuration, on the distribution of the dopants within the TiO_2 particles, as well as on the incident light flux [289]. (2) Doping TiO_2 with a low CuO content produced visible photocatalytic activity, which became smaller upon increasing the copper amount [298]. (3) Detailed study of Cr-doped TiO_2 shows absorption in the visible region with no visible photocatalytic oxidation of oxalic acid, propene, or 2-propanol, although UV photoactivity has diminished by 25–1,000 times upon doping [299]. (4) EPR study of TiO_2 aqueous colloids doped with Fe^{3+} or V^{4+} shows the growth of trapped electron signals upon UV illumination, which has been attributed to inhibition of hole–electron recombination by the dopants [300]. (5) Doping TiO_2 with Fe^{3+} improved the UV photoactivity of methanol oxidation [281, 301] and phenol degradation [302], while in the latter case visible activity has also been noticed with a considerable lower quantum yield. However, the enhanced photoactivity has been reported for low levels of Fe^{3+} while at high levels the photoactivity decreased. This has been attributed to the formation of conduction band electrons and the highly reactive h_T^+ at low Fe^{3+} levels when most light is absorbed by TiO_2, while excitation of the Fe^{3+} states gives rise to less reactive species [303]. It has been suggested that three-dimensional networks of Fe^{3+} doped in aqueous suspensions lead to improved photocatalytic activity through an "Antenna Mechanism," i.e., the energetic coupling throughout a long chain of TiO_2 nanoparticles enables energy and/or exciton transfer from the absorbing nanoparticle to a distant one [130].

Visible absorption induced by doping is not always expected to show enhanced photocatalytic activity as demonstrated in the following examples: (1) Both p-type doping with Al^{3+}, Cr^{3+}, or Ga^{3+} and n-type doping of TiO_2 with Nb^{5+}, Ta^{5+}, or Sb^{5+} had a rather detrimental effect on UV photocatalysis [304], which has been assigned to increased electron–hole recombination rate [304, 305]. (2) The p-type doping creates acceptor centers, which trap photo-electrons and then, once negatively charged, attract holes and create a recombination centers [299]. (3) The n-type doping agents create donor centers that increase the concentration of conduction electrons in the solid. Hence, electron–hole recombination is also favored, which is detrimental for photo-efficiency [299]. (4) Incorporation of transition metal ions such as Cr^{3+} and Mo^{5+} at low concentrations creates new trapping sites, which decrease the lifetime of the charge carriers compared to bare TiO_2. The proposed mechanism involves oxidation of Cr^{3+} or Mo^{5+} to Cr^{4+} or Mo^{6+} and subsequent reduction of these immobilized holes by TiO_2 electrons [306]. (5) Vanadium reduces the photoreactivity of TiO_2 by promoting charge-carrier recombination at surface VO_2^+ segments by electron trapping whereas V(IV) impurities in surface V_2O_5 promote charge-carrier recombination by hole trapping. Substitutional V(IV) in the lattice also act also as a charge-carrier recombination center [307]. These complexities are expected to be present in the mechanisms of other transition metal ions doped into TiO_2; (6) bare TiO_2 doped with Cu showed detrimental effect [257]. However, doping of TiO_2 with Cu under N-plasma produced N-doped TiO_2/Cu, which showed higher activity than bare TiO_2 under UV illumination, but decreased activity below the undoped level at high Cu–N content [308].

7.2 Bilayers and Mixed Oxides

Metal oxide bilayers [309–313] and mixed-phase bimetal oxides [314–319] have been studied to achieve charge separation and inhibit electron–hole recombination as demonstrated in the following examples: (1) Zirconia/titania showed visible and UV photocatalytic activity toward reduction of Pb^{2+} or Cd^{2+} in aqueous suspension particularly in the presence of HCO_2^-, which converts the holes into $CO_2^{\bullet -}$ [320]. The photocatalytic activity of this binary oxide assembly for the degradation of 4-chlorophenol is gradually enhanced with increased ZrO_2 content up to 12 wt%, which has been ascribed to electrons' transfer from the excited ZrO_2 to the conduction band of TiO_2 [321]. Inhibition of TiO_2 particle size by ZrO_2 in hollow binary ZrO_2/TiO_2 oxide fibers using mixed precursor solution and the formation of stronger surface acid sites has been proposed to account for the improved efficiency of ethylene and trichloromethane oxidation in the gas phase compared to P-25 [322]. A similar interpretation has been presented for enhanced oxidation of aqueous phenol by UV photolysis of TiO_2-ZrO_2 [323]. (2) An interesting concept of "composites" is the application of mixed rutile and anatase nanocrystals. Titania nanocrystals containing both rutile and anatase structures have shown higher photocatalytic activity with respect to degradation of *p*-coumaric acid compared to pure anatase or rutile. The most efficient catalyst contained 30% rutile and 70%

anatase. This was supported by photocatalytic oxidation of naphthalene in acetonitrile–water medium (maximum at 90% anatase) [324], photodegradation of phenol (85% anatase) [325] and 4-chlorophenol (77.4% anatase) [326], and oxidation of gaseous CH_3CHO [327]. These observations are explained by the different energy levels of rutile and anatase TiO_2 so that the electrons migrate from the rutile to the anatase and holes from anatase to rutile, thus achieving charge separation and inhibition of the electron–hole recombination. Mixed-phase TiO_2 photocatalyst has been recently shown to be superior to P-25 TiO_2 in both oxidation of methylene blue and reduction of CO_2 to CH_4 in the presence of 2-propanol as a hole scavenger [313]. (3) Electron migration from TiO_2 to WO_3 has been reported when the WO_3/TiO_2 composite is illuminated in water vapors or humid air. Subsequent energy storage is achieved by water oxidation at the TiO_2 surface and O_2 reduction to H_2O_2 at WO_3 [328]. WO_3/TiO_2 films in the form of mixtures of separate TiO_2 and WO_3 phases increase the UV activity of TiO_2 [329] as is the case with In_2O_3/TiO_2 [330] and SiO_2/TiO_2 [331]. The enhanced photoactivity has been attributed to a better charge separation where electrons are at WO_3 (or In_2O_3 or SiO_2) and holes are at the TiO_2 layer. On the other hand, mixed-phase study involving ZrO_2 and TiO_2 suggests that binary oxides are less active than the respective pure oxides [332]. Also, doping of Al_2O_3/TiO_2 composite films with Ru, Si, and Te shows lower photocatalytic activity in the UV region compared to the undoped material [333].

Photocatalytic layers are supported on different solids including quartz, glass, stainless steel, or Ti plates, which may affect the photocatalytic activity as demonstrated for the degradation of malic acid [334] and methylene blue [335]. In the latter case, the effect was attributed in part to inhibition of electron–hole recombination by the creation of electron sinks.

7.3 *Surface Modification*

The main purpose of surface modification is to achieve a better electron–hole separation, although composite particles such as binary oxides are also used to enlarge the TiO_2 surface area or enhance adsorption of substrates. Surface modification of TiO_2 and TiO_2 composites with Pt [132, 160, 163, 229, 336–348], Pd [342, 349, 350], Au [268, 342, 348, 349, 351–363], Ag [268, 342, 344, 349, 364–371], Cu [277, 342, 349, 370, 372], Ni [342, 349, 370], Co [342], V [373], metal oxides [374–381], and polyoxometalates [155, 382–384] often extends the optical absorption and enhances charge separation and catalytic interfacial electron transfer to oxygen or other substrates (Fig. 3).

Surface metal nanoclusters may act also as catalysts for O_2 reduction, which dictate the yield of substrate oxidation as has been thoroughly discussed [137, 385–389]. We shall focus below on systems that enhance charge separation and/or induce visible absorption: (1) Radiation-induced synthesis of mono- and multi-metallic clusters and nanocolloids enhances the photocatalytic oxidation of

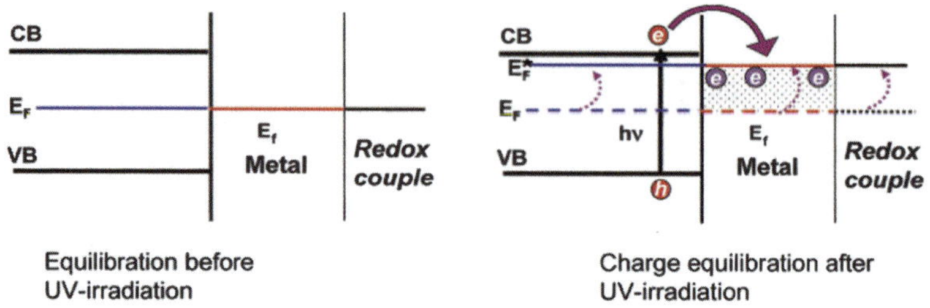

Fig. 3 Equilibration of semiconductor–metal nanocomposites with redox couple before and after UV irradiation. Reproduced with permission from [355]. Copyright 2004 American Chemical Society

phenol and rhodamine B under UV illumination due to inhibition of electron–hole recombination, which was observed by time-resolved microwave conductivity [390]. (2) POM ($H_3PW_{12}O_{40}$) placed between the TiO_2 and metal nanoparticles (Cu, Ag, Pt, and Au) improves the UV and visible-light photocatalytic activity. Short-circuit contact between the metal and TiO_2 was avoided by photo-deposition of the metals at POM, so that charge separation with an activation barrier for electron–hole recombination was introduced [383]. (3) Fe_xO_y/TiO_2 composite showed visible-light photocatalytic activity, although the Fe^{III} had a detrimental effect in the UV region probably due to an inner filter effect by Fe^{III} or formation of electron–hole recombination centers [391]. (4) Visible absorption has been reported for TiO_2 surface modification with noble metals (Pt, Au, and Pd). The metal clusters give rise to localized energy levels within the band gap of TiO_2 into which the valence band electrons of TiO_2 are excited at wavelengths above 400 nm. The quantum yields increase with increasing the metal loading up to a maximum value due to the decrease in the electron–hole recombination rate. At higher metal levels the metal clusters act as electron–hole recombination centers and lower the photonic efficiency [392].

7.4 TiO_2 Composites with Carbonaceous Nanomaterials

Combination of TiO_2 with carbonaceous nanomaterials such as carbon nanotubes (CNT), graphene, activated carbon, [60]-fullerenes, thin layer carbon coating, and nanometric carbon black have been recently reviewed [393, 394]. Carbon nanostructures possess unique electronic and catalytic properties. CNT has a large specific surface area, excellent mechanical strength, and capability to undergo chemical surface modifications including bonds formation with titania [395]. TiO_2/CNT composites show extended electron–hole lifetime, which is essential for enhanced photocatalytic activity. In addition, CNT absorbs visible light extending the TiO_2/CNT photoactivity into the visible range [394, 395].

The recent emergence of graphene, a stable two-dimensional carbon, has initiated a large interest in the photocatalytic activity of its nanocomposites.

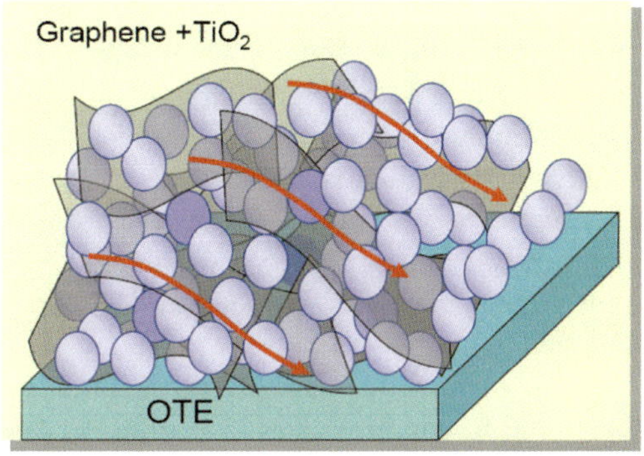

Fig. 4 Schematic diagram illustrating the graphene oxide sheets in transporting electrons in graphene oxide–semiconductor composite films. Reproduced with permission from [400]. Copyright 2010 American Chemical Society

Graphene sheets possess very high conductivity due to the existence of an extended sp2-bonded carbon network.

TiO_2/graphene hybrids are particularly promising for photocatalytic applications because of their enhanced charge separation due to the excellent electronic property and two-dimensional nanostructure of the grapheme [396–399] as illustrated in Fig. 4.

TiO_2/graphene composites are obtained when graphene is covered with a large number of TiO_2 nanoparticles. Transient photovoltage spectrum of anatase TiO_2/graphene demonstrates that the lifetime of the electron–hole pairs is longer compared to bare TiO_2 resulting from the transfer of TiO_2 electrons to the grapheme [401]. Enhanced photocatalytic activity of TiO_2/graphene hybrids has been observed, although no quantum yields or reaction rates under the same absorbed light density have been reported [402–404]. TiO_2/reduced-graphene-oxide nanocomposites have recently shown nearly 90% enhancement of the UV photocurrent and complete degradation of 2,4-dichlorophenoxyacetic acid [400].

8 Mechanism and Concluding Remarks

Doping TiO_2 nanoparticles has often achieved photocatalytic activity in the visible spectrum. However, doping may affect the adsorption properties of the photocatalyst and the dopants may act also as recombination centers and decrease the photocatalytic yields. The basic principles and mechanisms involving doped TiO_2 have much in common with undoped TiO_2. Irrespective of the mechanism by which the optical absorption becomes red-shifted, the energies and mobility of the electrons and/or holes are changed by doping. Hence, reactions of electrons and/or holes with the substrates may become thermodynamically unfeasible or may have different reaction rate constants. Evidently, the results may differ with different dopants and substrates.

Unfortunately, the visible photoactivity has been usually compared to undoped TiO_2 at the same visible wavelength range without reporting quantum yields or carrying out comparative tests with equal numbers of absorbed photons for the doped and undoped titania. Therefore, it is usually difficult to assess the relative quality of the new photocatalysts, although the preparation procedures and the absorption spectra may be of theoretical and practical interest. "Enhanced" activity is attributed to transfer of electrons to the doped metals, which is claimed to favor effective charge separation, larger surface area, creation of surface shallow traps, and substrate adsorption. However, when comparing under the same absorbed light density, doping may decrease the rate of substrate depletion compared to undoped TiO_2.

TiO_2 composites and modification of all kinds have been predominantly attempted to enhance charge separation. The principle is to design a bicomponent structure so that the electrons are more stable on one component and the holes on the other. Such a separation is believed to inhibit electron–hole recombination and therefore enhance trapping at the surface and subsequent reactions with substrates in solution or gas phase. Depending on the nature of the TiO_2 partner component, additional benefits can be expected including enhancement of O_2 reduction, e.g., by noble metals, where electrons are converted into superoxide radicals, which react relatively slowly with holes compared to TiO_2 electrons. The reduction of superoxide radicals by electrons produces H_2O_2 and subsequently $^{\bullet}OH$, which increases the overall photocatalytic yield. In addition, surface modification may catalyze other oxidation or reduction reactions with substrates and increase adsorption, although bare TiO_2 adsorbs many organic and inorganic compounds and the presence of a second adsorbing material may induce the opposite effect. Nanocomposites may also shift the excitation spectrum to the visible range while achieving improved charge separation throughout distribution of the electrons and holes between different components of the photocatalyst.

Since the mid-1980s, many works have been published concerning "sandwich" nanostructures particularly in colloid solutions [405]. In such a system, two different semiconductor parts are connected so that illumination of one part produces response from the second part. The first example has been observed in CdS sols where visible excitation induced reduction of MV^{2+} and oxidation of CH_3OH to formaldehyde, which increased ca. ten times to $\Phi_{HCHO} = 1$ upon the addition of TiO_2 sol [406]. TiO_2 and CdS nanoparticles produce a sandwich upon mixing, and the conduction band electrons migrate to the TiO_2 where reduction of MV^{2+} takes place. The holes remain at the CdS and apparently oxidize CH_3OH. TiO_2 also quenched the CdS fluorescence demonstrating unequivocally that effective charge separation took place [406]. Note, however, that the charge separation achieved in the "sandwich colloids" does not prevent electron–hole recombination at the interface between the two parts of the sandwich. The high quantum yield has been observed at moderately high scavenger concentrations at the semiconductors surfaces [406], yet $\Phi_{HCHO} = 1$ is only half of the maximum yield under the given experimental conditions.

Similarly, bilayered materials and surface-modified TiO_2 are not able to prevent electron–hole recombination at the boundary between the two components, although the recombination rate may be lower. This is particularly effective in photocatalytic systems for decontamination because such systems require photocatalytic reactions with relatively very low substrate concentrations.

The basic mechanisms of all systems remain similar. The electrons and holes are shared between the TiO_2 and the surface modifier or binary partner if the energy levels are such that the electron is more stable at one component and the hole at the other. The principles governing the reactions with substrates remain the same, but the energetic is inevitably changed, and this may dictate different reaction paths and reaction rates. This may have particular importance when mixed substrates are involved such as in decontamination systems.

Acknowledgment This work was supported by the Israel-USA BSF.

References

1. Bakardjieva S, Stengl V, Szatmary L, Subrt J, Lukac J, Murafa N, Niznansky D, Cizek K, Jirkovsky J, Petrova N (2006) Transformation of brookite-type TiO_2 nanocrystals to rutile: correlation between microstructure and photoactivity. J Mater Chem 16:1709–1716
2. Liu L, Zhao H, Andino JM, Li Y (2012) Photocatalytic CO_2 reduction with H_2O on TiO_2 nanocrystals: comparison of anatase, rutile, and brookite polymorphs and exploration of surface chemistry. ACS Catal 2:1817–1828
3. Rothenberger G, Moser J, Gratzel M, Serpone N, Sharma DK (1985) Charge carrier trapping and recombination dynamics in small semiconductor particles. J Am Chem Soc 107:8054–8059
4. Serpone N, Lawless D, Khairutdinov R, Pelizzetti E (1995) Subnanosecond relaxation dynamics in TiO_2 colloidal sols (particle sizes Rp = 1.0–13.4 nm). Relevance to heterogeneous photocatalysis. J Phys Chem 99:16655–16661
5. Kavan L, Graetzel M, Gilbert SE, Klemenz C, Scheel HJ (1996) Electrochemical and photoelectrochemical investigation of single-crystal anatase. J Am Chem Soc 118:6716–6723
6. Asahi R, Taga Y, Mannstadt W, Freeman AJ (2000) Electronic and optical properties of anatase TiO_2. Phys Rev B 61:7459–7465
7. Safrany A, Gao RM, Rabani J (2000) Optical properties and reactions of radiation induced TiO_2 electrons in aqueous colloid solutions. J Phys Chem B 104:5848–5853
8. Bahnemann D, Henglein A, Lilie J, Spanhel L (1984) Flash photolysis observation of the absorption spectra of trapped positive holes and electrons in colloidal titanium dioxide. J Phys Chem 88:709–711
9. Kolle U, Moser J, Graetzel M (1985) Dynamics of interfacial charge-transfer reactions in semiconductor dispersions. Reduction of cobaltocenium dicarboxylate in colloidal TiO_2. Inorg Chem 24:2253–2258
10. Kamat PV, Bedja I, Hotchandani S (1994) Photoinduced charge transfer between carbon and semiconductor clusters. One-electron reduction of Ca in colloidal TiO_2 semiconductor suspensions. J Phys Chem B 98:9137–9142
11. O'Regan B, Graetzel M, Fitzmaurice D (1991) Optical electrochemistry. 2. Real-time spectroscopy of conduction band electrons in a metal oxide semiconductor electrode. J Phys Chem 95:10525–10528
12. O'Regan B, Graetzel M, Fitzmaurice D (1991) Optical electrochemistry I. Steady-state spectroscopy of conduction-band electrons in a metal oxide semiconductor electrode. Chem Phys Lett 183:89–93

13. Tamaki Y, Hara K, Katoh R, Tachiya M, Furube A (2009) Femtosecond visible-to-IR spectroscopy of TiO_2 nanocrystalline films: elucidation of the electron mobility before deep trapping. J Phys Chem C 113:11741–11746
14. Yoshihara T, Katoh R, Furube A, Tamaki Y, Murai M, Hara K, Murata S, Arakawa H, Tachiya M (2004) Identification of reactive species in photoexcited nanocrystalline TiO_2 films by wide-wavelength-range (400–2500 nm) transient absorption spectroscopy. J Phys Chem B 108:3817–3823
15. Arbour C, Sharma DK, Langford CH (1990) Picosecond flash spectroscopy of titania colloids with adsorbed dyes. J Phys Chem 94:331–335
16. Skinner DE, Colombo DP Jr, Cavaleri JJ, Bowman RM (1995) Femtosecond investigation of electron trapping in semiconductor nanoclusters. J Phys Chem 99:7853–7856
17. Colombo J, Philip D, Bowman RM (1996) Does interfacial charge transfer compete with charge carrier recombination? A femtosecond diffuse reflectance investigation of TiO_2 nanoparticles. J Phys Chem 100:18445–18449
18. Tamaki Y, Furube A, Murai M, Hara K, Katoh R, Tachiya M (2007) Dynamics of efficient electron–hole separation in TiO_2 nanoparticles revealed by femtosecond transient absorption spectroscopy under the weak-excitation condition. Phys Chem Chem Phys 9:1453–1460
19. Howe RF, Gratzel M (1985) EPR observation of trapped electrons in colloidal TiO_2. J Phys Chem 89:4495–4499
20. Rabani J, Yamashita K, Ushida K, Stark J, Kira A (1998) Fundamental reactions in illuminated titanium dioxide nanocrystallite layers studied by pulsed laser. J Phys Chem B 102:1689–1695
21. Colombo DP Jr, Roussel KA, Saeh J, Skinner DE, Cavaleri JJ, Bowman RM (1995) Femtosecond study of the intensity dependence of electron-hole dynamics in TiO_2 nanoclusters. Chem Phys Lett 232:207–214
22. Bahnemann D, Henglein A, Spanhel L (1984) Detection of the intermediates of colloidal titania-catalyzed photoreactions. Faraday Discuss Chem Soc 78:151–163
23. Lawless D, Serpone N, Meisel D (1991) Role of OH radicals and trapped holes in photocatalysis. A pulse radlolysls study. J Phys Chem 95:5166–5170
24. Bahnemann DW, Hilgendorff M, Memming R (1997) Charge carrier dynamics at TiO_2 particles: reactivity of free and trapped holes. J Phys Chem B 101:4265–4275
25. Lepore GP, Langford CH, Vichova J, Vlcek A (1993) Photochemistry and picosecond absorption-spectra of aqueous suspensions of a polycrystalline titanium-dioxide optically transparent in the visible spectrum. J Photochem Photobiol A Chem 75:67–75
26. Jaeger CD, Bard AJ (1979) Spin trapping and electron spin resonance detection of radical intermediates in the photodecomposition of water at titanium dioxide particulate systems. J Phys Chem 83:3146–3152
27. Anpo M, Shima T, Kubokawa Y (1985) ESR and photoluminescence evidence for the photocatalytic formation of hydroxyl radicals on small titanium dioxide (TiO_2) particles. Chem Lett 168:1799–1802
28. Turchi CS, Ollis DF (1990) Photocatalytic degradation of organic-water contaminants – mechanisms involving hydroxyl radical attack. J Catal 122:178–192
29. Fujihira M, Satoh Y, Osa T (1981) Heterogeneous photocatalytic oxidation of aromatic compounds on titanium dioxide. Nature (London) 293:206–208
30. Ollis DF, Hsiao CY, Budiman L, Lee CL (1984) Heterogeneous photoassisted catalysis: conversions of perchloroethylene, dichloroethane, chloroacetic acids, and chlorobenzenes. J Catal 88:89–96
31. Al-Ekabi H, Serpone N, Pelizzetti E, Minero C, Fox MA, Draper RB (1989) Kinetic studies in heterogeneous photocatalysis. 2. Titania-mediated degradation of 4-chlorophenol alone and in a three-component mixture of 4-chlorophenol, 2,4-dichlorophenol, and 2,4,5-trichlorophenol in air-equilibrated aqueous media. Langmuir 5:250–255

32. Okamoto K, Yamamoto Y, Tanaka H, Tanaka M, Itaya A (1985) Heterogeneous photocatalytic decomposition of phenol over anatase powder. Bull Chem Soc Jpn 58:2015–2022
33. Minero C, Aliberti C, Pelizzetti E, Terzian R, Serpone N (1991) Kinetic studies in heterogeneous photocatalysis. 6. AM1 simulated sunlight photodegradation over titania in aqueous media: a first case of fluorinated aromatics and identification of intermediates. Langmuir 7:928–936
34. Matthews RW (1984) Hydroxylation reactions induced by near-ultraviolet photolysis of aqueous titanium dioxide suspensions. J Chem Soc Faraday Trans 1(80):457–471
35. Wei T-Y, Wan C (1992) Kinetics of photocatalytic oxidation of phenol on titanium oxide (TiO_2) surface. J Photochem Photobiol A 69:241–249
36. Mills A, Morris S, Davies R (1993) Photomineralization of 4-chlorophenol sensitized by titanium dioxide: a study of the intermediates. J Photochem Photobiol A Chem 70:183–191
37. Richard C (1993) Regioselectivity of oxidation by positive holes (H^+) in photocatalytic aqueous transformations. J Photochem Photobiol A Chem 72:179–182
38. Cunningham J, Srijaranai S (1988) Isotope-effect evidence for hydroxyl radical involvement in alcohol photooxidation sensitized by titanium dioxide in aqueous suspension. J Photochem Photobiol A 43:329–335
39. Goldstein S, Czapsky G, Rabani J (1994) Oxidation of phenol by radiolytically generated $^{\bullet}OH$ and chemically generated $SO_4^{\bullet -}$. A distinction between $^{\bullet}OH$ transfer and hole oxidation in the photolysis of TiO_2 colloid solution. J Phys Chem 98:6586–6591
40. Goldstein S, Meyerstein D, Czapski G (1993) The Fenton reagents. Free Radic Biol Med 15:435–445
41. Meyerstein D, Goldstein S (1999) Comments on the mechanism of "Fenton-like" process. Acc Chem Res 32:547–550
42. Rajh T, Saponjic ZV, Micic OI (1992) Reactions of hydrous titanium oxide colloids with strong oxidizing agents. Langmuir 8:1265–1270
43. Micic OI, Zhang Y, Cromack KR, Trifunac AD, Thurnauer MC (1993) Trapped holes on titania colloids studied by electron paramagnetic resonance. J Phys Chem 97:7277–7283
44. Salvador P (2007) On the nature of photogenerated radical species active in the oxidative degradation of dissolved pollutants with TiO_2 aqueous suspensions: a revision in the light of the electronic structure of adsorbed water. J Phys Chem C 111:17038–17043
45. Henderson MA (2002) The interaction of water with solid surfaces: fundamental aspects revisited. Surf Sci Rep 46:1–308
46. Kurtz RL, Stockbauer R, Madey TE, Roman E, De SJL (1989) Synchrotron radiation studies of water adsorption on titania(110). Surf Sci 218:178–200
47. Krischok S, Hofft O, Gunster J, Stultz J, Goodman DW, Kempter V (2001) H_2O interaction with bare and Li-precovered TiO_2. Studies with electron spectroscopies (MIES and UPS(HeI and II)). Surf Sci 495:8–18
48. Winter B, Weber R, Widdra W, Dittmar M, Faubel M, Hertel IV (2004) Full valence band photoemission from liquid water using EUV synchrotron radiation. J Phys Chem A 108:2625–2632
49. Nakamura R, Nakato Y (2004) Primary intermediates of oxygen photoevolution reaction on TiO_2 (rutile) particles, revealed by in situ FTIR absorption and photoluminescence measurements. J Am Chem Soc 126:1290–1298
50. Brookes IM, Muryn CA, Thornton G (2001) Imaging water dissociation on TiO_2(110). Phys Rev Lett 87:266103-1–266103-4
51. Ji Y, Wang B, Luo Y (2012) Location of trapped hole on rutile-TiO_2(110) surface and its role in water oxidation. J Phys Chem C 116:7863–7866
52. Siripala W, Tomkiewicz M (1982) Interactions between photoinduced and dark charge transfer across n-titania-aqueous electrolyte interface. J Electrochem Soc 129:1240–1245
53. Fan FR, Bard AJ (1990) Scanning tunneling microscopy and tunneling spectroscopy of the n-TiO_2 (001) surface. J Phys Chem 94:3761–3766

54. Redmond G, Fitzmaurice D, Graetzel M (1993) Effect of surface chelation on the energy of an intraband surface state of a nanocrystalline titania film. J Phys Chem 97:6951–6954
55. Boschloo GK, Goossens A (1996) Electron trapping in porphyrin-sensitized porous nanocrystalline TiO_2 electrodes. J Phys Chem 100:19489–19494
56. Boschloo G, Fitzmaurice D (1999) Spectroelectrochemical investigation of surface states in nanostructured TiO_2 electrodes. J Phys Chem B 103:2228–2231
57. Dung D, Ramsden J, Graetzel M (1982) Dynamics of interfacial electron-transfer processes in colloidal semiconductor systems. J Am Chem Soc 104:2977–2985
58. Dimitrijevic NM, Savic D, Micic OI, Nozik AJ (1984) Interfacial electron-transfer equilibria and flatband potentials of a-ferric oxide and titanium dioxide colloids studied by pulse radiolysis. J Phys Chem 88:4278–4283
59. Fitzmaurice DJ, Frei H (1991) Transient near-infrared spectroscopy of visible light sensitized oxidation of iodide at colloidal titania. Langmuir 7:1129–1137
60. Fitzmaurice DJ, Eschle M, Frei H, Moser J (1993) Time-resolved rise of iodine molecule (1-) upon oxidation of iodide at aqueous titania colloid. J Phys Chem 97:3806–3812
61. Liu L-M, Crawford P, Hu P (2009) The interaction between adsorbed OH and O_2 on TiO_2 surfaces. Prog Surf Sci 84:155–176
62. Henderson MA (1996) Structural sensitivity in the dissociation of water on TiO_2 single-crystal surfaces. Langmuir 12:5093–5098
63. Herrmann JM, Disdier J, Pichat P (1981) Oxygen species ionosorbed on powder photocatalyst oxides from room-temperature photoconductivity as a function of oxygen pressure. J Chem Soc Faraday Trans 1(77):2815–2826
64. Zhang L, Ji H, Lei Y, Xiao W (2011) Oxygen adsorption on anatase surfaces and edges. Appl Surf Sci 257:8402–8408
65. Herrmann J-M (2010) Photocatalysis fundamentals revisited to avoid several misconceptions. Appl Catal B 99:461–468
66. Serpone N (1995) Brief introductory remarks on heterogeneous photocatalysis. Sol Energy Mater Sol Cells 38:369–379
67. Ollis DF (2005) Kinetics of liquid phase photocatalyzed reactions: an illuminating approach. J Phys Chem B 109:2439–2444
68. Monllor-Satoca D, Gomez R, Gonzalez-Hidalgo M, Salvador P (2007) The "Direct-Indirect" model: an alternative kinetic approach in heterogeneous photocatalysis based on the degree of interaction of dissolved pollutant species with the semiconductor surface. Catal Today 129:247–255
69. Montoya JF, Velasquez JA, Salvador P (2009) The direct-indirect kinetic model in photocatalysis: a reanalysis of phenol and formic acid degradation rate dependence on photon flow and concentration in TiO_2 aqueous dispersions. Appl Catal B 88:50–58
70. Montoya JF, Peral J, Salvador P (2011) Commentary on the article: "A new kinetic model for heterogeneous photocatalysis with titanium dioxide: case of non-specific adsorption considering back reaction, by S. Valencia, F. Catano, L. Rios, G. Restrepo and J. Marin, published in Applied Catalysis B: Environmental, 104 (2011) 300–304". Appl Catal B 111–112:649–650
71. Montoya JF, Salvador P (2010) The influence of surface fluorination in the photocatalytic behavior of TiO_2 aqueous dispersions: an analysis in the light of the direct-indirect kinetic model. Appl Catal B 94:97–107
72. Jia C, Qin Q, Wang Y, Zhang C (2012) Photocatalytic degradation of bisphenol A in aqueous suspensions of titanium dioxide. Adv Mater Res 433–440:172–177
73. Kubacka A, Ferrer M, Fernandez-Garcia M (2012) Kinetics of photocatalytic disinfection in TiO_2-containing polymer thin films: UV and visible light performances. Appl Catal B 121–122:230–238
74. Moreira J, Serrano B, Ortiz A, de Lasa H (2012) A unified kinetic model for phenol photocatalytic degradation over TiO_2 photocatalysts. Chem Eng Sci 78:186–203

75. Ishibashi KI, Fujishima A, Watanabe T, Hashimoto K (2000) Generation and deactivation processes of superoxide formed on TiO_2 film illuminated by very weak UV light in air or water. J Phys Chem B 104:4934–4938
76. Rabani J, Klugroth D, Henglein A (1974) Pulse radiolytic investigations of $OHCH_2O_2$ radicals. J Phys Chem 78:2089–2093
77. von-Sonntag C, Dowideit P, Fang X, Mertens R, Pan X, Schuchmann MN, Shuchmannn H-P (1997) The fate of peroxyl radicals in aqueous solution. Water Sci Technol 35:9–15
78. Kormann C, Bahnemann DW, Hoffmann MR (1991) Photolysis of chloroform and other organic-molecules in aqueous TiO_2 suspensions. Environ Sci Technol 25:494–500
79. Curco D, Malato S, Blanco J, Gimenez J, Marco P (1996) Photocatalytic degradation of phenol: comparison between pilot-plant-scale and laboratory results. Sol Energy 56:387–400
80. Gimenez J, Curco D, Marco P (1997) Reactor modelling in the photocatalytic oxidation of wastewater. Water Sci Technol 35:207–213
81. Brandi RJ, Alfano OM, Cassano AE (2000) Evaluation of radiation absorption in slurry photocatalytic reactors. 2. Experimental verification of the proposed method. Environ Sci Technol 34:2631–2639
82. Christensen PA, Dilks A, Egerton TA, Temperley J (2000) Infrared spectroscopic evaluation of the photodegradation of paint – Part II: the effect of UV intensity & wavelength on the degradation of acrylic films pigmented with titanium dioxide. J Mater Sci 35:5353–5358
83. Fisher AC, Peter LM, Ponomarev EA, Walker AB, Wijayantha KGU (2000) Intensity dependence of the back reaction and transport of electrons in dye-sensitized nanacrystalline TiO_2 solar cells. J Phys Chem B 104:949–958
84. Kuo WS, Lin YT (2000) Photocatalytic oxidation of xenobiotics in water with immobilized TiO_2 on agitator. J Environ Sci Health B 35:61–75
85. Golego N, Studenikin SA, Cocivera M (2000) Sensor photoresponse of thin-film oxides of zinc and titanium to oxygen gas. J Electrochem Soc 147:1592–1594
86. Wang C, Bahnemann DW, Dohrmann JK (2001) Determination of photonic efficiency and quantum yield of formaldehyde formation in the presence of various TiO_2 photocatalysts. Water Sci Technol 44:279–286
87. Brandi RJ, Rintoul G, Alfano OM, Cassano AE (2002) Photocatalytic reactors – reaction kinetics in a flat plate solar simulator. Catal Today 76:161–175
88. Wang CY, Rabani J, Bahnemann DW, Dohrmann JK (2002) Photonic efficiency and quantum yield of formaldehyde formation from methanol in the presence of various TiO_2 photocatalysts. J Photochem Photobiol A Chem 148:169–176
89. Gao RM, Stark J, Bahnemann DW, Rabani J (2002) Quantum yields of hydroxyl radicals in illuminated TiO_2 nanocrystallite layers. J Photochem Photobiol A Chem 148:387–391
90. Vulliet E, Chovelon J-M, Guillard C, Herrmann J-M (2003) Factors influencing the photocatalytic degradation of sulfonylurea herbicides by TiO_2 aqueous suspension. J Photochem Photobiol A Chem 159:71–79
91. Du Y, Rabani J (2003) The measure of TiO_2 photocatalytic efficiency and the comparison of different photocatalytic titania. J Phys Chem B 107:11970–11978
92. Yoshikawa N, Kimura T, Kawase Y (2003) Oxidative degradation of nonionic surfactants with TiO_2 photocatalyst in a bubble column reactor. Can J Chem Eng 81:719–724
93. Nakashima T, Ohko Y, Kubota Y, Fujishima A (2003) Photocatalytic decomposition of estrogens in aquatic environment by reciprocating immersion of TiO_2-modified polytetrafluoroethylene mesh sheets. J Photochem Photobiol A Chem 160:115–120
94. Villarreal TL, Gomez R, Gonzalez M, Salvador P (2004) A kinetic model for distinguishing between direct and indirect interfacial hole transfer in the heterogeneous photooxidation of dissolved organics on TiO_2 nanoparticle suspensions. J Phys Chem B 108:20278–20290
95. Kimura T, Yoshikawa N, Matsumura N, Kawase Y (2004) Photocatalytic degradation of nonionic surfactants with immobilized TiO_2 in an airlift reactor. J Environ Sci Health A Tox Hazard Subst Environ Eng 39:2867–2881

96. Lim TH, Kim SD (2004) Trichloroethylene degradation by photocatalysis in annular flow and annulus fluidized bed photoreactors. Chemosphere 54:305–312
97. Heyd DV, Au B (2005) Fluorescence development during 514 nm irradiation of catechol adsorbed on nanocrystalline titanium dioxide. J Photochem Photobiol A Chem 174:62–70
98. Coleman HM, Abdullah MI, Eggins BR, Palmer FL (2005) Photocatalytic degradation of 17 beta-oestradiol, oestriol and 17 alpha-ethynyloestradiol in water monitored using fluorescence spectroscopy. Appl Catal B Environ 55:23–30
99. Lu Y, Spitler MT, Parkinson BA (2006) Photochronocoulometric measurement of the coverage of surface-bound dyes on titanium dioxide crystal surfaces. J Phys Chem B 110:25273–25278
100. Jin CQ, Christensen PA, Egerton TA, White JR (2006) Rapid measurement of photocatalytic oxidation of poly(vinyl chloride) by in situ FIR spectrometry of evolved CO_2. Mater Sci Technol 22:908–914
101. Jin CQ, Christensen PA, Egerton TA, Lawson EJ, White JR (2006) Rapid measurement of polymer photo-degradation by FTIR spectrometry of evolved carbon dioxide. Polym Degrad Stab 91:1086–1096
102. Sakanoue M, Kinoshita Y, Otsuka Y, Imai H (2007) Photocatalytic activities of rutile and anatase nanoparticles selectively prepared from an aqueous solution. J Ceramic Soc (Japan) 115:821–825
103. Li YJ, Sun SG, Ma MY, Ouyang YZ, Yan WB (2008) Kinetic study and model of the photocatalytic degradation of rhodamine B (RhB) by a TiO_2-coated activated carbon catalyst: effects of initial RhB content, light intensity and TiO_2 content in the catalyst. Chem Eng J 142:147–155
104. Goldstein S, Behar D, Rabani J (2008) Mechanism of visible light photocatalytic oxidation of methanol in aerated aqueous suspensions of carbon-doped TiO_2. J Phys Chem C 112:15134–15139
105. Satuf ML, Brandi RJ, Cassano AE, Alfano OM (2008) Photocatalytic degradation of 4-chlorophenol: a kinetic study. Appl Catal B Environ 82:37–49
106. Marugan J, van Grieken R, Cassano AE, Alfano OM (2009) Scaling-up of slurry reactors for the photocatalytic oxidation of cyanide with TiO_2 and silica-supported TiO_2 suspensions. Catal Today 144:87–93
107. Turchi CS, Ollis DF (1989) Mixed reactant photocatalysis: intermediates and mutual rate inhibition. J Catal 119:483–496
108. Luo Y, Ollis DF (1996) Heterogeneous photocatalytic oxidation of trichloroethylene and toluene mixtures in air: kinetic promotion and inhibition, time-dependent catalyst activity. J Catal 163:1–11
109. Upadhya S, Ollis DF (1998) A simple kinetic model for the simultaneous concentration and intensity dependencies of TCE photocatalyzed destruction. J Adv Oxid Technol 3:199–202
110. Chen J, Ollis DF, Rulkens WH, Bruning H (1999) Kinetic processes of photocatalytic mineralization of alcohols on metalized titanium dioxide. Water Res 33:1173–1180
111. Ollis DF (2002) Photocatalytic powder layer reactor: a uniformly mixed gas phase occurring in a catalytic fixed-bed flow reactor. Ind Eng Chem Res 41:6409–6412
112. Lewandowski M, Ollis DF (2003) Extension of a two-site transient kinetic model of TiO_2 deactivation during photocatalytic oxidation of aromatics: concentration variations and catalyst regeneration studies. Appl Catal B 45:223–238
113. Mills A, Wang J, Ollis DF (2006) Dependence of the kinetics of liquid-phase photocatalyzed reactions on oxygen concentration and light intensity. J Catal 243:1–6
114. Chin P, Roberts GW, Ollis DF (2007) Kinetic modeling of photocatalyzed soot oxidation on titanium dioxide thin films. Ind Eng Chem Res 46:7598–7604
115. Chin P, Ollis DF (2008) Design approaches for a cycling adsorbent/photocatalyst system for indoor air purification: formaldehyde example. J Air Waste Manage Assoc 58:494–501
116. Ollis D (2010) Kinetics of photocatalyzed film removal on self-cleaning surfaces: simple configurations and useful models. Appl Catal B 99:478–484

117. Khairutdinov RF, Burshtein KY, Serpone N (1996) Photochemical reactions on the surface of a circular disk: a theoretical approach to kinetics in restricted two-dimensional space. J Photochem Photobiol A Chem 98:1–14
118. Khairutdinov RF, Burshtein KY, Serpone N (1997) Theoretical study of the quenching of excited molecules on the surface of disk-shaped nanoparticles. Khim Fiz 16:20–30
119. Serpone N (2007) Some remarks on so-called heterogeneous photocatalysis and on the mechanical application of the Langmuir-Hinshelwood kinetic model. J Adv Oxid Technol 10:111–115
120. Ferguson MA, Hoffmann MR, Hering JG (2005) TiO_2-photocatalyzed As(III) oxidation in aqueous suspensions: reaction kinetics and effects of adsorption. Environ Sci Technol 39:1880–1886
121. Mladenova D, Dushkin C, Li PG (2008) Photocatalysis with TiO_2 and SnO_2/TiO_2 films examined by a kinetic model. J Adv Oxid Technol 11:477–485
122. Huang HH, Tseng DH, Juang LC (2008) Titanium dioxide mediated photocatalytic degradation of monochlorobenzene in aqueous phase. Chemosphere 71:398–405
123. Sun L, Hoy AR, Bolton JR (1996) Generation efficiency of the hydroxyl radical adduct of the DMPO spin trap in homogeneous and heterogeneous media. J Adv Oxid Technol 1:44–52
124. Chen J, Ollis DF, Rulkens WH, Bruning H (1998) Photocatalyzed oxidation of alcohols and organochlorides in the presence of native TiO_2 and metalized TiO_2 suspensions. Part (II): photocatalytic mechanisms. Water Res 33:669–676
125. Ferry JL, Glaze WH (1998) Photocatalytic reduction of nitro organics over illuminated titanium dioxide: role of the TiO_2 surface. Langmuir 14:3551–3555
126. Stark J, Rabani J (1999) Photocatalytic dechlorination of aqueous carbon tetrachloride solutions in TiO_2 layer systems: a chain reaction mechanism. J Phys Chem B 103:8524–8531
127. Tada H, Teranishi K, Ito S (1999) Additive effect of sacrificial electron donors on Ag/TiO_2 photocatalytic reduction of bis(2-dipyridyl)disulfide to 2-mercaptopyridine in aqueous media. Langmuir 15:7084–7087
128. Kawahara T, Konishi Y, Tada H, Tohge N, Ito S (2001) Patterned TiO_2/SnO_2 bilayer type photocatalyst. 2. Efficient dehydrogenation of methanol. Langmuir 17:7442–7445
129. Wang C-Y, Pagel R, Bahnemann DW, Dohrmann JK (2004) Quantum yield of formaldehyde formation in the presence of colloidal TiO_2-based photocatalysts: effect of intermittent illumination, platinization, and deoxygenation. J Phys Chem B 108:14082–14092
130. Wang CY, Pagel R, Dohrmann JK, Bahnemann DW (2006) Antenna mechanism and deaggregation concept: novel mechanistic principles for photocatalysis. C R Chim 9:761–773
131. Kitano M, Matsuoka M, Hosoda T, Ueshima M, Anpo M (2008) Effect of HF treatment on the activity of TiO_2 thin films for photocatalytic water splitting. Res Chem Intermediat 34:577–585
132. Chiarello GL, Ferri D, Selli E (2011) Effect of the CH_3OH/H_2O ratio on the mechanism of the gas-phase photocatalytic reforming of methanol on noble metal-modified TiO_2. J Catal 280:168–177
133. Kandiel TA, Dillert R, Robben L, Bahnemann DW (2011) Photonic efficiency and mechanism of photocatalytic molecular hydrogen production over platinized titanium dioxide from aqueous methanol solutions. Catal Today 161:196–201
134. Mohamed HH, Mendive CB, Dillert R, Bahnemann DW (2011) Kinetic and mechanistic investigations of multielectron transfer reactions induced by stored electrons in TiO_2 nanoparticles: a stopped flow study. J Phys Chem A 115:2139–2147
135. Sun L, Bolton JR (1996) Determination of the quantum yield for the photochemical generation of hydroxyl radicals in TiO_2 suspensions. J Phys Chem 100:4127–4134
136. Gao RM, Safrany A, Rabani J (2003) Reactions of TiO_2 excess electron in nanocrystallite aqueous solutions studied in pulse and gamma-radiolytic systems. Radiat Phys Chem 67:25–39
137. Du Y, Goldstein S, Rabani J (2011) The catalytic effects of copper ions on photooxidation in TiO_2 suspensions: the role of superoxide radicals. J Photochem Photobiol A Chem 225:1–7

138. Grabner G, Li GZ, Quint RM, Quint R, Getoff N (1991) Pulsed laser-induced oxidation of phenol in acid aqueous TiO_2 sols. J Chem Soc Faraday Trans 87:1097–1101
139. Fitzmaurice D, Frei H, Rabani J (1995) Time-resolved optical study on the charge-carrier dynamics in a TiO_2/AgI sandwich colloid. J Phys Chem 99:9176–9181
140. Fujihara K, Izumi S, Ohno T, Matsumura M (2000) Time-resolved photoluminescence of particulate TiO_2 photocatalysts suspended in aqueous solutions. J Photochem Photobiol A Chem 132:99–104
141. Tachikawa T, Tojo S, Fujitsuka M, Majima T (2003) One-electron oxidation of aromatic sulfides adsorbed on the surface of TiO_2 particles studied by time-resolved diffuse reflectance spectroscopy. Chem Phys Lett 382:618–625
142. Mytych P, Karocki A, Stasicka Z (2003) Mechanism of photochemical reduction of chromium(VI) by alcohols and its environmental aspects. J Photochem Photobiol A Chem 160:163–170
143. Tachikawa T, Tojo S, Fujitsuka M, Majima T (2004) Direct observation of the one-electron reduction of methyl viologen mediated by the CO_2^- radical anion during TiO_2 photocatalytic reactions. Langmuir 20:9441–9444
144. Tachikawa T, Tojo S, Fujitsuka M, Majima T (2004) Formation of the dimer radical cation of aromatic sulfide on the TiO_2 surface during photocatalytic reactions. Langmuir 20:4327–4329
145. Shkrob IA, Sauer MC, Gosztola D (2004) Efficient, rapid photooxidation of chemisorbed polyhydroxyl alcohols and carbohydrates by TiO_2 nanoparticles in an aqueous solution. J Phys Chem B 108:12512–12517
146. Chang BK, Jang BW, Dai S, Overbury SH (2005) Transient studies of the mechanisms of CO oxidation over Au/TiO_2 using time-resolved FTIR spectroscopy and product analysis. J Catal 236:392–400
147. Gundlach L, Ernstorfer R, Willig F (2006) Escape dynamics of photoexcited electrons at catechol: TiO_2(110). Phys Rev B 74:035324 (035310)
148. Min L, Wu X-Z, Tetsuya S, Inoue H (2007) Time-resolved chemiluminescence study of the TiO_2 photocatalytic reaction and its induced active oxygen species. Luminescence 22:105–112
149. Graetzel M, Frank AJ (1982) Interfacial electron-transfer reactions in colloidal semiconductor dispersions. Kinetic analysis. J Phys Chem 86:2964–2967
150. Rossetti R, Beck SM, Brus LE (1984) Direct observation of charge-transfer reactions across semiconductor: aqueous solution interfaces using transient Raman spectroscopy. J Am Chem Soc 106:980–984
151. Nosaka Y, Fox MA (1988) Kinetics for electron transfer from laser-pulse irradiated colloidal semiconductors to adsorbed methylviologen: dependence of the quantum yield on incident pulse width. J Phys Chem 92:1893–1897
152. Shkrob IA, Sauer MC Jr (2004) Hole scavenging and photostimulated recombination of electron-hole pairs in aqueous TiO_2 nanoparticles. J Phys Chem B 108:12497–12511
153. Green ANM, Chandler RE, Haque SA, Nelson J, Durrant JR (2005) Transient absorption studies and numerical modeling of iodine photoreduction by nanocrystalline TiO_2 films. J Phys Chem B 109:142–150
154. Tachikawa T, Tojo S, Fujitsuka M, Majima T (2006) One-electron oxidation pathways during beta-cyclodextrin-modified TiO_2 photocatalytic reactions. Chemistry 12:7585–7594
155. Chen T, Feng Z, Wu G, Shi J, Ma G, Ying P, Li C (2007) Mechanistic studies of photocatalytic reaction of methanol for hydrogen production on Pt/TiO_2 by in situ Fourier transform IR and time-resolved IR spectroscopy. J Phys Chem C 111:8005–8014
156. Rowley JG, Meyer GJ (2011) Di- and tri-iodide reactivity at illuminated titanium dioxide interfaces. J Phys Chem C 115:6156–6161
157. Roy SC, Varghese OK, Paulose M, Grimes CA (2010) Toward solar fuels: photocatalytic conversion of carbon dioxide to hydrocarbons. ACS Nano 4:1259–1278

158. Dimitrijevic NM, Shkrob IA, Gosztola DJ, Rajh T (2012) Dynamics of interfacial charge transfer to formic acid, formaldehyde, and methanol on surface of TiO_2 nanoparticles and role in methane production. J Phys Chem C 116:878–885
159. Inoue T, Fujishima A, Konishi S, Honda K (1979) Photoelectrocatalytic reduction of carbon dioxide in aqueous suspensions of semiconductor powders. Nature (London) 277:637–638
160. Abe R, Sayama K, Arakawa H (2003) Significant effect of iodide addition on water splitting into H_2 and O_2 over Pt-loaded TiO_2 photocatalyst: suppression of backward reaction. Chem Phys Lett 371:360–364
161. Gao RM, Safrany A, Rabani J (2002) Fundamental reactions in TiO_2 nanocrystallite aqueous solutions studied by pulse radiolysis. Radiat Phys Chem 65:599–609
162. Kasarevic-Popovic Z, Behar D, Rabani J (2004) Role of excess electrons in TiO_2 nanoparticles coated with Pt in reduction reactions studied in radiolysis of aqueous solutions. J Phys Chem B 108:20291–20295
163. Behar D, Rabani J (2006) Kinetics of hydrogen production upon reduction of aqueous TiO_2 nanoparticles catalyzed by Pd^0, Pt^0, or Au^0 coatings and an unusual hydrogen abstraction; steady state and pulse radiolysis study. J Phys Chem B 110:8750–8755
164. Mohamed HH, Dillert R, Bahnemann DW (2011) Reaction dynamics of the transfer of stored electrons on TiO_2 nanoparticles: a stopped flow study. J Photochem Photobiol A Chem 217:271–274
165. Mohamed HH, Dillert R, Bahnemann DW (2011) Growth and reactivity of silver nanoparticles on the surface of TiO_2: a stopped-flow study. J Phys Chem C 115:12163–12172
166. Mohamed HH, Dillert R, Bahnemann DW (2012) Kinetic and mechanistic investigations of the light induced formation of gold nanoparticles on the surface of TiO_2. Chemistry 18:4314–4321
167. Lide DR (ed) (1992–1993) Handbook of chemistry and physics, 73rd edn. CRC Press, London
168. Asahi R, Morikawa T, Ohwaki T, Aoki K, Taga Y (2001) Visible-light photocatalysis in nitrogen-doped titanium oxides. Science 293:269–271
169. Zhang H, Chen G, Bahnemann DW (2009) Photoelectrocatalytic materials for environmental applications. J Mater Chem 19:5089–5121
170. Kuznetsov VN, Serpone N (2009) On the origin of the spectral bands in the visible absorption spectra of visible-light-active TiO_2 specimens analysis and assignments. J Phys Chem C 113:15110–15123
171. Yang Y, Zhong H, Tian C (2011) Photocatalytic mechanisms of modified titania under visible light. Res Chem Intermediat 37:91–102
172. Grabowska E, Reszczynska J, Zaleska A (2012) Mechanism of phenol photodegradation in the presence of pure and modified-TiO_2: a review. Water Res 46:5453–5471
173. Pelaez M, Nolan NT, Pillai SC, Seery MK, Falaras P, Kontos AG, Dunlop PSM, Hamilton JWJ, Byrne JA, O'Shea K, Entezari MH, Dionysiou DD (2012) A review on the visible light active titanium dioxide photocatalysts for environmental applications. Appl Catal B Environ 125:331–349
174. Belver C, Bellod R, Stewart SJ, Requejo FG, Fernandez-Garcia M (2006) Nitrogen-containing TiO_2 photocatalysts. Part 2. Photocatalytic behavior under sunlight excitation. Appl Catal B 65:309–314
175. Emeline AV, Kuzmin GN, Serpone N (2008) Wavelength-dependent photostimulated adsorption of molecular O_2 and H_2 on second generation titania photocatalysts: the case of the visible-light-active N-doped TiO_2 system. Chem Phys Lett 454:279–283
176. Fu H, Zhang L, Zhang S, Zhu Y, Zhao J (2006) Electron spin resonance spin-trapping detection of radical intermediates in N-doped TiO_2-assisted photodegradation of 4-chlorophenol. J Phys Chem B 110:3061–3065
177. Asahi R, Morikawa T (2007) Nitrogen complex species and its chemical nature in TiO_2 for visible-light sensitized photocatalysis. Chem Phys 339:57–63

178. Emeline AV, Sheremetyeva NV, Khomchenko NV, Ryabchuk VK, Serpone N (2007) Photoinduced formation of defects and nitrogen stabilization of color centers in N-doped titanium dioxide. J Phys Chem C 111:11456–11462
179. Fang X, Zhang Z, Chen Q, Ji H, Gao X (2007) Dependence of nitrogen doping on TiO_2 precursor annealed under NH_3 flow. J Solid State Chem 180:1325–1332
180. Chen D, Yang D, Geng J, Zhu J, Jiang Z (2008) Improving visible-light photocatalytic activity of N-doped TiO_2 nanoparticles via sensitization by Zn porphyrin. Appl Surf Sci 255:2879–2884
181. Li Q, Li YW, Wu P, Xie R, Shang JK (2008) Palladium oxide nanoparticles on nitrogen-doped titanium oxide: accelerated photocatalytic disinfection and post-illumination catalytic "memory". Adv Mater 20:3717–3723
182. Jagadale TC, Takale SP, Sonawane RS, Joshi HM, Patil SI, Kale BB, Ogale SB (2008) N-doped TiO_2 nanoparticle based visible light photocatalyst by modified peroxide sol-gel method. J Phys Chem C 112:14595–14602
183. Huang D, Liao S, Quan S, Liu L, He Z, Wan J, Zhou W (2008) Synthesis and characterization of visible light responsive N-TiO_2 mixed crystal by a modified hydrothermal process. J Non-Cryst Solids 354:3965–3972
184. Yang S, Gao L (2008) Photocatalytic activity of nitrogen doped rutile TiO_2 nanoparticles under visible light irradiation. Mater Res Bull 43:1872–1876
185. Zhao L, Jiang Q, Lian J (2008) Visible-light photocatalytic activity of nitrogen-doped TiO_2 thin film prepared by pulsed laser deposition. Appl Surf Sci 254:4620–4625
186. Zhang J, Wang Y, Jin Z, Wu Z, Zhang Z (2008) Visible-light photocatalytic behavior of two different N-doped TiO_2. Appl Surf Sci 254:4462–4466
187. Kun R, Tarjan S, Oszko A, Seemann T, Zoellmer V, Busse M, Dekany I (2009) Preparation and characterization of mesoporous N-doped and sulfuric acid treated anatase TiO_2 catalysts and their photocatalytic activity under UV and Vis illumination. J Solid State Chem 182:3076–3084
188. Wang J, Tafen DN, Lewis JP, Hong Z, Manivannan A, Zhi M, Li M, Wu N (2009) Origin of photocatalytic activity of nitrogen-doped TiO_2 nanobelts. J Am Chem Soc 131:12290–12297
189. Gao X, Liu J, Chen P (2011) Nitrogen-doped titania photocatalysts induced by shock wave. Mater Res Bull 44:1842–1845
190. Bellardita M, Addamo M, Di PA, Palmisano L, Venezia AM (2009) Preparation of N-doped TiO_2: characterization and photocatalytic performance under UV and visible light. Phys Chem Chem Phys 11:4084–4093
191. Wang Y, Zhou G, Li T, Qiao W, Li Y (2009) Catalytic activity of mesoporous TiO_2-xNx photocatalysts for the decomposition of methyl orange under solar simulated light. Catal Commun 10:412–415
192. Wang J, Wang Z, Li H, Cui Y, Du Y (2010) Visible light-driven nitrogen doped TiO_2 nanoarray films: preparation and photocatalytic activity. J Alloys Compd 494:372–377
193. Cho HJ, Hwang PG, Jung D (2011) Preparation and photocatalytic activity of nitrogen-doped TiO_2 hollow nanospheres. J Phys Chem Solids 72:1462–1466
194. In S, Vesborg PCK, Abrams BL, Hou YD, Chorkendorff I (2011) A comparative study of two techniques for determining photocatalytic activity of nitrogen doped TiO_2 nanotubes under visible light irradiation: photocatalytic reduction of dye and photocatalytic oxidation of organic molecules. J Photochem Photobiol A Chem 222:258–262
195. Jie HS, Lee HB, Chae KH, Huh MY, Matsuoka M, Cho SH, Park JK (2012) Nitrogen-doped TiO_2 nanopowders prepared by chemical vapor synthesis: band structure and photocatalytic activity under visible light. Res Chem Intermediat 38:1171–1180
196. Choi H, Antoniou MG, Pelaez M, De la Cruz AA, Shoemaker JA, Dionysiou DD (2007) Mesoporous nitrogen-doped TiO_2 for the photocatalytic destruction of the cyanobacterial toxin microcystin-LR under visible light irradiation. Environ Sci Tech 41:7530–7535
197. Li X, Xiong R, Wei G (2008) S-N co-doped TiO_2 photocatalysts with visible-light activity prepared by sol-gel method. Catal Lett 125:104–109

198. Liu C, Tang X, Mo C, Qiang Z (2008) Characterization and activity of visible-light-driven TiO_2 photocatalyst codoped with nitrogen and cerium. J Solid State Chem 181:913–919
199. Liu G, Zhao Y, Sun C, Li F, Lu GQ, Cheng HM (2008) Synergistic effects of B/N doping on the visible-light photocatalytic activity of mesoporous TiO_2. Angew Chem Int Ed 47:4516–4520
200. Song K, Zhou J, Bao J, Feng Y (2008) Photocatalytic activity of (copper, nitrogen)-codoped titanium dioxide nanoparticles. J Am Ceram Soc 91:1369–1371
201. Fan JW, Liu JY, Hong J, Zhang J (2009) The synthesis of nanostructure TiO_2 co-doped with N and Fe and their application for micro-polluted source water treatment. Environ Technol 30:1447–1452
202. Livraghi S, Elghniji K, Czoska AM, Paganini MC, Giamello E, Ksibi M (2009) Nitrogen-doped and nitrogen-fluorine-codoped titanium dioxide. Nature and concentration of the photoactive species and their role in determining the photocatalytic activity under visible light. J Photochem Photobiol A Chem 205:93–97
203. Lv Y, Ding Y, Zhou J, Xiao W, Feng Y (2009) Preparation, characterization, and photocatalytic activity of N, S-codoped TiO_2 nanoparticles. J Am Ceram Soc 92:938–941
204. Sathish M, Viswanath RP, Gopinath CS (2009) N, S-co-doped TiO_2 nanophotocatalyst: synthesis, electronic structure and photocatalysis. J Nanosci Nanotechnol 9:423–432
205. Shen Y, Xiong T, Du H, Jin H, Shang J, Yang K (2009) Investigation of Br-N co-doped TiO_2 photocatalysts: preparation and photocatalytic activities under visible light. J Sol-Gel Sci Technol 52:41–48
206. Tan K, Zhang H, Xie C, Zheng H, Gu Y, Zhang WF (2010) Visible-light absorption and photocatalytic activity in molybdenum- and nitrogen-codoped TiO_2. Catal Commun 11:331–335
207. Wu M, Yang B, Lv Y, Fu Z, Xu J, Guo T, Zhao Y (2010) Efficient one-pot synthesis of Ag nanoparticles loaded on N-doped multiphase TiO_2 hollow nanorod arrays with enhanced photocatalytic activity. Appl Surf Sci 256:7125–7130
208. Kurtoglu ME, Longenbach T, Sohlberg K, Gogotsi Y (2011) Strong coupling of Cr and N in Cr-N-doped TiO_2 and its effect on photocatalytic activity. J Phys Chem C 115:17392–17399
209. Wang E, He T, Zhao L, Chen Y, Cao Y (2011) Improved visible light photocatalytic activity of titania doped with tin and nitrogen. J Mater Chem 21:144–150
210. Dolat D, Quici N, Kusiak-Nejman E, Morawski AW, Li PG (2012) One-step, hydrothermal synthesis of nitrogen, carbon co-doped titanium dioxide (N, C-TiO_2) photocatalysts. Effect of alcohol degree and chain length as carbon dopant precursors on photocatalytic activity and catalyst deactivation. Appl Catal B 115–116:81–89
211. Wu D, Long M (2012) Visible light assisted photocatalytic degradation of methyl orange using Ag/N-TiO_2 photocatalysts. Water Sci Technol 65:1027–1032
212. Xiao Q, Yao C (2012) Visible light photocatalytic activity of C, N, S-tridoped anatase TiO_2 nanosheets. Adv Mater Res 391–392:1117–1122
213. Han X, Shao G (2011) Electronic properties of rutile TiO_2 with nonmetal dopants from first principles. J Phys Chem C 115:8274–8282
214. Nguyen CK, Nguyen VK, Nguyen HA, Nga DT, Nguyen VM (2011) The origin of visible light photocatalytic activity of N-doped and weak ferromagnetism of Fe-doped TiO_2 anatase. Adv Nat Sci Nanosci Nanotechnol 2:15008
215. Xu L, Steinmiller EMP, Skrabalak SE (2011) Achieving synergy with a potential photocatalytic Z-scheme: synthesis and evaluation of nitrogen-doped TiO_2/SnO_2 composites. J Phys Chem C 116:871–877
216. Xie Y, Zhao X (2008) The effects of synthesis temperature on the structure and visible-light-induced catalytic activity of F, N-codoped and S, N-codoped titania. J Mol Catal A Chem 285:142–149
217. Umezawa N, Ye J (2012) Role of complex defects in photocatalytic activities of nitrogen-doped anatase TiO_2. Phys Chem Chem Phys 14:5924–5934

218. Zhang Z, Long J, Xie X, Zhuang H, Zhou Y, Lin H, Yuan R, Dai W, Ding Z, Wang X, Fu X (2012) Controlling the synergistic effect of oxygen vacancies and N dopants to enhance photocatalytic activity of N-doped TiO_2 by H_2 reduction. Appl Catal A 425–426:117–124
219. Pelaez M, Falaras P, Kontos AG, De la Cruz AA, O'Shea K, Dunlop PSM, Byrne JA, Dionysiou DD (2012) A comparative study on the removal of cylindrospermopsin and microcystins from water with NF-TiO_2-P25 composite films with visible and UV-vis light photocatalytic activity. Appl Catal B Environ 121–122:30–39
220. Kang QM, Yuan BL, Xu JG, Fu ML (2011) Synthesis, characterization and photocatalytic performance of TiO_2 codoped with bismuth and nitrogen. Catal Lett 141:1371–1377
221. Tachikawa T, Tojo S, Kawai K, Endo M, Fujitsuka M, Ohno T, Nishijima K, Miyamoto Z, Majima T (2004) Photocatalytic oxidation reactivity of holes in the sulfur- and carbon-doped TiO_2 powders studied by time-resolved diffuse reflectance spectroscopy. J Phys Chem B 108:19299–19306
222. Sakthivel S, Kisch H (2003) Daylight photocatalysis by carbon-modified titanium dioxide. Angew Chem Int Ed 42:4908–4911
223. Neumann B, Bogdanoff P, Tributsch H, Sakthivel S, Kisch H (2005) Electrochemical mass spectroscopic and surface photovoltage studies of catalytic water photooxidation by undoped and carbon-doped titania. J Phys Chem B 109:16579–16586
224. Zabek P, Eberl J, Kisch H (2009) On the origin of visible light activity in carbon-modified titania. Photochem Photobiol Sci 8:264–269
225. Wang H, Wu Z, Liu Y (2009) A simple two-step template approach for preparing carbon-doped mesoporous TiO_2 hollow microspheres. J Phys Chem C 113:13317–13324
226. Wu Z, Dong F, Liu Y, Wang H (2009) Enhancement of the visible light photocatalytic performance of C-doped TiO_2 by loading with V_2O_5. Catal Commun 11:82–86
227. Dong F, Guo S, Wang H, Li X, Wu Z (2011) Enhancement of the visible light photocatalytic activity of C-doped TiO_2 nanomaterials prepared by a green synthetic approach. J Phys Chem C 115:13285–13292
228. Xue LM, Zhang FH, Fan HJ, Bai XF (2011) Preparation of C doped TiO_2 photocatalysts and their photocatalytic reduction of carbon dioxide. Adv Mater Res 183–185:1842–1846
229. Jung D, Kim G, Kim MS, Kim BW (2012) Evaluation of photocatalytic activity of carbon-doped TiO_2 films under solar irradiation. Korean J Chem Eng 29:703–706
230. Li LH, Lu J, Wang ZS, Yang L, Zhou XF, Han L (2012) Fabrication of the CN co-doped rod-like TiO_2 photocatalyst with visible-light responsive photocatalytic activity. Mater Res Bull 47:1508–1512
231. Goldstein S, Behar D, Rabani J (2009) Nature of the oxidizing species formed upon UV photolysis of C-TiO_2 aqueous suspensions. J Phys Chem C 113:12489–12494
232. Li Y, Hwang DS, Lee NH, Kim SJ (2005) Synthesis and characterization of carbon-doped titania as an artificial solar light sensitive photocatalyst. Chem Phys Lett 404:25–29
233. Wu G, Nishikawa T, Ohtani B, Chen A (2007) Synthesis and characterization of carbon-doped TiO_2 nanostructures with enhanced visible light response. Chem Mater 19:4530–4537
234. Wang Y, Wang Y, Meng Y, Ding H, Shan Y, Zhao X, Tang X (2008) A highly efficient visible-light-activated photocatalyst based on bismuth- and sulfur-codoped TiO_2. J Phys Chem C 112:6620–6626
235. Cui Y, Du H, Wen L (2009) Origin of visible-light-induced photocatalytic properties of S-doped anatase TiO_2 by first-principles investigation. Solid State Commun 149:634–637
236. Yu J, Liu S, Xiu Z, Yu W, Feng G (2009) Synthesis of sulfur-doped TiO_2 by solvothermal method and its visible-light photocatalytic activity. J Alloys Compd 471:L23–L25
237. Rockafellow EM, Stewart LK, Jenks WS (2009) Is sulfur-doped TiO_2 an effective visible light photocatalyst for remediation? Appl Catal B 91:554–562
238. Yu CL, Cai DJ, Yang K, Yu JC, Zhou Y, Fan CF (2010) Sol-gel derived S, I-co-doped mesoporous TiO_2 photocatalyst with high visible-light photocatalytic activity. J Phys Chem Solids 71:1337–1343

239. Umebayashi T, Yamaki T, Yamamoto S, Miyashita A, Tanaka S, Sumita T, Asai K (2003) Sulfur-doping of rutile-titanium dioxide by ion implantation: photocurrent spectroscopy and first-principles band calculation studies. J Appl Phys 93:5156–5160
240. Jiang Y, Liu Y, Yang L, Li G, Liu H (2011) Preparation and photocatalysis of sulfur-doped nano-TiO_2/Ti film. Adv Mater Res 177:281–283
241. Dozzi MV, Livraghi S, Giamello E, Selli E (2011) Photocatalytic activity of S- and F-doped TiO_2 in formic acid mineralization. Photochem Photobiol Sci 10:343–349
242. Han C, Pelaez M, Likodimos V, Kontos AG, Falaras P, Dionysiou DD (2011) Innovative visible light-activated sulfur doped TiO_2 films for water treatment. Appl Catal B Environ 107:77–87
243. Peng Y, He J, Liu Q, Sun Z, Yan W, Pan Z, Wu Y, Liang S, Cheng W, Wei S (2011) Impurity concentration dependence of optical absorption for phosphorus-doped anatase TiO_2. J Phys Chem C 115:8184–8188
244. Wang Y-W, Li Y-F, Yang P-H (2011) Preparation, characterization of nonmetal doped TiO_2 nanoparticles with their excellent photocatalytic properties. Adv Mater Res 183–185:2254–2257
245. Tojo S, Tachikawa T, Fujitsuka M, Majima T (2008) Iodine-doped TiO_2 photocatalysts: correlation between band structure and mechanism. J Phys Chem C 112:14948–14954
246. Xu H, Zheng Z, Zhang L, Zhang H, Deng F (2008) Hierarchical chlorine-doped rutile TiO_2 spherical clusters of nanorods: large-scale synthesis and high photocatalytic activity. J Solid State Chem 181:2516–2522
247. Deng P, Hu J, Wang H, Sun B (2010) Hydrothermal preparation and comparative study of halogen-doping TiO_2 photocatalysts. J Adv Oxid Technol 13:200–205
248. Li D, Haneda H, Hishita S, Ohashi N, Labhsetwar NK (2005) Fluorine-doped TiO_2 powders prepared by spray pyrolysis and their improved photocatalytic activity for decomposition of gas-phase acetaldehyde. J Fluorine Chem 126:69–77
249. Li D, Haneda H, Hishita S, Ohashi N (2005) Visible-light-driven N-F-codoped TiO_2 photocatalysts. 2. Optical characterization, photocatalysis, and potential application to air purification. Chem Mater 17:2596–2602
250. Ho W, Yu JC, Lee S (2006) Synthesis of hierarchical nanoporous F-doped TiO_2 spheres with visible light photocatalytic activity. Chem Commun 10:1115–1117
251. Dozzi MV, Ohtani B, Selli E (2011) Absorption and action spectra analysis of ammonium fluoride-doped titania photocatalysts. Phys Chem Chem Phys 13:18217–18227
252. Kuznetsov VN, Serpone N (2006) Visible light absorption by various titanium dioxide specimens. J Phys Chem B 110:25203–25209
253. Serpone N (2006) Is the band gap of pristine TiO_2 narrowed by anion- and cation-doping of titanium dioxide in second-generation photocatalysts? J Phys Chem B 110:24287–24293
254. Miyauchi M, Ikezawa A, Tobimatsu H, Irie H, Hashimoto K (2004) Zeta potential and photocatalytic activity of nitrogen doped TiO_2 thin films. Phys Chem Chem Phys 6:865–870
255. Irie H, Watanabe Y, Hashimoto K (2003) Nitrogen-concentration dependence on photocatalytic activity of TiO_2-xNx powders. J Phys Chem B 107:5483–5486
256. Long R, English NJ (2009) First-principles calculation of nitrogen-tungsten codoping effects on the band structure of anatase-titania. Appl Phys Lett 94:132102-1–132102-3
257. Jiang Y, Scott J, Amal R (2012) Exploring the relationship between surface structure and photocatalytic activity of flame-made TiO_2-based catalysts. Appl Catal B 126:290–297
258. Jia L, Wu C, Li Y, Han S, Li Z, Chi B, Pu J, Jian L (2011) Enhanced visible-light photocatalytic activity of anatase TiO_2 through N and S codoping. Appl Phys Lett 98:211903-1–211903-3
259. Dvoranová D, Brezová V, Mazúra M, Malati MA (2002) Investigations of metal-doped titanium dioxide photocatalysts. Appl Catal B Environ 37:91–105
260. Khan R, Kim SW, Kim TJ, Nam CM (2008) Comparative study of the photocatalytic performance of boron-iron co-doped and boron-doped TiO_2 nanoparticles. Mater Chem Phys 112:167–172

261. Jiang H, Song H, Zhou Z, Liu X, Meng G (2008) Characterization of LiF-doped TiO_2 and its photocatalytic activity for decomposition of trichloromethane. Mater Res Bull 43:3037–3046
262. Xu J, Ao Y, Chen M, Fu D (2009) Low-temperature preparation of boron-doped titania by hydrothermal method and its photocatalytic activity. J Alloys Compd 484:73–79
263. Chen Q, Shi W, Xu Y, Wu D, Sun Y (2010) Ag-Si co-doped TiO_2 photocatalyst synthesized via a nonaqueous method. J Nanosci Nanotechnol 10:7221–7225
264. Thomas J, Yoon M (2012) Facile synthesis of pure TiO_2(B) nanofibers doped with gold nanoparticles and solar photocatalytic activities. Appl Catal B 111–112:502–508
265. Guo L, Fu F, Wang D, Qiang X, Wei QB, Wu Y (2011) Preparation of silver-doped TiO_2 photocatalyst via a simple sol-hydrothermal and their visible light photocatalytic activity. Mater Sci Forum 694:824–830
266. Hamal DB, Klabunde KJ (2011) Valence state and catalytic role of cobalt ions in cobalt TiO_2 nanoparticle photocatalysts for acetaldehyde degradation under visible light. J Phys Chem C 115:17359–17367
267. Thind SS, Wu G, Chen A (2012) Synthesis of mesoporous nitrogen-tungsten co-doped TiO_2 photocatalysts with high visible light activity. Appl Catal B 111–112:38–45
268. Zielinska A, Kowalska E, Sobczak JW, Lacka I, Gazda M, Ohtani B, Hupka J, Zaleska A (2010) Silver-doped TiO_2 prepared by microemulsion method: surface properties, bio- and photoactivity. Sep Purif Technol 72:309–318
269. Grabowska E, Zaleska A, Sobczak JW, Gazda M, Hupka J (2009) Boron-doped TiO_2: characteristics and photoactivity under visible light. Procedia Chem 1:1553–1559
270. Zaleska A, Grabowska E, Sobczak JW, Gazda M, Hupka J (2009) Photocatalytic activity of boron-modified TiO_2 under visible light: the effect of boron content, calcination temperature and TiO_2 matrix. Appl Catal B 89:469–475
271. Gracia F, Holgado JP, Caballero A, Gonzalez-Elipe AR (2004) Structural, optical, and photoelectrochemical properties of Mn^{2+}-TiO_2 model thin film photocatalysts. J Phys Chem B 108:17466–17476
272. Hao H, Zhang J (2009) The study of Iron (III) and nitrogen co-doped mesoporous TiO_2 photocatalysts: synthesis, characterization and activity. Microporous Mesoporous Mater 121:52–57
273. Lin X, Rong F, Ji X, Fu D (2011) Visible light photocatalytic activity and photoelectrochemical property of Fe-doped TiO_2 hollow spheres by sol-gel method. J Sol-Gel Sci Technol 59:283–289
274. Serpone N, Lawless D, Disdier J, Herrmann J-M (1994) Spectroscopic, photoconductivity, and photocatalytic studies of TiO_2 colloids: naked and with the lattice doped with Cr^{3+}, Fe^{3+}, and V^{5+} cations. Langmuir 10:643–652
275. Diamandescu L, Vasiliu F, Tarabasanu-Mihaila D, Feder M, Vlaicu AM, Teodorescu CM, Macovei D, Enculescu I, Parvulescu V, Vasile E (2008) Structural and photocatalytic properties of iron- and europium-doped TiO_2 nanoparticles obtained under hydrothermal conditions. Mater Chem Phys 112:146–153
276. Vu AT, Nguyen QT, Bui THL, Tran MC, Dang TP, Tran TKH (2010) Synthesis and characterization of TiO_2 photocatalyst doped by transition metal ions (Fe^{3+}, Cr^{3+} and V^{5+}). Adv Nat Sci Nanosci Nanotechnol 1:015009
277. Di PA, Garcia-Lopez E, Ikeda S, Marci G, Ohtani B, Palmisano L (2002) Photocatalytic degradation of organic compounds in aqueous systems by transition metal doped polycrystalline TiO_2. Catal Today 75:87–93
278. Di PA, Marci G, Palmisano L, Schiavello M, Uosaki K, Ikeda S, Ohtani B (2002) Preparation of polycrystalline TiO_2 photocatalysts impregnated with various transition metal ions: characterization and photocatalytic activity for the degradation of 4-nitrophenol. J Phys Chem B 106:637–645
279. Yu J, Xiang Q, Zhou M (2009) Preparation, characterization and visible-light-driven photocatalytic activity of Fe-doped titania nanorods and first-principles study for electronic structures. Appl Catal B 90:595–602

280. Teoh WY, Amal R, Maedler L, Pratsinis SE (2007) Flame sprayed visible light-active Fe-TiO_2 for photomineralization of oxalic acid. Catal Today 120:203–213
281. Wang CY, Bahnemann DW, Dohrmann JK (2000) A novel preparation of iron-doped TiO_2 nanoparticles with enhanced photocatalytic activity. Chem Commun 16:1539–1540
282. Wang Q, Xu S, Shen F (2011) Preparation and characterization of TiO_2 photocatalysts co-doped with iron (III) and lanthanum for the degradation of organic pollutants. Appl Surf Sci 257:7671–7677
283. Suciu R-C, Corina RM, Silipas TD, Indrea E, Popescu V, Popescu GL (2011) Fe_2O_3–TiO_2 thin films prepared by sol-gel method. Environ Eng Manag J 10:187–192
284. Ji T, Yang F, Lv Y, Zhou J, Sun J (2009) Synthesis and visible-light photocatalytic activity of Bi-doped TiO_2 nanobelts. Mater Lett 63:2044–2046
285. Liu S, Guo E, Yin L (2012) Tailored visible-light driven anatase TiO_2 photocatalysts based on controllable metal ion doping and ordered mesoporous structure. J Mater Chem 22:5031–5041
286. Song YT, Shao WN, Cao WB (2011) Preparation of nanocrystalline W-doped TiO_2 powders and their photocatalytic properties under visible light irradiation. Mater Sci Forum 695:489–492
287. Deng QR, Xia XH, Guo ML, Gao Y, Shao G (2011) Mn-doped TiO_2 nanopowders with remarkable visible light photocatalytic activity. Mater Lett 65:2051–2054
288. Li J, Wang D, Liu H, Zhu Z (2012) Multilayered Mo-doped TiO_2 nanofibers and enhanced photocatalytic activity. Mater Manuf Process 27:631–635
289. Choi W, Termin A, Hoffmann MR (1994) The role of metal ion dopants in quantum-sized TiO_2: correlation between photoreactivity and charge carrier recombination dynamics. J Phys Chem 98:13669–13679
290. Devi LG, Murthy BN, Kumar SG (2009) Photocatalytic activity of V^{5+}, Mo^{6+} and Th^{4+} doped polycrystalline TiO_2 for the degradation of chlorpyrifos under UV/solar light. J Mol Catal A Chem 308:174–181
291. Dutta SS, Singh D, Saini KK, Kant C, Sharma V, Jain SC, Sharma CP (2006) Sol-gel-derived super-hydrophilic nickel doped TiO_2 film as active photo-catalyst. Appl Catal A 314:40–46
292. Gao H, Liu W (2012) La and/or Y doped TiO_2: facile synthesis and enhanced photocatalysis. Adv Mater Res 463–464:290–294
293. Yao S, Sui C, Shi Z (2011) Preparation and characterization of visible-light-driven europium doped mesoporous titania photocatalyst. J Rare Earth 29:929–933
294. Fan X, Chen X, Zhu S, Li Z, Yu T, Ye J, Zou Z (2008) The structural, physical and photocatalytic properties of the mesoporous Cr-doped TiO_2. J Mol Catal A Chem 284:155–160
295. Lin WC, Yang WD (2012) Synthesis, characterization and photocatalytic activity of copper (II)-doped titanium dioxide powders. Adv Mater Res 391–392:728–731
296. Devi LG, Murthy BN, Kumar SG (2010) Photocatalytic activity of TiO_2 doped with Zn^{2+} and V^{5+} transition metal ions: influence of crystallite size and dopant electronic configuration on photocatalytic activity. Mater Sci Eng B 166:1–6
297. Choudhury B, Borah B, Choudhury A (2012) Extending photocatalytic activity of TiO_2 nanoparticles to visible region of illumination by doping of cerium. J Photochem Photobiol A Chem 88:257–264
298. Li G, Dimitrijevic NM, Chen L, Rajh T, Gray KA (2008) Role of surface/interfacial Cu^{2+} sites in the photocatalytic activity of coupled CuO-TiO_2 nanocomposites. J Phys Chem C 112:19040–19044
299. Herrmann JM, Disdier J, Pichat P (1984) Effect of chromium doping on the electrical and catalytic properties of powder titania under UV and visible illumination. Chem Phys Lett 108:618–622
300. Gratzel M, Howe RF (1990) Electron paramagnetic resonance studies of doped TiO_2 colloids. J Phys Chem 94:2566–2572

301. Wang CY, Boettcher C, Bahnemann DW, Dohrmann JK (2003) A comparative study of nanometer sized Fe(III)-doped TiO_2 photocatalysts: synthesis, characterization and activity. J Mater Chem 13:2322–2329
302. Wang XH, Li JG, Kamiyama H, Moriyoshi Y, Ishigaki T (2006) Wavelength-sensitive photocatalytic degradation of methyl orange in aqueous suspension over iron(III)-doped TiO_2 nanopowders under UV and visible light irradiation. J Phys Chem B 110:6804–6809
303. Yu J, Yu H, Ao CH, Lee SC, Yu JC, Ho W (2006) Preparation, characterization and photocatalytic activity of in situ Fe-doped TiO_2 thin films. Thin Solid Films 494:273–280
304. Mu W, Herrimann J-M, Pichat P (1989) Room temperature photocatalytic oxidation of liquid cyclohexane into cyclohexanone over neat and modified TiO_2. Catal Lett 3:73–84
305. Radecka M, Wierzbicka M, Komornicki S, Rekas M (2004) Influence of Cr on photoelectrochemical properties of TiO_2 thin films. Physica B Condens Matter 348:160–168
306. Wilke K, Breuer HD (1999) The influence of transition metal doping on the physical and photocatalytic properties of titania. J Photochem Photobiol A Chem 121:49–53
307. Martin ST, Morrison CL, Hoffmann MR (1994) Photochemical mechanism of size-quantized vanadium-doped TiO_2 particles. J Phys Chem 98:13695–13704
308. Trejo-Tzab R, Alvarado-Gil JJ, Quintana P, Bartolo-Perez P (2012) N-doped TiO_2 P25/Cu powder obtained using nitrogen (N_2) gas plasma. Catal Today 193:179–185
309. Kaleji BK, Sarraf-Mamoory R (2012) Nanocrystalline sol-gel TiO_2-SnO_2 coatings: preparation, characterization and photocatalytic performance. Mater Res Bull 47:362–369
310. Kim C, Choi M, Jang J (2010) Nitrogen-doped SiO_2/TiO_2 core/shell nanoparticles as highly efficient visible light photocatalyst. Catal Commun 11:378–382
311. Wang D-H, Jia L, Wu X-L, Lu L-Q, Xu A-W (2012) One-step hydrothermal synthesis of N-doped TiO_2/C nanocomposites with high visible light photocatalytic activity. Nanoscale 4:576–584
312. Ye F, Ohmori A (2002) The photocatalytic activity and photoabsorption of plasma sprayed TiO_2-Fe_3O_4 binary oxide coatings. Surf Coat Technol 160:62–67
313. Li G, Ciston S, Saponjic ZV, Chen L, Dimitrijevic NM, Rajh T, Gray KA (2008) Synthesizing mixed-phase TiO_2 nanocomposites using a hydrothermal method for photooxidation and photoreduction applications. J Catal 253:105–110
314. Leon-Ramos JA, Kibanova D, Santiago-Jacinto P, Mar-Santiago Y, Trejo-Valdez M (2011) Synthesis, characterization and photocatalytic properties of tungsten-doped hydrothermal TiO_2. J Sol-Gel Sci Technol 57:43–50
315. Zhang P, Xu M, Fang H, Li L (2009) Low-temperature synthesis of $InVO_4$ doped TiO_2 sol and visible-light photocatalytic activities of $InVO_4$-TiO_2 films. Mater Lett 63:2146–2148
316. Ang TP, Toh CS, Han YF (2009) Synthesis, characterization, and activity of visible-light-driven nitrogen-doped TiO_2-SiO_2 mixed oxide photocatalysts. J Phys Chem C 113:10560–10567
317. Hou YD, Wang XC, Wu L, Chen XF, Ding ZX, Wang XX, Fu XZ (2008) N-doped SiO_2/TiO_2 mesoporous nanoparticles with enhanced photocatalytic activity under visible-light irradiation. Chemosphere 72:414–421
318. Fuerte A, Hernandez-Alonso MD, Maira AJ, Martinez-Arias A, Fernandez-Garcia M, Conesa JC, Soria J (2001) Visible light-activated nanosized doped-TiO_2 photocatalysts. Chem Commun 24:2718–2719
319. Xu L, Steinmiller EMP, Skrabalak SE (2012) Achieving synergy with a potential photocatalytic Z-scheme: synthesis and evaluation of nitrogen-doped TiO_2/SnO_2 composites. J Phys Chem C 116:871–877
320. Mishra T, Hait J, Aman N, Jana RK, Chakravarty S (2007) Effect of UV and visible light on photocatalytic reduction of lead and cadmium over titania based binary oxide materials. J Colloid Interface Sci 316:80–84
321. Neppolian B, Wang Q, Yamashita H, Choi H (2007) Synthesis and characterization of ZrO_2-TiO_2 binary oxide semiconductor nanoparticles: application and interparticle electron transfer process. Appl Catal A 333:264–271

322. Wu B, Yuan R, Fu X (2009) Structural characterization and photocatalytic activity of hollow binary ZrO_2/TiO_2 oxide fibers. J Solid State Chem 182:560–565
323. Kambur A, Pozan GS, Boz I (2012) Preparation, characterization and photocatalytic activity of TiO_2-ZrO_2 binary oxide nanoparticles. Appl Catal B 115–116:149–158
324. Ohno T, Tokieda K, Higashida S, Matsumura M (2003) Synergism between rutile and anatase TiO_2 particles in photocatalytic oxidation of naphthalene. Appl Catal A 244:383–391
325. Kolen'ko YV, Churagulov BR, Kunst M, Mazerolles L, Colbeau-Justin C (2004) Photocatalytic properties of titania powders prepared by hydrothermal method. Appl Catal B 54:51–58
326. Bakardjieva S, Subrt J, Stengl V, Dianez MJ, Sayagues MJ (2005) Photoactivity of anatase-rutile TiO_2 nanocrystalline mixtures obtained by heat treatment of homogeneously precipitated anatase. Appl Catal B 58:193–202
327. Kawahara T, Ozawa T, Iwasaki M, Tada H, Ito S (2003) Photocatalytic activity of rutile-anatase coupled TiO_2 particles prepared by a dissolution-reprecipitation method. J Colloid Interface Sci 267:377–381
328. Tatsuma T, Saitoh S, Ngaotrakanwiwat P, Ohko Y, Fujishima A (2002) Energy storage of TiO_2-WO_3 photocatalysis systems in the gas phase. Langmuir 18:7777–7779
329. Rampaul A, Parkin IP, O'Neill SA, DeSouza J, Mills A, Elliott N (2003) Titania and tungsten doped titania thin films on glass; active photocatalysts. Polyhedron 22:35–44
330. Shchukin D, Poznyak S, Kulak A, Pichat P (2004) TiO_2-In_2O_3 photocatalysts: preparation, characterizations and activity for 2-chlorophenol degradation in water. J Photochem Photobiol A Chem 162:423–430
331. Enriquez R, Beaugiraud B, Pichat P (2004) Mechanistic implications of the effect of TiO_2 accessibility in TiO_2-SiO_2 coatings upon chlorinated organics photocatalytic removal in water. Water Sci Technol 49:147–152
332. Navio JA, Colon G, Herrmann JM (1997) Photoconductive and photocatalytic properties of $ZrTiO_4$. Comparison with the parent oxides TiO_2 and ZrO_2. J Photochem Photobiol A Chem 108:179–185
333. Chu SZ, Inoue S, Wada K, Li D, Suzuki J (2005) Fabrication and photocatalytic characterizations of ordered nanoporous X-doped (X = N, C, S, Ru, Te, and Si) TiO_2/Al_2O_3 films on ITO/glass. Langmuir 21:8035–8041
334. Fernandez A, Lassaletta G, Jimenez VM, Justo A, Gonzalez-Elipe AR, Herrmann JM, Tahiri H, Ait-Ichou Y (1995) Preparation and characterization of TiO_2 photocatalysts supported on various rigid supports (glass, quartz and stainless steel). Comparative studies of photocatalytic activity in water purification. Appl Catal B 7:49–63
335. Kundu S, Kafizas A, Hyett G, Mills A, Darr JA, Parkin IP (2011) An investigation into the effect of thickness of titanium dioxide and gold-silver nanoparticle titanium dioxide composite thin-films on photocatalytic activity and photoinduced oxygen production in a sacrificial system. J Mater Chem 21:6854–6863
336. Sato S (1988) Effects of surface modification with silicon oxides on the photochemical properties of powdered titania. Langmuir 4:1156–1159
337. Ryu J, Choi W (2004) Effects of TiO_2 surface modifications on photocatalytic oxidation of arsenite: the role of superoxides. Environ Sci Technol 38:2928–2933
338. Hidalgo MC, Maicu M, Navio JA, Colon G (2007) Photocatalytic properties of surface modified platinized TiO_2: effects of particle size and structural composition. Catal Today 129:43–49
339. Kozlova EA, Lyubina TP, Nasalevich MA, Vorontsov AV, Miller AV, Kaichev VV, Parmon VN (2011) Influence of the method of platinum deposition on activity and stability of Pt/TiO_2 photocatalysts in the photocatalytic oxidation of dimethyl methylphosphonate. Catal Commun 12:597–601
340. Khnayzer RS, Thompson LB, Zamkov M, Ardo S, Meyer GJ, Murphy CJ, Castellano FN (2012) Photocatalytic hydrogen production at titania-supported Pt nanoclusters that are derived from surface-anchored molecular precursors. J Phys Chem C 116:1429–1438

341. Kowalska E, Remita H, Colbeau-Justin C, Hupka J, Belloni J (2008) Modification of titanium dioxide with platinum ions and clusters: application in photocatalysis. J Phys Chem C 112:1124–1131
342. Ikeda S, Sugiyama N, Pal B, Marci G, Palmisano L, Noguchi H, Uosaki K, Ohtani B (2001) Photocatalytic activity of transition-metal-loaded titanium(IV) oxide powders suspended in aqueous solutions: correlation with electron-hole recombination kinetics. Phys Chem Chem Phys 3:267–273
343. Ohtani B, Iwai K, Nishimoto SI, Sato S (1997) Role of platinum deposits on titanium (IV) oxide particles: structural and kinetic analyses of photocatalytic reaction in aqueous alcohol and amino acid solutions. J Phys Chem B 101:3349–3359
344. Crittenden JC, Liu J, Hand DW, Perram DL (1997) Photocatalytic oxidation of chlorinated hydrocarbons in water. Water Res 31:429–438
345. Sun B, Smirniotis PG, Boolchand P (2005) Visible light photocatalysis with platinized rutile TiO_2 for aqueous organic oxidation. Langmuir 21:11397–11403
346. Colon G, Maicu M, Hidalgo MC, Navio JA, Kubacka A, Fernandez-Garcia M (2010) Gas phase photocatalytic oxidation of toluene using highly active Pt doped TiO_2. J Mol Catal A Chem 320:14–18
347. Wang Y, Jing M, Zhang M, Yang J (2012) Facile synthesis and photocatalytic activity of platinum decorated TiO_2-xNx: perspective to oxygen vacancies and chemical state of dopants. Catal Commun 20:46–50
348. Li X, Zhuang Z, Li W, Pan H (2012) Photocatalytic reduction of CO_2 over noble metal-loaded and nitrogen-doped mesoporous TiO_2. Appl Catal A 429–430:31–38
349. Kim J, Monllor-Satoca D, Choi W (2012) Simultaneous production of hydrogen with the degradation of organic pollutants using TiO_2 photocatalyst modified with dual surface components. Energ Environ Sci 5:7647–7656
350. Alaoui QT, Herissan A, Le QC, MM Z, Sorgues S, Remita H, Colbeau-Justin C (2012) Elaboration, charge-carrier lifetimes and activity of Pd-TiO_2 photocatalysts obtained by gamma radiolysis. J Photochem Photobiol A Chem 242:34–43
351. Lee WI, Choi GJ, Do YR (1997) Effect of Au and WO_3 on the surface structure and photocatalytic activity of TiO_2. Bull Korean Chem Soc 18:667–670
352. Dawson A, Kamat PV (2001) Semiconductor-metal nanocomposites. Photoinduced fusion and photocatalysis of gold-capped TiO_2 (TiO_2/gold) nanoparticles. J Phys Chem B 105:960–966
353. Wu CG, Tzeng LF, Kuo YT, Shu CH (2002) Enhancement of the photocatalytic activity of TiO_2 film via surface modification of the substrate. Appl Catal A 226:199–211
354. Arabatzis IM, Stergiopoulos T, Andreeva D, Kitova S, Neophytides SG, Falaras P (2003) Characterization and photocatalytic activity of Au/TiO_2 thin films for azo-dye degradation. J Catal 220:127–135
355. Subramanian V, Wolf EE, Kamat PV (2004) Catalysis with TiO_2/gold nanocomposites. Effect of metal particle size on the Fermi level equilibration. J Am Chem Soc 126:4943–4950
356. Kowalska E, Abe R, Ohtani B (2009) Visible light-induced photocatalytic reaction of gold-modified titanium(iv) oxide particles: action spectrum analysis. Chem Commun 241–243
357. Dozzi MV, Chiarello GL, Selli E (2010) Effects of surface modification on the photocatalytic activity of TiO_2. J Adv Oxid Technol 13:305–312
358. Lee M, Amaratunga P, Kim J, Lee D (2010) TiO_2 nanoparticle photocatalysts modified with monolayer-protected gold clusters. J Phys Chem C 114:18366–18371
359. Kochuveedu ST, Kim DP, Kim DH (2012) Surface-plasmon-induced visible light photocatalytic activity of TiO_2 nanospheres decorated by Au nanoparticles with controlled configuration. J Phys Chem C 116:2500–2506
360. Kowalska E, Mahaney OOP, Abe R, Ohtani B (2010) Visible-light-induced photocatalysis through surface plasmon excitation of gold on titania surfaces. Phys Chem Chem Phys 12:2344–2355

361. Zielinska-Jurek A, Kowalska E, Sobczak JW, Lisowski W, Ohtani B, Zaleska A (2011) Preparation and characterization of monometallic (Au) and bimetallic (Ag/Au) modified-titania photocatalysts activated by visible light. Appl Catal B 101:504–514
362. Tanaka A, Ogino A, Iwaki M, Hashimoto K, Ohnuma A, Amano F, Ohtani B, Kominami H (2012) Gold-titanium(IV) oxide plasmonic photocatalysts prepared by a colloid-photodeposition method: correlation between physical properties and photocatalytic activities. Langmuir 28:13105–13111
363. Kowalska E, Rau S, Ohtani B (2012) Plasmonic titania photocatalysts active under UV and visible-light irradiation: influence of gold amount, size, and shape. J Nanotechnol 2012:361853–361864
364. Cai C, Zhang J, Pan F, Zhang W, Zhu H, Wang T (2008) Influence of metal (Au, Ag) micro-grid on the photocatalytic activity of TiO_2 film. Catal Lett 123:51–55
365. Cao Y, Tan H, Shi T, Tang T, Li J (2008) Preparation of Ag-doped TiO_2 nanoparticles for photocatalytic degradation of acetamiprid in water. J Chem Technol Biotechnol 83:546–552
366. Priya R, Baiju KV, Shukla S, Biju S, Reddy MLP, Patil K, Warrier KGK (2009) Comparing ultraviolet and chemical reduction techniques for enhancing photocatalytic activity of silver oxide/silver deposited nanocrystalline anatase titania. J Phys Chem C 113:6243–6255
367. Wang E, Liu S, Lu Q, Xiu Z, Li T, Song L (2011) Photocatalytic property of surface-modified TiO_2 nanobelts under visible light irradiation. J Sol-Gel Sci Technol 58:705–710
368. Zhang S, Peng F, Wang H, Yu H, Zhang S, Yang J, Zhao H (2011) Electrodeposition preparation of Ag loaded N-doped TiO_2 nanotube arrays with enhanced visible light photocatalytic performance. Catal Commun 12:689–693
369. Subrahmanyam A, Biju KP, Rajesh P, Jagadeesh KK, Raveendra KM (2012) Surface modification of sol gel TiO_2 surface with sputtered metallic silver for sun light photocatalytic activity: initial studies. Sol Energy Mater Sol Cells 101:241–248
370. Korzhak AV, Ermokhina NI, Stroyuk AL, Bukhtiyarov VK, Raevskaya AE, Litvin VI, Kuchmiy SY, Ilyin VG, Manorik PA (2008) Photocatalytic hydrogen evolution over mesoporous TiO_2/metal nanocomposites. J Photochem Photobiol A Chem 198:126–134
371. Mills A, Hill G, Stewart M, Graham D, Smith WE, Hodgen S, Halfpenny PJ, Faulds K, Robertson P (2004) Characterization of novel Ag on TiO_2 films for surface-enhanced Raman scattering. Appl Spectrosc 58:922–928
372. Irie H, Miura S, Kamiya K, Hashimoto K (2008) Efficient visible light-sensitive photocatalysts: grafting Cu(II) ions onto TiO_2 and WO_3 photocatalysts. Chem Phys Lett 457:202–205
373. Higashimoto S, Tanihata W, Nakagawa Y, Azuma M, Ohue H, Sakata Y (2008) Effective photocatalytic decomposition of VOC under visible-light irradiation on N-doped TiO_2 modified by vanadium species. Appl Catal A 340:98–104
374. Zhou W, Liu H, Wang J, Liu D, Du G, Cui J (2010) Ag_2O/TiO_2 nanobelts heterostructure with enhanced ultraviolet and visible photocatalytic activity. ACS Appl Mater Interfaces 2:2385–2392
375. Jin Q, Fujishima M, Tada H (2011) Visible light-active iron oxide-modified anatase titanium(IV) dioxide. J Phys Chem C 115:6478–6483
376. Jin Q, Ikeda T, Fujishima M, Tada H (2011) Nickel(II) oxide surface-modified titanium(IV) dioxide as a visible-light-active photocatalyst. Chem Commun 47:8814–8816
377. Liu P, Li W, Zhang J, Lin Y (2011) Photocatalytic activity enhancement of TiO_2 porous thin film due to homogeneous surface modification of RuO_2. J Mater Res 26:1532–1538
378. Tada H, Jin Q, Nishijima H, Yamamoto H, Fujishima M, Okuoka S-I, Hattori T, Sumida Y, Kobayashi H (2011) Titanium(IV) dioxide surface-modified with iron oxide as a visible light photocatalyst. Angew Chem Int Ed 50:3501–3505
379. Fujishima M, Jin Q, Yamamoto H, Tada H, Nolan M (2012) Tin oxide-surface modified anatase titanium(iv) dioxide with enhanced UV-light photocatalytic activity. Phys Chem Chem Phys 14:705–711

380. Jin Q, Fujishima M, Nolan M, Iwaszuk A, Tada H (2012) Photocatalytic activities of Tin (IV) oxide surface-modified titanium(IV) dioxide show a strong sensitivity to the TiO_2 crystal form. J Phys Chem C 116:12621–12626
381. Nolan M, Iwaszuk A, Tada H (2012) Molecular metal oxide cluster-surface modified titanium(iv) dioxide photocatalysts. Aust J Chem 65:624–632
382. Ozer RR, Ferry JL (2001) Investigation of the photocatalytic activity of TiO_2-polyoxometalate systems. Environ Sci Technol 35:3242–3246
383. Pearson A, Bhargava SK, Bansal V (2011) UV-switchable polyoxometalate sandwiched between TiO_2 and metal nanoparticles for enhanced visible and solar light photococatalysis. Langmuir 27:9245–9252
384. Chen C, Lei P, Ji H, Ma W, Zhao J, Hidaka H, Serpone N (2004) Photocatalysis by titanium dioxide and polyoxometalate/TiO_2 cocatalysts. Intermediates and mechanistic study. Environ Sci Technol 38:329–337
385. Schwitzgebel J, Ekerdt JG, Gerischer H, Heller A (1995) Role of the oxygen molecule and of the photogenerated electron in TiO_2-photocatalyzed air oxidation reactions. J Phys Chem 99:5633–5638
386. Wang CM, Heller A, Gerischer H (1992) Palladium catalysis of O_2 reduction by electrons accumulated on TiO_2 particles during photoassisted oxidation of organic compounds. J Am Chem Soc 114:5230–5234
387. Gerischer H, Heller A (1991) The role of oxygen in photooxidation of organic molecules on semiconductor particles. J Phys Chem 95:5261–5267
388. Serpone N, Maruthamuthu P, Pichat P, Pelizzetti E, Hidaka H (1995) Exploiting the interparticle electron transfer process in the photocatalyzed oxidation of phenol, 2-chlorophenol and pentachlorophenol: chemical evidence for electron and hole transfer between coupled semiconductors. J Photochem Photobiol A Chem 85:247–255
389. Kesselman JM, Shreve GA, Hoffmann MR, Lewis NS (1994) Flux-matching conditions at TiO_2 photoelectrodes: is interfacial electron transfer to O_2 rate-limiting in the TiO_2-catalyzed photochemical degradation of organics? J Phys Chem 98:13385–13395
390. Belloni J, Mostafavi M, Remita H, Marignier JL, Delcourt MO (1998) Radiation-induced synthesis of mono- and multimetallic clusters and nanocolloids. New J Chem 22:1239–1255
391. Libera JA, Elam JW, Sather NF, Rajh T, Dimitrijevic NM (2010) Iron(III)-oxo centers on TiO_2 for visible-light photocatalysis. Chem Mater 22:409–413
392. Sakthivela S, Shankarb MV, Palanichamyb M, Arabindoob B, Bahnemanna DW, Murugesanb V (2004) Enhancement of photocatalytic activity by metal deposition: characterisation and photonic efficiency of Pt, Au and Pd deposited on TiO_2 catalyst. Water Res 38:3001–3008
393. Kamat PV (2010) Graphene-based nanoarchitectures. Anchoring semiconductor and metal nanoparticles on a two-dimensional carbon support. J Phys Chem Lett 1:520–527
394. Leary R, Westwood A (2011) Carbonaceous nanomaterials for the enhancement of TiO_2 photocatalysis. Carbon 49:741–772
395. Woan K, Pyrgiotakis G, Sigmund W (2009) Photocatalytic carbon-nanotube-TiO_2 composites. Adv Mater 21:2233–2239
396. Novoselov KS, Geim AK, Morozov SV, Jiang D, Zhang Y, Dubonos SV, Grigorieva IV, Firsov AA (2004) Electric field effect in atomically thin carbon films. Science 306:666–669
397. Ding S, Chen JS, Luan D, Boey FYC, Madhavi S, Lou XW (2011) Graphene-supported anatase TiO_2 nanosheets for fast lithium storage. Chem Commun 47:5780–5782
398. Katsnelson MI, Novoselov KS (2007) Graphene: new bridge between condensed matter physics and quantum electrodynamics. Solid State Commun 143:3–13
399. Schedin F, Geim AK, Morozov SV, Hill EW, Blake P, Katsnelson MI, Novoselov KS (2007) Detection of individual gas molecules adsorbed on graphene. Nat Mater 6:652–655
400. Ng YH, Lightcap IV, Goodwin K, Matsumura M, Kamat PV (2010) To what extent do graphene scaffolds improve the photovoltaic and photocatalytic response of TiO_2 nanostructured films? J Phys Chem Lett 1:2222–2227

401. Jiang B, Tian C, Pan Q, Jiang Z, Wang JQ, Yan W, Fu H (2011) Enhanced photocatalytic activity and electron transfer mechanisms of graphene/TiO_2 with exposed 001 facets. J Phys Chem C 115:23718–23725
402. Dong P, Wang Y, Guo L, Liu B, Xin S, Zhang J, Shi Y, Zeng W, Yin S (2012) A facile one-step solvothermal synthesis of graphene/rod-shaped TiO_2 nanocomposite and its improved photocatalytic activity. Nanoscale 4:4641–4649
403. Xiang Q, Yu J, Jaroniec M (2011) Enhanced photocatalytic H_2-production activity of graphene-modified titania nanosheets. Nanoscale 3:3670–3678
404. Wang Y, Shi R, Lin J, Zhu Y (2010) Significant photocatalytic enhancement in methylene blue degradation of TiO_2 photocatalysts via graphene-like carbon in situ hybridization. Appl Catal B 100:179–183
405. Henglein A (1989) Small-particle research: physicochemical properties of extremely small colloidal metal and semiconductor particles. Chem Rev 89:1861–1873
406. Spanhel L, Weller H, Henglein A (1987) Photochemistry of semiconductor colloids. 22. Electron ejection from illuminated cadmium sulfide into attached titanium and zinc oxide particles. J Am Chem Soc 109:6632–6635

UV LED Sources for Heterogeneous Photocatalysis

Oluwatosin Tokode, Radhakrishna Prabhu, Linda A. Lawton, and Peter K.J. Robertson

Abstract This review article presents an overview of the application of ultraviolet light-emitting diode (UV LED) sources in heterogeneous photocatalysis within the context of artificial UV sources. The feasibility of UV LEDs as a source of UV irradiation in heterogeneous photocatalysis was first demonstrated almost a decade ago; however, for the most part, photocatalytic experimental set-ups utilise artificial light sources in the form of conventional UV lamps to initiate the desired photocatalytic transformations. A look at all sources of UV irradiation used in heterogeneous photocatalysis is taken with a focus on the growing importance of solid-state lighting devices such as UV LEDs. UV LEDs have higher external quantum efficiency and a lifetime of over 100,000 h; they are small in size and produce directional UV light which can be of the desired wavelength. In recent times, these UV LED sources have become widely applied in heterogeneous photocatalysis studies in the research literature and are fast becoming a viable alternative to conventional UV lamps.

Keywords Photocatalysis, Photoreactors, UV LED, UV light

Contents

O. Tokode (✉), R. Prabhu, and L.A. Lawton
Institute for Innovation, Design and Sustainability (IDEAS), Robert Gordon University, Garthdee Road, Aberdeen AB10 7GJ, UK
e-mail: o.i.tokode@rgu.ac.uk

P.K.J. Robertson
School of Chemistry and Chemical Engineering, Queen's University Belfast, David Keir Building, Stranmillis Road, Belfast BT9 5AG, Northern Ireland

D.W. Bahnemann and P.K.J. Robertson (eds.), *Environmental Photochemistry Part III*, Hdb Env Chem (2015) 35: 159–180, DOI 10.1007/698_2014_306,

1 Introduction

The demonstration of photocatalytic water splitting on titanium dioxide electrodes in 1972 by Fujishima and Honda [1] marked the beginning of heterogeneous photocatalysis as an active field of research. Four broad areas of research have emerged over the years; these are environmental remediation, kinetics and mechanistic studies, organic synthesis and energy applications. Several comprehensive review articles have been published to cover these broad areas and other specific areas such as principles and mechanisms [2–5], history and future prospects [6, 7], water splitting and hydrogen generation [8–10], water purification [11–14], air purification [15–17], reactor development [18–21], solar technologies [22] and catalyst synthesis, modification and immobilisation [23–25]. A keyword search in the Web of Science returns results showing the total number of publications on photocatalysis in the literature now exceeds 12,000.

2 Heterogeneous Photocatalysis

Heterogeneous photocatalysis can be divided into two kinds with respect to where the initial excitation takes place; when photoexcitation first occurs in the adsorbate molecule which then interacts with the catalyst substrate in the ground state, it is called a catalysed photoreaction; when the catalyst substrate is photoexcited and then transfers an electron to a molecule in the ground state, it is referred to as a sensitised photoreaction. The vast majority of studies in the literature on photocatalytic reactions are sensitised reactions; thus, photocatalysis takes place when light (photons) having energy (hv) equal to or greater than the bandgap (E_{bg}) of a semiconductor photocatalyst is absorbed by the catalyst particle. The primary steps after photon absorption are shown in reactions (1–8) [26].

$$TiO_2 + hv \rightarrow e_{CB}{}^{-} + h_{VB}{}^{+}, \tag{1}$$

$$e_{CB}{}^{-} \rightarrow e_{tr}{}^{-}, \tag{2}$$

$$h_{VB}{}^{+} \rightarrow h_{tr}{}^{+}, \tag{3}$$

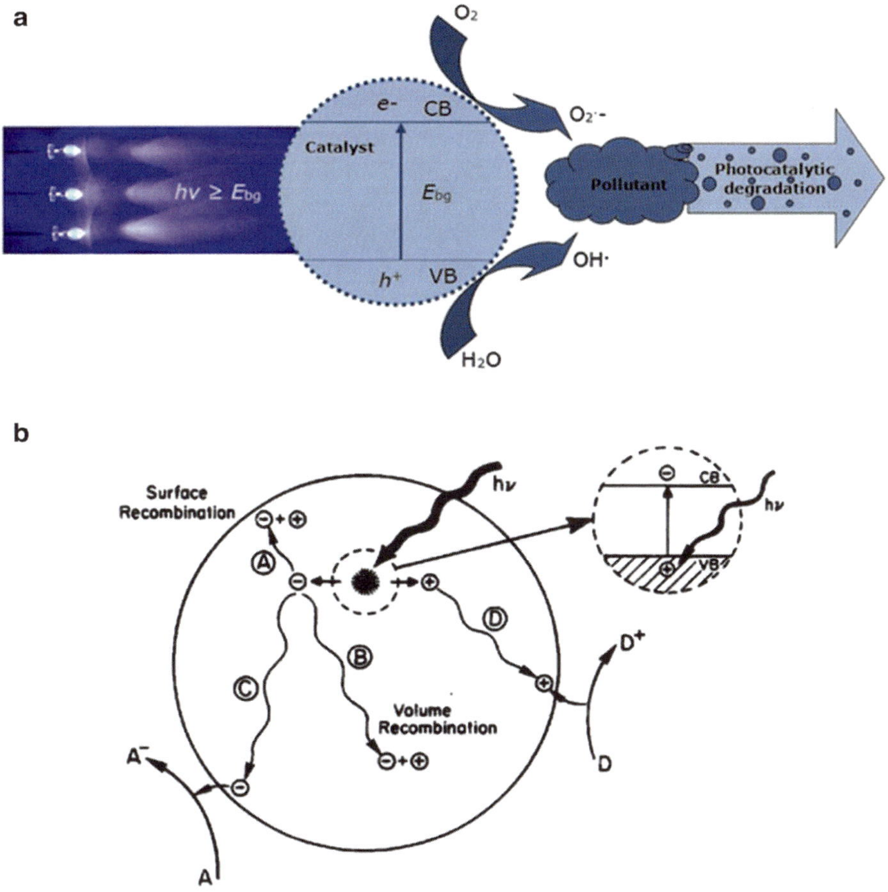

Fig. 1 (**a**) Illustration of pollutant degradation through heterogeneous photocatalysis. (**b**) Electron-hole pair pathways after photogeneration (Reprinted with permission from [4] Copyright 1995, American Chemical Society)

$$h_{VB}{}^{+} + H_2O \rightarrow OH^{\bullet}, \tag{4}$$

$$e_{CB}{}^{-} + O_2 \rightarrow O_2{}^{\bullet -}, \tag{5}$$

$$e_{CB}{}^{-} + h_{VB}{}^{+} \rightarrow TiO_2, \tag{6}$$

$$e_{tr}{}^{-} + h_{VB}{}^{+} \rightarrow TiO_2, \tag{7}$$

$$e_{tr}{}^{-} + h_{tr}{}^{+} \rightarrow TiO_2. \tag{8}$$

Upon bandgap excitation, charge separation due to the promotion of an electron ($e_{cb}{}^{-}$) from the valence to the conduction band takes place, generating a hole ($h_{vb}{}^{+}$) at the valence band in the process. The resulting electron-hole pair has several pathways which include surface and bulk recombination (Fig. 1), but for productive photocatalysis to occur, trapping of the $h_{vb}{}^{+}/e_{cb}{}^{-}$ or both is necessary.

This takes place on the surface of the photocatalyst in traps located below the edge of the conduction band [27, 28]; the redox reactions may also occur in the bulk volume as well. The highly reactive photogenerated h_{vb}^+ and e_{cb}^- can be directly involved in the oxidation and reduction of organic molecules, respectively, or indirectly through $OH^•$ and $O_2^{•-}$ which are equally reactive intermediate species with high-standard redox potentials [29, 30]. Photocatalysis is however not catalysis in the real sense of the word as the photons which initiate the desired photocatalytic reaction are also reactants and become consumed in the process. The recombination of the photogenerated electron-hole pair results in inefficient use of photons and this result in low quantum yields or photonic efficiencies that have been reported in the research literature on photocatalytic oxidation in both aqueous and gaseous phases [31].

2.1 Irradiation Sources for Heterogeneous Photocatalysis

The use of several catalysts have been reported in the literature on heterogeneous photocatalysis; they include metal oxides such as titanium dioxide (TiO_2), zinc oxide (ZnO), tin oxide (SnO_2) and metal chalcogenides such as cadmium sulphide (CdS) and zinc sulphide (ZnS). TiO_2 is by far the most investigated and widely used semiconductor catalyst in heterogeneous photocatalysis [32] not just because of its pioneering role in the Honda-Fujishima effect [1] but also for its superior activity [33, 34] when in the anatase polymorphic form and its photoinduced superhydrophilicity [35–38]. One of the factors determining the photocatalytic activity of a semiconductor catalyst is its light absorption properties. The bandgap energy of these commonly used catalysts fall within photon energies found in the UV range of the solar electromagnetic radiation reaching the earth with the exception of catalysts such as CdS having bandgap energies corresponding to photon energies within the visible range. Hence, CdS has been reported as a suitable visible light catalyst for heterogeneous photocatalysis [39, 40], while other semiconductor catalysts can also absorb visible light upon modification [41].

2.1.1 Natural Light Sources

The solar radiation reaching the earth is composed mainly of visible and infrared radiation (Fig. 2). At sea level, about 50% of this radiation is visible radiation, infrared radiation makes up 40%, while UV radiation accounts for <10%. Heterogeneous photocatalysis using natural light sources involves the activation of a photocatalyst using photons from sunlight having adequate photon energies to generate h_{vb}^+ and e_{cb}^- and subsequently, the intermediate $OH^•$ and $O_2^{•-}$. This area of photocatalysis is generally referred to as solar heterogeneous photocatalysis in the literature. The utilisation of cheaply available solar radiation in heterogeneous photocatalysis is advantageous when the high cost of artificial

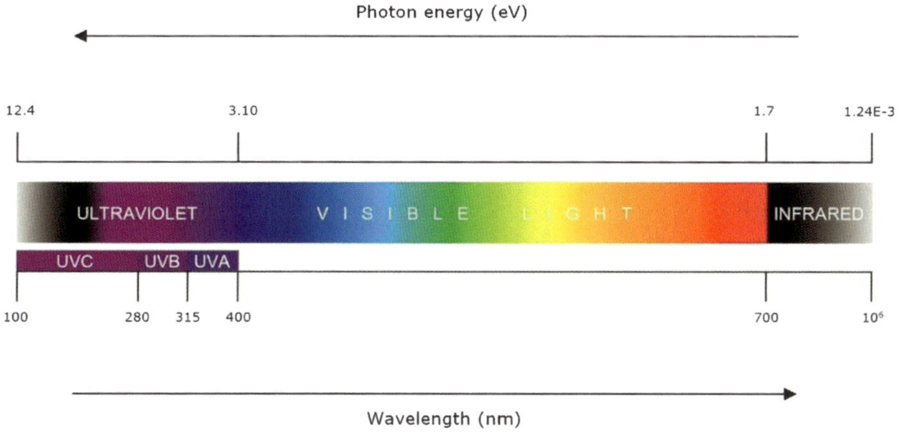

Fig. 2 Solar electromagnetic spectrum reaching the earth showing wavelengths (nm) and corresponding photon energies (eV)

light sources (lamps and electricity) is to be avoided; however, highly efficient solar photocatalysis systems are cost intensive and require complex designs and components.

The design of solar photocatalysis systems has generally followed conventional solar thermal designs especially the collectors used by both designs, but at this point the components for solar detoxification systems begin to look similar to those of other water treatment systems. The major design consideration in solar photoreactors with regard to irradiation source is whether to use non-concentrated or concentrated sunlight. Bockelmann et al. [42] compared solar water detoxification systems using concentrating and non-concentrating reactors and provide a detailed discussion of the advantages and disadvantages, while the design considerations and configurations for solar reactors have been reviewed by Alfano et al. [43]. Sunlight as a natural source of UV irradiation is cost-effective but also insufficient and not readily available in all geographic locations; however, it remains an area of active research with several studies particularly on water detoxification [22, 44, 45].

2.1.2 Artificial Light Sources

When artificial light sources have been preferred over natural light in heterogeneous photocatalysis, artificial light devices in the form of UV lamps, lasers, or light-emitting diodes (LEDs) have been employed. These devices typically emit either longwave (315–400 nm) or shortwave (200–315 nm) UV radiation. A survey of the literature on heterogeneous photocatalysis shows conventional UV lamps have been utilised in most experimental designs of photocatalytic reactions and reactors.

UV Lamps

The technology used for conventional UV lamps employed in heterogeneous photocatalytic reactions is of two types: incandescence and gas discharge. Gas discharge lamps can be divided into low-pressure lamps such as fluorescent lamps which have a working pressure below atmospheric pressure, high-pressure lamps and high-intensity discharge (HID) lamp. These lamps can also be divided into two broad categories: black lights and germicidal lamps. This division is based on the emission wavelength of the UV lamps; black lights emit longwave UV radiation and can be but are not limited to fluorescent and mercury vapour or incandescent lamps. Germicidal lamps emit shortwave UV radiation and find greater application in water disinfection because of the DNA disruption capacity of the high energy photons from the emitted UV light [46]. Germicidal lamps are better described as either low-pressure lamps or medium-pressure lamps [47], and they have secondary emissions in the longwave range which become dominant after the deterioration of the germicidal output towards the end of the rated life of the lamp [46, 47]. Unlike solar irradiation, UV lamps have a narrower emission spectrum which falls within the UV and visible region (Fig. 3); however, this emission is very broad when compared to UV lasers and LEDs.

A comparison between solar and artificial UV irradiation in the photocatalytic degradation of nitrobenzene at equivalent catalyst loading by Bhaktende et al. [48] showed a much faster photocatalytic reaction rate under artificial UV radiation even though concentrated sunlight was used. This is as a result of the percentage useful component (UV) present in the artificial lamp used being higher than that in sunlight; however, solar irradiation can show a higher efficiency when visible radiation as well as UV is utilised by the catalyst in the photocatalytic reaction [49].

UV Lasers

UV lasers produce coherent, monochromatic, high-intensity UV light with low beam divergence which in the case of tunable lasers can have their wavelengths altered to the desired range. The use of UV lasers in heterogeneous photocatalysis has been very limited to date due to the high cost of lasers, specialist training required for their operation, unsuitability for reactor design and the safety concerns associated with the use of lasers. Where they have been used, specific applications like pulsed laser deposition (PLD) [50–52], controlled periodic illumination (CPI) [53] and laser flash photolysis [54–57] have been investigated. However, there is also general interest in laser irradiated heterogeneous photocatalysis because of the unique properties of UV light produced from a laser beam which have been previously reported to enhance photocatalytic efficiency and conversion [58, 59].

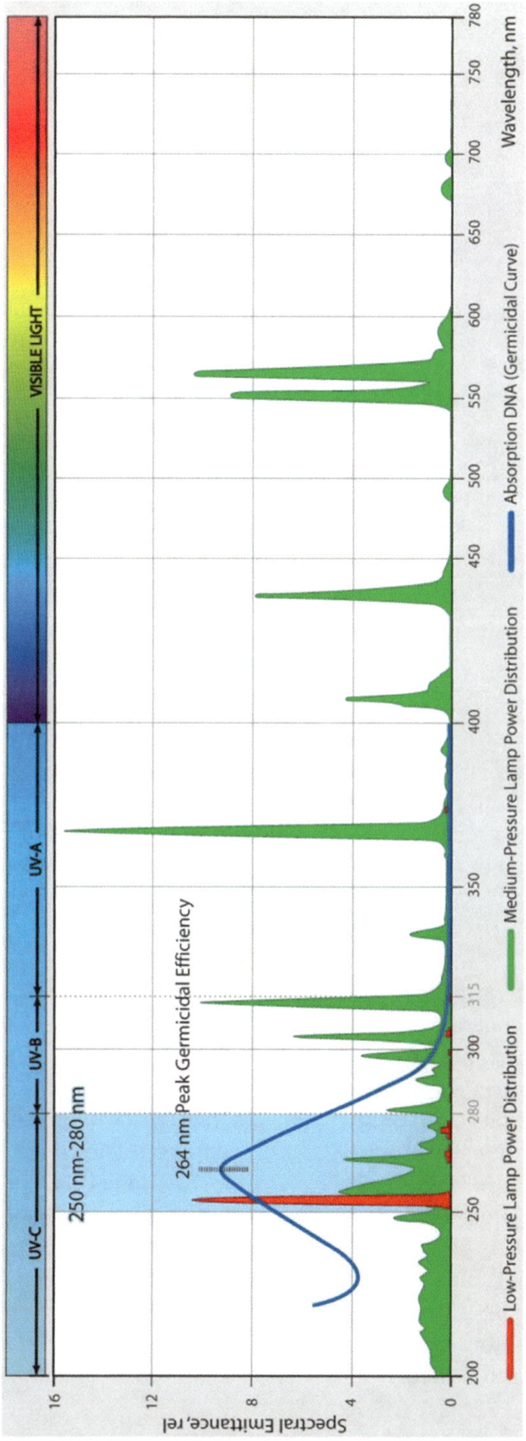

Fig. 3 Universal emission spectrum of a UV lamp showing wavelength of germicidal effectiveness (Copyright 2014, Emperor Aquatics, Inc.)

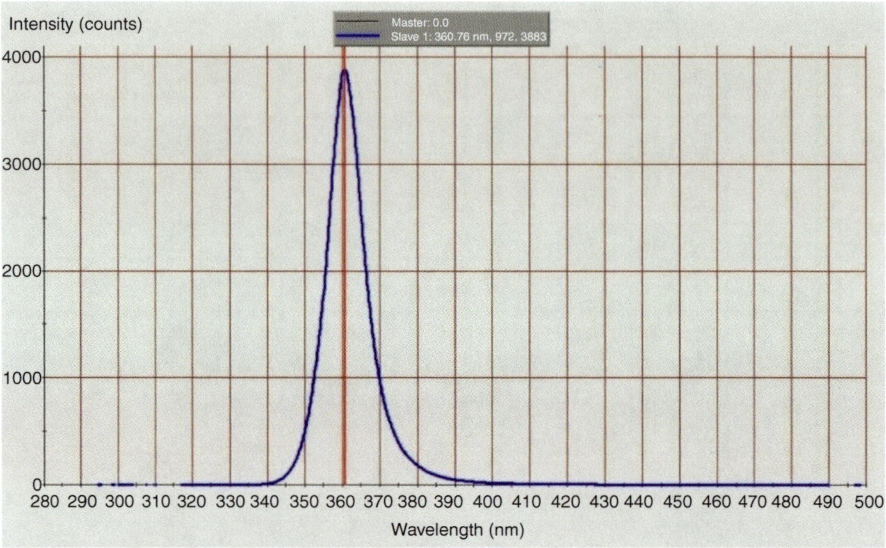

Fig. 4 Narrow emission spectra of a UV LED at 360 nm (Copyright 2014, The Fox Group Inc.)

UV LEDs

LEDs are solid-state light (SSL) sources which use inorganic semiconductors having a junction with hole-carrying p-layer and an electron-carrying n-layer to generate photons [60]. The application of a forward voltage to these layers ejects an electron and a hole from the n-layer and p-layer, respectively, and the recombination of the ejected electrons and holes within the device releases energy in the form of photons. The colour of light produced is determined by the bandgap energy of the semiconductor material. A UV LED is a light source which produces light of narrow band emission wavelengths within the ultraviolet range; hence, they are able to produce monochromatic light having a narrow emission spectra (Fig. 4).

The intensity of the produced light is strongly dependent on the forward voltage, while the energy of generated photons is dependent on the wavelength of emission. UV LED light gets brighter with increasing forward voltage, while the generated photons become more energetic with decreasing wavelength. These UV LEDs are cheap, rugged, compact and lightweight and have lower operating temperature compared to conventional lamps (Fig. 5). While the efficiency of fluorescent and incandescent lamps is limited by energy loss incurred in photon energy conversion and heat loss in the lamp filament, respectively, the efficiency of SSL sources like UV LEDs increases exponentially in parallel with advances in semiconductor technologies, material science and optics in accordance with Haitz's law [61].

UV LEDs are highly efficient because the materials used generally ensure the electron-hole pair is not trapped before recombination and the generated photons exit the device without being absorbed [62]. Technological progress in the

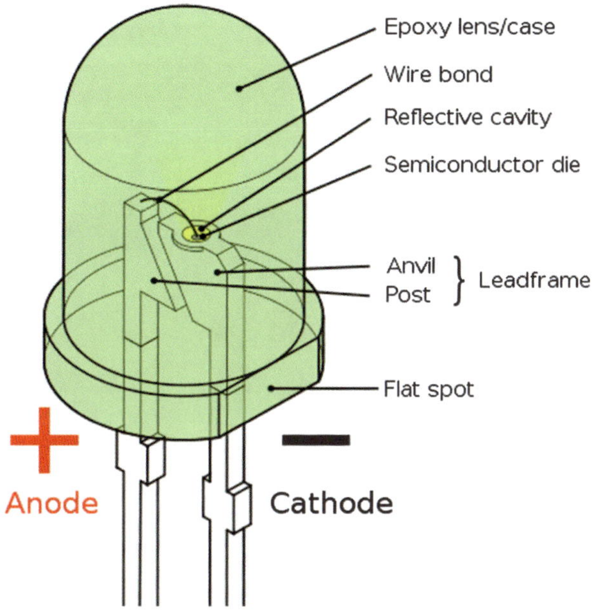

Fig. 5 Internals of the construction of a typical UV LED light bulb used in heterogeneous photocatalysis

performance of LEDs in general especially in the areas of operating efficiency (electrical input to UV output), power output and shorter wavelengths as well as cost has been rapid in recent years. However, this progress is faster in white light LEDs which are now used in general lighting applications [63, 64] than UV LEDs which have niche applications in sensors and sensing systems [65, 66], water disinfection [67, 68], phototherapy [69], forensic analysis [70], UV curing [71] and recently, heterogeneous photocatalysis.

3 UV LEDs in Heterogeneous Photocatalysis

The utilisation of UV LEDs as an irradiation source in heterogeneous photocatalysis began several years after such solid-state lighting devices had become mainstream alternatives in other areas of application. Chen et al. [72] claim to have carried out the first ever study employing UV LEDs in heterogeneous photocatalysis. They investigated the photo-oxidation of perchloroethylene (PCE) in a designed UV LED (375 nm) illuminated photocatalytic reactor. A PCE conversion of 43% was reported at very weak UV illumination of 49 μW cm^{-2}. Their study demonstrated the viability of UV LEDs as alternative irradiation sources for heterogeneous photocatalysis and concluded with a prediction of the replacement of conventional UV lamps by UV LEDs as the favoured source of irradiation in heterogeneous photocatalysis studies and applications. A search of the literature, however, revealed studies by Li et al. [73] who investigated visible light-

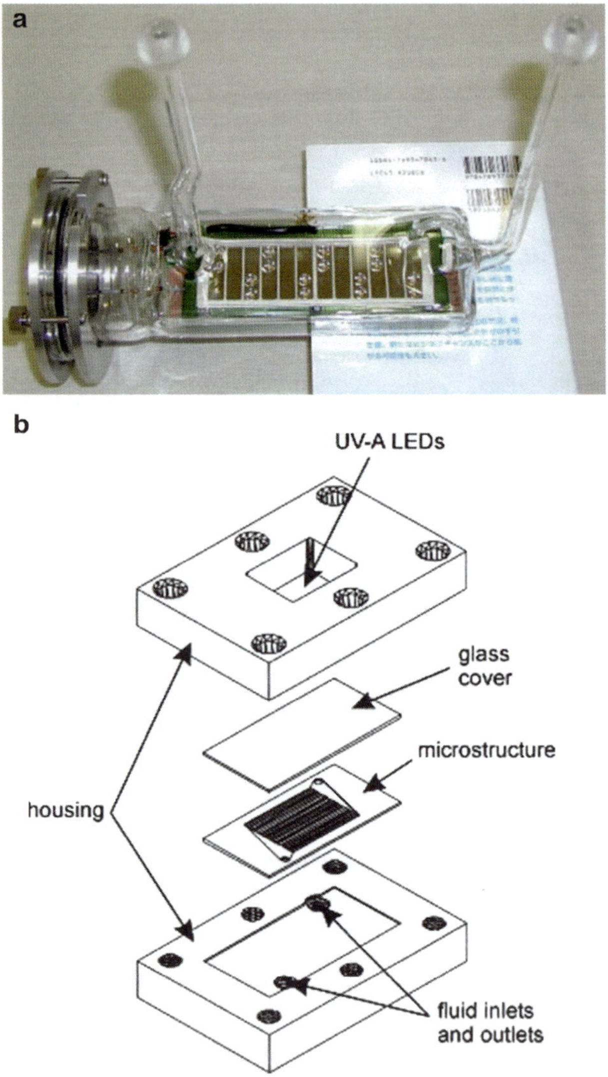

Fig. 6 Pioneering LED-based photocatalytic reactors: (**a**) VIS LED reactor by Li et al. (Reprinted from [73] Copyright 2014, with permission from Elsevier) and (**b**) UV LED reactor by Gorges et al. (Reprinted from [74] Copyright 2004, with permission from Elsevier)

driven photocatalytic decomposition of acetaldehyde in the gas phase using VIS LEDs (475 nm) and Gorges et al. [74] who studied photocatalysis in microreactors employing UV LED (385 nm) sources (Fig. 6) predating that by Chen et al. [72].

To date, studies employing conventional UV lamps as the UV irradiation source greatly outnumber those employing UV LEDs; nevertheless, the frequency of photocatalysis studies involving LEDs in the literature is on the increase (Fig. 7). UV LEDs have now gained recognition in testing and standardisation of photocatalytic irradiation processes. In a recent technical specification [75], the European Committee for Standardization (CEN) prescribed conditions under which

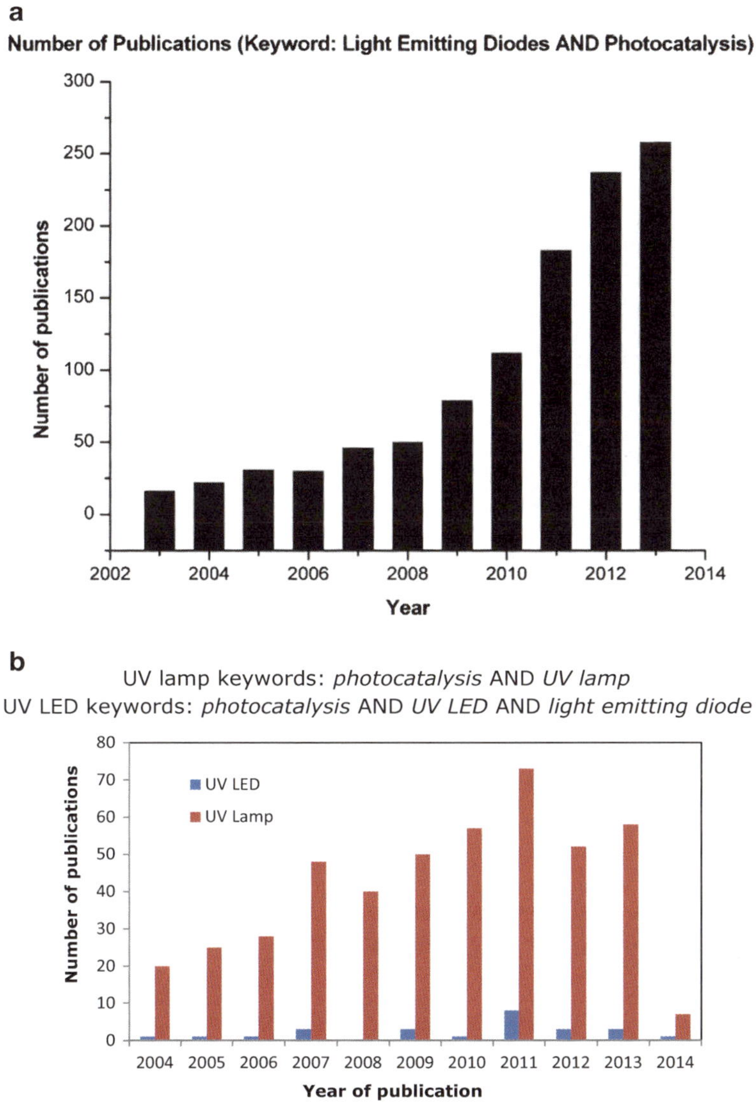

Fig. 7 Number of (**a**) publications involving LEDs and photocatalysis (Reprinted with permission from [76] Copyright 2014, American Chemical Society) and (**b**) peer-reviewed journal papers published on heterogeneous photocatalysis using UV LEDs or UV lamps as source of UV illumination. *Source*: Web of Science (Robert Gordon University database subscription)

photocatalytic surfaces should be irradiated during the performance of tests on photocatalytic efficiency. LEDs were included among other lamp types as sources of irradiation and their unique properties were adequately highlighted.

A startling number of these studies are still devoted to investigating the feasibility of UV LED sources in the heterogeneous photocatalysis, while several others

Table 1 Some heterogeneous photocatalysis studies in the literature employing UV LED sources

Primary investigation	Authors	References
Feasibility of UV LED sources	Chen et al.	[72]
	Shie et al.	[77]
	Levine et al.	[78]
	Tayade et al.	[79]
	Natarajan et al.	[80]
	Jo et al.	[81]
Visible light photocatalysis[a]	Li et al.	[73]
	Wang et al.	[82]
	Wang et al.	[83]
Energy efficiency of UV LED sources	Natarajan et al.	[84]
	Repo et al.	[85]
Microreactors	Gorges et al.	[74]
	Daniel and Gutz	[86]
	Matsushita et al.	[87]
Controlled periodic illumination	Chen et al.	[88]
	Tokode et al.	[89, 90]
Modelling/irradiance distribution	Wang et al.	[91]
	Jamali et al.	[92]
Pathogenic microorganisms	Xiong and Hu	[93]
	McCullagh et al.	[94]

[a]Involved VIS LEDs as opposed to UV LEDs

exploit specific advantages of UV LEDs in the photocatalytic degradation of various pollutants (Table 1).

3.1 UV LED Reactor Configurations

In the design of reactors for heterogeneous photocatalysis, the most important considerations to be made include mass transfer of target molecule to catalyst surface, maximising illuminated catalyst surface area and enhancing kinetic rate of reaction. With respect to the orientation of the light source and reaction chamber, reactors can be classified into immersion, external and distributive types. Immersion-type reactors have the light source internally situated, centrally or peripherally within the reaction chamber. When designed in the immobilised configuration, the lamp surface can also provide additional surface area for catalyst immobilisation in immersion-type reactors [95]. The light sources in external type reactors are situated outside the reaction chamber; this may result in decreased efficiency if the light source and the reactor are no longer a singular unit. Distributive type reactors are reactors in which the light for irradiation is distributed from the light source to the catalyst by means of reflectors or optical fibre. Optical fibre reactors (OFR) are distributive-type reactors which have the advantages of a fixed-bed reactor configuration while achieving equivalent efficiencies of a slurry system [96]. The advantage of distributed type reactors is that the light is incident on the catalyst surface without therefore limiting absorption by reactants and product phases and ensuring a large illuminated catalyst surface area. UV LED reactors can fall under any of immersion, external or distributive types and the various configurations reported in the literature are shown in Fig. 8.

Most of the UV LED reactors developed to date are experimental set-ups which do not fully integrate the UV LED array and reaction vessel into a single unit, thereby failing to take advantage of the small sizes and ease of coupling provided by this light source. Generally in reactor development, mass transfer limitations are greatest in the immobilised configuration because of the fixed catalyst and reduced surface area-to-volume ratio [99]. This reduces the efficiency of the photocatalytic process in immobilised reactors and makes them less efficient when compared with slurry reactors which employ suspended catalysts having the greatest surface area-to-volume ratio. The total illuminated catalyst surface area in contact with the reacting volume, κ, varies not only with the reactor configuration but with the illumination source employed as well. Ray and Beenackers [100] determined κ for different reactor configurations and the comparison is shown in Table 2. The miniature size of UV LEDs and the ease with which they can be coupled into an array can be exploited to ensure maximum illuminated surface area of the catalyst.

3.2 *Comparison of UV LEDs and Conventional UV Lamps*

UV LEDs fall under the category of solid-state light sources (SSL). These light sources differ greatly from incandescent and fluorescent sources and have revolutionised the lighting industry within the last decade. The benefits and potential of solid-state light sources have been well discussed by several authors [101–103], but with regard to heterogeneous photocatalysis, limited effort has been made to highlight the advantages of these light sources. An efficient source of irradiation for heterogeneous photocatalysis is one that has uniform light intensity and narrow emission spectra, minimises heat loss during operation, emits in a wavelength that is not absorbed by the substrate/pollutant, provides maximum illumination of the catalyst surface and is easily incorporated into reactor design to form a singular reactor unit. A comparison of conventional UV lamps and UV LEDs and their technological differences are highlighted in Table 3.

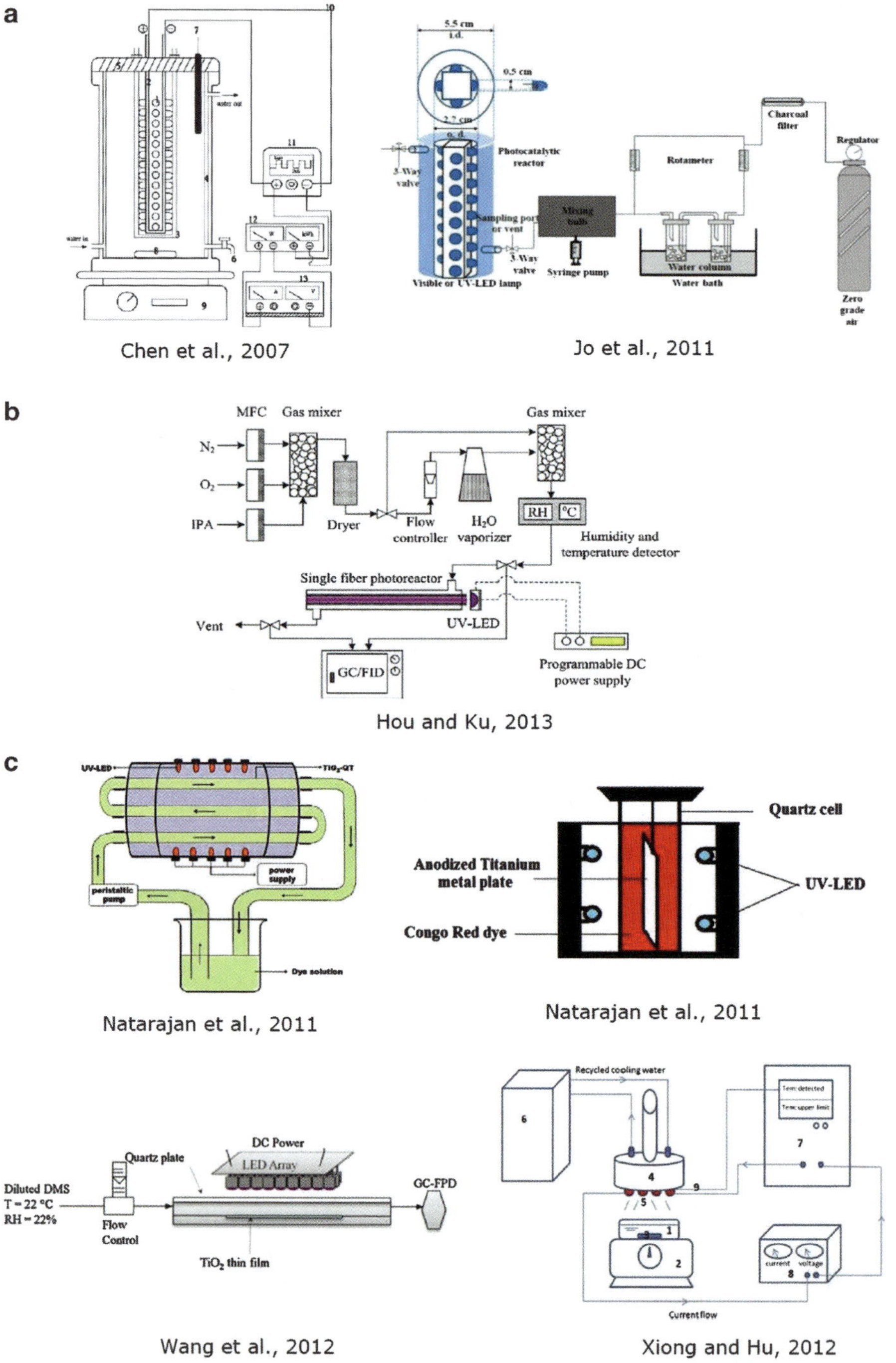

Fig. 8 UV LED reactor configurations in the heterogeneous photocatalysis literature: (**a**) immersive type, Chen et al. (Reprinted from [88] Copyright 2007, with permission from Elsevier) and Jo et al. (Reprinted from [81] Copyright 2011, with permission from Wiley); (**b**) distributive

Table 2 Comparison of illuminated surface area per unit treated volume, κ, of some common reactor configurations

Reactor configuration	κ (m^2/m^3)	κ (m^{-1})
Slurry reactor	$[6C_c/\rho_c]\ 1/d_p$	2,631
Immersion type (annular) with classical lamps	$[4\varepsilon/(1-\varepsilon)]\ 1/d_0$	133
Immersion type (annular) with modern lamps	$[4\varepsilon/(1-\varepsilon)]\ 1/d_0$	2,667
External type (annular)	$4d_0/(d_0^2 - d_l^2)$	27
Distributive type with hollow tubes	$[4\varepsilon/(1-\varepsilon)]\ 1/d_0$	2,000
Microreactors with UV LEDs[a]	$(2h+w)/hw$	11,667

[a]See [74] for further details

Table 3 Comparison of conventional UV lamps and UV LEDs

Area of comparison	UV LEDs	Conventional UV lamps
Rated life time	100,000 h	1,000–17,000 h
Emission spectra	Quasi-monochromatic (full width at half maximum)	Polychromatic
Toxicity	Non-toxic	Toxic (may contain substances like mercury)
Power output	Up to 1 W	Up to 1,000 W
Failure mode	Gradual failure	Abrupt failure
Instant start/on-off cycling	Achieve full brightness instantly	Achieve full brightness in several seconds
Wall plug efficiency	As high as 35%	As high as 23%
Size	Small (as small as 2 mm)	Large
External quantum efficiency (EQE)	As high as 40%	7–15%
Design possibilities	Easily integrated with reactors	Difficult to integrate with reactors
Durability	Rugged and durable	Fragile
Operating technology	Semiconductor technology and optics	Plasma physics and optics
Purchase cost	High (reducing rapidly with improving technology)	Low
Thermal radiation	Negligible	High

Fig. 8 (continued) type, Hou and Ku (Reprinted from [97] Copyright 2013, with permission from Elsevier); and (**c**) external type, Natarajan et al. (Reprinted from [98] Copyright 2011, with permission from Elsevier), Natarajan et al. (Reprinted with permission from [84] Copyright 2011, American Chemical Society), Wang et al. (Reprinted from [91] Copyright 2012, with permission from Elsevier) and Xiong and Hu (Reprinted from [93] Copyright 2013, with permission from Elsevier)

3.3 Efficiency of Heterogeneous Photocatalysis Using UV LED Sources

Heterogeneous photocatalysis is one of the advanced oxidation processes (AOPs) [104, 105] which exploit the in situ generation of highly reactive OH radical ($OH^•$) in the degradation of a variety of organic compounds. Among the various AOPs, OH radical generation through the H_2O_2/UV process (9) is the most efficient because it has a quantum yield of $\Phi\ (OH^•) = 1.11 \pm 0.07$ in the UV excitation range of 205–280 nm [106].

$$H_2O_2 + h\nu \rightarrow 2OH^{\bullet} \tag{9}$$

Generally, the efficiency of the photocatalytic process is referred to as the quantum yield [107] which is defined as the number of molecules changed divided by the number of absorbed photons (assuming all photons are absorbed by the catalyst and losses due to light scattering are negligible). For a species i, the quantum yield Φ is

$$\Phi x_i \equiv \frac{\pm(\mathrm{d}[X_i]/\mathrm{d}t)_0}{\mathrm{d}[h\nu]_{abs}/\mathrm{d}t}, \tag{10}$$

where Φx_i is the quantum yield for x_i, $\mathrm{d}[x_i]/\mathrm{d}t$ is the initial rate of formation or degradation of x_i and $\mathrm{d}[h\nu]/\mathrm{d}t$ is the rate of photon absorption by the catalyst. From an energy efficiency standpoint, however, H_2O_2 has a very strong absorption below 280 nm and a relatively low molar extinction coefficient, therefore requiring highly energetic photons while having a weak absorption of these incident photons. This increases the electrical energy demand of the H_2O_2/UV process and reduces its efficiency in photon utilisation. In a study by Muñoz et al. [108], the International Organization for Standardization (ISO) 14040 was used to apply a life cycle assessment (LCA) tool in determining the environmental impact of heterogeneous photocatalysis. The results show heterogeneous photocatalysis to have the greatest environmental impact among AOPs primarily because of the high electrical energy (power) consumption required by the conventional UV lamps which provided UV irradiation of the catalyst. For heterogeneous photocatalysis to be energy efficient, the photon generation and utilisation processes must be both efficient and cost-effective since photons are the single most expensive component required in any heterogeneous photocatalytic reaction.

The use of UV LEDs provides a greater overall efficiency for heterogeneous photocatalysis in regard to photon generation and utilisation. This can be attributed to the high energy conversion efficiency with which electrical power is converted into optical power in UV LED bulbs as well as the generation of photons of the appropriate wavelength for catalyst absorption. Bolton et al. [109] previously developed figures of merit called electric energy per order (E_{EO}) and electric energy per mass (E_{EM}) for evaluating energy and cost efficiency by relating the electrical energy consumption to the fundamental efficiency factors of photon absorption and quantum yield. These figures of merit can be used for comparison of different AOPs [110], reactors and UV sources in heterogeneous photocatalysis. Studies evaluating the efficiency of heterogeneous photocatalysis using these figures of merit can be found in the literature for UV LEDs [84, 85, 111] and conventional UV lamp [112, 113] sources; however, no direct comparison of the efficiency of a UV LED and conventional UV lamp irradiated heterogeneous photocatalytic reaction has been reported to date.

4 Lamp Emission Modelling for UV LED Sources

Most reactor configurations employed in heterogeneous photocatalysis are already well known and used for thermal and thermal-catalytic reactions, but the differentiating feature of photocatalytic reactions lies in the initiation of the reaction by light absorption. As a result of this, the importance of the characteristics of the illumination source and radiation field in reactors cannot be overemphasised. Generally in designing photocatalytic reactors, knowledge of the intrinsic kinetics of the desired reaction, lamp characteristics, reactor geometry and radiation distribution within the chosen reactor geometry is required. In order to obtain these details, the local volumetric rate of energy absorption (LVREA) must be known as the photocatalytic reaction will only proceed if the energy released when photons are absorbed is available to the reactants. The LVREA can be obtained through the solution of the radiative transport equation (RTE) which is an integro-differential equation comprising several equations, hence requiring an iterative solution. Its solution has been approached using a variety of methods found in the literature [114, 115].

Within the reactor geometry, the boundary condition for the RTE equation is provided by the lamp emission model (Fig. 9). Presently in the heterogeneous photocatalysis literature, the available models have been exclusively those of conventional, tubular UV lamps for the following reasons: (1) UV LEDs have just began to receive widespread application in reactor design, (2) modelling of UV LEDs is a more difficult process than conventional UV lamps, and (3) there is a wide range of UV LED radiation patterns [116].

Cassano et al. [117] have reviewed the various lamp emission models used in the solution of the RTE equation for various UV lamps, while Wang et al. [91] developed a model predicting the radiation pattern of UV LEDs in a continuous reactor. The suitability of such UV LED radiation field model for reactor design remains to be seen, while a general lack of studies on the solution of the RTE equation using UV LED specific lamp emission models presents a challenge for the design and analysis of reactors employing UV LED sources.

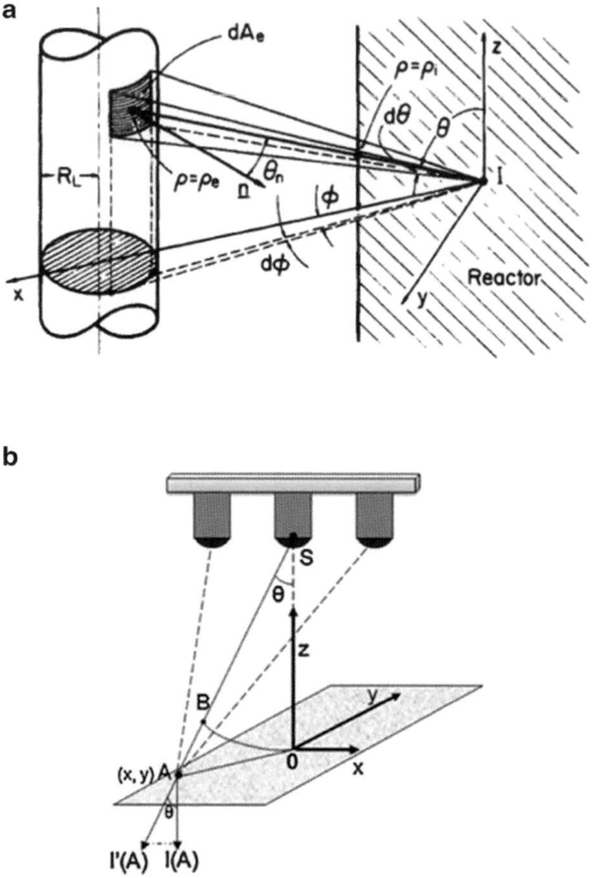

Fig. 9 Schematic of (**a**) the extended, superficial, diffuse emission (E-SDE) lamp emission model by Cassano et al. (Reprinted with permission from [117] Copyright 1995, American Chemical Society) and (**b**) UV LED radiation field model by Wang et al. (Reprinted from [91] Copyright 2012, with permission from Elsevier)

5 Future Prospects and Trends

The current surge in the number of studies utilising UV LED sources in heterogeneous photocatalysis studies coincides with the overall expansion of the UV LED market which is expected to exceed $280 million by 2017. Presently, about 90% of UV LED applications require ultraviolet light in the UVA/UVB wavelength range, and while UV curing applications currently dominate these applications, photocatalytic air and water purification applications are on the increase [118]. This market expansion signifies growing demand for UV LEDs which will prompt falling prices, innovation and new areas of application.

6 Conclusion

The use of UV LED sources in heterogeneous photocatalysis is currently experiencing a great deal of attention after several feasibility studies proved their viability in photocatalytic applications. A variety of compounds such as dyes, phenols, volatile organic compounds (VOCs) and toxins such as cyanotoxins have been photocatalytically treated using UV LED sources. There are also results in the literature which have shown the application of UV LEDs in photocatalysis studies where conventional lamps were unsuitable such as controlled periodic illumination, design of microreactors, energy efficient treatment and the use of monochromatic light. Despite the low power output of UV LEDs which limits their applications to pollutants at low concentration, these sources now present a practically competitive alternative light source in heterogeneous photocatalysis. Provided progress in LED technology continues to advance according to Haitz's law and suitable emission models required for reactor design are developed, UV LEDs have the potential to replace conventional UV lamps in photocatalytic applications.

References

1. Fujishima A, Honda K (1972) Nature 238:37
2. Hoffmann MR, Martin ST, Choi W, Bahnemann DW (1995) Chem Rev 95:69–96
3. Mills A, Le Hunte S (1997) J Photochem Photobiol A 108:1
4. Linsebigler LA, Lu G, Yates TJ (1995) Chem Rev 95:735–758
5. Fox AM, Dulay TM (1993) Chem Rev 93:341–357
6. Hashimoto K, Irie H, Fujishima A (2005) Jpn J Appl Phys 14:8269–8285
7. Fujishima A, Zhang X (2006) C R Chim 9:750
8. Chen X, Shen S, Guo L, Mao SS (2010) Chem Rev 110:6503
9. Kudo A, Miseki Y (2009) Chem Soc Rev 38:253
10. Ni M, Leung MK, Leung DY, Sumathy K (2007) Renew Sustain Energy Rev 11:401
11. Mills A, Davies HR, Worsley D (1993) Chem Soc Rev 22:417–425
12. Legrini O, Oliveros E, Braun AM (1993) Chem Rev 93:671–698
13. Friedmann D, Mendive C, Bahnemann D (2010) Appl Catal B 99:398
14. Bahnemann D, Bockelmann D, Goslich R (1991) Sol Energy Mater Sol Cells 24:564
15. Zhao J, Yang X (2003) Build Environ 38:645
16. Mo J, Zhang Y, Xu Q, Lamson JJ, Zhao R (2009) Atmos Environ 43:2229
17. Khan FI, Ghoshal AK (2000) J Loss Prev Process Ind 13:527
18. McCullagh C, Skillen N, Adams M, Robertson PKJ (2011) J Chem Technol Biotechnol 86:1002
19. Mozia S (2010) Sep Purif Technol 73:71
20. Bouchy M, Zahraa O (2003) Int J Photoenergy 5:191
21. Birnie M, Riffat S, Gillott M (2006) Int J Low Carbon Technol 1:47
22. Malato S, Fernández-Ibáñez P, Maldonado M, Blanco J, Gernjak W (2009) Catal Today 147:1
23. Chen X, Mao SS (2007) Chem Rev 107:2891
24. Li Puma G, Bono A, Krishnaiah D, Collin JG (2008) J Hazard Mater 157:209
25. Shan AY, Ghazi TIM, Rashid SA (2010) Appl Catal A 389:1
26. Goldstein S, Behar D, Rabani J (2009) J Phys Chem C 113:12489

27. Szczepankiewicz SH, Moss JA, Hoffmann MR (2002) J Phys Chem B 106:2922–2927
28. Wang H, He J, Gerrit Boschloo G, Lindstrom H, Hagfeldt A, Lindquist S (2001) J Phys Chem B 105:2529–2533
29. Sawyer DT, Valentine JS (1981) Acc Chem Res 14:393
30. Wood PM (1988) Biochem J 253:287
31. Zhang Z, Wang C, Zakaria R, Ying JY (1998) J Phys Chem B 102:10871
32. Diebold U (2003) Surf Sci Rep 48:53
33. Ohtani B, Prieto-Mahaney OO, Li D, Abe R (2010) J Photochem Photobiol A 216:179
34. Pelizzetti E (1995) Sol Energy Mater Sol Cells 38:453
35. Mills A, Crow M (2008) Int J Photoenergy 2008:Article ID 470670
36. Nakajima A, Koizumi S, Watanabe T, Hashimoto K (2000) Langmuir 16:7048
37. Miyauchi M, Nakajima A, Watanabe T, Hashimoto K (2002) Chem Mater 14:2812
38. Langlet M, Permpoon S, Riassetto D, Berthomé G, Pernot E, Joud JC (2006) J Photochem Photobiol A 181:203
39. Shiragami T, Pac C, Yanagida S (1990) J Phys Chem 94:504
40. Shiragami T, Ankyu H, Fukami S, Pac C, Yanaglda S, Mori H, Fujita H (1992) J Chem Soc Faraday Trans 88:1055
41. Chatterjee D, Dasgupta S (2005) J Photochem Photobiol C 6:186
42. Bockelmann D, Weichgrebe D, Goslich R, Bahnemann D (1995) Sol Energy Mater Sol Cells 38:441
43. Alfano OM, Bahnemann D, Cassano AE, Dillert R, Goslich R (2000) Catal Today 58:199
44. Sichel C, De Cara M, Tello J, Blanco J, Fernández-Ibáñez P (2007) Appl Catal B 74:152
45. Fernández P, Blanco J, Sichel C, Malato S (2005) Catal Today 101:345
46. Guo M, Hu H, Bolton JR, El-Din MG (2009) Water Res 43:815
47. Schalk S, Adam V, Arnold E, Brieden K, Voronov A, Witzke H (2006) IUVA News 8:32
48. Bhatkhande DS, Kamble SP, Sawant SB, Pangarkar VG (2004) Chem Eng J 102:283
49. Kuo W, Ho P (2001) Chemosphere 45:77
50. Zhao L, Han M, Lian J (2008) Thin Solid Films 516:3394
51. Suda Y, Kawasaki H, Ueda T, Ohshima T (2004) Thin Solid Films 453:162
52. Zhao L, Jiang Q, Lian J (2008) Appl Surf Sci 254:4620
53. Stewart G, Fox AM (1995) Res Chem Intermed 21:933–938(6)
54. Williams G, Seger B, Kamat PV (2008) ACS Nano 2:1487
55. Tanielian C, Duffy K, Jones A (1997) J Phys Chem B 101:4276
56. Tachikawa T, Tojo S, Fujitsuka M, Majima T (2006) Chem A Eur J 12:3124
57. Navío JA, Marchena FJ, Roncel M, De la Rosa MA (1991) J Photochem Photobiol A 55:319
58. Gondal M, Sayeed M, Seddigi Z (2008) J Hazard Mater 155:83
59. Yahaya AH, Gondal MA, Hameed A (2004) Chem Phys Lett 400:206
60. Bergh AA, Dean PJ (1976) Light emitting diodes. Clarendon, Oxford, 598p
61. Haitz R, Tsao JY (2011) Phys Status Solidi A 208:17
62. Bettles T, Schujman S, Smart JA, Liu W, Schowalter L (2007) IUVA News 9:11
63. Krames MR, Shchekin OB, Mueller-Mach R, Mueller GO, Zhou L, Harbers G, Craford MG (2007) J Disp Technol 3:160
64. Pimputkar S, Speck JS, DenBaars SP, Nakamura S (2009) Nat Photonics 3:180
65. O'Toole M, Diamond D (2008) Sensors 8:2453
66. de Lacy Costello BPJ, Ewen RJ, Ratcliffe NM, Richards M (2008) Sens Actuators B Chem 134:945
67. Vilhunen S, Särkkä H, Sillanpää M (2009) Environ Sci Pollut Res Int 16:439
68. Würtele MA, Kolbe T, Lipsz M, Külberg A, Weyers M, Kneissl M, Jekel M (2011) Water Res 45:1481
69. Hao Z, Zhang J, Zhang X, Ren X, Luo Y, Lu S, Wang X (2008) J Phys D 41:182001
70. Wawryk J, Odell M (2005) J Clin Forensic Med 12:296
71. Ollett SH, Lampe Jr RW (2005) US Patent 6,880,954
72. Chen DH, Ye X, Li K (2005) Chem Eng Technol 28:95

73. Li D, Haneda H, Ohashi N, Hishita S, Yoshikawa Y (2004) Catal Today 93–95:895
74. Gorges R, Meyer S, Kreisel G (2004) J Photochem Photobiol A 167:95
75. Eurpean Committee for Standardization (2014) Technical Committee CEN/TC 386 “Photocatalysis”, CEN/TS 16599
76. Jo W, Tayade RJ (2014) Ind Eng Chem Res 53:2073
77. Shie J, Lee C, Chiou C, Chang C, Chang C, Chang C (2008) J Hazard Mater 155:164
78. Levine LH, Richards JT, Coutts JL, Soler R, Maxik F, Wheeler RM (2011) J Air Waste Manage Assoc 61:932
79. Tayade RJ, Natarajan TS, Bajaj HC (2009) Ind Eng Chem Res 48:10262
80. Natarajan TS, Thomas M, Natarajan K, Bajaj HC, Tayade RJ (2011) Chem Eng J 169:126
81. Jo W, Eun S, Shin S (2011) Photochem Photobiol 87:1016
82. Wang X, Lim T (2010) Appl Catal B 100:355
83. Wang P, Fane AG, Lim T (2013) Chem Eng J 215–216:240
84. Natarajan TS, Natarajan K, Bajaj CH, Tayade JR (2011) Ind Eng Chem Res 50:7753
85. Repo E, Rengaraj S, Pulkka S, Castangnoli E, Suihkonen S, Sopanen M, Sillanpää M (2013) Sep Purif Technol 120:206
86. Daniel D, Gutz IGR (2007) Electrochem Commun 9:522
87. Matsushita Y, Ohba N, Kumada S, Sakeda K, Suzuki T, Ichimura T (2008) Chem Eng J 135 (Supplement 1):S303
88. Chen H, Ku Y, Irawan A (2007) Chemosphere 69:184
89. Tokode O, Prabhu R, Lawton LA, Robertson PKJ (2014) Chem Eng J 246:337
90. Tokode OI, Prabhu R, Lawton LA, Robertson PKJ (2012) J Catal 290:138
91. Wang Z, Liu J, Dai Y, Dong W, Zhang S, Chen J (2012) J Hazard Mater 215–216:25
92. Jamali A, Vanraes R, Hanselaer P, Van Gerven T (2013) Chem Eng Process Process Intensif 71:43
93. Xiong P, Hu J (2013) Water Res 47:4547
94. McCullagh C, Robertson JM, Bahnemann DW, Robertson PK (2007) Res Chem Intermed 33:359
95. Ray AK (1998) Catal Today 44:357
96. Peill NJ, Hoffmann MR (1995) Environ Sci Technol 29:2974
97. Hou W, Ku Y (2013) J Mol Catal A Chem 374–375:7
98. Natarajan K, Natarajan TS, Bajaj HC, Tayade RJ (2011) Chem Eng J 178:40
99. Turchi CS, Ollis DF (1988) J Phys Chem 92:6852
100. Ray AK, Beenackers AA (1998) AIChE J 44:477
101. Bergh A, Craford G, Duggal A, Haitz R (2001) Phys Today 54:42
102. Schubert EF, Kim JK (2005) Science 308:1274
103. Tsao JY (2004) IEEE Circuits Devices Mag 20:28
104. Topudurtir K, Tay S, Monschein E (1998) Advanced Photochemical Oxidation Processes Handbook ISBN-13: 978-1249248996
105. Andreozzi R, Caprio V, Insola A, Marotta R (1999) Catal Today 53:51
106. Goldstein S, Aschengrau D, Diamant Y, Rabani J (2007) Environ Sci Technol 41:7486
107. Tokode O, Prabhu R, Lawton LA, Robertson PKJ (2014) Appl Catal B 156–157:398
108. Muñoz I, Rieradevall J, Torrades F, Peral J, Domènech X (2006) Chemosphere 62:9
109. Bolton JR, Bircher KG, Tumas W, Tolman CA (2001) Pure Appl Chem 73:627
110. Muruganandham M, Selvam K, Swaminathan M (2007) J Hazard Mater 144:316
111. Chen H, Ku Y, Wu C (2007) J Chem Technol Biotechnol 82:626
112. Daneshvar N, Aber S, Seyed Dorraji MS, Khataee AR, Rasoulifard MH (2007) Sep Purif Technol 58:91
113. Khataee AR, Pons MN, Zahraa O (2009) J Hazard Mater 168:451
114. Romero RL, Alfano OM, Cassano AE (1997) Ind Eng Chem Res 36:3094
115. Pasquali M, Santarelli F, Porter JF, Yue P (1996) AIChE J 42:532
116. Moreno I, Sun C (1808) Opt Express 16(2008)
117. Cassano EA, Martin AC, Brandi JR, Alfano MO (1995) Ind Eng Chem Res 34:2155–2201
118. Mukish P (2013) Market & technology reports. Yole Développement

Semiconductor Photocatalysis for Atom-Economic Reactions

Horst Kisch

Abstract Based on preceding work on photoelectrochemistry at semiconductor single crystal electrodes the field of photocatalysis at semiconductor powders has experienced a tremendous growth in the last three decades. The reason for this is the genuine property of inorganic semiconductor surfaces to photocatalyze concerted reduction and oxidation reactions of a great variety of electron donor and acceptor substrates. Whereas high attention was paid to water splitting and exhaustive aerobic degradation of pollutants, only a small part of research explored synthetic aspects. After introducing the basic mechanistic principles, standard experiments for the preparation and characterization of visible light active photocatalysts and for the investigation of reaction mechanisms are discussed. Novel atom-economic C–C and C–N coupling reactions illustrate the relevance for organic synthesis. They exemplify that the multidisciplinary field of semiconductor photocatalysis combines classical photochemistry with electrochemistry, solid state chemistry, and heterogeneous catalysis.

Keywords Addition reaction, Catalysis, Photocatalysis, Semiconductor, Solar energy

Contents

H. Kisch (✉)
Department Chemie und Pharmazie, Universität Erlangen-Nürnberg, Egerlandstr.1, D-91058 Erlangen, Germany
e-mail: Horst.Kisch@fau.de

D.W. Bahnemann and P.K.J. Robertson (eds.), *Environmental Photochemistry Part III*, Hdb Env Chem (2015) 35: 181–220, DOI 10.1007/698_2013_251,

1 Introduction

Semiconductor photocatalysis research predominantly deals with the splitting of water into hydrogen and oxygen, the "holy grail" of photochemistry. The use of this process was already proposed in 1874 by Jules Verne in his book "The Mysterious Island": *Yes, my friends, I believe that water will one day be employed as fuel, that hydrogen and oxygen which constitute it will furnish an inexhaustible source of heat and light. Water will be the coal of the future.* Different from this artificial water cleavage, in photosynthesis instead of hydrogen, organic matter is produced by C–C coupling via carbon dioxide fixation. This complicated heterogeneous photocatalytic process can be decomposed into three basic reaction steps: photoinduced charge separation, charge trapping, and interfacial electron exchange with water. However, similar processes may occur upon irradiation of semiconductor surfaces in contact with liquid or gaseous substrates. The trapped charges can undergo interfacial electron transfer (IFET) with donor and acceptor substrates like air pollutants and oxygen, respectively. These reaction steps are the basis for the use of semiconductor photocatalysis for the exhaustive aerobic photooxidation of pollutants in air and water. This field of photocatalysis is already practically utilized for air purification [1, 2], but not for water splitting which is still a topic of academic research. Contrary to these degradation and cleavage reactions, only a small part of the literature is concerned with bond-forming processes of synthetic value analogous to C–C coupling in photosynthesis. The use of solar energy to drive organic syntheses was proposed already by G. D. Ciamician in 1912: *On the arid lands there will spring up industrial colonies without smoke and without smoke-stacks; forests of glass tubes will extend over the plants and glass buildings will rise everywhere; inside of these will take place the photochemical processes that hitherto have been the guarded secret of the plants, but that will have been mastered by human industry which will know how to make them bear even more abundant fruit than nature, for nature is not in a hurry and mankind is* [3, 4].[1]

Recent review articles deal with general photocatalysis of organic reactions and to a minor part also with semiconductor-photocatalyzed organic transformations,

[1] It is noted that the spectral composition of solar light arriving at the earth surface is approximately 3% UV, 45% Vis, and 52% IR.

all affording well-known products [5–9]. This review is primarily concerned with synthetically useful bond formations occurring by visible light excitation of semiconductor powders suspended in a liquid solution of substrates. After a brief introduction to semiconductor photocatalysis, novel atom-economic organic reactions and the activation of alkanes are discussed. All these processes contain as basic reaction steps the formation of C–C, C–N, and C–S bonds. The mechanistic discussion is based on experimentally proving the postulated key steps relevant to chemical synthesis. An attempt is also made to categorize the manifold of previously observed photocatalytic processes into two simple reaction classes, which are independent of the detailed mechanism.

2 Mechanistic Aspects

A general principle of conversion of light to chemical energy is schematically summarized by Eqs. (1)–(7) for the sensitization of the redox reaction $A + D = A_{red} + D_{ox}$ by a transition metal complex. Three key steps are photoinduced charge separation [Eq. (1)] and electron exchange with donor and acceptor substrates D and A affording primary redox products [Eqs. (3) and (4)]. The latter are converted to stable final products [Eqs. (5) and (6)]. In the case of *solar energy storage*, the overall reaction has to be endergonic, whereas for *solar energy utilization* it is in general exergonic. Whereas many systems undergo the first reaction step, only a few enable also the crucial electron exchange steps due to the highly favored charge recombination [Eq. (2)]. In most cases, an efficient back electron transfer (BET) between the primary redox products [Eq. (7)] prevents successive reactions to the final redox products [Eqs. (5) and (6)]. Thus, the basic problem of conversion of light to chemical energy is how to inhibit charge recombination and BET processes. In homogeneous systems, this can be solved to some extent by making one of the redox steps so fast that it successfully competes with charge recombination. In summary, the function of the transition metal complex is to photosensitize two *consecutive homogeneous* electron exchange reactions with a donor and acceptor substrate.

The reaction sequence discussed above differs significantly from photosensitization by a semiconductor, in general just named *photocatalysis*. In this case, a solid photocatalyst *simultaneously* sensitizes two *heterogeneous* redox reactions. By analogy with

$$|M|\text{—}L \xrightarrow{h\nu} |M|^{\cdot+}\text{—}L^{\cdot-} \tag{1}$$

$$|M|^{\cdot+}\text{—}L^{\cdot-} \rightarrow |M|\text{—}L + h\nu/\text{Wärme} \tag{2}$$

$$|M|^{\cdot+}\text{—}L^{\cdot-} + A \rightarrow |M|^{\cdot+}\text{—}L + A^{\cdot-} \tag{3}$$

$$|M|^{\cdot+}\text{—}L + D \rightarrow |M|\text{—}L + D^{\cdot+} \tag{4}$$

$$A^{\cdot-} \rightarrow A_{red} \quad (5)$$

$$D^{\cdot+} \rightarrow D_{ox} \quad (6)$$

$$A^{\cdot-} + D^{\cdot+} \rightarrow A + D \quad (7)$$

Eqs. (1)–(7), the basic reaction steps may be summarized in a simplified way according to Eqs. (8)–(10). Light absorption generates, inter alia, reactive electron–hole pairs trapped at the semiconductor surface. It is expected that the distance between these redox centers should be larger than in a molecular sensitizer and therefore charge recombination may become slow enough to allow the desired IFET at the solid/liquid or solid/gas interface (IFET). Note that the consecutive electron exchange reactions in the homogeneous case [Eqs. (3) and (4)] become concerted in the heterogeneous system [Eq. (10)]. The subsequent reaction steps are described by Eqs. (5)–(7):

$$SC \xrightarrow{h\nu} SC(e_r^-, h_r^+) \quad (8)$$

$$SC(e_r^-, h_r^+) \rightarrow SC + h\nu/\text{heat} \quad (9)$$

$$SC(e_r^-, h_r^+) + A + D \rightarrow SC + A^{-\cdot} + D^{+\cdot}. \quad (10)$$

In principle, one expects that similar to a molecule, the exited semiconductor may undergo also energy transfer. Indeed, recent experimental findings evidence formation of singlet oxygen according to Eqs. (11) and (12) [10–12]. However, it cannot completely be ruled out that singlet oxygen may be also produced by hole oxidation of superoxide [Eq. (13)] generated through reduction of oxygen by the reactive electron [10, 11]:

$$SC \xrightarrow{h\nu} SC(e_r^-, h_r^+) \quad (11)$$

$$SC(e_r^-, h_r^+) + O_2 \rightarrow SC + {}^1O_2 \quad (12)$$

$$O_2^- + h_r^+ \rightarrow {}^1O_2. \quad (13)$$

The bandgap E_g is defined as the energy difference between the edges of conduction and valence band. It determines the minimum energy necessary for optical excitation. It amounts to 3.2 eV for the frequently used anatase crystal phase of titania. Accordingly, only light of wavelengths shorter than 390 nm can be absorbed (Fig. 1). For cadmium and zinc sulfide, the corresponding values are 2.4 eV/520 nm and 3.6 eV/345 nm, respectively.

The above discussion applies for an idealized semiconductor crystal without any impurities or surface defects. In reality however, this condition is never met in practical photocatalysis. As a consequence thereof, new and *localized* electronic states located within the bandgap, generally called *surface states,* are always present in a real material. They may be generated also by adsorption of solvent and substrates onto the semiconductor surface. The surface states can act as charge

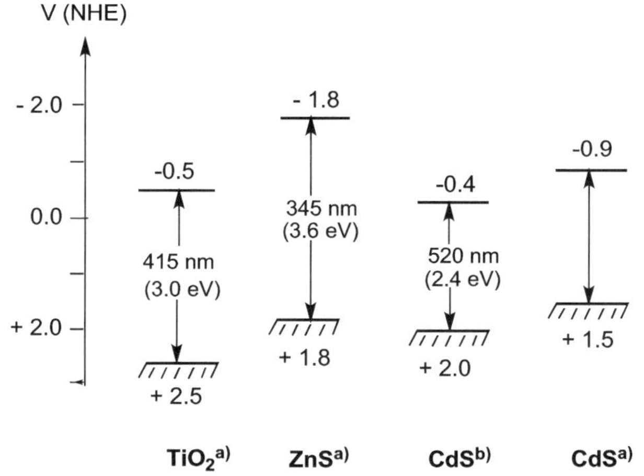

Fig. 1 Light absorption onsets, bandgap energies, and band edge positions of some semiconductor powders in contact with neutral water for titania [21], ZnS [157], CdS (*a*) [158]; (*b*) [43]. (*a*) and (*b*) refer to single crystal and powder, respectively

recombination centers but also as reactive electron–hole pairs, thus controlling the IFET and therefore the photocatalytic activity.

2.1 *Primary Processes*

Scheme 1 schematically summarizes the primary processes occurring after light absorption of a semiconductor powder in contact with an aqueous solution of an electron donor D and an acceptor A. Dissociation of the light-generated exciton and relaxation affords electrons and holes energetically located close to the corresponding edges of the energy bands. From there, they may undergo nonradiative and radiative primary recombination and trapping at nonreactive (e^-_{tr}, h^+_{tr}) or reactive (e^-_r, h^+_r) surface sites. The corresponding energy levels are located at the surface and not in the bulk as may be assumed from the simplified Scheme 1. Depicted life times were taken from results obtained for colloidal titania [13, 14]. It is noted that band bending, responsible for charge separation at semiconductor single crystal electrodes, is probably not relevant for powders discussed in this review. They contain in general crystallites of 20–100 nm size, which are too small for invoking the presence of a charge depletion layer. This is supported by the fact that the *incident photon to current efficiency* of a titania powder electrode does not depend on the applied potential, which determines the band bending, but on the ease of donor oxidation [15]. Therefore, the efficiency of charge separation is determined by competition between recombination and IFET reactions.

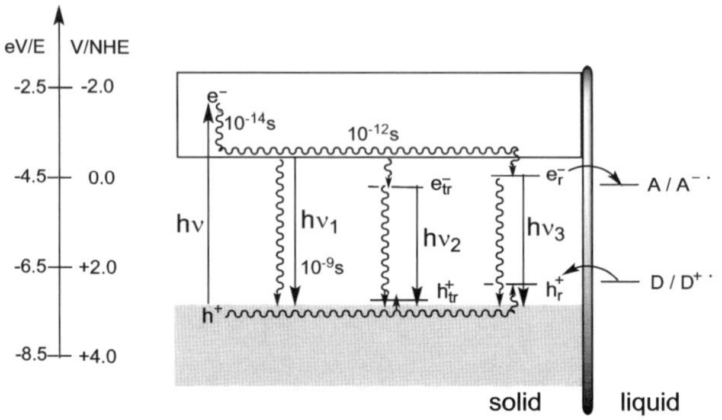

Scheme 1 Simplified scheme of the primary processes occurring after light absorption at a semiconductor–liquid interface. *Straight* and *waved lines* correspond to radiative and nonradiative processes, respectively. The *thick vertical bar* symbolizes the solid/liquid interface. In this scheme, electrons and holes gain stability when moving down and up, respectively. Throughout this article, all band edge and Fermi potentials refer to a semiconductor in contact with water of pH 7 unless otherwise noted. The depicted band positions apply for titania. The electron energy (eV) is given relative to the vacuum level

Experimental differentiation of the various surface sites may be possible by combining luminescence quenching with product inhibition studies, a standard procedure in homogeneous molecular photochemistry. The reactive electron–hole pair can undergo the IFET reactions with A and D [Eqs. (14) and (15)] if the corresponding reduction potentials are located at *appropriate* positions *within* the bandgap as depicted in Scheme 1. In some cases, IFET may be coupled with proton transfer (see Chaps. 3 and 4). Consecutive reactions of the primary redox products in general afford reduced and oxidized products [Eq. (16)], and only in a few cases, intermolecular bond formation may lead to one unique addition product [Eq. (17)]. Whereas the first case is equivalent to photoelectrochemical reaction schemes, the second case is novel and may include more than two addition partners (see Chaps. 3 and 4). We have proposed to term the first and second reaction modes as *semiconductor photocatalysis type A and type B*, respectively [16]. It is generally assumed that D and A have to be adsorbed at the semiconductor surface in order to enable high IFET rates. However, redox transformations of substrates having no contact with the surface at all have been also observed. Rare examples are aerobic oxidation reactions. In this case, the IFET with adsorbed oxygen and water generates short-lived reactive oxygen compounds like OH and HO_2 radicals (vide infra) capable of diffusing to remote substrates [17–19].

The simplified Scheme 1 does not contain electron transfer reactions with the material itself leading to surface destruction and therefore to photocatalyst deactivation, the process of so-called photocorrosion. Whereas titania is quite photostable, zinc and cadmium sulfide in general suffer from photocorrosion to elemental metal and sulfur as a result of redox reactions with the lattice components in the absence of air. In its presence, metal and sulfate ions are the final products. The tendency to photocorrosion often is highest for very pure materials suggesting that surface states of less pure semiconductors facilitate not only the undesired recombination but also the desired IFET in favor of photocorrosion:

$$A + e_r^- \rightarrow A^{-\cdot} \tag{14}$$

$$D + h_r^+ \rightarrow D^+ \tag{15}$$

$$A + D \underset{SC}{\overset{h\nu}{\rightarrow}} B_{red} + C_{ox} \tag{16}$$

$$A + D \underset{SC}{\overset{h\nu}{\rightarrow}} A - D. \tag{17}$$

The thermodynamics of the two IFET reactions [Eqs. (14) and (15)] can be estimated by comparing the substrates reduction potentials with the potentials of the reactive electron–hole pair, i.e., the quasi-Fermi levels of electrons and holes. Since the latter are not known exactly, the positions of the corresponding band edges are taken. It is recalled that the quasi-Fermi level is defined as the Fermi level measured under irradiation [20]. In general, the difference between the two levels is rather small [21].

Aerobic photooxidation reactions catalyzed by semiconductor surfaces are already technically applied for air purification [22] by solar light. Different from conventional air cleaning based on adsorption, which requires a final chemical degradation of the pollutant, photocatalysis removes the pollutant by an exhaustive aerobic photooxidation. In most cases, the final products are harmless compounds like carbon dioxide, water, and nitrate or sulfate. The two mechanisms operating in these reactions are used to exemplify two basic actions of a semiconductor surface during a photocatalytic reaction. In *direct semiconductor photocatalysis*, light ($h\nu_1$) absorbed by the semiconductor and charge recombination is prevented by the faster IFET reactions [Eqs. (18) and (19)] as schematically summarized at the left part of Scheme 2. Oxidation of the pollutant D may occur also via intermediate OH radicals generated by competitive water oxidation. In *indirect* photocatalysis, light absorption ($h\nu_2$) occurs by the substrate D having an excited state reduction potential negative enough for electron injection into the conduction band under formation of the radical cation $D^{+\cdot}$ The resulting spatial separation of the two generated charges slows down the BET according to Eq. (20) enabling successful competition via oxygen reduction and oxidative decomposition of D^+ by superoxide or its secondary products [Eqs. (18) and (21)]. Thus, no valence band hole is formed in *indirect* photocatalysis, and the semiconductor functions as an electron relay preventing the undesired back reaction:[2]

$$O_2 + e_r^- \rightarrow O_2^{\bullet -} \tag{18}$$

[2] Many authors prefer the term "photosensitized photocatalysis" for such a type of reaction. This is not correct, since according to its definition a sensitizer cannot simultaneously function as substrate.

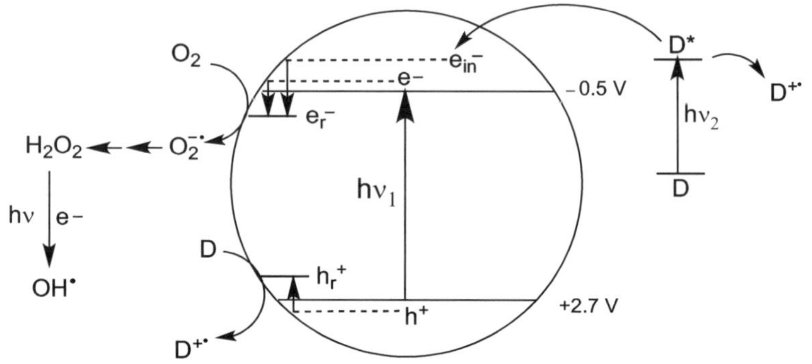

Scheme 2 Pictorial view of the primary processes for direct ($h\nu_1$) and indirect ($h\nu_2$) semiconductor photocatalysis exemplified for titania-catalyzed aerobic photooxidations. The *circle* symbolizes a large semiconductor crystal or an aggregate of nanocrystals, and the *two horizontal lines* represent the band edges at pH 7

$$D + h_r^+ \rightarrow D^{+\cdot} \tag{19}$$

$$D^{+\cdot} + e_r^- \rightarrow D \tag{20}$$

$$D^{+\cdot} + O_2^{\bullet -}/HO_2^{\bullet}/OH^{\bullet} \rightarrow C_{ox}. \tag{21}$$

In some cases, the substrates may interact with the semiconductor surface and induce appearance of a new absorption band having charge-transfer (CT) character. Typical examples are various aromatic compounds and sulfur dioxide [23]. In the case of aromatic 1,2-diols like catechol, stable red chelate complexes are formed. They exhibit a broad absorption band with a maximum at 420 nm extending down to 600 nm [24–26]. Thus, different to aliphatic 1,2-diols, aromatic ones can be photooxidized in the presence of titania not only by UV but also by visible light. Also in this case, excitation generates a conduction band electron and a donor radical cation [Eqs. (22) and (23)], as discussed above. But this occurs via an optical electron transfer and not through a photoinduced electron transfer involving the excited state of D (see Scheme 2). Irrespective of this difference in the mechanism of charge generation, in both cases, no valence band hole is formed and also the CT mechanism can be classified as *indirect* semiconductor photocatalysis:

$$SC—D \overset{h\nu_{CT}}{\rightarrow} SC^{\cdot -}—D^{\cdot +} \tag{22}$$

$$SC^{\cdot -}—D^{\cdot +} + O_2 \rightarrow SC + D^{\cdot +} + O_2^{\cdot -}. \tag{23}$$

Both direct and indirect photocatalysis transform the initially generated superoxide to reactive oxygen compounds of high oxidative power [Eqs. (24)–(29)] having reduction potentials of 0.94 V (O_2^-/H_2O_2), 1.29 V (H_2O_2/H_2O), and 1.90 V (OH/OH^-). (All values apply for pH 7; all reduction potentials apply for aqueous solutions relative to NHE as defined and summarized by [27].) They are involved in the thermal degradation steps of the radical cation and can induce auto-oxidation reactions:

$$O_2^{\bullet-} + H^+ \rightarrow HO_2^{\bullet} \quad (24)$$

$$HO_2^{\bullet} + HO_2^{\bullet} \rightarrow H_2O_2 + O_2 \quad (25)$$

$$O_2^{\bullet-} + HO_2^{\bullet} \rightarrow O_2 + HO_2^- \quad (26)$$

$$HO_2^- + H^+ \rightarrow H_2O_2 \quad (27)$$

$$H_2O_2 + O_2^{\bullet-} \rightarrow {}^{\bullet}OH + OH^- \quad (28)$$

$$H_2O_2 + e_r^- \rightarrow {}^{\bullet}OH + OH^- \quad (29)$$

2.2 *Characterization of Semiconductor Powders*

In addition to standard characterization methods of heterogeneous catalysts like bulk and surface elemental analysis, X-ray powder diffraction (XRD), specific surface area measurements, diffuse reflectance spectroscopy (DRS), emission spectroscopy, and electron paramagnetic resonance [28, 29] especially photoelectrochemical experiments are of basic importance. Only the latter may proof the presence of semiconductor properties in order to exclude a general heterogeneous photocatalysis mechanism, which does not involve simultaneous reductive and oxidative IFET reactions [30]. Lifetimes of photogenerated charges may be obtained by time-resolved DRS, photoluminescence [31–33], and surface photovoltage measurements (Dember voltage). In the latter, the semiconductor is often embedded in a polymer film of a few micrometer thickness. The obtained photovoltage decay curve in general consists of a slow and fast component representing bulk and surface recombination, respectively. Typical lifetimes of surface charges, relevant for photocatalysis, are in the range of micro- to nanoseconds. The sign of the photovoltage may indicate the presence of an *n*- or *p*-type material [34, 35].

Bandgap values are easily obtained by DRS measurements. In this method, the diffuse reflectance of the powder is recorded relative to a white standard like alumina or barium sulfate. The Kubelka–Munk function $F(R_\infty)$ is equivalent to absorbance and is obtained from Eq. (30) wherein R_∞ is the diffuse reflectivity [Eq. (31)] of an infinitely thick sample layer and k and S are the absorption and scattering coefficients, respectively. It is noted that Eq. (30) applies only for (1) monochromatic irradiation,

$$F(R_\infty) = (1 - R_\infty)^2/2R_\infty = k/S \quad (30)$$

$$R_\infty = R_{\text{sample}}/R_{\text{standard}}; \quad (31)$$

(2) infinite thick sample layer (is for most powders reached at about 5 mm), (3) low sample concentrations, (4) uniform distribution, and (5) absence of fluorescence. Although dilution with the white standard improves resolution, this is

Scheme 3 Dependence of driving force of the IFET reduction of methylviologen on the pH value. (**a**) pH < H_o, (**b**) pH = pH_o, (**c**) pH > pH_o

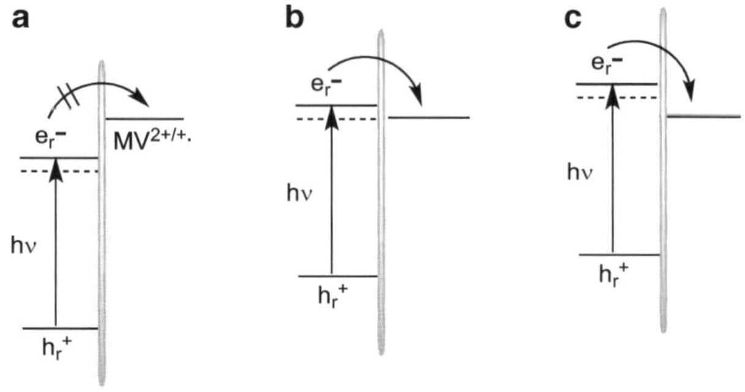

only rarely done in the literature [36]. To obtain the bandgap, a modified Kubelka-Munk function is plotted as function of the energy

$$(F(R_\infty)h\nu)^{1/n} \propto h\nu - E_g \tag{32}$$

of the exciting light according to Eq. (32). Extrapolation of the linear part of the resulting curve affords the bandgap. The number n depends on the nature of the electronic transition and is 1 for a direct and 0.5 for an indirect semiconductor of crystalline structure [37]. It is noted that an unambiguous conclusion may not be possible since the extrapolation is connected with a rather large error [38].

The *quasi-Fermi level* of electrons ($_nE_F^*$) may be obtained from the photocurrent onset [39, 40] or from the light intensity saturation of the photocurrent [41, 42], both measured with an electrode having the semiconductor as a thin powder layer on conducting glass. When the quasi-Fermi level is located not too far from the pH-independent reduction potential of a dissolved redox couple like methylviologen, the so-called suspension method can be used in the case that the Fermi level is pH-dependent. The latter requirement is usually met for oxidic and sulfidic semiconductors exhibiting dependence according to Eq. (33), wherein the constant k is usually in the range of 0.059 V. Thus, in alkaline suspensions, the Fermi level of electrons is located at more negative potentials than in neutral or acidic ones. According to thermodynamics, the IFET reduction of MV^{2+} is feasible only when the quasi-Fermi level becomes equal (Scheme 3, pH = pH_o) or more negative than the methylviologen potential of −0.44 V. This is also visually recognized by the appearance of the blue color of the viologen radical cation upon increasing the pH value. Depending on the reduction potential of h_r^+, it may oxidize water, another donor, or the semiconductor itself (photocorrosion) [21, 43, 44]. Initially the photocurrent of the powder suspension in water was measured as function of the pH value with a standard three-electrode setup using a platinum flag as working electrode. Depending on the irradiation equipment, $_nE_F^*$ values may slightly depend on the light intensity. Upon a tenfold increase of the latter, a cathodic shift of 27 mV was observed for CdS [21]. The onset of the photocurrent corresponds to the quasi-Fermi level of electrons. All values cited in this review apply for

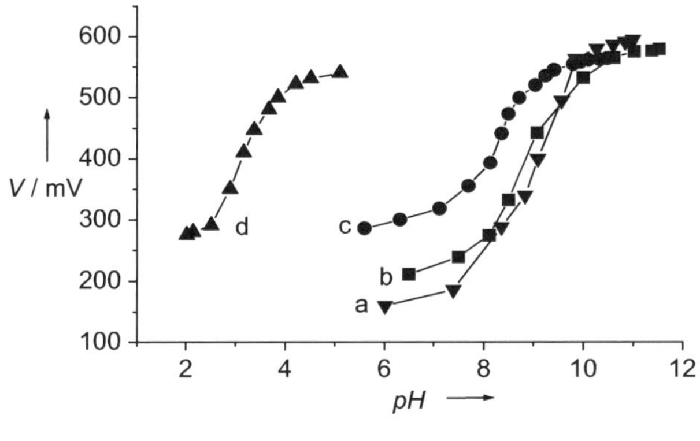

Fig. 2 Dependence of photovoltage on suspension pH value of CdS (*a*) and CdS–O–SiO_2 containing 50% (*b*), 30% (*c*), and 12% (*d*) CdS

pH 7 as calculated from Eq. (33) and are given relative to NHE, unless otherwise mentioned:

$$E_F(\text{pH}) = E_F(\text{pH} = 0) - k\text{pH}_o. \tag{33}$$

When in the suspension method instead of photocurrent, the photovoltage between the platinum flag and a reference electrode is measured, one obtains a sigmoidal type of voltage vs. pH value titration curve [44]. Its shape depends on the potential of the reference electrode, the $[MV^{2+}]/[MV^{+\bullet}]$ ratio, the pH value, k, and on ${}_nE^*_F$. At the pH value of the inflection point (pH_o), the quasi-Fermi level becomes equal to −0.44 V, the potential of methylviologen. Figure 2 illustrates the use of this convenient method for the investigation of electronic interactions in supported or "coupled" semiconductor photocatalysts. A considerable cathodic shift of 0.21 V is observed upon grafting 12% of CdS onto silica as evidenced by the shift of pH_o from −0.38 to −0.59 V upon grafting CdS onto silica [43]. As mentioned above, the factor k has to be known in order to refer quasi-Fermi levels to the same pH values. For suspensions, it can be obtained from the slope of the voltage-current plot above the inflection point [44] or from the slope of the onset of photocurrent vs. pH value [21]. In our hands, both methods did not exhibit good reproducibility due to considerable voltage fluctuations and very low photocurrents. Alternatively, the pH_o point can be measured not only for one but also for a series of pH-independent redox couples [45]. Figure 3 depicts the linear dependence of pH_o values on the electron acceptor reduction potentials. From the slopes, k values of 0.050 and 0.060 V are obtained for titania and a chloroplatinate (IV)-modified titania.

Knowing the quasi-Fermi level of electrons, also the level of holes can be estimated by adding the bandgap energy. This rough but helpful procedure is based on the assumption that both levels are located very close to the corresponding band edges. Since most of the employed powders represent highly doped *n*-semiconductors, this is a reliable approximation. It is noted that the position of the Fermi level depends not only on the presence of impurities and modifiers but also on the

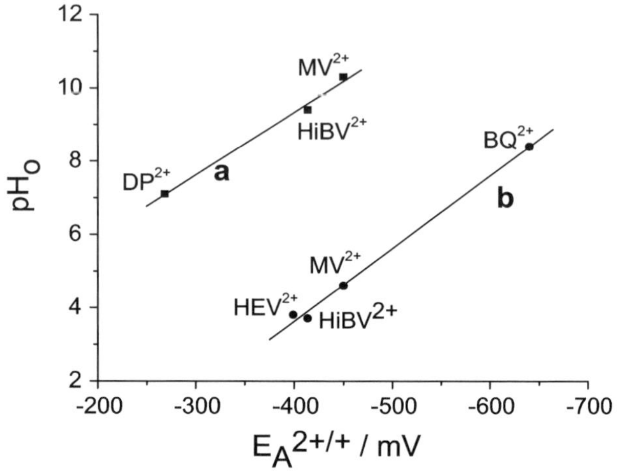

Fig. 3 Dependence of the pH_o value of titania (*a*) and of a chloroplatinate(IV)-modified titania on the reduction potential of various bipyridinium electron acceptors. MV^{2+} = methylviologen, $HiBV^{2+}$ = *N*,*N*′-bis(2-methyl-3-hydroxypropyl)-4,4′-dipyridinium, HEV^{2+} = *N*,*N*′-bis (2-hydroxyethyl)-1,1′-dipyridinium, BQ^{2+} = *N*,*N*′-1,4-butandiyl-2,2′-byridinium, DP^{2+} = *N*,*N*′-1,2-ethandiylphenanthrolinium [45]

nature of adsorbed substrates and solvent. Thus, cathodic shifts of up to 1.0 V were reported upon cleaning the CdS surface of a single crystal electrode from elemental sulfur and cadmium [46]. When titania was suspended in acetonitrile instead of water, the Fermi level shifted by 0.4 V cathodically [47].

When the suspension photovoltage measurements are performed wavelength dependent, detailed information on the electronic structure of the semiconductor powder and on primary processes at the solid–solution interface can be gained [48].

2.3 Rates and Quantum Yields

The rate of any photoreaction is given by the product between absorbed photon flux (I_a is the number of photons absorbed at a given wavelength per time and volume) and quantum yield (Φ). Since I_a depends on the intensity of the light source, only the quantum yield, i.e., rate divided by I_a, is independent of the intensity and therefore can be used to compare the

$$\Phi_p = \frac{\text{rate}}{I_a} \tag{34}$$

efficiencies of various photoreactions. It has to be specified if the substrate disappearance or product formation rate [Eq. (34)] was measured. Except for photoinduced chain reactions, the maximum value of the *product* quantum yield (Φ_p) is

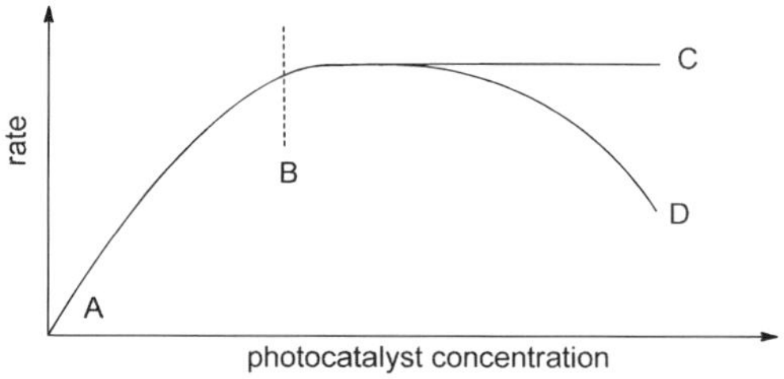

Fig. 4 Dependence of reaction rate on photocatalyst concentration

one, meaning that each quantum of light absorbed by the reacting system generates one product molecule.[3] Whereas the amount of light absorbed by the reacting system can easily be measured for homogeneous solutions, it is

$$\zeta_p = \frac{\text{rate}}{I_0} \qquad (35)$$

extremely difficult for suspensions of powders as generally employed in semiconductor photocatalysis [49, 50]. In this case, light is not only absorbed but also scattered and reflected by the suspended semiconductor particles. Up to 13–76% of the light arriving at the powder surface may be lost [51–54].

To solve this intrinsic problem of the quantitative comparison of heterogeneous photoreactions, it was proposed just to replace I_a by the easily measurable number of photons of a given wavelength per time and volume arriving at the *inside* of a flat front window of the photoreactor (incident photon flux I_0). The apparent quantum yield obtained by this procedure was termed *photonic efficiency* [Eq. (35)] [52, 55–57]. Comparison of the resulting numbers, however, is meaningful only if the fraction of light absorbed is the same in each experiment. This is a rather unlikely assumption since the amount of scattered and reflected light may change considerably from experiment to experiment. This is a serious problem when comparing photonic efficiencies obtained by various research groups, since usually different photoreactors are employed. But a comparison seems meaningful when referring a set of reactions performed in the same photoreactor to a standard system obtaining thereby "relative photonic efficiencies" [56, 57]. A more simple and practicable procedure allows comparing the rate of reactions performed in the same photoreactor. For this the rate is measured as function of increasing photocatalyst concentration. As for a *homogeneous* system [58] (Fig. 4), also for a *heterogeneous* reaction [51] the rate initially increases linearly with photocatalyst concentration (Fig. 4, curve A–B) due to the increase of absorbed photon flux and

[3] This is based on the Stark–Einstein law that one quantum of light can convert only one molecule. Higher values are observable only in photoinduced chain reactions wherein a photogenerated intermediate acts a thermal catalyst.

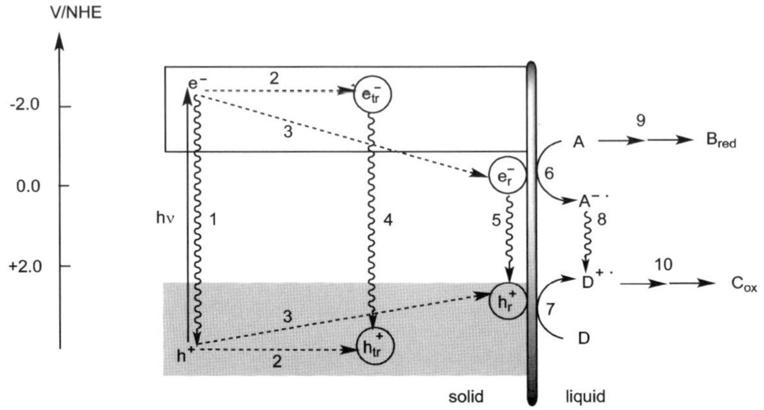

Scheme 4 Schematic description of primary processes occurring during a semiconductor-photocatalyzed redox reaction. For the sake of simplicity, *wavy arrows* symbolize nonradiative, radiative, and photocorrosion processes. *Dashed arrows* indicate charge trapping

then stays constant in the region of B to C representing concentrations of constant and optimal light absorption. This corresponds to optimum light absorption and the resulting optimum rate corresponds to a pseudo-photon efficiency enabling comparison of "photocatalytic activities."[4] In some cases, a rate decrease is observed in region from B to D due to reduced penetration depth and increased scattering of the incident light beam [59, 60]. For the often employed titania powders of specific surface areas within 50–200 m^2/g, the optimum concentration is in the range of 0.5–3.0 g/L depending on details of the irradiation system. In general, a semiquantitative comparison of "photocatalytic activities" is meaningful only if the reactions have been conducted in the plateau region of Fig. 4. *However, this simple requirement is quite often not met in the literature and therefore the reported conclusions are doubtful, especially for small differences.*[5] A recent proposal coming from the field of thermal heterogeneous catalysis does not present a reliable alternative [61, 62].

The following qualitative discussion of the parameters determining the quantum yield illustrates the problems in more detail. Scheme 4 summarizes the most important primary processes determining the efficiency of charge generation at the solid/solution interface. It is noted that due to the small crystallites (10–50 nm) the powder consists of, no band bending has to be considered. Light absorption

[4] The latter term is commonly used and in general is based on yield or rate of the catalytic photoreaction.

[5] Even when the reactions are performed in the same photoreactor, the reproducibility of optimum rates is usually in the range of at least ±10%. It is noted that the "rate constants" published in some papers also depend on the absorbed photon flux and therefore are no constants.

generates an electron–hole pair delocalized throughout the crystal. It may recombine and undergo trapping at unreactive (e_{tr}^-, h_{tr}^+) and reactive (e_r^-, h_r^+)

$$\eta_r = \frac{k_3}{k_1 + k_2 + k_3 + k_4} \tag{36}$$

$$\eta_{ifet} = \frac{k_{6,7}}{k_5 + k_{6,7}} \tag{37}$$

$$\eta_p = \frac{k_{9,10}}{k_8 + k_{sb} + k_{9,10}} \tag{38}$$

surface sites (Scheme 4, processes 1, 2, 3, respectively) followed by nonradiative and radiative charge recombination (processes 4 and 5). IFET from the reactive electron–hole pair affords the primary redox products A^- and D^+ (processes 6 and 7) which may suffer back electron transfer to A and D (process 8) or undergo the desired conversion to final products (processes 9, 10). By analogy with homogeneous systems, the efficiency of formation of the reactive electron–hole pair (η_r), the efficiency of the IFET (η_{ifet}), and the efficiency of product formation from the primary redox products (η_p) are given by the ratio of the rate constants for formation and consumption of the corresponding intermediates present in a quasi-stationary state [Eqs. (36)–(38); see Scheme 4]. In Eq. (38), the rate constant k_{sb} belongs to the secondary back electron transfer according to Eqs. (39) and (40):

$$A^{-\bullet} + h_r^+ \rightarrow A \tag{39}$$

$$D^{+\bullet} + e_r^- \rightarrow D. \tag{40}$$

The efficiency of the overall reaction, i.e., the quantum yield of product formation (Φ_p), can then be formulated as the product of $\eta_r \times \eta_{ifet} \times \eta_p$ [Eq. (41)]. Thus, Φ_p depends on the rate constants of various elementary reactions, and it is difficult to correlate its variations with one

$$\Phi_p = \eta_r \eta_{ifet} \eta_p \tag{41}$$

unique process. However, the efficiency of product formation from the primary redox products (η_p) should only weakly depend on small changes of intrinsic semiconductor properties, whereas the efficiencies of formation of the reactive electron–hole pair and of the IFET reactions (η_r and η_{ifet}) may change considerably. Minor alterations during synthesis of the powder and adsorption of reaction components can strongly affect the photophysical properties of the semiconductor–liquid interface. Furthermore, since the semiconductor powder in general consists of micrometer-sized aggregates built up by nano-sized crystallites, the detailed nature of the intercrystallite interaction may strongly affect the efficiency of charge generation and therefore the value of η_r. The solid–solid interface, especially of mixed-phase oxides like anatase/rutile materials, may improve charge generation by preventing recombination through an intercrystallite electron transfer as supported by EPR spectroscopy [29]. It was recently reported that coagulation of colloidal titania nanoparticles increases the photocurrent density at a platinum electrode measured upon photoreduction of methylviologen by methanol in the presence of titania nanoparticles. The effect was rationalized by the assumption that an intercrystallite charge transfer may improve charge separation [63]. However, it cannot be excluded that other effects like a faster methanol oxidation by the reactive hole may be the reason. Similarly, colloidal CdS is inactive whereas micrometer aggregates photocatalyze organic addition reactions (see Chaps. 3 and 4) [64]. For titania covered by a few weight percent of metallic silver as a type of *nanocathode* (Ag/TiO_2), a strong electronic coupling between properly aligned crystallites was proposed to generate an "antenna effect" inducing improved charge separation [65, 66]. Similar effects may rationalize the higher yield of superoxide observed upon UV excitation of ZrO_2–TiO_2 nanoparticles networks as compared to TiO_2–TiO_2 and ZrO_2–ZrO_2 systems [67, 68]. Very recently it was proven that thermal electron transfer takes place from small to large ZnO nanocrystals. Due to the quantum size effect, the conduction band edge of the smaller particle is located at more negative potential as compared to the larger particle [69].

Thus, when comparing photonic efficiencies or optimum reaction rates induced by various substrates, it is not obvious which of the three efficiencies of the multistep reaction is responsible for the observed changes. In most cases, it is even unknown if the electron–hole pairs sometimes observable by emission spectroscopy are identical with the reactive electron–hole pairs involved in the IFET reactions. In general, good emitters are poor photocatalysts because radiative charge recombination is faster than the IFET process. This relationship, however, does not proof that the emitting electron–hole pairs are identical with the reactive ones since the latter may be produced via the former. A combination of emission quenching and reaction inhibition studies offers a simple possibility to answer this question (see Chap. 3).

3 C–C and C–N Coupling

After the early work of Bard et al. on the TiO_2-catalyzed photo-Kolbe reaction [70], many papers appeared dealing with *photocatalytic* organic reactions in the presence of colloidal or suspended semiconductor particles. They include cis-trans isomerizations [71–74], valence isomerizations [75, 76], substitution and cycloaddition reactions [77–82], oxidations [83, 84], and reductions [85–87]. In general, UV excitation was indispensable and in all cases well-known compounds were formed, which were not isolated but only identified by spectroscopic methods. One reason for that is that in most cases photocorrosion of the photocatalyst prevents efficient product formation during reaction times long enough for preparative chemistry. This is very true for colloidal metal sulfide semiconductors, which are photochemically too unstable for synthetic processes [64, 88–90], although the pseudohomogeneous nature of their solutions allows classical mechanistic investigations.

The structure of almost all products can be rationalized within the mechanistic scheme of semiconductor photocatalysis type A. In general, it is the oxidative part of the reaction wherein the organic substrate is involved. That means that the reactivity of the initially formed radical cation determines the kind of products finally formed. Typical reactions of radical cations are deprotonation, bond cleavage, and electron transfer [91, 92]. In addition to deprotonation via C–H cleavage, radical cations may be transformed to the corresponding radicals also by C–C, C–H, and other bond cleavages. All the reactions mentioned above do not lead to novel products nor do they introduce new aspects to organic synthesis. Different from that, the reactions discussed in the next two chapters afford new compounds isolated on a gram scale. The major part belongs to addition reactions representing atom-economic processes. They demonstrate that semiconductor photocatalysis may be a valuable new method for preparative chemistry. Recent reviews on organic photochemistry also mention the use of semiconductors as photocatalysts [8, 93–97].

3.1 C–C Coupling Through Semiconductor Photocatalysis Type A

Irradiation of an aqueous 2,5-dihydrofuran (2,5-DHF) suspension of ZnS or platinized CdS (Pt/CdS) with UV or visible light affords a few liters of hydrogen and gram amounts of hitherto unknown dehydrodimers in isolated yields of 60% (Scheme 5) [88, 98–101]. No reaction occurs in the absence of water, and the initially evolved hydrogen gas contains about 90% of D_2 when D_2O is employed. Colloidal zinc or cadmium sulfide and high-purity single crystals do not catalyze the reaction. Structure and statistical ratio of the three regioisomeric dehydrodimers suggest that the products are formed by dimerization of an intermediate dihydrofuryl radical. Product formation can be rationalized within the scheme for *semiconductor photocatalysis type A* as depicted in Schemes 1 and 5. In the reductive reaction step, water is reduced to hydrogen [Eq. (42)], whereas in the oxidative part, 2,5-DHF is oxidized to the allylic dihydrofuryl radical and a proton [Eq. (43)]. 3-Hydroxytetrahydrofuran, the product of addition of water to a mutual 2,5-DHF radical cation, could not be detected. According to this basic reaction scheme, although water is reduced, it is not consumed, since it is reformed as indicated by the sum of Eqs. (42) and (43). In fact, the water concentration did not decrease although about two liters of hydrogen were produced. The net reaction can therefore be summarized according to Eq. (44). Especially noteworthy is that the initially observed D_2 content of 90% drops to 40% after evolution of one

$$2e_r^- + 2H_2O \rightarrow H_2 + 2OH^- \tag{42}$$

$$2h_r^+ + 2RH \rightarrow 2R^\bullet + 2H^+ \tag{43}$$

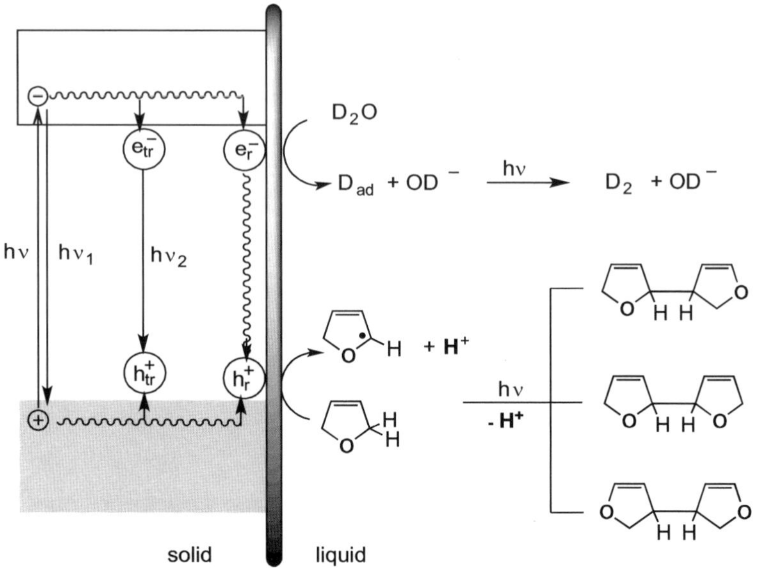

Scheme 5 Photodehydrodimerization of 2,5-dihydrofuran ($2e^-/2h^+$ process). Band edges are positioned at −1.8 and +1.8 V or −0.9 and +1.5 V for ZnS and CdS, respectively

$$2RH \xrightarrow[H_2O]{h\nu,\,ZnS} H_2 + R - R \tag{44}$$

liter of hydrogen, whereas the sum of HD and H_2 increased from 10% to 60%. From these results, it is obvious that the formation of D_2 from D_2O in "sacrificial" systems is a necessary but not sufficient criterion for "permanent" water reduction. This result is of basic importance since the *consumption* of water was never proven in the enormously growing field of "sacrificial" photochemical hydrogen production. Typical examples are primary and secondary alcohols as reducing agents, also generating protons according to Eq. (45). Therefore, water is not consumed and the reaction can be classified as a water-assisted dehydrogenation of the alcohol:

$$h_r^+ + R_2CHOH \rightarrow R_2C^{\bullet}OH + H^+. \tag{45}$$

Since the dehydrodimers were unknown before, this C–C coupling reaction is the first example for the preparation of a novel compound through semiconductor photocatalysis. The analogous products were obtained from 3,4-dihydropyran (3,4-DHP), 3-methyl-2,3-dihydropyran (3-MeDHP), and cyclohexene in isolated yields of 30–60%. The saturated ether tetrahydrofuran is also dehydrodimerized whereas 1,4-dioxane does not react. In situ prepared zinc sulfide photocatalyzes the same reactions but exhibits a different chemoselectivity since 1,4-dioxane is also dehydrodimerized [102, 103].

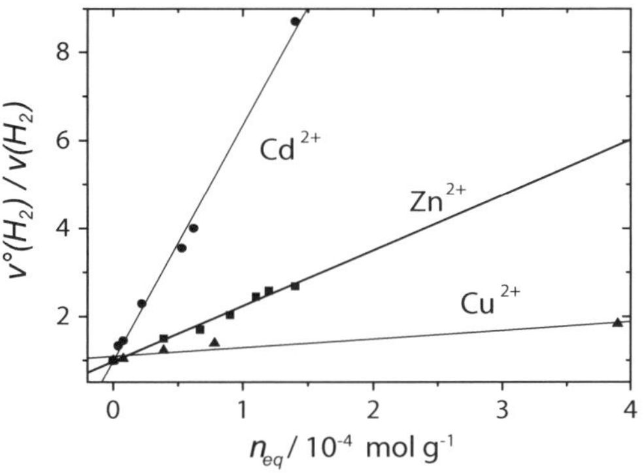

Fig. 5 Stern–Volmer plot for the ZnS-photocatalyzed dehydrodimerization of 2,5-DHF

To unravel the detailed mechanism, substrate adsorption, quenching, inhibition, and kinetic studies were conducted for the ZnS-catalyzed photodehydrodimerization of 2,5-DHF. A plot of the amount of 2,5-DHF adsorbed (n_{eq}) against the residual concentration in solution exhibits two saturation plateaus at $n_{eq(max)}$ of 2.8×10^{-3} and 65×10^{-3} mol g^{-1}. The first plateau is due to formation of a mixed solvent–solute surface monolayer, and the second corresponds to multilayer adsorption. Assuming that the formation of the monolayer can be described by competitive adsorption between water and 2,5-DHF, the data were analyzed according to the Hiemenz model [104, 105]. The average area occupied by 2,5-DHF in the surface–solute monolayer is obtained from this model as 10.2 Å^2. From the surface density of zinc sites (11.4×10^{-6} mol m^{-2}) of cubic zinc sulfide [106] and the specific surface area (100–170 m^2 g^{-1}) of the ZnS employed, one estimates that the surface concentration of 2,5-DHF in the saturated monolayer is in the range of (1–2) $\times 10^{-3}$ mol g^{-1}. This good agreement with the experimental value of 2.8×10^{-3} mol g^{-1} suggests that each zinc site is occupied by one 2,5-DHF molecule. The rather small downfield shift of 1.5 ppm as observed for the C(sp^3) atoms of adsorbed 2,5-DHF suggests that the oxygen atom does not directly interact with zinc sites but rather indirectly through hydrogen bonding to coordinated water.

To find out if emissive (e_{tr}^-, h_{tr}^+) and reactive (e_r^-, h_r^+) electron–hole pairs are identical or not, emission quenching and product inhibition studies were performed. Addition of zinc or cadmium sulfate slightly increases or does not alter the two emission bands of an aqueous ZnS suspension at 366 and 430 nm. Also 2,5-DHF has no significant influence. Contrary to emission, the reaction is strongly inhibited when cadmium or zinc salts are added. This indicates that emitting and reacting surface sites are different. A Stern–Volmer plot of the reduced reaction rate as function of the initial inhibitor concentration affords a straight line only, when the concentration of *adsorbed* ions is employed (Fig. 5). From the corresponding slopes, Stern–Volmer constants of 13×10^3 and 50×10^3 M^{-1} are calculated for Zn^{2+} and Cd^{2+}, respectively. Copper(II) ions exert only a very weak effect.

Inhibition by Cd^{2+} proceeds via competitive IFET [Eq. (46)], since even at the very low concentration of 3.9×10^{-6} mol g^{-1} formation of elemental cadmium is observable; complete inhibition occurs at 6×10^{-4} mol g^{-1}. This differs significantly from the effect of zinc ions

$$2e_r^- + Cd^{2+} \rightarrow Cd \tag{46}$$

in which the expected elemental zinc [107, 108] could not be detected, even at the high concentration of 0.8 mol g^{-1}. Therefore, zinc ions either prevent formation of the reactive electron–hole pair or efficiently promote its radiationless deactivation. Inhibition studies with various metal ions reveal that Fe^{2+} and Ni^{2+} accelerate the reaction up to a concentration of about 0.7×10^{-3} M, but inhibit the reaction at concentrations above 2×10^{-3} and 6×10^{-3} M, respectively. At a given surface concentration of these ions ($n_{eq} = 3 \times 10^{-5}$ mol g^{-1}), there is no simple relation with the reduction potential of the metal ion. However, the reaction rate increases approximately linearly with the electrochemical exchange current density of hydrogen evolution at the corresponding metal electrode [109]. This strongly suggests that water reduction at the photoexcited ZnS/M_{ad}^{2+} surface occurs at small metal islands generated by photoreduction.

From these results, the primary events at the semiconductor surface can be summarized as schematically depicted in Scheme 5. The light-generated electron–hole pair has a lifetime of 0.1–20 ns and either recombines through radiative or nonradiative processes or is trapped at emitting (e_{tr}^-, h_{tr}^+) or reacting surface (e_r^-, h_r^+) sites. Whereas the former are detected by their emission at 440 nm, the latter could not be observed directly but their existence is evidenced through the inhibition experiments. The reduction of water most likely is located at surface states generated by metal islands. For the formation of the 2,5-dihydrofuryl radical in the oxidative part of the reaction, three pathways may be considered. First is hydrogen abstraction by a surface sulfur radical. Although such radicals have been detected at zinc sulfides, this reaction path is rather unlikely since THF is also dehydrodimerized but does not undergo H-abstraction with sulfur-centered radicals in homogeneous solution [110]. Second is a stepwise formation through an initially formed radical cation followed by deprotonation. And third is a dissociative IFET in which electron transfer and deprotonation proceed concerted as indicated in Scheme 5. All experimental evidence is in favor of the concerted pathway.

Taking a redox potential of 1.6–2.0 V for the reactive hole and for 2,5-DHF an oxidation potential of 2.6 V, oxidation to the radical cation is endergonic by at least 0.6 eV. Furthermore, there is no simple relation between apparent product quantum yields and ether redox potentials. On the other hand, a similar estimation for the concerted process of a dissociative electron transfer [Eq. (47)] reveals that the reaction is exergonic by at least 0.9 eV:

$$RH \rightarrow R^{\bullet} + H^+ + e^-. \tag{47}$$

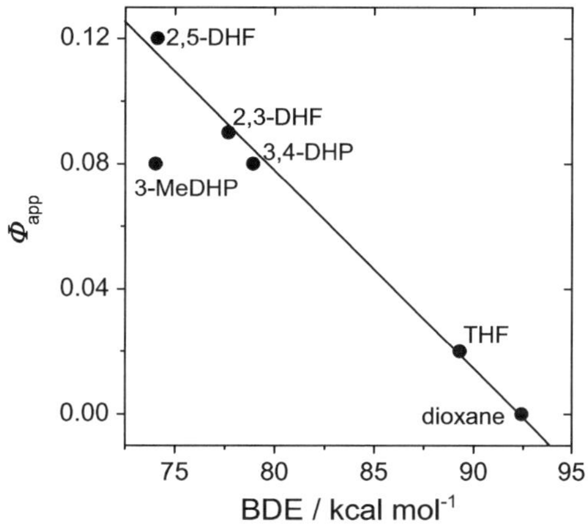

Fig. 6 Variation of apparent product quantum yield with the calculated bond dissociation energy of the allylic C–H bond

Since the driving force of this reaction is the difference between the free enthalpy of C–H bond homolysis and the potential of the hydrogen electrode, the former value should be decisive when comparing apparent quantum yields of various substrates. Figure 6 displays the relation between quantum yield and calculated bond dissociation energy of the corresponding C–H bond [88]. The expected increase with decreasing bond strength favors the concerted oxidation pathway. The deviation of 3-MeDHP most likely arises from steric hindrance of the radical C–C coupling step by the adjacent methyl group.

It is noted that the intermediate allylic radicals may suffer disproportionation, in addition to double bonds, electron transfer, and dimerization, as well known from their chemistry in homogeneous solution. Surprisingly, the latter pathway is followed to about 90% as indicated by a complete material balance. This unexpected high chemoselectivity strongly suggests that C–C coupling does not proceed via fully solvated radicals but in the H_2O/2,5-DHF surface layer. Evidence for that comes also from the quadratic dependence of the initial rate on the amount of adsorbed 2,5-DHF, which is characteristic for a heterogeneous catalytic dimerization by a modified Langmuir–Hinshelwood mechanism affording easily desorbable products [111]. C–C coupling between radicals adsorbed in the water 2,5-DHF surface layer is further supported by competition experiments with THF. Although the unsaturated ether reacts only ten times faster than the saturated one, no THF dehydrodimers or cross-products are detected when THF is present in tenfold excess over 2,5-DHF. Only at a 500-fold higher concentration the expected products are observed.

From the results discussed above, one can conclude that hydrogen formation and C–C coupling occur via subsequent absorption of two photons ($2e^-/2h^+$ process). The question arises why the C–C homocoupling between two radicals is so highly favored over C–H heterocoupling with an adsorbed hydrogen atom to reform 2,5-DHF. One possibility is that the first electron does not produce an adsorbed hydrogen atom, but is stored at the metallic zinc or Pt/CdS center and water is subsequently reduced in a two-electron step. As another possibility, the adsorbed hydrogen atom may not undergo coupling to the dihydrofuryl radical because of unknown kinetic barriers. Therefore, it seemed worthwhile to replace water by an organic acceptor, which could produce a more stable one-electron reduction intermediate, perhaps capable to undergo the postulated heterocoupling with the one-electron oxidation intermediate. The following chapter illustrates that this concept led to the discovery of a novel type of linear photoaddition reaction.

3.2 C–N Coupling Through Semiconductor Photocatalysis Type B

When azobenzene is added to a running ZnS- or CdS-catalyzed photodehydrodimerization experiment of 2,5-DHF, hydrogen evolution is completely inhibited. Instead, the novel allylhydrazine **3c**, a linear addition product of 2,5-DHF to azobenzene, and small amounts of hydrazobenzene (**4**) are formed (Scheme 6) [64, 89, 112]. When all the azobenzene is consumed and some excess 2,5-DHF is still present, hydrogen evolution starts again. CdS or CdS grafted onto silica (CdS–O–SiO_2) allows conducting the reaction with visible light. The photoaddition exhibits a significant solvent dependence. No reaction occurs in dry *n*-hexane or THF, but upon addition of water or methanol, the reaction is as fast as in pure methanol. When colloidal CdS is employed, no photoaddition but efficient photocorrosion occurs. Scheme 6 summarizes the addition of olefins 2**a**–**f** to 1,2-diazenes **1** affording hitherto unknown allylhydrazines **3a**–**f** on a gram scale. Due to the poor crystallization properties, isolated yields are only in the range of 10–40% whereas HPLC yields are about twice larger.

When either the 1,2-diazene [113–115] or the olefin [116–118] is substituted by electron-withdrawing groups, the same reaction type was observable when the substrates were irradiated in homogeneous solution in the *absence* of a photocatalyst. However, these reactions are of very limited preparative utility. Surprisingly, only a few other allylhydrazines have been prepared in the literature by conventional thermal procedures [119].

From the discussion at the end of the previous chapter and the experimental results presented above, a simplified reaction scheme is depicted in Scheme 7. Since the presence of the diazene completely inhibits hydrogen evolution and the reaction proceeds only in protic solvents or in the presence of water, the reductive IFET is formulated as a proton-coupled reduction of the diazene to a hydrazyl radical [Eq. (48)]. The oxidative IFET is assumed to proceed as described for

$$\mathrm{ArN} = \mathrm{NAr} + \mathrm{e}_\mathrm{r}^- + \mathrm{H}^+ \rightarrow \mathrm{ArN}^\bullet - \mathrm{N(H)Ar} \qquad (48)$$

the photodehydrodimerization. Heterocoupling of the hydrazyl and allyl radicals affords the allylhydrazine (path B). Thus, formation of the addition product is a

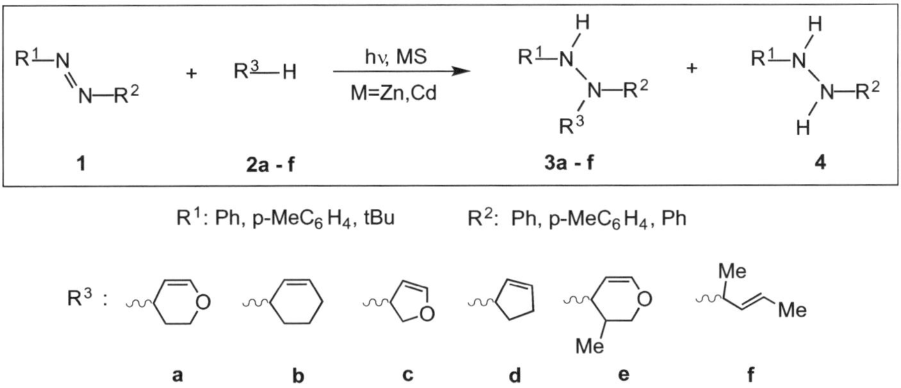

Scheme 6 Preparation of allylhydrazines through addition of cyclic allyl/enol ethers and olefins to 1,2-diazenes photocatalyzed by ZnS or CdS suspended in methanol

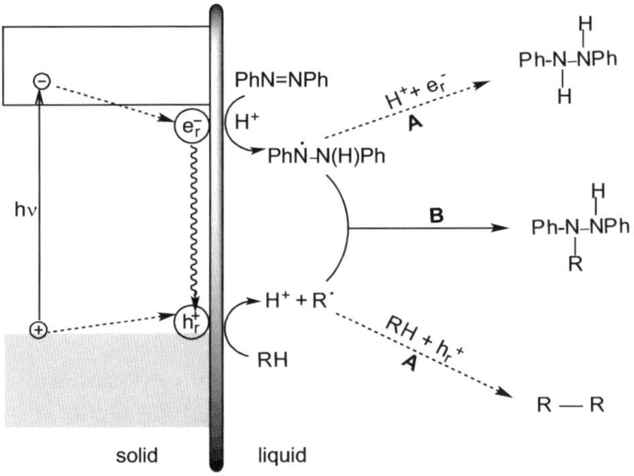

Scheme 7 Simplified scheme for the CdS- or ZnS-photocatalyzed addition of cyclic unsaturated ethers or olefins to 1,2-diazenes ($1e^-/1h^+$ process)

$1e^-/1h^+$ process, whereas the by-products are formed via a $2e^-/2h^+$ process, irrespective whether the hydrazobenzene derivative **4** is formed by subsequent disproportionation or reduction of the hydrazyl radical [Eqs. (49) and (50)]. The energetic relations between band positions and redox potentials are summarized in Fig. 7:

$$2\mathrm{ArN}^{\bullet} - \mathrm{N(H)Ar} \rightarrow \mathrm{ArN(H)} - \mathrm{N(H)Ar} + \mathrm{ArN} = \mathrm{NAr}, \tag{49}$$

$$\mathrm{ArN(H)} - \mathrm{NAr} + e_r^- + \mathrm{H}^+ \rightarrow \mathrm{ArN(H)} - \mathrm{N(H)Ar}. \tag{50}$$

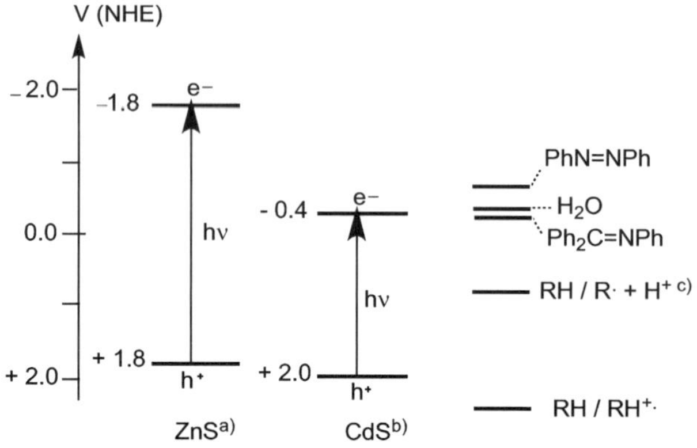

Fig. 7 Metal sulfide band edge and substrate redox potential positions. (*a*) Single crystal, (*b*) self-prepared powder, (*c*) RH = 2,5-DHF

Formation of the hydrazobenzene product is strongly favored when platinized zinc or cadmium sulfide is used as the photocatalyst. In both cases, the rate decreases considerably and hydrazobenzene becomes the major product. It is known that the presence of platinum favors multi-electron processes [120].

Adsorption studies were conducted with CdS in methanol, the solvent employed in the photoaddition reaction. From ^{13}C NMR spectra of 2,5-DHF adsorbed from the gas phase onto the dry powders, it is concluded that 2,5-DHF is adsorbed parallel to the surface. From the maximum surface concentration of 0.4×10^{-3} mol g^{-1} found for 2,5-DHF and the maximum number of 1.54×10^{-3} mol g^{-1} calculated for the Cd^{2+} surface concentration in cubic CdS [121], it follows that 2,5-DHF adsorbs at about every fourth Cd^{2+} center, in agreement with a parallel orientation. The maximum surface concentration $n_{eq(max)}$ for azobenzene of about 10^{-5} mol g^{-1} is two orders of magnitude lower, whereas the adsorption constants are much higher. It is estimated that in the case of CdS, only every 220th Cd^{2+} site interacts with an azobenzene molecule, corresponding to a surface coverage of only about 0.7%. This suggests that the more polar methanol ($\mu = 1.7$ D) efficiently competes with the less polar trans-azobenzene ($\mu_{trans} \cong 0$ D) for adsorption sites at the polar CdS surface. It seems likely that interaction between surface Brϕnsted acid sites [112] and the basic nitrogen lone pairs is the driving force for adsorption (vide infra).

The apparent quantum yields of allylhydrazine formation (Φ_{app}) were measured at 366 nm, the wavelength at which light absorption by the diazene is minimized. In the system CdS/olefin/1,2-diphenyldiazene/MeOH Φ_{app} increases from 0.02 (2,5-DHF) over 0.03 (cyclohexene) and 0.04 (3,4-DHP) to 0.05 (2,3-DHF). As also observed for the ZnS-catalyzed photodehydrodimerization, there is no simple relation with the redox potentials of the olefins.

The postulated C–N heterocoupling requires diffusion of the two radicals either in the solvent–solute surface layer or in the bulk solution. In both cases, one expects that the reaction rate should decrease upon increasing solvent viscosity through the application of high pressure. To achieve the latter, the CdS–O–SiO_2-catalyzed photoaddition of 2,5-DHF to azobenzene was conducted at pressures ranging

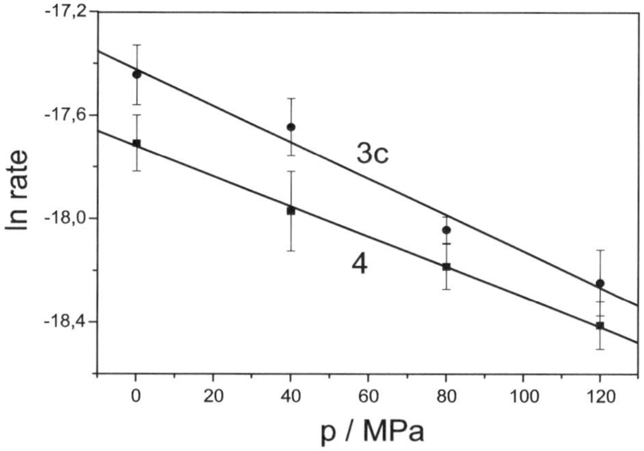

Fig. 8 Pressure dependence of the **3c** and **4** formation rates

from 0.1 to 120 MPa [112]. Both the formation rate of addition and reduction product **3c** and **4** ($R^1 = R^2 =$ Ph) decrease with increasing pressure. From a plot of ln(rate) vs. pressure, activation volumes $\Delta V^{\ddagger}$ are obtained as 17.4 ± 3.4 and 15.8 ± 2.3 cm^3 mol^{-1} for **3c** and **4**, respectively, (Fig. 8). However, since with increasing pressure also the dielectric constant increases, the observed effects may originate from the change of this property [122–125]. In order to differentiate between these two possibilities, the rates were measured in a series of alcohols for which viscosity and dielectric constant change in an opposite fashion. Whereas the rates again decrease with increasing viscosity, they increase when plotted as function of increasing dielectric constant. This indicates that the rate decrease at higher pressure is a viscosity effect.

It is unlikely that the activation volume is connected with substrate adsorption and product desorption [126] or with the IFET steps. Usually interfacial collision rates depend on molecular mass but not on diffusion rates [127]. Most likely, the activation volume measured for the formation of **3c** originates from the diffusion of the intermediary radicals to each other or from the subsequent C–N coupling step itself. The latter case can be excluded since bond formation between neutral organic species in homogeneous solution in general has a negative activation volume [128–131]. The only exception is radical recombination in the termination step of polymerizations [132, 133]. These reactions possess $\Delta V^{\ddagger}$ values in the range of 13 to 25 cm^3 mol^{-1}which are composed of the large and positive contribution of diffusion and the small and negative part of radical C–C coupling. Hence, the activation volume found for **3c** most likely originates primarily from diffusion of the intermediate radicals to each other and only to a minor part from C–N coupling. Therefore, it should resemble the activation volume for the viscous flow of methanol. The fact that the latter value of 8 cm^3 mol^{-1} [131] is significantly smaller suggests that the radicals do not diffuse in the bulk homogeneous solution but in the solvent–solute surface layer. The latter should have a higher viscosity, and consequently the activation volume should become more positive. In accordance with this interpretation are also the small activation energies of 2.8 ± 0.3 kcal/mol and

2.5 ± 0.2 kcal/mol observed for **3c** and **4**, respectively. Since the same activation parameters as for **1c** were also found for the formation of the reduction product **4**, the disproportionation pathway [Eq. (49)], which involves radical diffusion, is favored over the secondary reduction step [Eq. (50)]. However, the latter may be partly involved as suggested by the slightly smaller pressure effect as compared to **3c**.

3.3 C–C Coupling Through Semiconductor Photocatalysis Type B

According to the proposed mechanism for this novel photoaddition reaction, other substrates capable of forming radicals upon CdS-photoinduced one-electron oxidation or reduction should undergo similar C–C couplings. Replacing the 1,2-diazene by an aromatic imine, the expected reactions were observed [16, 112, 134–140]. Trisubstituted imines **5** afford the new homoallylamines **6a–g** in isolated yields of 30–75% (Scheme 8).

When a disubstituted imine **(7a–d)** is employed instead of the trisubstituted one, in addition to the homoallylamine (**8**), also the hydrodimer (**9**) of the imine, i.e., the dimer of an anticipated α-aminobenzyl radical, is isolated (Scheme 9). The observation that the hydrodimer is produced only from the disubstituted but not from the trisubstituted imine parallels the electrochemical reduction which affords hydrodimers from aldimines but not from ketimines [141, 142]. Thus, product formation can be rationalized by assuming that the allylic radical generated in the oxidative IFET as discussed above undergoes C–C heterocoupling with the α-aminodiphenylmethyl radical produced according to Scheme 10. In no cases a product arising from C–N heterocoupling could be observed. Thus, different from mutual thermal routes, which usually involve the use of organometallic reagents [143–145], the reaction is regioselective and much easier to perform.

When the CdS surface is alkylated with 3-bromopropyltrimethoxysilane, the resulting powder is completely inactive. However, it becomes very active, when its iminium salt substitutes the imine. This indicates that the surface OH and SH groups of cadmium sulfide protonate the imine to render its redox potential more positive [137]. The reductive IFET is therefore formulated according to Eq. (51):

$$[\mathrm{Ar_2C = N(H)Ar}]^+ + e_r^- \rightarrow \mathrm{Ar_2C^\bullet - N(H)Ar}. \tag{51}$$

As observed for 2,5-DHF, also the imine **7a** exhibits an adsorption isotherm indicating the presence of mono- and multilayer adsorption. From the former a maximum surface concentration of 20 × 10^{-7} mol g^{-1} can be estimated. Application of the Hiemenz model suggests that only 1–2% of the surface is covered by **7a** in competition with the solvent. The results resemble those obtained for the adsorption of azobenzene. Assuming a size of 8 Å^2 for methanol, one arrives at

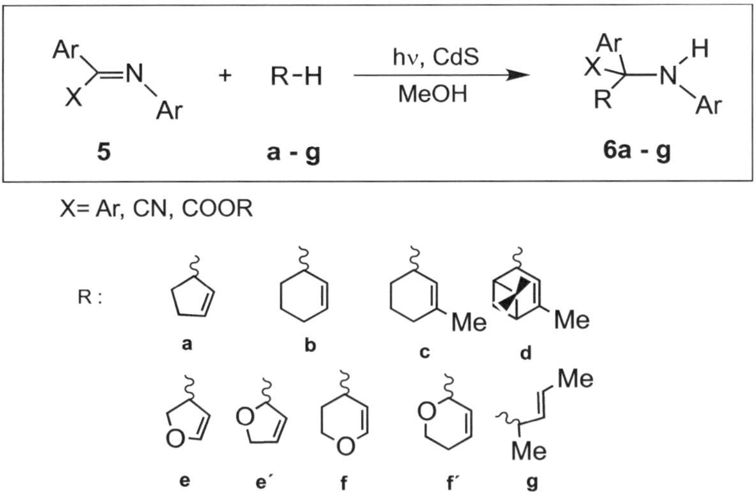

Scheme 8 Preparation of homoallylamines through addition of allyl/enol ethers and olefins to disubstituted imines. Imine hydrodimers **9** are formed as by-products

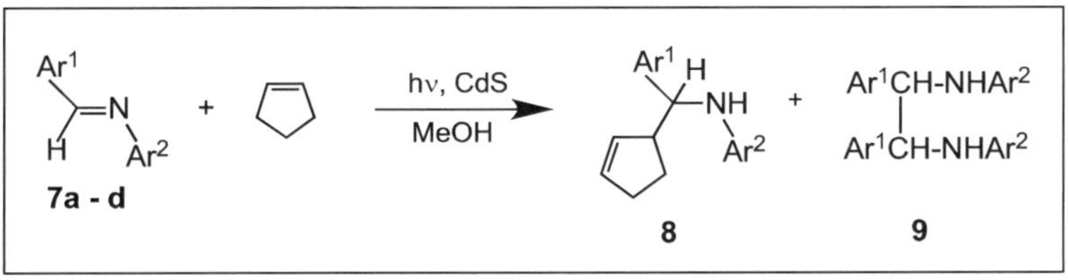

	a	b	c	d
Ar^1	4-ClC_6H_4	2,6-$Cl_2C_6H_3$	4-ClC_6H_4	4-$MeOC_6H_4$
Ar^2	4-ClC_6H_4	C_6H_5	3,5-$Me_2C_6H_3$	4-MeC_6H_4
8[a)]	60	40	55	80
9[a)]	20	10	40	--

Scheme 9 Addition of cyclopentene to monosubstituted imines. a) Yield of isolated product

the conclusion that methanol is present in a 500-fold excess over the imine in the methanol–imine surface monolayer. As expected, methanol should adsorb much stronger than **7a** onto the hydrous CdS surface [146]. Therefore, it is rational that **7c** does not influence the photocurrent of a CdS electrode whereas methanol induces current doubling (Weiß and Kisch, unpublished).

Control experiments with **7a** showed that in the absence of olefins, hydrodimers were also formed but the reaction rate was decreased by about 90%. Whereas the reductive reaction step can proceed as depicted in Scheme 10, the solvent must be

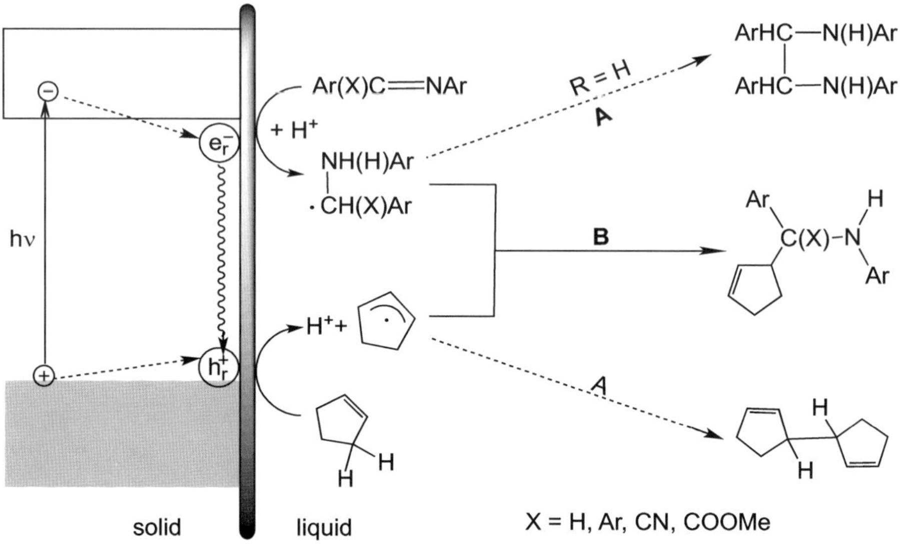

Scheme 10 Simplified mechanistic scheme for the CdS-photocatalyzed addition of cyclopentene to imines

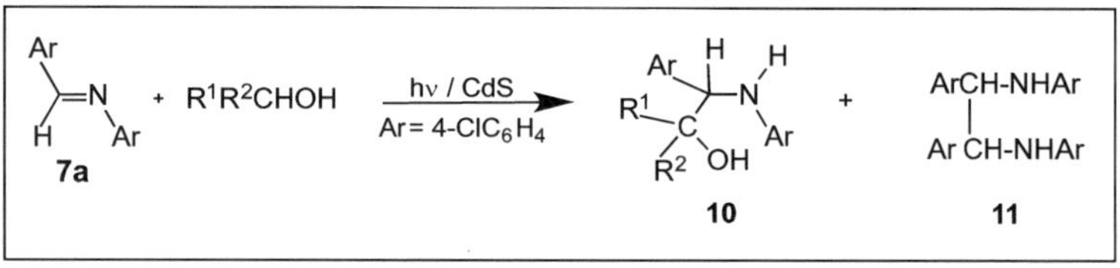

	a	b	c	d	e
R^1	H	Me	Et	Pr	Me
R^2	H	H	H	H	Me

Scheme 11 Preparation of β-hydroxyamines through addition of alcohols to imine **7a**

involved in the oxidative step since no significant oxidative photocorrosion occurs. Accordingly, irradiation of CdS in a solution of **7a** in different alcohols transforms the imine at different rates to the corresponding addition products **10a–e** and hydrodimers **11a–e** (Scheme 11).

Except for methanol and 2-propanol, the products are racemic diastereomeric mixtures, which are isolated in low yields (5–20%); they are often mixed with the two-electron reduction product N-4-chlorobenzyl-4-chloroaniline. The major product in all reactions is the hydrodimer **11**, obtained in yields of 10% (MeOH), 28% (BuOH), 29% (PrOH), 42% (EtOH), and 60% (*i*PrOH). The structure of **10** indicates that in all cases the α-CH bond of the alcohol is added to the imine in agreement with the preferred formation of α-hydroxyalkyl radicals. These results show that the solvent can be directly involved in the oxidative step. Formation of hydrodimers in the absence of olefins thus can be explained by the oxidation of the alcohols. It is noted that in the presence of olefins, no alcohol addition products could be detected by HPLC analysis, although methanol is present in a 500-fold molar excess. This nicely reflects the high chemoselectivity of the semiconductor–liquid interface.

Increasing the light intensity results in a linear increase of the reaction rate. Above an incident intensity of about 10^{18} quanta s^{-1}, a saturation effect is observed. This is in accord with other photoreactions catalyzed by semiconductor powders [71, 147]. Noteworthy, the product ratio of 0.9 observed for **8a**:**9a** is not influenced by changing the light intensity. This suggests that the rates of aminobenzyl radical dimerization and addition to the allyl radical exhibit the same dependence on the concentration of the light-generated electron–hole pairs.

The reaction rate increases approximately linearly on CdS concentration and reaches a plateau at about 3 g L^{-1}. Surprisingly, in the same concentration range, the ratio of addition to hydrodimer product (**8a**:**9a**) decreases from 2 to about 1. In the same direction, the surface concentration of the intermediate radicals should decrease although the ratio of α-aminobenzyl to cyclopentenyl radical concentration should not change. Therefore, the product ratio is expected to stay constant. However, a lower concentration of the radicals increases their lifetime, assuming that they undergo only second-order decay reactions. This effect should favor hydrodimerization, which is a $2e^{-}/2h^{+}$ process and therefore requires that a second radical pair be generated during the lifetime of the first one. Furthermore, one can make the plausible assumption that there is still some weak interaction within a reactive electron–hole pair and therefore the distance between the charges in a pair should be smaller than the distance between neighboring pairs. This means that the radical homocoupling most likely requires a longer diffusion path than heterocoupling. Accordingly, a longer radical lifetime should also enable a more efficient diffusion and therefore favor the hydrodimer formation.

To obtain information on the stereochemistry of the radical C–C coupling, chiral imines were employed in addition reactions with α-pinene (Scheme 12). The (+)-menthylester affords the C2(R)C1(S) diastereomer, whereas both diastereomers are produced with the (−)-menthylester [148].

To investigate how steric pressure at the imine nitrogen atom influences the reaction, the aryl group Ar^2 was replaced by the bulky 1-adamantyl group. In this case, CdS-grafted alumina was employed as the photocatalyst. Using cyclopentene, cyclohexene, and α-pinene and various N-adamantylimines hitherto, unknown homoallyladamantylamines were obtained in isolated yields of 21–85% (Scheme 13) [149]. Unsaturated adamantylamines are of pharmaceutical interest since this class of compounds has antibacterial, antitumor, antipyretic, and anti-inflammatory properties. Some of them were discussed as promising candidates for the treatment of Alzheimer's [134, 139] and Parkinson's diseases [150].

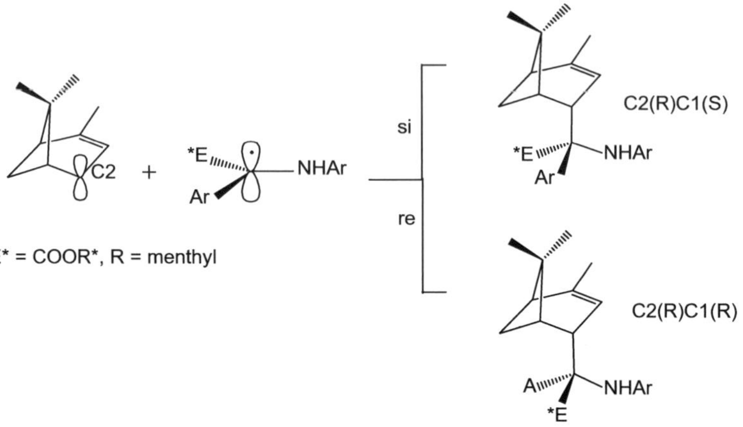

Scheme 12 Diastereoselectivity of C–C coupling

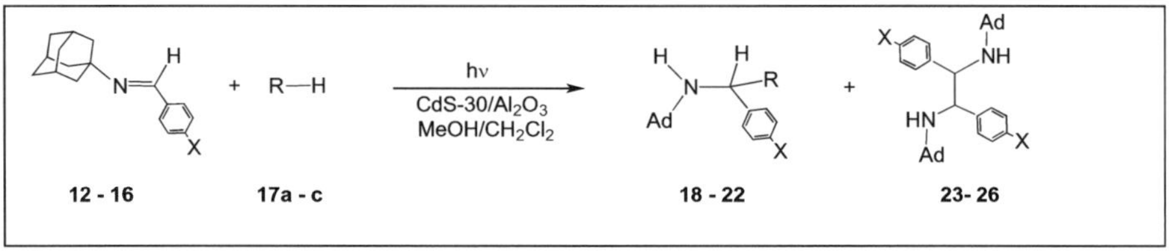

X	H	F	Cl	Br	OCH_3
imine	**12**	**13**	**14**	**15**	**16**
	18	**19**	**20**	**21**	**22**
		(23)	**24**	**(25)**	**(26)**

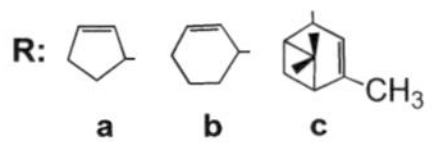

Scheme 13 Preparation of homoallyl-*N*-adamantylimines. Numbers in parentheses refer to products that were not isolated but their formation was evidenced by ^{1}H-NMR

Whereas the diastereoselectivity of C–C heterocoupling is rather low (Scheme 10, path B), the homocoupling between two α-aminobenzyl radicals (Scheme 10, path A) is a diastereospecific process as exemplified by the hydrodimerization of the *p*-chlorophenyl derivative **14**. According to HPLC and X-ray structural analysis, only the diastereomer **24** is formed in the reactions with cyclopentene and α-pinene, whereas **24′** is produced in the case of cyclohexene (Fig. 9). Surprisingly, the stereochemistry is controlled by the nature of the olefin, although it is not directly involved in the C–C homocoupling. However, this effect can be rationalized by recalling that the radicals have to diffuse to each other within a solvent–solute surface layer consisting inter alia of olefins adsorbed to CdS via hydrogen bonding to surface SH and OH groups. It is expected that steric interaction with the olefin should occur during this diffusion process. Thus, the olefin plays a dual role being substrate for the addition and stereodirecting spectator for the hydrodimerization reaction.

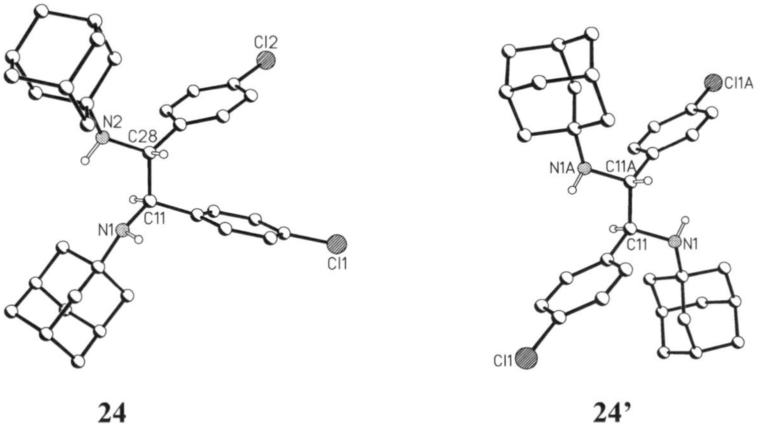

Fig. 9 Molecular structures of diastereomers isolated from reactions with imine **14** (nonrelevant hydrogen atoms omitted for clarity)

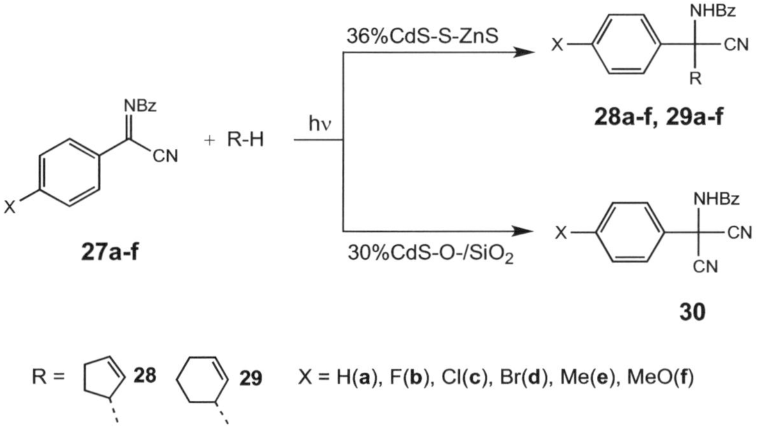

Scheme 14 Support-controlled chemoselectivity

To explore the general applicability of the olefin–imine addition reaction for the synthesis of valuable organic compounds, the *N*-aryl substituent in the imine **5** (X = CN) was replaced by an *N*-benzoyl group, which may be easily converted to an amino group. The resulting unsaturated amino acids could be of pharmaceutical relevance [151, 152]. Surprisingly, in the presence of CdS–O–SiO_2, the addition reactions with cyclopentene and cyclohexene were completely inhibited in favor of a novel *thermal* transhydrocyanation of the imine component affording novel malononitriles **30** in isolated yields of 40–50% (Scheme 14). However, in the presence of CdS, ZnS, or 36% CdS–S–ZnS, this dark reaction was completely inhibited in favor of the desired addition products (65–85% isolated yield).

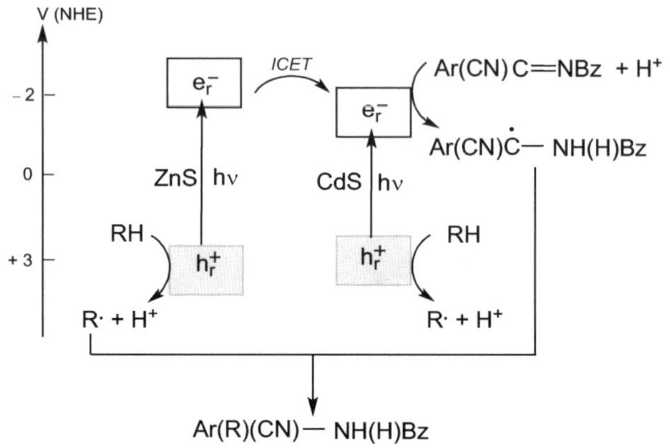

Scheme 15 In 36% CdS–S–ZnS, the photoinduced charge separation may be improved by an intercrystallite electron transfer (ICET)

Unexpectedly, the compounds **28** and **29** were unknown in the literature. Whereas in the case of 36% CdS–S–ZnS complete conversion of **27c** was observed already after 3 h of irradiation time, 29 and 19 h were required for self-prepared ZnS and CdS, respectively. Commercially available CdS and ZnS were inactive.

A speculative mechanism for the 36% CdS–S–ZnS-catalyzed addition reaction is summarized in Scheme 15. It is noted that this grafted photocatalyst induces about six times higher reaction rates than pristine CdS. According to time-resolved photovoltage measurements, 36% CdS–S–ZnS may be considered as a photochemical diode of the type *n*-CdS−S−*p*-ZnS [140]. The increased reactivity is in accord with the longer charge carrier lifetime of 4 μs as compared to 3 μs measured for pristine CdS. This slower charge recombination can be rationalized by assuming an intercrystallite electron transfer at the CdS–S–ZnS interface.

As observed for the radical C–C and C–N couplings, the driving force of the IFET reactions is not the rate-determining parameter. Thus, whereas the reduction potential of the *p*-chlorophenyl-substituted imine **27c** is 150 mV more positive than that of the bromo derivative **27d**, the pseudo rate constants do not differ significantly. To investigate if the rate-determining diffusion of the radicals contains also some contribution of C–C bond formation, the addition of cyclohexene and cyclopentene to a series of *p*-substituted imines was analyzed in terms of the Hammett equation [Eq. (52)]. Therein the parameter σ is a constant for a given substituent X in the p-XC_6H_4 group of the imine and k_X, k_H are the corresponding rate constants. The value of ρ depends on the specific reaction. Generally, positive ρ-values indicate that enhanced electron-withdrawing substituents X increase the reaction rate:

$$\log(k_X/k_H) = \sigma\rho. \tag{52}$$

A corresponding plot of the left term of Eq. (52) vs. the σ-parameters reveals a linear relationship (Fig. 10). Only the fluorophenyl imine **27b** does not follow this

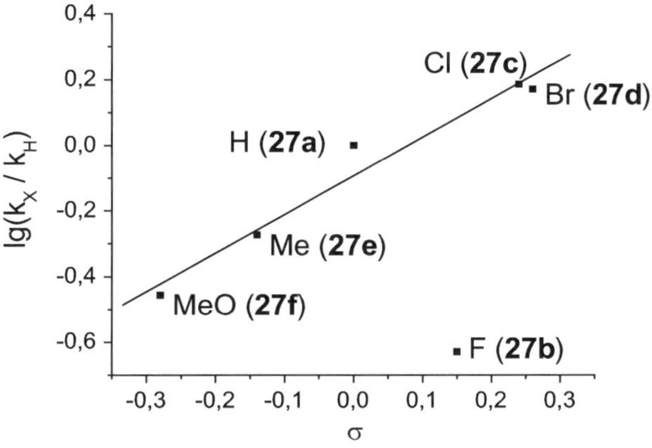

Fig. 10 Hammett plot for the addition of cyclopentene to imines **27a–f**

general trend. The positive ρ-values of $\rho = 1.18$ and $\rho = 1.44$ for cyclopentene and cyclohexene additions, respectively, suggest a nucleophilic attack of the allyl radical at the α-aminobenzyl radical.

4 C–H Activation of Alkanes

The activation and functionalization of alkanes is one of the major challenges in chemistry [153, 154]. The only industrially applied process is the photosulfoxidation of liquid alkanes by sulfur dioxide and oxygen in the presence of UV light [Eq. (53)]. For C_{16-20} chain alkanes, the resulting linear alkanesulfonic acids are used as biodegradable surfactants:

$$R - H + SO_2 + 1/2\ O_2 + h\nu \rightarrow RSO_3H. \tag{53}$$

The alkane activation step consists of hydrogen abstraction from the alkane by triplet-excited sulfur dioxide. Subsequent addition of SO_2 to the generated alkyl radical affords an alkylpersulfonyl radical, which by a further hydrogen abstraction produces another alkyl starter radical and the persulfonic acid (Scheme 16). Fragmentation and hydrogen abstraction [Eqs. (54) and (55)] produce the alkanesulfonic acid [155]. Accordingly, the overall reaction is a photoinduced radical chain reaction, and product formation continues even after turning off the light. In general, regioisomeric alkyl radicals are formed in the hydrogen abstraction step except in the

$$RSO_2 - O - O - H \rightarrow RSO_2 - O^{\cdot} + OH^{\cdot}, \tag{54}$$

$$RSO_2 - O^{\cdot} + R - H \rightarrow RSO_3H + R^{\cdot} \tag{55}$$

case of adamantane photosulfoxidation in the presence of hydrogen peroxide affording regioselectively 1-adamantanesulfonic acid.

Scheme 16 Mechanistic scheme of the UV photosulfoxidation of alkanes

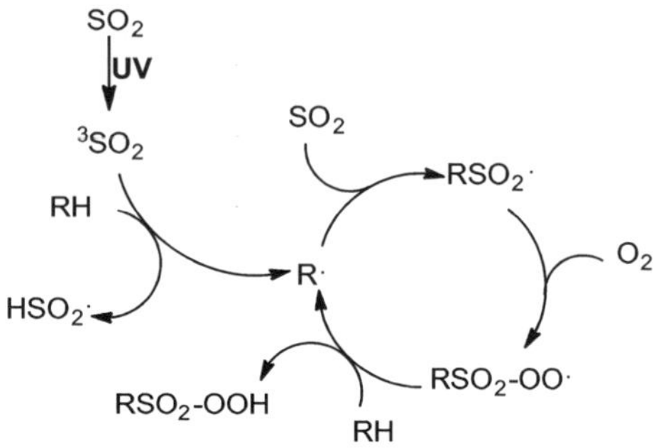

Fig. 11 Diffuse reflectance spectra of titania P25 in (*a*) the absence and (*b*) presence of SO_2. Curve (*c*) corresponds to the difference spectrum (*b*)–(*a*)

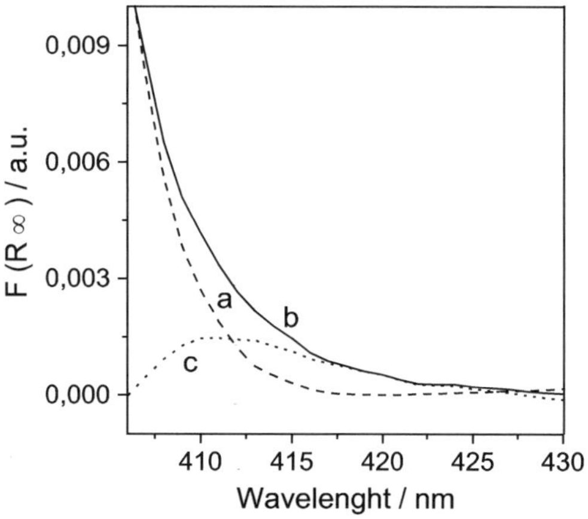

Since hydroxyl and hydroperoxyl radicals are generated at the semiconductor surface in the presence of oxygen [Eqs. (18), (24)–(29)], they may also undergo hydrogen abstraction with an alkane inducing a similar chain reaction. Accordingly, suspensions of various titania powders in *n*-heptane were irradiated with visible light ($\lambda \geq 400$ nm) under an atmosphere of $SO_2/O_2 = 1{:}1$ (v/v). Surprisingly, formation of *n*-heptanesulfonic acid was observed not only with modified titania absorbing visible light but also with the unmodified sample. This suggested formation of a charge-transfer complex between titania and one of the reaction components. In fact, exposure of P25 to sulfur dioxide results in a yellowish coloration of the powder originating from a broad absorption maximum in the diffuse reflectance spectrum at 410–420 nm (Fig. 11).

Under the given experimental conditions, product formation stopped after 6 h of irradiation time. However, separating the catalyst powder and washing with methanol restored the activity. Repeating this procedure three times, the photocatalyst still retained its original activity (Fig. 12). This behavior suggested that the reaction is inhibited by strong product adsorption and that washing desorbs the sulfonic acid. Accordingly, no product was formed when heptanesulfonic acid was added to the suspension prior to irradiation. A similar deactivation and activation were observed

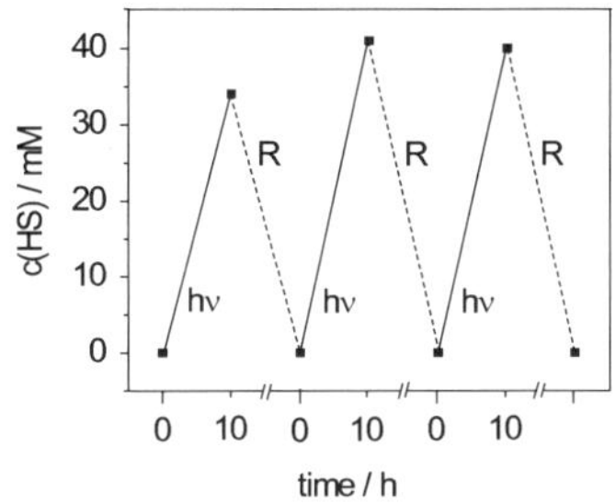

Fig. 12 Sequential photosulfoxidation of n-heptane. $\lambda_{irr} \geq 400$ nm. HS = n-heptanesulfonic acid, R = regeneration

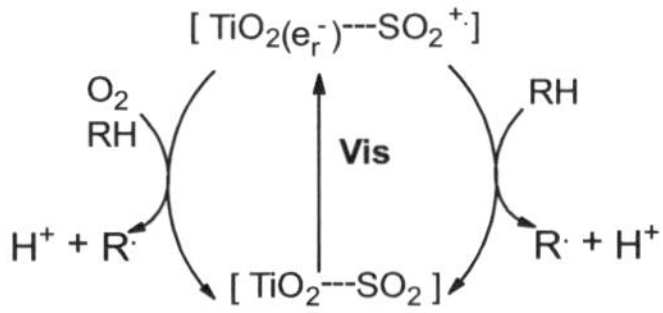

Scheme 17 Proposed mechanism of visible light induced generation of alkyl radicals

during photooxidation of sulfur dioxide in the presence of gaseous n-heptane at UV-irradiated titania [156]. Product formation was also inhibited when small amounts of water like 0.3 vol% were present in the suspension. This may be due to blocking the reactive surface centers for heptane oxidation by preferential adsorption. When after 2 h of irradiation, resulting in a product concentration of 15 mM, irradiation was stopped and the reaction left for 3 days in the dark at room temperature, product formation continued affording 50 mM of the sulfonic acid. However, when the radical scavenger hydroquinone was present during the dark phase, product formation did not continue. All these observations suggest that the new photosulfoxidation is a radical chain reaction. However, the alkyl starter radical is generated not via UV excitation of sulfur dioxide but through visible light absorption of the TiO_2/n-heptane/SO_2/O_2 system. Accordingly, a preliminary mechanism for alkyl radical generation is proposed as schematically depicted in Scheme 17.

Visible light excitation of the charge-transfer complex generates a reactive conduction band electron [$TiO_2(e_r^-)$] and an adsorbed sulfur dioxide radical cation. Oxygen reduction by $TiO_2(e_r^-)$ produces superoxide whereas the adsorbed radical cation may oxidize the alkane to the alkyl radical and a proton. A reduction potential of about 1.8 V is estimated for the latter reaction step. Superoxide may also generate an alkyl radical through protonation by adsorbed water or surface OH groups to the hydroperoxyl radical [see Eqs. (24)–(29)] and subsequent hydrogen abstraction from the alkane. The alkyl radical thus produced is expected to initiate a radical chain reaction as formulated for the stoichiometric UV photosulfoxidation [Scheme 16, Eqs. (54) and (55)]. In agreement with the mechanistic proposal is the complete inhibition observed in the presence of only 10 vol% of 2-propanol, which should be much faster oxidized than the alkane and which is also an efficient OH radical scavenger.

The general applicability of the presented C–H activation is demonstrated by the successful photosulfoxidation of cyclohexane and solid adamantane. In the latter case, glacial acetic acid was employed as a solvent. In summary, this novel visible light induced C–H activation can be classified as a *semiconductor photocatalysis type B* reaction, extending the two-substrate addition [Eq. (17)] [16] to the three-substrate addition scheme A + B + C = D.

5 Summary

The unique charge-separating properties of the semiconductor–liquid interface enable micrometer aggregates of nano-scaled semiconductor powders to photocatalyze interfacial electron exchange with dissolved substrates. In most cases, the primary products are radicals, which undergo selective chemical bond formation in secondary reactions. The high reactivity of such heterogeneous systems is exemplified by the visible light induced functionalization of alkanes. Improvement of the photocatalytic activity of a semiconductor can be accomplished by grafting it onto an insulating or semiconducting support. It is likely that the resulting chemical bonds between the crystallites of an aggregate induce an intercrystallite electron transfer slowing down the undesired charge recombination. Very recent results suggest that such types of hybrid systems are a promising approach for improving the efficiency of charge generation. In general, reduced and oxidized products are obtained (semiconductor photocatalysis type A), in complete analogy with electrochemical synthesis. An example is the anaerobic dehydrodimerization of unsaturated ethers in aqueous solution producing stoichiometric amounts of hydrogen. The reaction represents a combination of "sacrificial" hydrogen production and organic synthesis. It proves that formation of D_2 upon using D_2O as the solvent does not unambiguously indicate that hydrogen is produced from water as generally assumed in the literature for "sacrificial hydrogen production."

In a few cases, the semiconductor photocatalyzes an addition reaction between two or more substrates affording one single reaction product (semiconductor photocatalysis type B), a reaction type unknown in classical electrochemistry. Typical examples are the addition of olefins or unsaturated ethers to 1,2-diazenes or imines. These preparative photoreactions are easily conducted, and the heterogeneous photocatalyst can be conveniently separated. They are promising examples for a "green chemistry" since they do not produce waste materials and they can be driven by solar light. In both photocatalysis types, the semiconductor action is at least bifunctional. It enables a proper assembling of substrates through adsorption at the surface–solvent layer, and it catalyzes photoinduced IFET to and from substrates, often coupled to proton transfer. The generated radicals undergo regio- and stereoselective C–C and C–N coupling to valuable products. In a formal way, the overall process resembles natural photosynthesis, wherein light absorption also generates reducing and oxidizing centers, which finally induce the synthesis of organic matter through C–C coupling reactions. The few previous preparative results in "artificial photosynthesis" reveal that simple semiconductor powders photocatalyze atom-economic reactions driven by visible light. In summary, the multifunctional nature of a semiconductor surface offers promising aspects for selective photochemical transformations related to environmental chemistry and chemical solar energy utilization.

References

1. Fujishima A, Rao TN (1998) Pure Appl Chem 70:2177–2187
2. Pichat P (2010) Appl Catal B 99:428–434
3. Ciamician G (1912) Science (Washington, DC, US) 36:385–394
4. Albini A, Fagnoni M (2008) ChemSusChem 1:63–66
5. Esser P, Pohlmann B, Scharf H-D (1994) Angew Chem 106:2093–2108
6. Esser P, Pohlmann B, Scharf H-D (1994) Angew Chem Int Ed Engl 2033(2020):2009–2023
7. Palmisano G, Augugliaro V, Pagliaro M, Palmisano L (2007) Chem Commun (Cambridge, UK):3425–3437
8. Fagnoni M, Dondi D, Ravelli D, Albini A (2007) Chem Rev (Washington, DC, US) 107:2725–2756
9. Protti S, Fagnoni M (2009) Photochem Photobiol Sci 8:1499–1516
10. Daimon T, Hirakawa T, Kitazawa M, Suetake J, Nosaka Y (2008) Appl Catal A 340:169–175
11. Naito K, Tachikawa T, Cui S-C, Sugimoto A, Fujitsuka M, Majima T (2006) J Am Chem Soc 128:16430–16431
12. Janczyk A, Krakowska E, Stochel G, Macyk W (2006) J Am Chem Soc 128:15574–15575
13. Kamat PV (1993) Chem Rev (Washington, DC, US) 93:267–300
14. Tachikawa T, Fujitsuka M, Majima T (2007) J Phys Chem C 111:5259–5275
15. Hodes G, Howell IDJ, Peter LM (1992) J Electrochem Soc 139:3136–3140
16. Schindler W, Kisch H (1997) J Photochem Photobiol A Chem 103:257–264
17. Tatsuma T, Tachibana S-I, Fujishima A (2001) J Phys Chem B 105:6987–6992
18. Haick H, Paz Y (2003) ChemPhysChem 4:617–620
19. Paz Y (2010) Diffus Defect Data Pt B 162:135–162
20. Gerischer H (1990) Electrochim Acta 35:1677–1699
21. Ward MD, White JR, Bard AJ (1983) J Am Chem Soc 105:27–31
22. Tryk DA, Fujishima A, Honda K (2000) Electrochim Acta 45:2363–2376
23. Parrino F, Ramakrishnan A, Kisch H (2008) Angew Chem Int Ed 47:7107–7109
24. Seo YS, Lee C, Lee KH, Yoon KB (2005) Angew Chem Int Ed 44:910–913, S910/911-S910/910
25. Manzhos S, Jono R, Yamashita K, Fujisawa J-i, Nagata M, Segawa H (2011) J Phys Chem C 115:21487–21493
26. Macyk W, Szacilowski K, Stochel G, Buchalska M, Kuncewicz J, Labuz P (2010) Coord Chem Rev 254:2687–2701
27. Wardman P (1989) J Phys Chem Ref Data 18(4):1637
28. Hurum DC, Agrios AG, Crist SE, Gray KA, Rajh T, Thurnauer MC (2006) J Electron Spectrosc Relat Phenom 150:155–163
29. Li G, Gray KA (2007) Chem Phys 339:173–187
30. Anpo M, Moon SC, Chiba K, Martra's G, Coluccia S (1993) Res Chem Intermediat 19:495–519
31. Tachikawa T, Yamashita S, Majima T (2011) J Am Chem Soc 133:7197–7204
32. Tachikawa T, Yoshida A, Tojo S, Sugimoto A, Fujitsuka M, Majima T (2004) Chem Eur J 10:5345–5353
33. Furube A, Asahi T, Masuhara H, Yamashita H, Anpo M (1999) J Phys Chem B 103:3120–3127
34. Schiller M, Muller FW, Damm C (2002) J Photochem Photobiol A 149:227–236

35. Israel G, Muller FW, Damm C, Harenburg J (1997) J Inf Rec 23:559–584
36. Kortuem G, Braun W, Herzog G (1963) Angew Chem 75:653–661
37. Tauc J, Grigorovici R, Vanuc A (1966) Phys Status Solidi 15:627
38. Ohtani B, Mahaney OOP, Amano F, Murakami N, Abe R (2010) J Adv Oxid Technol 13:247–261
39. Morrison SR (1980) Electrochemistry at semiconductor and oxidized metal electrodes. Plenum, New York
40. Memming R (2001) Semiconductor electrochemistry. Wiley-VCH, Weinheim
41. Pleskov YV, Mazin VM, Evstefeeva YE, Varnin VP, Teremetskaya IG, Laptev VA (2000) Electrochem Solid State Lett 3:141–143
42. Beranek R, Kisch H (2007) Electrochem Commun 9:761–766
43. Weiss H, Fernandez A, Kisch H (2001) Angew Chem Int Ed 40:3825–3827
44. Roy AM, De GC, Sasmal N, Bhattacharyya SS (1995) Int J Hydrogen Energy 20:627–630
45. Burgeth G, Kisch H (2002) Coord Chem Rev 230:41–47
46. Meissner D, Lauermann I, Memming R, Kastening B (1988) J Phys Chem 92:3484–3488
47. Enright B, Redmond G, Fitzmaurice D (1994) J Phys Chem 98:6195–6200
48. Mitoraj D, Kisch H (2009) J Phys Chem C 113:20890–20895
49. Satuf ML, Brandi RJ, Cassano AE, Alfano OM (2007) Ind Eng Chem Res 46:43–51
50. Sagawe G, Satuf ML, Brandi RJ, Muschner JP, Federer C, Alfano OM, Bahnemann D, Cassano AE (2010) Ind Eng Chem Res 49:6898–6908
51. Schiavello M, Augugliaro V, Palmisano L (1991) J Catal 127:332–341
52. Hoffmann MR, Martin ST, Choi W, Bahnemann DW (1995) Chem Rev 95:69–96
53. Ollis DF (2008) Top Catal 35:217–223
54. Ohtani B (2008) Chem Lett 37:217–229
55. Serpone N, Terzian R, Lawless D, Kennepohl P, Sauve G (1993) J Photochem Photobiol 73:11–16
56. Serpone N, Sauvé G, Koch R, Tahiri H, Pichat P, Piccinini P, Pelizzetti E, Hidaka H (1996) J Photochem Photobiol A Chem 94:191–203
57. Tahiri H, Serpone N, Le van Mao R (1996) J Photochem Photobiol A Chem 93:199–203
58. Einaga H, Misono M (1996) Bull Chem Soc Jpn 69:3435–3441
59. Johne P, Kisch H (1997) J Photochem Photobiol A Chem 111:223–228
60. Curco D, Gimenez J, Addardak A, Cervera-March S, Esplugas S (2002) Catal Today 76:177–188
61. Maschmeyer T, Che M (2010) Angew Chem Int Ed 49:1536–1539
62. Kisch H (2010) Angew Chem Int Ed 49:9588–9589
63. Lakshminarasimhan N, Kim W, Choi W (2008) J Phys Chem C 112:20451–20457
64. Kuenneth R, Feldmer C, Knoch F, Kisch H (1995) Chem Eur J 1:441–448
65. Wang C-Y, Pagel R, Dohrmann JK, Bahnemann DW (2007) Mater Sci Forum 544–545:17–22
66. Zhang H, Chen G, Bahnemann DW (2009) J Mater Chem 19:5089–5121
67. Siedl N, Elser MJ, Bernardi J, Diwald O (2009) J Phys Chem C 113:15792–15795
68. Baumann SO, Elser MJ, Auer M, Bernardi J, Husing N, Diwald O (2011) Langmuir 27:1946–1953
69. Hayoun R, Whitaker KM, Gamelin DR, Mayer JM (2011) J Am Chem Soc 133:4228–4231
70. Kraeutler B, Bard AJ (1978) J Am Chem Soc 100:5985–5992
71. Al-Ekabi H, De Mayo P (1985) J Phys Chem 89:5815–5821
72. Yanagida S, Mizumoto K, Pac C (1986) J Am Chem Soc 108:647–654
73. Anpo M, Sunamoto M, Che M (1989) J Phys Chem 93:1187–1189
74. Kodama S, Yagi S (1989) J Phys Chem 93:4556–4561
75. Ikezawa H, Kutal C (1987) J Org Chem 52:3299–3303
76. Al-Ekabi H, De Mayo P (1986) J Phys Chem 90:4075–4080
77. Draper AM, Ilyas M, De Mayo P, Ramamurthy V (1984) J Am Chem Soc 106:6222–6230
78. Al-Ekabi H, De Mayo P (1986) Tetrahedron 42:6277–6284

79. Maldotti A, Amadelli R, Bartocci C, Carassiti V (1990) J Photochem Photobiol A 53:263–271
80. Wang CM, Mallouk TE (1990) J Am Chem Soc 112:2016–2018
81. Barber RA, De Mayo P, Okada K (1982) J Chem Soc Chem Commun:1073–1074
82. Ilyas M, De Mayo P (1985) J Am Chem Soc 107:5093–5099
83. Fox MA, Pettit TL (1985) J Org Chem 50:5013–5015
84. Boarini P, Carassiti V, Maldotti A, Amadelli R (1998) Langmuir 14:2080–2085
85. Joyce-Pruden C, Pross JK, Li Y (1992) J Org Chem 57:5087–5091
86. Mahdavi F, Bruton TC, Li Y (1993) J Org Chem 58:744–746
87. Ohtani B, Osaki H, Nishimoto S, Kagiya T (1986) J Am Chem Soc 108:308–310
88. Horner G, Johne P, Kunneth R, Twardzik G, Roth H, Clark T, Kisch H (1999) Chem Eur J 5:208–217
89. Kuenneth R, Feldmer C, Kisch H (1992) Angew Chem 104:1102–1103
90. Kuenneth R, Feldmer C, Kisch H (1992) Angew Chem Int Ed Engl 1131(1108):1039–1140
91. Schmittel M, Burghart A (1997) Angew Chem Int Ed Engl 36:2551–2589
92. Linker T, Schmittel M (1998) Radikale und radikalionen in der organischen synthese. Wiley-VCH, Weinheim
93. Fox MA (1987) Top Curr Chem 142:71–99
94. Fox MA, Dulay MT (1993) Chem Rev 93:341–357
95. Marinkovic S, Hoffmann N (2004) Eur J Org Chem 3102–3107
96. Hoffmann N (2007) Pure Appl Chem 79:1949–1958
97. Su F, Mathew SC, Moehlmann L, Antonietti M, Wang X, Blechert S (2011) Angew Chem Int Ed 50:657–660, S657/651-S657/653
98. Buecheler J, Zeug N, Kisch H (1982) Angew Chem 94:792–793
99. Hetterich W, Kisch H (1988) Chem Ber 121:15–20
100. Zeug N, Buecheler J, Kisch H (1985) J Am Chem Soc 107:1459–1465
101. Kuenneth R, Twardzik G, Emig G, Kisch H (1993) J Photochem Photobiol A Chem 76:209–215
102. Yanagida S, Azuma T, Midori Y, Pac C, Sakurai H (1985) J Chem Soc Perkin Trans 2:1487–1493
103. Yanagida S, Kawakami H, Midori Y, Kizumoto H, Pac C, Wada Y (1995) Bull Chem Soc Jpn 68:1811–1823
104. Adamson AW (1982) Physical chemistry of surfaces, 4th edn. Wiley, New York
105. Hiemenz PC (1986) Principles of colloid and surface chemistry, 2nd edn. Marcel Dekker, New York
106. Zhang Q, Xu Z, Finch JA (1995) J Colloid Interface Sci 169:414–421
107. Spanhel L, Haase M, Weller H, Henglein A (1987) J Am Chem Soc 109:5649–5655
108. Henglein A (1982) Ber Bunsenges Phys Chem 86:301–305
109. Hoerner G (2000)
110. Lunazzi L, Placucci G, Grossi L (1983) Tetrahedron 39:159–163
111. Wilkinson F (1980) Chemical kinetics and reaction mechanism. Van Nostrand Reinhold Co. Ltd., Workinghan
112. Reinheimer A, Van Eldik R, Kisch H (2000) J Phys Chem B 104:1014–1024
113. Schenck GO, Formanek H (1958) Angew Chem 70:505
114. Cookson RC, Stevens IDR, Watts CT (1965) Chem Commun (Lond) 259–260
115. Askani R (1965) Chem Ber 98:2551–2555
116. Rosenthal I, Elad D (1967) Tetrahedron 23:3193–3204
117. Ahlgren G (1973) J Org Chem 38:1369–1374
118. Ninomiya I, Naito T (1989) Photochemical synthesis. Academic, New York
119. Al-Sader BH, Crawford RJ (1970) Can J Chem 48:2745–2754
120. Memming R (1988) Top Curr Chem 143:79–112
121. Chan D, Perram JW, White LR, Healy TW (1975) J Chem Soc Faraday Trans 1 71:1046–1057

122. Schäfer K (1969) Landolt-Börnstein, Zahlenwerte und Funktionen, vol 6. Springer, Berlin
123. Srinivasan KR, Kay RL (1977) J Solution Chem 6:357–367
124. Brazier DW, Freeman GR (1969) Can J Chem 47:893–899
125. Gonikberg MG (1963) Chemical equilibria and reaction rates at high pressure. Israel Program for Scientific Translations, Jerusalem
126. Miyahara M, Iwasaki S, Kotera T, Kawamura T, Okazaki M (1995) J Colloid Interface Sci 170:335–339
127. Andersen OS, Feldberg SW (1996) J Phys Chem 100:4622–4629
128. Asano T, Le Noble WJ (1978) Chem Rev 78:407–489
129. Van Eldik R, Asano T, Le Noble WJ (1989) Chem Rev 89:549–688
130. Drljaca A, Hubbard CD, Van Eldik R, Asano T, Basilevsky MV, Le Noble WJ (1998) Chem Rev (Washington, DC) 98:2167–2289
131. Isaacs NS (1981) Liquid phase high pressure chemistry. Wiley, Chichester
132. Nicholson AE, Norrish RGW (1956) Discuss Faraday Soc 22:104–113
133. Yokawa M, Ogo Y (1976) Makromol Chem 177:429–436
134. Keck H, Schindler W, Knoch F, Kisch H (1997) Chem Eur J 3:1638–1645
135. Schindler W, Knoch F, Kisch H (1996) Chem Ber 129:925–932
136. Reinheimer A, Fernandez A, Kisch H (1999) Zeitschrift fuer Phys Chem (Muenchen) 213:129–133
137. Hopfner M, Weiss H, Meissner D, Heinemann FW, Kisch H (2002) Photochem Photobiol Sci 1:696–703
138. Kisch H, Sakthivel S, Janczarek M, Mitoraj D (2007) J Phys Chem C 111:11445–11449
139. Pehlivanugullari HC, Sumer E, Kisch H (2007) Res Chem Intermediat 33:297–309
140. Gaertner M, Ballmann J, Damm C, Heinemann FW, Kisch H (2007) Photochem Photobiol Sci 6:159–164
141. Takaki K, Tsubaki Y, Tanaka S, Beppu F, Fujiwara Y (1990) Chem Lett:203–204
142. Thies H, Schoenenberger H, Bauer KH (1960) Arch Pharm 293/65:67–73
143. Jin SJ, Araki S, Butsugan Y (1993) Bull Chem Soc Jpn 66:1528–1532
144. Mauze B, Miginiac ML (1973) Bull Soc Chim Fr 1832–1838
145. Arous-Chtara R, Moreau JL, Gaudemar M (1980) J Soc Chim Tunis 3:1–11
146. Mills A, Williams G (1987) J Chem Soc Faraday Trans 1 83:2647–2661
147. Egerton TA, King CJ (1979) J Oil Colour Chem Assoc 62:386–391
148. Kohl S (2007) Ph.D. Thesis, University of Erlangen-Nürnberg
149. Aldemir M, Heinemann FW, Kisch H (2012) Photochem Photobiol Sci 11:908–913
150. Arndt T, Guessregen B, Hohl A, Reis J (2005) Clin Chim Acta 359:125–131
151. Kollonitsch J, Barash L, Kahan FM, Kropp H (1973) Nat (Lond) 243:346–347
152. Kollonitsch J, Perkins LM, Patchett AA, Doldouras GA, Marburg S, Duggan DE, Maycock AL, Aster SD (1978) Nature 274:906–908
153. Bergman RG (2007) Nat (Lond, UK) 446:506
154. Graening T (2007) Nachr Chem 55:836–840
155. Ramloch H, Taeuber G (1979) Chem Unserer Zeit 13:157–162
156. Shang J, Zhu Y, Du Y, Xu Z (2002) J Solid State Chem 166:395–399
157. Fan FRF, Leempoel P, Bard AJ (1983) J Electrochem Soc 130:1866–1875
158. Finlayson MF, Wheeler BL, Kakuta N, Park K-H, Bard AJ, Campion A, Fox MA, Webber SE, White JM (1985) J Phys Chem 89:5676–5681

Efficient Mesoporous Semiconductor Materials for Environmental Applications

Adel A. Ismail and Detlef W. Bahnemann

Abstract Wastewater effluents from industry, at times, contain toxic organic chemicals that need to be treated prior to the effluent disposal. Advanced oxidation processes (AOPs) are efficient techniques that can potentially destroy a wide range of organic molecules. The choice of these techniques is based on their great potential for the complete mineralization of organic compounds present in these effluents that is readily explained by the generation of strongly oxidizing free radical intermediates, e.g., hydroxyl radicals. Photocatalysis is one of these AOPs where these radicals are generated on the surface of the illuminated semiconductor particles. Mesoporous metal oxides and mixed metal oxides have been receiving considerable attention in recent years due to their scientifically interesting properties and possible industrial applications in the fields of catalysis, adsorption, separation, ion exchange, and chemical sensing. This chapter covers recent developments in the syntheses of mesoporous semiconductor materials and presents applications of mesoporous semiconductors materials as efficient photocatalysts. The underlying reaction mechanisms will be explained and discussed.

Keywords Doped photocatalysts, Environmental remediation, Heterojunction photocatalysts, Mesoporous semiconductors, Photocatalyst networks

A.A. Ismail (✉)
Advanced Materials Department, Central Metallurgical R&D Institute, CMRDI, P.O. Box 87, Helwan 11421, Cairo, Egypt

Advanced Materials and NanoResearch Centre, Najran University, P.O. Box 1988, Najran 11001, Saudi Arabia
e-mail: adelali141@yahoo.com

D.W. Bahnemann
Institut für Technische Chemie, Leibniz Universität Hannover, Callinstrasse 3, 30167 Hannover, Germany

D.W. Bahnemann and P.K.J. Robertson (eds.), *Environmental Photochemistry Part III*,
Hdb Env Chem (2015) 35: 221–266, DOI 10.1007/698_2015_331,

Contents

1 Introduction

Mesoporous metal oxides and mixed metal oxides have been receiving considerable attention in recent years due to their scientifically interesting properties and their potential industrial applications in the fields of catalysis, adsorption, separation, ion exchange, and chemical sensing. They exhibit excellent surface properties, such as high surface area, large pore volume, and well-organized pore channels constituted by the network elements [1–4]. Porous semiconductor photocatalysts that are active upon illumination have attracted extensive attention recently due to their remarkable well-designed physical and chemical properties. Mesoporous TiO_2 is an interesting material for photocatalytic applications due to its continuous particle network, which may be beneficial compared to separate individual nanoparticles, in particular for catalyst recovery. Mesoporous metal oxides are prepared by template-based methods using soft templates (surfactant and block polymers) and hard templates (porous silica, polystyrene spheres, porous carbon) [5–9]. Metal oxides and mixed oxides with acidic, basic, and/or redox properties are used as catalysts in all fields of heterogeneous catalysis, including oil refining, fine chemical synthesis, and environmental catalysis [10, 11]. This research has produced good results and hence made it even more interesting to prepare ordered mesoporous crystalline metal oxides. Finding new catalysts remains a necessity, notably to answer the ever-increasing demand for sustainable and environmentally benign chemical processes. TiO_2 has been widely used as a photocatalyst for the removal of hazardous organic substances [12, 13] due to its strong oxidizing and reducing ability under UV-light irradiation. Two of the most important factors affecting the photocatalytic activity of mesoporous metal oxides and mixed metal oxides are their specific surface area in a continuous structure rather than in discrete particles and crystallinity. This continuity can be expected to make the electron transfer within the material easier, resulting in higher activity. If either mesoporous metal oxides or mixed metal oxides could be prepared with crystalline walls, it would be a useful material applicable as high-performance photocatalyst. In the last decade, research efforts have been directed to enhance the photoactivity of the mesoporous metal oxides using various methods such as increasing catalyst surface-to-volume ratio, sensitization of the catalyst using dye molecules [14, 15], and particle contacts of the photocatalyst with nonmetals such as nitrogen, carbon, fluoride, and iodine [16–19] and transition metals [20–22]. Wastewater effluents from industry, at times, contain toxic organic chemicals which need to be treated prior to effluent disposal. Advanced oxidation processes (AOPs) are efficient techniques which are characterized by the generation of hydroxyl radicals ($^{\bullet}OH$) that can potentially destroy a wide range of organic molecules. Among the numerous AOPs, heterogeneous photocatalysis using semiconductors has been found to be a highly effective treatment technology. The choice of this technique was based on its great potential for the complete mineralization of organic effluents, while the catalyst itself is nontoxic, cost-effective, and readily available. Semiconductor photocatalysis is emerging as a potent technique for treating such effluents. The main advantage of photocatalysis lies in the fact that organic contaminants are completely mineralized without requiring secondary treatment of concentrated wastes. Furthermore, photocatalysis has been reported to have the potential to be an effective method for treating a wide range of pollutants both in water and air [23–26]. The following chapter focuses on the synthesis of mesoporous semiconductor materials as efficient photocatalysts highlighting synthetic methods, architectural concepts, and fundamental principles that govern the rational design and synthesis. In this chapter, synthesis mechanisms and the corresponding pathways are first demonstrated for mesoporous materials prepared by the surfactant-templating approach. The continuing breakthroughs in the synthesis and modification of mesoporous metal oxides have resulted in new properties with improved photocatalysts. The chapter covers recent developments in the synthesis of mesoporous semiconductor materials, semiconductor/semiconductor contacts, and their application as efficient photocatalysts. The photocatalysis reaction mechanisms will be presented and discussed.

2 Mesoporous TiO_2 and Its Environmental Applications

Mesoporous TiO_2 as active photocatalysts can be prepared by different techniques such as sol–gel, hydrothermal, sonochemical, microwave, and electrodeposition. They have been well synthesized with or without the use of organic surfactant templates. These templates are used as structure-directing agents for organizing networks formed of metal oxide species in nonaqueous solutions. The most commonly used organic templates are amphiphilic poly(alkylene oxide) block copolymers, such as HO-$(CH_2CH_2O)_{20}(CH_2CH(CH_3)O)_{70}$ $(CH_2CH_2O)_{20}H$ (designated $EO_{20}PO_{70}EO_{20}$, called Pluronic P-123) [27–39] and HO$(CH_2CH_2O)_{106}$-$(CH_2CH$ $(CH_3)O)_{70}$ $(CH_2CH_2O)_{106}H$ (designated $EO_{106}PO_{70}$-EO_{106}, called Pluronic F-127) [38, 40–46]. Also, diblock polymers were used such as $[C_nH_{2n-1}(OCH_2CH_2)_yOH$, Brij 56 (B56, n/y)16/10) [47]. Other surfactants employed to direct the formation of mesoporous TiO_2 include tetradecyl phosphate by Antonelli and Ying [5] and commercially available dodecyl phosphate by Stone and Davis [48], Tween 80 [49], and cetyltrimethylammonium bromide (CTAB) TiO_2 [44, 50–59]. Semiconductor photocatalysts are interesting materials for photocatalytic applications and are regarded as the most efficient and environmentally benign photocatalyst being most widely used for the photodegradation of various pollutants [60]. The principle of the semiconductor photocatalytic reaction is straightforward. When photons with energies exceeding the bandgap energy of semiconductor metal oxide are absorbed by the particles in the mesoporous metal oxide photocatalyst framework, electrons are rapidly promoted from the valence band to the conduction band leaving holes behind in the valence band [60, 61]. The thus formed electrons and holes participate in redox processes at the semiconductor/water interface. The valance band holes migrate to the surface of the particles where they react with adsorbed hydroxide ions (or water molecules), generating adsorbed $^{\bullet}OH$ radicals. This photodecomposition process usually involves one or more radicals or intermediate species such as $^{\bullet}OH$, $O_2^{\bullet-}$, H_2O_2, or O_2, which play important roles in the photocatalytic reaction mechanisms [60]. The photocatalytic activity of a semiconductor is largely controlled by (1) the light absorption properties, e.g., light absorption spectrum and coefficient, (2) reduction and oxidation rates on the surface by the electron and hole, (3) and the electron–hole recombination rates.

2.1 Sol–Gel Method

In a sol–gel process, initially a colloidal suspension, or a sol, is formed from the hydrolysis and polymerization reactions of the precursors, which are usually inorganic metal salts or metal organic compounds such as metal alkoxides. Complete polymerization and loss of solvent leads to the transition from the liquid sol into a solid gel phase. First, a homogeneous solution is obtained by dissolving the surfactant (s) in a solvent. TiO_2 precursors are then added into the solution where they undergo hydrolysis catalyzed by an acid catalyst and transform to a sol of Ti–O–Ti chains [27–29, 50–53, 62, 63]. Chen et al. [27] prepared mesoporous TiO_2 using $TiCl_4$ and Ti $(OBu)_4$ as the precursors and P123 as the template by the nonhydrolytic evaporation-induced self-assembly (EISA) method with ordered 3D TiO_2 with a uniform pore size and high surface area being obtained. Photodegradation of phenol shows that the sample has a better photoactivity than commercial TiO_2 P25. This is attributed to the well-ordered 3D open-pore structure, which, combined with its relatively large surface area and pore volume, can facilitate the mass transport of the organic pollutants. Therefore, both the regular open-pore morphology and the biphasial structure are playing crucial roles in determining the sample's photoactivity. Beyers et al. [50] reported the preparation of mesoporous TiO_2 using CTAB as surfactant. The photocatalytic activity of mesoporous TiO_2 for the decomposition of rhodamine 6G could be increased by changing the synthesis medium from basic to acidic conditions. The slower condensation of the TiO_2 precursor leads to improved mesoporous structure, with better accessibility for photocatalysis. However, TiO_2

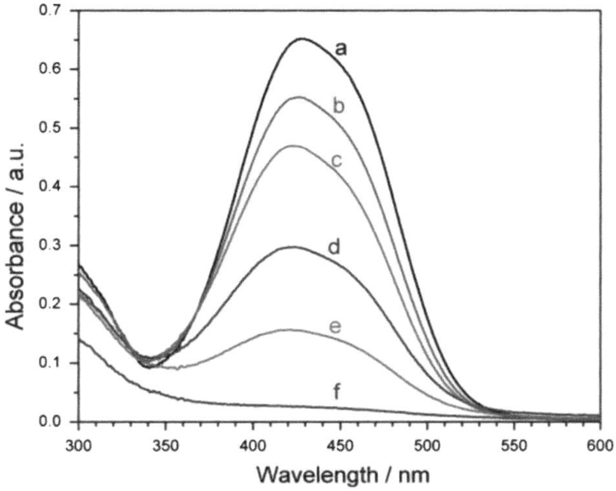

Fig. 1 UV–vis absorption spectra of methyl orange aqueous solution recorded at different time intervals following mesoporous anatase-catalyzed photodegradation. The spectra were recorded at time intervals of (**a**) 0 min, (**b**) 5 min, (**c**) 10 min, (**d**) 20 min, (**e**) 30 min, and (**f**) 45 min. The initial concentration of methyl orange was 30.0 mg/L, and the loading of mesoporous anatase catalyst was 3.0 g/L (adapted from Wang [51])

anatase was prepared by using the sol–emulsion gel method in the presence of both CTAB [51–53] and cyclohexane. The as-prepared anatase powders exhibited high photocatalytic activity and could be effectively used as catalysts for the photodegradation of methyl orange, bromopyrogallol red, and methylene blue [51]. Figure 1 shows the UV–vis absorption spectra of methyl orange aqueous solution recorded at different time intervals during the mesoporous anatase-catalyzed photodegradation [51]. The initial concentration of methyl orange was 30.0 mg/L, and the loading of the mesoporous anatase catalyst was 3.0 g/L. The absorption peak centered at 426 nm is the characteristic peak of methyl orange, and the absorption at 426 nm follows the Beer–Lambert law in the concentration range from 0 to 50.0 mg/L. The intensity of the absorption peak kept decreasing as the reaction time increased. After 45 min, this peak completely disappeared, indicating the complete degradation of methyl orange in the solution. The ordered mesoporous structure of the as-prepared anatase catalyst was of key importance to the high photocatalytic activity.

Shiraishi et al. [28] have developed highly selective methods for photochemical organic syntheses, driven by a mesoporous TiO_2, which enable the transformation of benzene into phenol with very high selectivity (>80%). Briefly, the preparation involves the dissolution of 1 g of P123 in dry ethanol (20 g). $TiCl_4$, 0.6 g, and TTIP, 2.5 g, were added to the solution, and the mixture was stirred for 2 h at room temperature [63]. The proposed system exhibits significant advantages for organic syntheses: (1) additive-free, (2) cheap source of oxidant (H_2O), and (3) mild reaction conditions. Liu et al. [64] have prepared nanostructure anatase TiO_2 monoliths using 1-butyl-3-methylimidazolium tetrafluoroborate ($BMIM^+BF_4^-$) ionic liquids as template solvents by a simple sol–gel method with a peptization process at ambient temperature. The as-prepared products showed wormhole-like mesoporous structures with a surface area of ca. 260 $m^2\ g^{-1}$ that was retained upon calcination up to 450°C, showing excellent thermal stability. The products revealed improved photodegradation ability toward rhodamine B as compared with that of the commercially available TiO_2 Degussa P25. Lu et al. [65] have synthesized well-defined, crystalline TiO_2 nanoparticles at room temperature by using spherical polyelectrolyte brush particles as a template. The template particles consisted of a polystyrene core onto which long chains of poly(styrene sodium sulfonate) (PS–NaSS) had been grafted. $Ti(OC_2H_5)_4$ was hydrolyzed in the presence of these brush particles leading to the formation of well-dispersed TiO_2 nanoparticles. The as-prepared TiO_2 nanocomposites exhibit high photocatalytic activity for the degradation of rhodamine B under UV irradiation (Fig. 2). The apparent rate constant k_{app} can therefore be expected to be proportional to the total surface S of the TiO_2

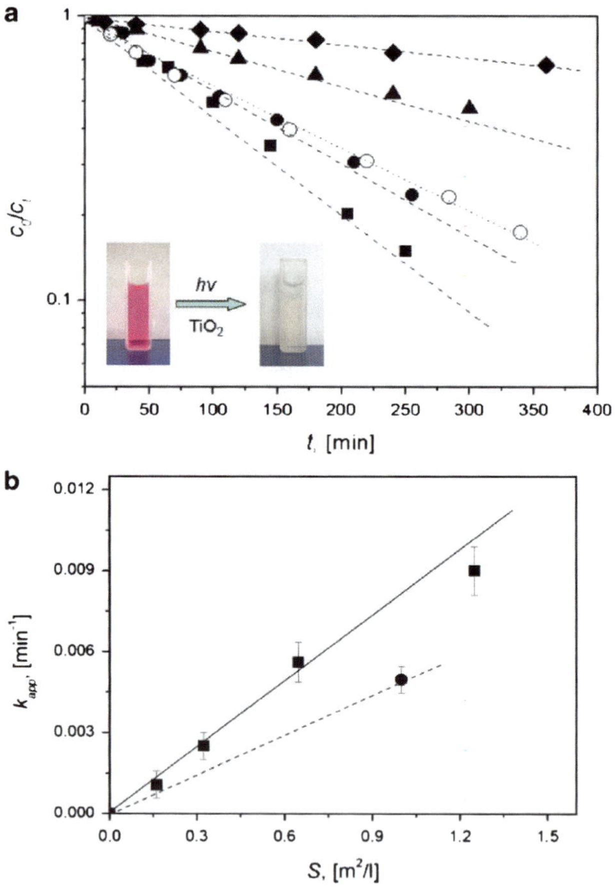

Fig. 2 (**a**) Photodegradation of RhB in the presence of PS–NaSS–TiO_2-3 composite particles (*solid symbols*) and P25 (Evonik Degussa) particles (*open circle*, 0.02 $g.L^{-1}$), respectively. Parameter of the different curves is the concentration of TiO_2 nanoparticles in the solution. *Quadrangles*, 0.01 $g.L^{-1}$; *circles*, 0.005 g/L^{-1}; *triangles*, 0.0025 $g.L^{-1}$; *diamonds*, 0.00125 $g.L^{-1}$. The inset demonstrates the decolorizing effect of photodegradation. (**b**) Rate constant k_{app} as a function of the surface area S of the TiO_2 nanoparticles (*quadrangles*, PS–NaSS–TiO_2-3; *circles*, P25) normalized to the unit volume of the system. The concentration of the reactant was as follows: [RhB] = 0.02 $mmol.L^{-1}$, $T = 20°C$ (adapted from Lu et al. [65])

nanoparticles present in the system. Figure 2b demonstrates that the apparent rate constant k_{app} indeed depends linearly on the specific surface area of the PS–NaSS–TiO_2 nanocomposite particles as expected.

TiO_2 hollow fibers with mesoporous long walls were fabricated with a sol–gel combined two-capillary spinneret electrospinning technique using P123 [29, 30]. The photodegradation rate of methylene blue and gaseous formaldehyde for the TiO_2 hollow fibers was found to be higher than that for P25 and mesoporous TiO_2 powders. Yu et al. [66] have prepared TiO_2 hollow microspheres based on template-directed deposition and in situ template-sacrificial dissolution. This synthesis method was developed in pure water by using SiO_2 microspheres as templates and TiF_4 as the precursor at 60°C. It is found that the thus prepared TiO_2 hollow microspheres show a stronger absorption in the UV–vis region (310–700 nm) than P25. Hence, this also leads to an enhanced photocatalytic activity of the TiO_2 hollow spheres. They exhibit hierarchically nanoporous structures and a high photocatalytic activity. Mesoporous titania with an amorphous wall can be prepared by sol–gel reactions of titanium oxysulfate sulfuric acid hydrate ($TiOSO_4.xH_2SO_4.xH_2O$) in the presence of $C_{16}TAB$ at room temperature [55]. Figure 3 shows the changes in concentration of 2-propanol and acetone observed during the respective illumination time. The 2-propanol concentration initially decreased due to its adsorption onto the mesoporous TiO_2 particles, which as mentioned have a high specific surface area. Upon UV irradiation, the concentration of 2-propanol decreased and acetone was generated. The results suggest that the obtained mesoporous material has both a good absorbability and a high photocatalytic activity.

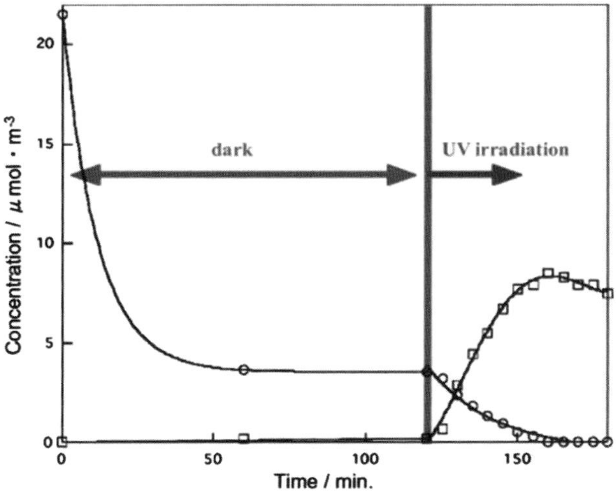

Fig. 3 Changes in the concentrations of 2-propanol and acetone under UV-light irradiation. Open circles and squares represent the concentrations of 2-propanol and acetone, respectively (adapted from Shibata et al. [55])

2.2 Hydrothermal Method

Hydrothermal syntheses are normally conducted in aqueous solutions employing steel pressure vessels called autoclaves equipped with Teflon liners under temperature and/or pressure control. The temperature can be elevated above the boiling point of water until the pressure of vapor saturation is reached. The temperature and the amount of solution added to the autoclave largely determine the internal pressure produced. Many research groups have used the hydrothermal method to prepare mesoporous TiO_2 nanoparticles [47, 54–57, 67–73].

2.2.1 Formation Mechanism of Stable Porous TiO_2

TiO_2 crystallites with low crystallinity are first prepared by a hydrolysis process in an acid system, and it can be expected that they easily form agglomerations because of the existence of some amorphous phase resulting from $Ti(OH)_n$ [67]. Subsequently, the CTAB introduction during the hydrothermal process under basic condition can effectively disperse the agglomeration and further induce the assembly of the as-prepared crystallites. Under basic condition, the CTA^+ groups are always positively charged while the nanoparticles are negatively charged. Thus, the strong electrostatic attraction between the nanoparticles and CTAB results in the dispersion of the nanoparticles together with their transformation from the amorphous phase to TiO_2 crystallites and further induces the assembly of the as-prepared crystallites to form mesoporous TiO_2 [54, 57, 68, 69, 71]. Mesoporous TiO_2 with amorphous walls can be prepared by the sol–gel reaction of titanium oxysulfate sulfuric acid hydrate ($TiOSO_4.xH_2SO_4.xH_2O$) [55] and $Ti(SO_4)_2$ [54] in the presence of the cationic surfactant CTAB at room temperature. The results suggest that the obtained mesoporous material has good adsorbability as well as high photocatalytic activity for the conversion of 2-propanol to acetone. The large surface area, small crystalline size, and well-crystallized anatase mesostructure can explain the high photocatalytic activity of mesoporous TiO_2 nanoparticles calcined at 400°C for the degradation of rhodamine B [71]. Mixed-phase TiO_2 nanocrystals with tunable brookite-to-rutile ratios can be synthesized using titanium tetrachloride as the titanium source in the presence of triethylamine [56] and CTAB as the template, followed by a posttreatment in the presence of ethylenediamine [57]. The high crystallinity, large surface area, and heterojunction microstructure between anatase and brookite may be responsible for the high photocatalytic activity in terms of the degradation of phenol and rhodamine B under UV irradiation. Interestingly, the thus prepared mesoporous TiO_2 photocatalysts showed higher photocatalytic efficiency than Degussa P25. Trimodal spongelike macro-/mesoporous TiO_2 was prepared by the hydrothermal treatment of precipitates of ($Ti(OC_4H_9)_4$) in pure water [70]. The resulting TiO_2 samples exhibit a disordered wormlike macroporous framework structure with continuous nanocrystalline TiO_2 particles. The hierarchically porous TiO_2 prepared at 180°C for 24 h displayed an especially high photocatalytic activity for acetone decomposition probably due to its special pore-wall structure, and its photocatalytic activity was about three times higher than that of Evonik Degussa P25. Wang et al. [47] have synthesized mesoporous TiO_2 using TTIP and the nonionic poly (alkylene oxide)-based surfactant (decaoxyethylene cetyl ether, $C_{16}(EO)_{10}$, Brij56) as the structural-directing agent. In a typical synthesis, TTIP (4.8 mL) was added dropwise to 30 mL of an aqueous solution containing $C_{16}(EO)_{10}$ (10 wt%) under very gentle stirring. The mixture was transferred to a Teflon-lined autoclave to age at 80°C for 24 h. The catalyst which was calcined at 350°C possessed an intact macro-/mesoporous structure and showed a photocatalytic reactivity for ethylene oxidation 60% higher than that of P25. Further heating at temperatures above 600°C destroyed both macro- and mesoporous structures, accompanied by a loss in photocatalytic activity [47]. The high photocatalytic performance of the intact macro-/mesoporous TiO_2 may be explained by the existence of macrochannels that increase their photoabsorption efficiency and allow efficient diffusion of gaseous molecules. Liu et al. [72] have synthesized hollow porous TiO_2 aggregates on a large scale by means of a simple hydrothermal method without using any templates. The photocatalytic rhodamine B degradation rate of the porous TiO_2 hollow aggregates was found to be more than twice as high as that of P25. The higher photocatalytic activity of the porous TiO_2 hollow aggregates has been explained by considering several factors: (1) The surface area of porous TiO_2 hollow was observed (porous TiO_2 hollow aggregates ca. 168 m^2g^{-1} versus P25 powder ca. 45 m^2g^{-1}); hence, there are more reactant adsorption/desorption sites for catalytic reaction. (2) The prevention of the unwanted aggregation of the nanoparticle clusters, which is also helpful in maintaining the high active surface area. (3) The highly porous structure, which allows rapid diffusion of various reactants and products during the reaction. (4) The smaller crystal sizes can also be beneficial for the separation of the photogenerated hole/electron pairs [72]. Mesoporous TiO_2 microspheres were synthesized by a facile solvothermal method [73]. The photocatalytic activities of the as-prepared TiO_2 microspheres were evaluated by the photodegradation of methyl orange (MO) and phenol in aqueous solutions, respectively, under simulated sunlight irradiation, and compared with those of commercial TiO_2 (P25). The TiO_2 microspheres exhibit excellent photodegradation activities for both MO and phenol, exceeding those of P25 by a factor of two under identical test conditions. These remarkable photoactivities have been attributed to higher surface area, stronger light-harvesting capability, lower recombination rate of photogenerated carriers, and hierarchical mesoporous microspherical structure packed by numerous interconnected nanoparticles.

2.3 Microwave Method

A dielectric material can be processed with energy in the form of high-frequency electromagnetic waves. The principal frequencies of microwave heating are between 900 and 2,450 MHz. Microwave radiation is applied to prepare various mesoporous TiO_2 nanoparticles. Crystalline anatase mesoporous nanopowders 100–300 nm in size with wormhole-like pore sizes of 3–5 nm were prepared by a modified sol–gel synthesis route starting from TTIP, accelerated by a microwave hydrothermal process [74]. The organic surfactant, tetradecylamine, which is used as a self-assembly micelle in the sol–gel and microwave hydrothermal process, enables to harvest crystallized mesoporous anatase nanoparticles with a high surface area. Mesoporous wormhole-like and crystalline powders with surface areas of 243–622 m^2/g are obtained. It is shown that crystallization by calcination at 400°C for 3 h inevitably reduces the surface area, while the microwave hydrothermal process demonstrated a rapid formation of crystalline mesoporous TiO_2 nanopowders with a high surface area and excellent photocatalytic activity for the methylene blue degradation.

2.4 Sonochemical Method

Ultrasound has been shown to be very useful for the synthesis of active mesoporous TiO_2 photocatalysts [31, 75]. Yu et al. [31] applied the sonochemical method for the preparation of highly photoactive TiO_2 nanoparticle with anatase and brookite phases using the hydrolysis of TTIP in pure water or in a 1:1 EtOH–H_2O solution under ultrasonic radiation. Mesoporous TiO_2 with a bicrystalline (anatase and brookite) framework was synthesized directly under high-intensity ultrasound irradiation. The photocatalytic activity of mesoporous TiO_2 synthesized in the presence of a triblock copolymer was about two times higher than that of P25. The high activities of the mesoporous TiO_2 with a bicrystalline framework can be attributed to the combined effect of three factors: high brookite content, high surface area, and the existence of mesopores. Yu et al. [75] prepared mesoporous TiO_2 nanocrystalline powders by the ultrasonic-induced hydrolysis reaction of (Ti $(OC_4H_9)_4$) in pure water without using any templates or surfactants. It was found that the as-prepared products are composed of anatase and brookite phases. The photocatalytic activity of the samples prepared by this ultrasonic method is reported to be higher than that of commercial P25 and of samples prepared by a conventional hydrolysis method (Fig. 4).

2.5 Synthesis of TiO_2 Using Electrodeposition

Electrodeposition is usually employed to produce metallic coatings on electrode surface by reduction processes at the cathode. On the other hand, TiO_2/benzoquinone hybrid films have been successfully electrodeposited anodically from basic Ti(IV)-alkoxide solutions containing hydroquinone in the presence of tetramethylammonium hydroxide [76]. The photodegradation of methylene blue (MB) as a

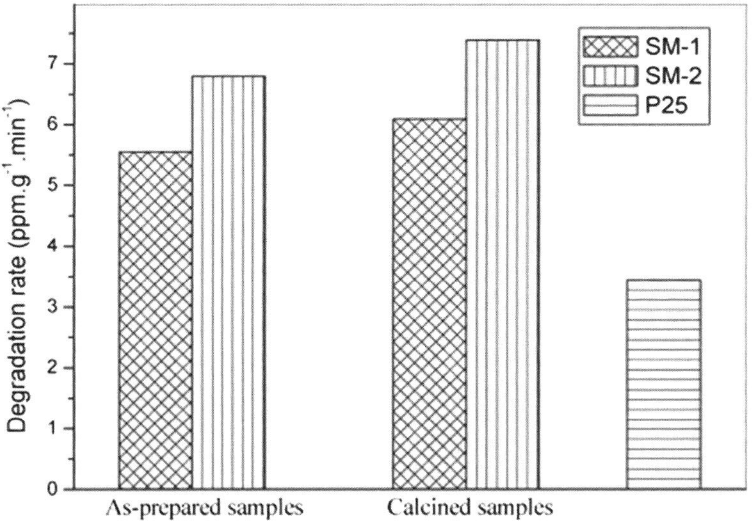

Fig. 4 Photocatalytic activities of as-prepared and calcined SM-1, SM-2, and P25 (adapted from Yu et al. [75])

representative for organic pollutants has been studied. The results revealed that the amorphous film calcined at 350°C is quite inactive, but the activity increases with increasing calcination temperature with the exception of the film calcined at 500°C. Matsumoto et al. reported a new method to prepare a mesoporous TiO_2 photocatalyst film onto alumite using an electrochemical technique [77], where the initial electrodeposition was carried out by electrolysis in $(NH_4)_2[TiO(C_2O_4)_2]$ solution, followed by pulse electrolysis in $TiCl_3$. This film exhibited a high photocatalytic activity for the decomposition of acetaldehyde with the corresponding concentration of CO_2 detected in the cell even after 45 min fluorescent lamp illumination.

2.6 Mesoporous TiO_2 Photocatalyst Films

Various titania sols containing poly(oxyethylenesorbitan monooleate) (Tween 80) surfactant to tailor-design the porous structure of TiO_2 have been successfully employed to prepare thin mesoporous TiO_2 films by dip coating at different molar ratios of Tween 80/isopropyl alcohol/acetic acid/TTIP=R:45:6:1 [49]. The thus prepared photocatalytic TiO_2 membranes have great potential for the development of highly efficient water treatment and reuse systems, for example, for the decomposition of organic pollutants, the inactivation of pathogenic microorganisms, the physical separation of contaminants, and the self-antifouling action because of their multifunctional capability. Wang et al. [78] have obtained 3D ordered mesoporous sulfated TiO_2 reacting a cubic mesoporous amorphous TiO_2 film with sulfuric acid at high temperature to produce sulfur-containing mesoporous TiO_2 with nanocrystalline frameworks. The resulting 3D ordered mesoporous sulfated TiO_2 superacids were found to be attractive photocatalysts for the degradation of bromomethane. High-quality mesostructured TiO_2 films have been prepared on silicon substrates by spin coating [79]. Posttreatment of the films in supercritical CO_2, in the presence of small amounts of either TTIP, tetramethoxysilane, or other precursors, greatly improved the thermal stability of the mesoporous coatings without affecting their optical transparency or integrity. A 0.02 M solution of stearic acid in methanol was first coated on the TiO_2-coated silicon wafers by spin coating. The photocatalytic degradation of stearic acid employing the sc-CO_2/TTIP-treated film is considerably faster compared to the untreated and sc-CO_2/TMOS-treated films. The higher efficiency was explained by the film's high porosity and by the presence of highly photoactive nanocrystalline anatase structures. Without sc-CO_2 treatment, the calcination at temperatures above 550°C inevitably leads to the collapse of the mesoporous structure and to the formation of rutile crystallites in the walls of the TiO_2 films by the transformation of anatase nanocrystals [79].

Kim et al. [80] prepared monodisperse spherical mesoporous TiO_2 with a morphology size of approximately 800 nm via the sol–gel approach using a triblock copolymer surfactant and TTIP with 2,4-pentanedione in aqueous solution. This precursor was then coated onto glass substrates without cracking by using the doctor blade method with various amounts of polyethylene oxide (PEO) and polyethylene glycol (PEG). The results revealed that the efficacy of the photocatalytic disinfection measured with the film adhesion method is strongly dependent on surface area and crystallite size. The inactivation rate constant was 0.118 min^{-1}, which is ca. 11 times faster than that of P25 (0.010 min^{-1}). This may be attributed to a larger surface area likely resulting in a better photocatalytic disinfection, because it provides more active sites. The mesoporous TiO_2 material coated on glass has a larger surface area (214 $m^2\ g^{-1}$) than the P25 TiO_2 (50 $m^2\ g^{-1}$). Therefore, mesoporous TiO_2 is likely to exhibit better photocatalytic inactivation than P25. Another factor that influences photocatalytic activity is crystallite size. Since the mesoporous TiO_2 was formed by the aggregation of nanocrystalline anatase particles having a smaller crystallite size (10 nm) than P25 (25 nm), its photocatalytic activity is expected to be better.

3 Semiconductor Heterojunction Photocatalysts

In recent years, a lot of efforts have been done to fabricate and design heterojunctions for enhancing the photocatalytic performances under UV and visible light [81]. In general, heterojunction photocatalysts are divided into four typical categories: (1) the semiconductor–semiconductor heterojunction(S–S), (2) the semiconductor–metal heterojunction (S–M), (3) the semiconductor–carbon

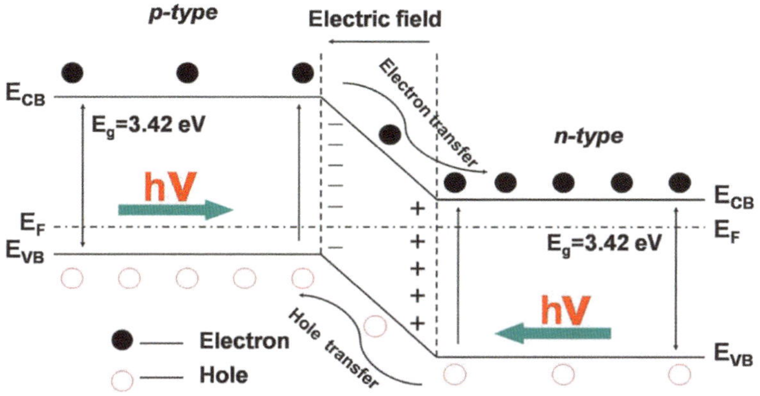

Scheme 1 Schematic diagram showing the energy band structure and electron–hole pair separation in the p–n heterojunction (adapted from Wang et al. [81])

group heterojunction (S–C), and (4) the multicomponent heterojunction [82]. There are two different S–S heterojunction systems: p–n semiconductor heterojunction and non-p–n heterojunction systems (Schemes 1 and 2). The proposed mechanism of p–n junction photocatalysts is described as follows: when the p- and n-type semiconductors are in contact and are irradiated by photons, the photogenerated electron–hole pairs can be quickly separated by the built-in electric field within the space charge region, due to the diffusion of electrons and holes in the opposite direction [83] (Scheme 1). Driven by the electric field, the electrons are transferred to the CB of the n-type semiconductors and the holes to the VB of the p-type semiconductors.

The advantages of p–n-type heterostructure are (1) a more effective charge separation, (2) a rapid charge transfer to the catalyst, (3) a longer lifetime of the charge carriers, and (4) a separation of locally incompatible reduction and oxidation reactions in nanospace [81]. On the other hand, there are non-p–n-type heterojunction systems, where the most suitable for photocatalytic applications is the staggered bandgap type (Scheme 2). In this type, the semiconductors A and B with matching band potentials are tightly bonded to construct the efficient heterostructure. When the CB level of semiconductor-B is lower than that of semiconductor-A, electrons in the CB of semiconductor-A can be transferred to that of semiconductor-B under visible-light irradiation. If the VB level of semiconductor-B is lower than that of semiconductor-A, holes in the VB of semiconductor-B can be transferred to that of semiconductor-A. The probability of electron–hole recombination can be reduced, and thus, the photocatalytic reaction can be enhanced greatly. We illustrate different examples of heterojunction semiconductor photocatalysts as detailed below.

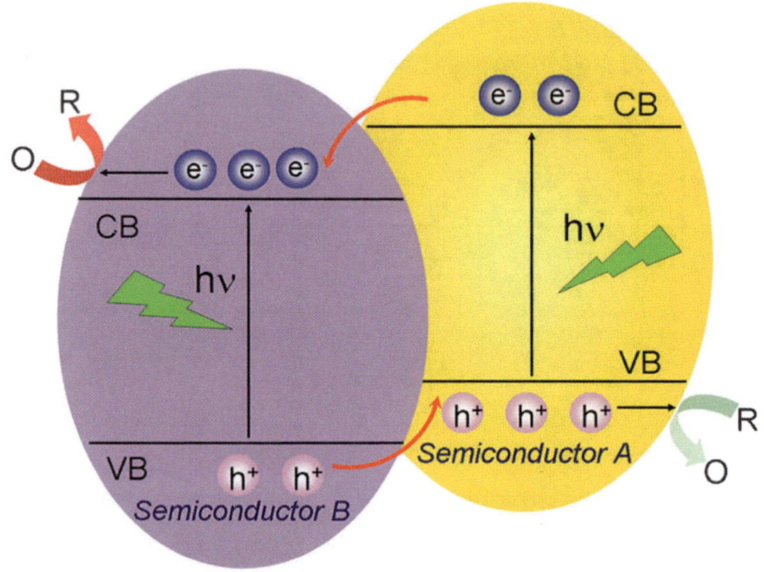

Scheme 2 Schematic diagram showing the energy band structure and electron–hole pair separation in the non-p–n heterojunction (adapted from Wang et al. [81])

3.1 Heterojunction of Anion (Nonmetal)/TiO$_2$ as Active Photocatalysts

One of the endeavors to improve the performance of TiO_2 is to increase its optical activity by shifting the onset of its response from the UV to the visible region. A promising approach is the particle contacts of TiO_2 with nonmetals. The rationale behind this approach is to sensitize TiO_2 toward visible light either by generating newly created midgap states or by narrowing the bandgap. The observed bandgap shift from the UV into the visible region has been attributed to (1) substitution of lattice oxygen by the anion or (2) formation of interstitial species in vacancies or micro-voids that give rise to surface or near-surface states [84–86]. The enhanced photocatalytic activity of heterojunction of nonmetal/TiO_2 is due to several possible factors. First, the presence of nonmetal ions is playing important roles such as S^{6+} serving as the surface center for the capture of photoinduced electrons and S^{4+} acting as the center for the capture of photoinduced hole; this sample may have the most suitable ratio of S^{6+} to S^{4+} to effectively suppress the recombination of photoinduced electrons and holes [87–89]. Also, carbon contact with TiO_2 can enhance the photocatalytic activity due to the conductivity of the TiO_2, allowing efficient charge transfer to the external site of the TiO_2 nanoparticles, where the desired oxidation reactions take place. Moreover, the elemental carbon can act as photosensitizer by injecting an electron into the conduction band of TiO_2 and is then transferred to oxygen absorbed on the TiO_2 surface, producing the O_2^- which is capable of degrading pollutants [90]. On the other hand, N/TiO_2 contact enhanced the photocatalytic activity due to the presence of the active N-species concentration and the lower number of Ti^{3+} sites [17].

3.1.1 Nitrogen/Mesoporous TiO_2

Visible-light-active mesoporous heterojunction of N/TiO_2 photocatalysts by the precipitation of the titanyl oxalate complex was prepared for photodegradation of MO [91]. The nitrogen species locates at the interstitial sites in TiO_2, which leads to the bandgap narrowing of TiO_2. A novel and interesting result is that N/TiO_2 calcined at 400°C has Bronsted acid sites arising from covalently bonded dicarboxyl groups, which greatly enhances the adsorption capacity for MO. Chi et al. [92] prepared mesoporous N/TiO_2 microspheres a template-free solvothermal method. The N/TiO_2 mesoporous spheres show higher visible-light photocatalytic activity than TiO_2. The enhanced photocatalytic activities of N/TiO_2 in UV light may be due to the increase of the surface deficiency after the introduction of nitrogen into the TiO_2 structure. It could be found that with the increase of the nitrogen amount in the spheres, the visible-light photocatalytic activity would also be enhanced, which could be the evidence to confirm the role of nitrogen in the lattice for improvement of the visible-light response of N/TiO_2 contact. Nitrogen-containing surfactant dodecylammonium chloride was introduced as a pore-templating material for tailor-designing the structural properties of TiO_2 and as a N/TiO_2 for its visible-light response [17]. Due to its narrow bandgap at 2.65 eV, N/TiO_2 efficiently degraded cyanobacterial toxin microcystin-LR (MC-LR) under visible spectrum above 420 nm. Acidic condition (pH 3.5) was more favorable for the adsorption and photocatalytic degradation of MC-LR on N/TiO_2 due to electrostatic attraction forces between negatively charged MC-LR and charged N/TiO_2. Even under UV light, MC-LR was decomposed 3–4 times faster using N/TiO_2 than control TiO_2. The degradation pathways and reaction intermediates of MC-LR were not directly related to the energy source for TiO_2 activation (UV and visible) and nature of TiO_2 [82]. Cong et al. [93] have synthesized N-doped TiO_2 nanocatalysts with a homogeneous anatase structure through a microemulsion–hydrothermal method by using triethylamine, urea, thiourea, and hydrazine hydrate. The results of the photodegradation of rhodamine B and 2,4-dichlorophenol in the visible-light irradiation ($\lambda > 420$ nm) suggested that the TiO_2 photocatalysts after nitrogen doping were greatly improved compared with the undoped TiO_2 photocatalysts and Degussa P25. Nitrogen doping could inhibit the recombination of the photoinduced electron and thereafter increase the efficiency of the photocurrent carrier. Wang et al. [32] prepared mesoporous $TiO_{2-x}N_x/ZrO_2$ visible-light photocatalysts by a sol–gel method. Results revealed that nitrogen was doped into the lattice of TiO_2 by the thermal treatment of NH_3-adsorbed TiO_2 hydrous gels, converting the TiO_2 into a visible-light responsive catalyst. The introduction of ZrO_2 into $TiO_{2-x}N_x$ considerably inhibits the undesirable crystal growth during calcination. The photocatalytic activity of the samples was evaluated by the decomposition of ethylene in air under visible-light ($\lambda > 450$ nm) illumination. The activity of the $TiO_{2-x}N_x$ is initially quite high and decreases rapidly with increasing calcination temperature. At the sintering temperature of 400°C, the conversion of C_2H_4 on $TiO_2{-}_xN_x$ is 28%, but it drops to 7% at 500°C. Mesoporous N/TiO_2 thin films have been prepared, containing anatase nanocrystallites that exhibited photocatalytic activity in the blue region of the visible spectrum [40]. Multiple coated thin films having different thicknesses were prepared to improve the efficiency of N/TiO_2 thin films. The photocatalytic tests for the degradation of methylene blue give the best results under visible-light excitation for the film nitrided at 500°C. At this temperature the concentration of nitrogen in the structure is optimal since oxygen vacancies are still not important enough to promote the recombination of the photogenerated electrons and holes [94–96].

Heterojunction of N/TiO_2 photocatalysts with noble metal for CO_2 photoreduction by water in the gas phase was investigated [41]. The optimum loading amount of Pt was 0.2 wt%, and the optimum N amount was 0.84% on the basis of the lattice oxygen atoms. With unique properties, such as the mesoporous structure, light absorption, and the electron-transfer character, the mesoporous N/TiO_2 samples showed good activity for CO_2 photoreduction to methane under visible light. Also, visible-light-responsive mesoporous Cu/N/TiO_2 photocatalysts were synthesized by a template-free homogeneous coprecipitation method that was followed by an impregnation method [97]. The Cu/N/TiO_2 photocatalysts had a well-defined mesoporous structure and large surface area and were responsive to visible light. The photocatalytic activities were evaluated via the photodegradation of gaseous xylene under UV- and visible-light irradiation. The photocatalytic activity of Cu/N/TiO_2 was considerably greater than that of both the N/TiO_2 sample and commercial P25, which has low Cu content. As shown in Fig. 5, the 0.6 mol% Cu/N/TiO_2 catalyst showed the maximum activity: the extent of xylene degradation was 82% under UV light and 78% under visible light, while the degrees of degradation using P25 catalyst were 32 and 10% under UV and visible light, respectively. When the

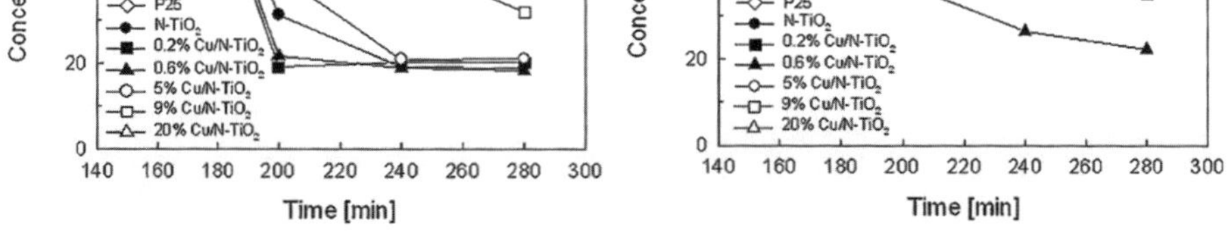

Fig. 5 Photodegradation of xylene under (**a**) UV- and (**b**) visible-light irradiation using P25, TiO_2, N/TiO_2, and the various Cu/N/TiO_2 samples (adapted from Kim et al. [97])

Cu content exceeded the optimum value, the photocatalytic activity decreased dramatically and was smaller than that for the N/TiO_2. There are two possible reasons for the decrease of photocatalytic activity for Cu contents higher than 0.6% mol. Firstly, active sites on the catalyst are covered with the excessive Cu. Secondly, the charge recombination rates increase with an increase in the amount of Cu loaded. Nitrogen doped TiO_2 was improved the photocatalytic activity by enhancing the strong absorption in the visible-light region and lowering the bandgap energy, and the addition of small amounts of Cu did to enhance the absorption in the visible-light range additionally and to decrease the recombination rate by absorbing the excited electron.

Heterojunction of mesoporous N, W/TiO_2 photocatalysts that contained various percentages of atomic W were synthesized for photodegradation rate of rhodamine B under visible light ($\lambda > 420$) [98]. Photodegradation studies of rhodamine B on the different samples revealed that an enhancement factor of up to 14 times in the reaction rate was observed with the 1.5 at% W/TiO_2 sample in contrast to commercial P25. This significant improvement in photocatalytic activity may be attributed to the synergistic effect of the red shift in absorption combined with a high surface area. Consequently, when these N, W/TiO_2 are irradiated, the electrons may be promoted from the valence band to the impurity level introduced by the metal atom contact/TiO_2 or from the lower to higher impurity level.

3.1.2 Carbon–Mesoporous TiO_2

Mesoporous C–TiO_2 photocatalysts through a direct solution-phase carbonization using $TiCl_4$ and diethanolamine as precursors were investigated [16]. The photocatalytic activities of the as-prepared samples were tested in a flow system on the degradation of NO at typical indoor air levels under simulated solar-light irradiation. The samples showed a more effective removal efficiency of NO than P25 on the degradation of the common indoor pollutant NO. However, for the C–TiO_2 calcined at 500°C, the removal rate reached the highest value after being irradiated for 30 min. Compared to the C–TiO_2 calcined at 600°C, the C–TiO_2 calcined at 500°C showed superior photocatalytic activity on the degradation of NO at parts per billion levels, which can be explained by the bandgap energy, the surface properties, as well as the mesoporous architecture. The inhabitation of the undesirable electron–hole pair recombination is important to enhance the photocatalytic activity because it can improve the ability to produce $^{\bullet}OH$, which is possibly beneficial for oxidation of NO. A recent study revealed that the holes formed for C–TiO_2 photocatalysts under visible-light irradiation were less reactive than those formed under UV-light irradiation for pure TiO_2 [99]. For the C–TiO_2 nanocomposites, the holes were trapped at midgap levels and showed less mobility, which was beneficial for the capture of surface hydroxyl to produce $^{\bullet}OH$. However, the density and nature of the localized states in the bandgap were significantly influenced by the carbon contents [99], which may be used to explain the difference in photocatalytic activity of C–TiO_2 calcined at 500 and 600°C on the degradation of NO at typical parts per billion levels. Liu et al. [100] reported highly ordered mesoporous C–TiO_2 nanocomposites with nanocrystal–glass frameworks via the organic–inorganic–amphiphilic coassembly followed by the in situ crystallization technology. The resol precursor (Mw < 500) used as a carbon precursor was prepared accordingly [101]. The C–TiO_2 nanocomposites with frameworks exhibit highly ordered hexagonal mesostructure and high thermal stability up to 700°C. The C–TiO_2 nanocomposites show good photocatalytic activity for the photodegradation of rhodamine B in an aqueous suspension, which may be attributed to the highly crystallized frameworks and high adsorptive capacity from the large surface areas. Zhang et al. [102] have prepared hollow TiO_2 microparticles about 20–60 μm in size and hollow TiO_2/carbon composite microparticles about 30–90 μm in size by employing commercial Sephadex G-100 beads as the template as well as the carbon precursor. In both cases, the product calcined at an intermediate temperature exhibited the highest photocatalytic activity for photodegradation of rhodamine B possibly because of a compromise between the anatase crystallinity and the surface area. Compared with the hollow TiO_2 microparticles, the hollow TiO_2/carbon composite microparticles exhibit remarkably enhanced photocatalytic activity. Lei et al. [48] prepared a 3D ordered macroporous TiO_2/graphitized carbon. It was found that the TiO_2/graphitized carbon showed higher activity in terms of degradation of rhodamine B and eosin Y than TiO_2/ amorphous carbon and P25.

3.1.3 Fluoride–Mesoporous TiO_2

Pan et al. [103] synthesized monodisperse F–TiO_2 hollow microspheres by hydrothermal treatment of TiF_4 in H_2SO_4 aqueous solution at 160°C for 4 h. The removal rates of MB over the course of the photocatalytic degradation reaction are shown in Fig. 6, which indicates that with identical UV-light exposure of 6 h, the mesoporous F–TiO_2 hollow microspheres show higher photocatalytic activity in the degradation of MB than that of P25.

F^-–TiO_2 has been prepared by hydrothermal treatment of TiF_4 in an HCl solution [104, 105]. The flowerlike F–TiO_2 hollow microspheres synthesized at 180°C showed the highest photocatalytic activity for the degradation of methylene blue under visible-light irradiation. Yu et al. [18] have used a novel and simple method for preparing highly photoactive F–TiO_2 photocatalyst with anatase, and brookite phase was developed by hydrolysis of TTIP in a mixed NH_4F–H_2O solution. The photocatalytic activity for the oxidation of acetone in air by F–TiO_2 photocatalysts exceeded that of P25 when the molar ratio of NH_4F to H_2O was kept in the range of 0.5–3. Yu et al. [106] have prepared mesoporous surface-fluorinated TiO_2 anatase phase by a one-step hydrothermal strategy in a NH_4HF_2–H_2O–C_2H_5OH mixed solution with TBOT as precursor. The photocatalytic activity of F–TiO_2 powders for the decomposition of acetone is obviously higher than that of pure TiO_2 and P25 by a factor of more than three times due to the fact that the strong electron-withdrawing ability of the surface Ti–F groups reduces the recombination

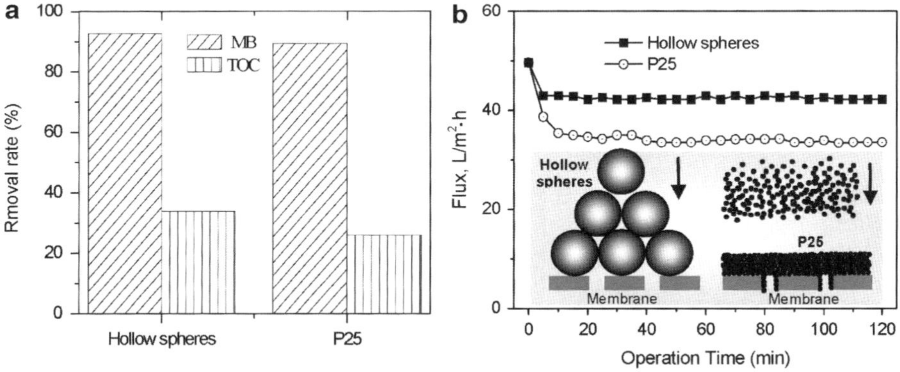

Fig. 6 (**a**) MB, TOC removal, and (**b**) membrane flux over mesoporous F–TiO_2 hollow microspheres and P25. Insert of panel *b*: schematic diagram of membrane fouling caused by photocatalysts (adapted from Pan et al. [103])

of photogenerated electrons and holes and enhances the formation of free OH radicals. Yu et al. [107] prepared TiO_2 film using TTIP in the presence of a P123. The surface modification of the films was conducted by dipping the as-prepared TiO_2 films in an aqueous 0.25 M trifluoroacetic acid (TFA) solution at room temperature. The photocatalytic activity of modified TiO_2 films for acetone oxidation in air is higher than that of unmodified TiO_2 thin films, and the modified film treated at 250°C shows the highest activity. This is ascribed to the fact that the TFA complex bound on the surface of TiO_2 acts as an electron scavenger and, thus, reduces the recombination of photogenerated electrons and holes. The enhancement is only temporary, however, as the TFA eventually decomposes under the strong oxidizing environment of photocatalysis.

3.2 *Particle/Particle Contacts*

3.2.1 Semiconductor/TiO_2 Heterostructures

TiO_2 anatase can only be excited by UV irradiation ($\lambda < 380$ nm) because of its large bandgap energy of 3.2 eV. Moreover, the rapid recombination of photoinduced electrons and holes greatly lowers the quantum efficiency [60]. Therefore, it is of great interest to improve the generation and separation of photoinduced electron–hole pairs in TiO_2 for further applications. In recent years it has been shown that the formation of semiconductor heterostructures is one of the effective methods to improve the photoinduced electron–hole generation and separation [108, 109]. Multiple semiconductor devices can absorb a larger fraction of the solar spectrum, which is beneficial for the excitation of the semiconductor and thus the photoinduced generation of electrons and holes. Moreover, the coupling of two

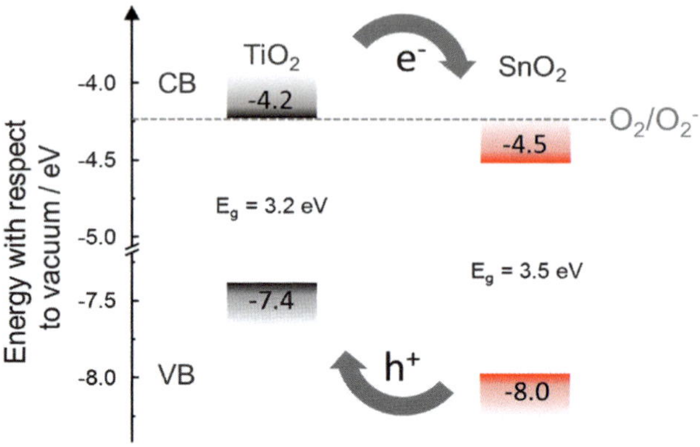

Fig. 7 Schematic energy diagram showing the positions of conduction and valence band edges of TiO_2 and SnO_2. Expected vectorial charge transfer directions are indicated for electrons (*top arrow*) and holes (*bottom arrow*) (adapted from Siedl et al. [110])

different semiconductors enables, e.g., the transfer electrons from an excited small bandgap semiconductor into an attached one provided that the latter has the appropriate position of its conduction band potential. This favors the separation of photoinduced charge carriers and thus improves the photocatalytic efficiency of such semiconductor heterostructures dramatically. For the exploration of surface charge-induced heteroaggregation, TiO_2–SnO_2 system was selected [110]. As a result of composite formation upon generation of heterointerfaces, light-induced charge separation and vectorial charge transfer (Fig. 7) were found to be facilitated in composites [111]. The conduction band position of SnO_2 is lower than that of TiO_2 and such that it is incapable of reducing oxygen molecules to form superoxide anions. The band offsets between TiO_2 and SnO_2 promoted charge separation across the interfaces. The quantitative analysis of the yield of photogenerated charges clearly shows that the adjustment of surface charge during particle network formation allows for the achievement of high mixing qualities and enables the realization of a high concentration of heterojunctions that are vital for the separation of photogenerated electrons and holes.

Fe_2O_3/TiO_2

Xuan et al. [112] have prepared well-defined magnetic separable, hollow spherical Fe_3O_4/TiO_2 hybrid photocatalysts through a poly(styrene-acrylic acid) template method. Fe_3O_4/TiO_2 hybrid with hollow spherical nature exhibits good photocatalytic activity for the degradation of RhB under UV light and can be recycled six times by magnetic separation without major loss of activity. Kim et al. [113] have synthesized mesoporous iron oxide-layered titanate nanohybrids through a reassembling reaction between exfoliated titanate nanosheets and iron hydroxide nanoclusters, in which an electrostatic attraction between both nanosized species could be achieved. The photocatalytic activity revealed that the present nanohybrids could induce the photodegradation of MB and DCA under visible-light illumination ($\lambda > 420$ nm). Fe/nanocrystalline TiO_2 with a mesoporous structure was prepared via a facile nonhydrolytic sol–gel route [114, 115]. During the photodegradation of MB under visible-light irradiation, as-prepared Fe/TiO_2 exhibited a higher activity than either the pure TiO_2 or the Fe/TiO_2 obtained via the traditional hydrolytic sol–gel route. The promoting effect of the heterojunction of Fe/TiO_2 on the photocatalytic activity for MB decomposition could be attributed to the formation of intermediate energy levels that allow Fe/TiO_2 to be activated easily in the visible area. The nonhydrolytic sol–gel method is superior owing to the controllable reaction rate and lack of surface tension, which ensures the formation of mesopores and well-crystallized anatase in the Fe/TiO_2 sample, leading to a higher activity since the reactant molecules are easily adsorbed and the recombination between the photoelectrons and the holes is effectively inhibited [114]. A new multifunctional nanocomposite (Fe_xO_y@Ti–hexagonal mesoporous silica (HMS)) involving superparamagnetic iron oxide nanoparticles and ordered mesoporous channels has been developed via the coating of as-synthesized iron oxide nanoparticles with an amorphous silica layer followed by the sol–gel polymerization using TEOS, tetrapropyl orthotitanate (TPOT), and a structure-directing reagent [116]. The Fe_xO_y@Ti–HMS acted as an efficient heterogeneous catalyst for the liquid-phase selective oxidation reactions of organic compounds using H_2O_2 as an oxidant. The meso-TiO_2/α-Fe_2O_3 composites possess synergy of the photocatalytic ability of meso-TiO_2 for oxidation of As (III) to As (V) and the adsorption performance of α-Fe_2O_3 for As(V) [117]. The results show that the meso-TiO_2/α-Fe_2O_3 composites can oxidize higher toxic As(III) to lower toxic As (V) with high efficiency at various pH values in the photocatalysis reaction (Fig. 8), and As(V) is effectively removed by adsorption onto the surface of composites. Mesoporous TiO_2 and α-Fe_2O_3, meso-TiO_2/α-Fe_2O_3 composites can possess more sufficient adsorption property for As(III) because of its special surface property and high surface area. When meso-TiO_2/α-Fe_2O_3 composites are added into the water containing arsenite, arsenite is adsorbed onto the surface of the composites. In the presence of UV irradiation, photocatalytic oxidation As(III) to As(V) occurs.

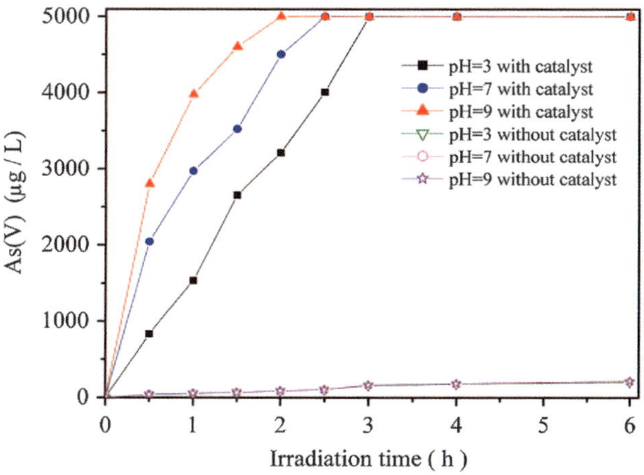

Fig. 8 Time profiles of photocatalytic oxidation As (III) to As (V) with and without *meso*-TiO_2/α-Fe_2O_3 composites (sample A) at pH 3, 7, and 9, respectively (adapted from Zhou et al. [117])

Furthermore, As(III) adsorbed onto the surface of this bifunctional composite is more advantageous of the photocatalytic oxidation, so the excellent adsorption property of α-Fe_2O_3 combining with good photocatalytic ability of meso-TiO_2 is an important factor for the photocatalytic oxidation of As(III) and the removal of As(V).

The bandgap of the TiO_2 was modified with transition metal ions Ag, Co, Cu, Fe, and Ni having different work functions by the wet impregnation method [22]. The investigations were carried out to demonstrate the effect of ionic radius and work function of metal ions on photocatalytic activity of mesoporous TiO_2 for the degradation of acetophenone (AP) and nitrobenzene (NB) in aqueous medium under UV irradiation. The initial rate of the photocatalytic degradation of AP and NB varies due to the change in bandgap of the catalyst, work function, ionic radii, and position of the impregnated metal ion on the TiO_2 lattice. The silver-impregnated catalysts showed the highest initial rate of photocatalytic degradation for both compounds due to the interstitial position of impregnated silver metal ion in the TiO_2 lattice.

Bi(III)–TiO_2

The Bi_2O_3 photosensitization of TiO_2 could extend the spectral response from UV to visible area, making the Bi_2O_3/TiO_2 photocatalyst easily activated by visible lights for the degradation of chlorophenol [33]. The ordered mesoporous channels facilitate the diffusion of reactant molecules. Meanwhile, the high surface area could enhance the Bi_2O_3 dispersion, the light harvesting, and the reactant adsorption. Furthermore, the highly crystallized anatase may promote the transfer of photoelectrons from bulk to surface and thus inhibit their recombination with photoholes, leading to enhanced quantum efficiency. Kong et al. [34] prepared visible-light-driven mesoporous bismuth titanate photocatalyst, which possesses wormlike channels, mixed-phase mesostructured frameworks, large pore diameter (~6.1 nm), and low bandgap energy (2.5 eV). The calcined sample exhibited visible-light photocatalytic reactivity valued by the degradation of 2, 4-DCP in aqueous media. However, Zhang et al. [118] synthesized $BiOI/TiO_2$ heterostructures with different Bi to Ti molar ratios through a simple soft chemical method at a temperature as low as 80°C. The photocatalytic activities of these $BiOI/TiO_2$ were evaluated on the degradation of MO under visible-light irradiation ($\lambda > 420$ nm). The results revealed that the $BiOI/TiO_2$ heterostructures exhibited much higher photocatalytic activities than pure BiOI and TiO_2, respectively, and 50% $BiOI/TiO_2$ showed the best activity among all these heterostructured photocatalysts. The visible-light photocatalytic activity enhancement of $BiOI/TiO_2$ heterostructures could be attributed to its strong absorption in the visible region and low recombination rate of the electron–hole pairs because of the heterojunction formed between BiOI and TiO_2.

Cr(III)–TiO_2

Cubic Im3m mesoporous Cr–TiO_2 was fabricated with ordered and well-crystallized [42]. The mesoporous TiO_2 is ineffective, but the mesoporous Cr–TiO_2 shows a very high decomposition rate for the photodegradation of organic pollutants [42, 119]. This must be due to the heterojunction of Cr^{3+}–TiO_2, which allows the activation of the mesoporous TiO_2 sample in the visible-light region. Cr^{3+} ions promote the separation of photogenerated holes and electrons and hence increase the photocatalytic reactivity of TiO_2. The excellent photocatalytic performance is related to the open mesoporous architecture with a large surface area, good anatase crystallinity, and a 3D-connected pore system [120, 121]. The 3D-interconnected mesochannels in the cubic mesoporous Cr–TiO_2 composite serve as efficient transport paths for reactants and products in photocatalytic reactions [122]. However, the Ti–Cr–MCM-48 photocatalyst prepared in a single step exhibits far superior photocatalytic activity for the degradation of acetaldehyde in gas phase compared to the TiO_2–Cr–MCM-48 prepared by a post-impregnation method. The high activity of the Ti–Cr–MCM-48 photocatalyst is attributed to a synergistic interaction between Cr ions dispersed in the silica framework and the nanocrystalline nature of TiO_2 crystallites anchored onto the pore walls. Ti–Cr–MCM-48 prepared in a single step showed the highest activity for CO_2 production. The high activity of Ti–Cr–MCM-48 arises from the synergistic interaction of the Cr ions dispersed in the MCM-48 framework and the TiO_2 nanocrystallites anchored onto the pore walls of MCM-48. The highly dispersed chromium ions can be excited by visible-light radiation to form a CT excited state, involving an electron transfer from O^{2-} to Cr^{6+} [123].

WO_3 and ZrO_2/TiO_2

Pan et al. [43] prepared highly ordered cubic mesoporous WO_3/TiO_2 thin films. The photocatalytic activity of WO_3/TiO_2 thin films in decomposing 2-propanol in the gas phase was optimized at 4 mol% of WO_3 concentration. Its photocatalytic activity was 2.2 times that of a mesoporous TiO_2 film and 6.1 times that of a nonporous TiO_2 film derived from a typical sol–gel method. The enhanced photocatalytic activity of WO_3/TiO_2 is ascribed to the increase in surface acidity. Liu et al. [124] have produced codoped Zr^{4+} and F^- ions within anatase hollow microspheres by a fluoride-mediated self-transformation strategy. Urea was used to catalyze the hydrolysis of aqueous mixtures of $Ti(SO_4)_2$ and $ZrOCl_2$ in the presence of NH_4F under hydrothermal conditions. The concomitant participation of F^- promotes lattice substitution of Ti^{4+} ions by Zr^{4+} and facilitates the transformation of surface-segregated amorphous ZrO_x clusters into Zr–F species. The better photocatalytic activity of fluorinated samples may be at least partially attributed to the presence of well-crystallized anatase with retention of small grain size, high surface area, and porosity, as well as hollow microarchitecture. Zr^{4+}, F^{-}–TiO_2 is associated with electron-transfer-mediated charge compensation between the Zr–F impurities, which reduces the number of both bulk and surface defects and provides a stabilizing effect on the local structure. Moreover, these synergetic interactions influence the textural characteristics and surface states of the TiO_2 host, such that the photocatalytic activity with regard to the decomposition of gaseous toluene is enhanced. Also the formation of porous TiO_2/ZrO_2 networks was achieved by a polymer gel templating technique [21]. The mixed TiO_2/ZrO_2 network structures exhibit higher surface areas than a corresponding pure TiO_2 network and in a certain range of metal oxide compositions. The photocatalytic efficiencies of the TiO_2 and TiO_2/ZrO_2 networks have been assessed by monitoring the photodecomposition of two organic molecules: salicylic acid and 2-chlorophenol. The TiO_2 network was found to exhibit an efficiency of ~60 and ~65% of the standard P25 for the salicylic acid and 2-chlorophenol reactions, respectively. For both photocatalytic reactions, the presence of ZrO_2 in the TiO_2 network resulted in enhanced photocatalytic activity relative to the TiO_2 network, which is believed to be due to a number of factors including an increased surface area and a decrease of the anatase to rutile crystal phase transformation.

Ce(III) and Zn(II)/TiO_2

Visible-light-induced metal-to-metal charge transfer (MMCT) for hetero-bimetallic Ti(IV)–O–Ce(III) assemblies on the pore of mesoporous silica, MCM-41, has been achieved [125]. It was concluded that the catalytic oxidation of 2-propanol is driven by the visible-light-induced MMCT of Ti(IV)–O–Ce(III) assemblies. A thermally stable mesoporous ceria–titania using hexadecylamine as structure-directing reagent and triethanolamine as an additive in mixed propanol–water medium was prepared [20]. These novel mesoporous CeO_2–TiO_2 materials showed high performance for the removal of toluene. The toluene removal performance was further enhanced for Pt-impregnated mesoporous CeO_2–TiO_2. Ismail and Bouzid [9] reported that Ce^{4+} (4f level) plays an important role in interfacial charge transfer and elimination of electron–hole recombination (Scheme 3). It could act an effective electron scavenger to trap the CB electrons of TiO_2 and Ce^{4+}, as Lewis acid, which apparently was superior to molecular oxygen (O_2) in the capability of trapping CB electrons [126]. In CeO_2–TiO_2 photocatalyst under UV illumination, when photons with energies larger than the bandgap are absorbed by TiO_2 particles, electrons are promoted from the valence band to the conduction band leaving holes behind in the valence band. The conduction band electrons ($E_{CB} = -0.5$ V vs. NHE at pH 7) migrate through the three-dimensional TiO_2 network until they reach Ce^{4+}. The standard reduction potential of Ce^{4+}/Ce^{3+}($E^{\circ} = +1.76$ V) was more positive than that of adsorbed molecular oxygen, which is reduced to form $O_2^{\bullet-}$ radicals (redox potential $O_2/O_2^{\bullet-} = -0.33$ V). The photoinduced holes at the valence band of TiO_2 (E°(ox) = 2.94 V vs. NHE) migrate to the surfaces of the particles where they react with adsorbed hydroxide ions yielding surface-adsorbed $^{\bullet}OH$ radicals

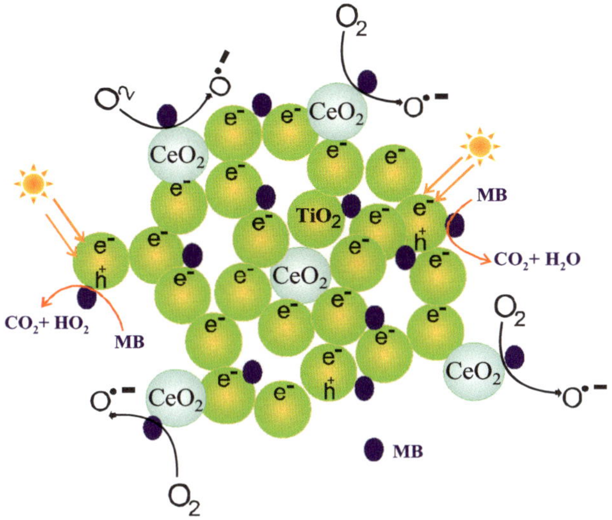

Scheme 3 Schematic illustration of the proposed antenna mechanism to explain the photonic efficiency of mesostructured CeO_2–TiO_2 photocatalyst for photodegradation of methylene blue; absorption of UV light by the semiconducting nanoparticle promotes an electron from the valence band to the conduction band. This charge carrier can recombine in bulk or migrate to the surface and react with the adsorbed species, which leads to their decomposition by direct oxidation on the holes or by $O_2^{\bullet -}$ and $^{\bullet}OH$ radicals (adapted from Ismail and Bouzid [9])

with the redox potential ($OH^- = {}^{\bullet}OH + e^-$; $E^o = -2.8$ V) [60] and then $^{\bullet}OH$ radicals oxidized MB ox/red ($E^o = 0.01$ V).

Kim et al. [127] have investigated the chemical bonding character and physicochemical properties of mesoporous zinc oxide-layered titanate nanocomposites synthesized by an exfoliation–restacking route. Upon hybridization with ZnO nanoparticles, the photocatalytic activity of layered titanate is enhanced with respect to the oxidative photodegradation of phenol and dichloroacetate. But of greater importance is that the chemical stability of guest ZnO against acidic corrosion is greatly improved by hybridization with layered titanate.

Nb(V) and PO_4(III)/TiO_2

Stone et al. [128] have prepared mesoporous titania and niobia molecular sieves by a ligand-assisted templating method. The transition metal oxides were tested as photocatalysts in the liquid-phase oxidative dehydrogenation of 2-propanol to acetone. The observed quantum yield of the reaction was 0.45 over P25. However, mesoporous TiO_2 converted 2-propanol with a very low quantum yield of 0.0026. A very low quantum yield was also found for the mesoporous niobia sample compared to a crystalline standard. Apparently, the surface reactivities of the poorly crystallized samples were suppressed by defects that act as electron–hole traps. The relative inactivity of mesoporous TiO_2 samples can be attributed to a high surface concentration of defects which can act as surface electron–hole recombination sites and/or the poisoning of catalytic surface sites by the phosphorus remaining from the surfactant. Yu et al. [35] synthesized a phosphated mesoporous TiO_2 by incorporating phosphorus from phosphoric acid directly into the framework of TiO_2 via a surfactant-templated approach. Both pure and phosphated mesoporous TiO_2 show significant activities on the oxidation of *n*-pentane. The higher photocatalytic activity of phosphated mesoporous TiO_2 can be explained by the extended bandgap energy, the large surface area, and the existence of Ti ions in a tetrahedral coordination.

Ni(III) and La(III)–TiO_2

Jing et al. [23] have prepared Ni–TiO_2 by using TBOT and acetylacetone in the presence of laurylamine. The results of photocatalytic hydrogen evolution in aqueous methanol solution under UV–vis-light irradiation showed that the activity of hydrogen production strongly depended on the Ni contents. The highest activity was achieved with 1% Ni–TiO_2. The results were rationalized by assuming that Ni^{2+} serves as shallow trapping sites, greatly enhancing the activity of the mesoporous photocatalyst. Also, photocatalytic degradation of commercial phoxim emulsion in aqueous suspension was investigated by using La–TiO_2 as the photocatalyst under UV irradiation [58]. The photocatalytic activity of the obtained La–TiO_2 nanopowders was detected by the degradation of RB aqueous solution under UV irradiation and compared with pure TiO_2 samples and commercial P25. La–TiO_2 with mesostructures showed much better photoactivity than that of pure TiO_2 ones and the P25 due to its large surface area, highly crystallized mesoporous wall, and more active sites for concentrating the substrate [129]. However, it is probably not just a consequence of enrichment of RB at the surface of mesoporous nanosized TiO_2. It can be found that the photocatalytic activity increased when the calcination temperature is increased to 300°C (Fig. 9). Compared with the as-synthesized sample, the sample calcined at 300°C possesses reduced surface area but a more excellent crystallinity. Therefore, the crystallinity also played an important role in the photoactivity of TiO_2.

Multimetal Ion/TiO_2

Up to date, although a variety of approaches have been developed to prepare many types of visible-light-driven semiconductor heterojunction photocatalysts, there are still some disadvantages, such as the limited region of visible-light photo-response. To avoid these problems, multicomponent heterojunction systems have been developed [36, 130–132], in which two or more visible-light active components and an electron-transfer system are spatially integrated. Yang et al. [36] prepared Ag/In_2O_3–TiO_2 nanocomposites by a one-step sol–gel–solvothermal method in

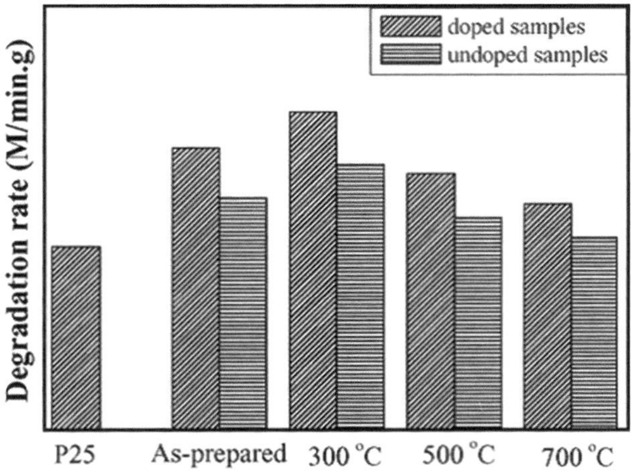

Fig. 9 The photocatalytic properties of doped and undoped TiO_2 samples as-prepared and calcined at different temperatures as well as the P25 nanoparticles (RB, $c_0 = 1.0 \times 10^{-5}$ M, pH 6.0) after 90 min UV-light radiation (adapted from Peng et al. [129])

the presence of P123. The resulting $Ag/In_2O_3–TiO_2$ three-component systems mainly exhibited an anatase-phase structure, high crystallinity, and extremely small particle sizes with Ag particles well-distributed on the surface. At 2.0% Ag and 1.9% $In_2O_3–TiO_2$, the $Ag/In_2O_3–TiO_2$ system exhibited the highest UV-light photocatalytic activity for the degradation of rhodamine B and methyl ter-butyl ether after 120 min UV-light irradiation. In addition, the UV-light photocatalytic activity of three-component systems exceeded that of pure TiO_2 and two-component (Ag/TiO_2 or $In_2O_3–TiO_2$) systems as well as the commercial photocatalyst, P25. These results indicate that (1) two-particle contact TiO_2 system is more photoactive than single-particle contact TiO_2 system, (2) single-particle contact system is more photoactive than pure TiO_2, and (3) $Ag/In_2O_3–TiO_2$ with P123 is more photoactive than that prepared without P123. Chu et al. [130] have synthesized 3D highly porous TiO_2-4%SiO_2-1%$TeO_2/Al_2O_3/TiO_2$ composite nanostructures (30–120 nm) directly fixed on glass substrates by anodization of a superimposed Al/Ti layer sputter deposited on glass and a sol–gel process. The porous composite nanostructures exhibited enhanced photocatalytic performances in decomposing CH_3CHO gas under UV illumination. Specially, the composite nanostructure showed the highest photocatalytic activity that is 6–10 times higher than commercial P25. The interfacial charge carrier dynamics of the three-component semiconductor (TiO_2)–semiconductor (In_2O_3)–metal (Pt) heterojunction system were investigated [131]. To examine the photoinduced charge carriers for the $In_2O_3–TiO_2$–Pt system, MB, a cationic dye that can be decomposed by accepting electrons following the irradiation on photocatalysts. $In_2O_3–TiO_2$ NBs performed better toward MB photodegradation than pristine TiO_2 NBs and pure In_2O_3 nanocrystals, which can be accounted for by the effective charge separation that occurred at the interface of In_2O_3 and TiO_2 (Fig. 10). The photocatalytic efficiency of In2O3–TiO_2 NBs was significantly improved upon the deposition of 1.0 wt% Pt. This improvement mainly emerged from the deposited Pt that can promote the overall charge separation of NBs by readily accepting

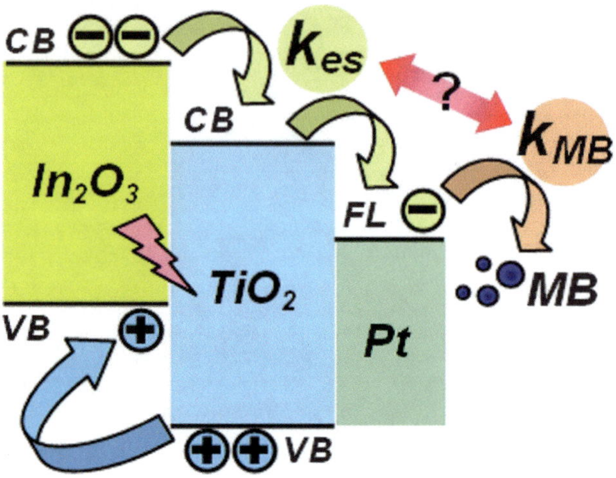

Fig. 10 Correlations of electron-scavenging rate constant (k_{es}) and rate constant of MB photodegradation (kMB) with the content of Pt for In_2O_3–TiO_2–Pt NBs (adapted from Chen et al. [131])

photoexcited electrons from TiO_2, thereby providing a considerable amount of free electrons for participation in MB degradation. For In_2O_3–TiO_2–Pt NBs with higher Pt content (5.0 wt%), an even better performance in MB photodegradation was attained, probably due to the much more pronounced charge separation of NBs caused by the increasing amount of Pt. Nevertheless, a depressed efficiency of MB photodegradation was observed for In_2O_3–TiO_2–Pt NBs as the content of Pt was further increased to 10.0 wt%. It is believed that the substantially abundant electrons trapped at excess Pt would encourage the electron–hole recombination to deplete the photoinduced charge carriers of NBs [132], thus leading to the depression in the resultant photocatalytic efficiency.

3.2.2 Highly Order Mesostructured TiO_2–SiO_2

Dong et al. [37] have prepared highly ordered mesoporous crystalline TiO_2–SiO_2 nanocomposites. They exhibit excellent photocatalytic activities more than P25 for the degradation of rhodamine B in aqueous suspension due to the bifunctional effect of highly crystallized anatase nanoparticles and high porosity (Fig. 11). The adsorption amount of RhB on the mesoporous materials is higher than that on P25, because the former have much larger surface areas. With the decrease in the Ti/Si ratio, the adsorption amount of RhB increases, which agrees with the increase in the surface area. After the light is on, the concentration of RhB decreases fast with the irradiation time and the pseudo-first-order reaction is observed.

TiO_2/SBA-15 composites through a post-synthetic approach with the assistance of ethylenediamine were prepared [133, 134]. The excellent photocatalytic activity of the composites is evaluated via the photodecomposition of phenol in the liquid phase under visible- and UV-light illumination. The conversion of phenol varies with the content of TiO_2 in the composites, and the optimal value is up to 46.2%

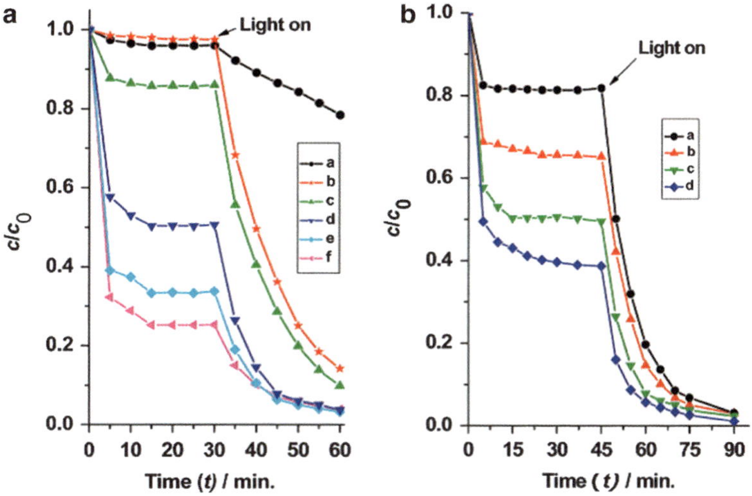

Fig. 11 Photocatalytic degradation of RhB monitored as the normalized concentration change versus irradiation time in the presence of (*A*) mesoporous TiO_2–SiO_2 composites prepared with different Ti–Si ratios. (**a**) Mesoporous TiO_2 calcined at 400°C for 2 h; (**b**) commercial photocatalyst P25; (**c**) mesoporous $90TiO_2$–$10SiO_2$ composite calcined at 700°C for 2 h; (**d**) mesoporous $80TiO_2$–$20SiO_2$, (**e**) $70TiO_2$–$30SiO_2$, and (**f**) $60TiO_2$-$40SiO_2$ composites calcined at 850°C for 2 h. (*B*) Photocatalytic degradation for mesoporous $80TiO_2$–$20SiO_2$ composites calcined at 700°C for 4 h (**a**), 800°C for 2 h (**b**), 850°C for 2 h (**c**), and 900°C for 2 h (**d**) (adapted from Dong et al. [37])

under illumination in the visible region. Li et al. [135] have prepared a core–shell structure of Ti–O–Si species modified TiO_2 embedded in mesoporous silica by the sol–gel method. The as-synthesized TiO_2–$xSiO_2$ composites exhibit both much higher absorption capability of organic pollutants and better photocatalytic activity for the photooxidation of benzene than pure TiO_2 and P25. The better photocatalytic activity of as-synthesized TiO_2–$xSiO_2$ composites than pure TiO_2 is attributed to their high surface area, higher UV absorption intensity, and easy diffusion of absorbed pollutants on the absorption sites to photogenerated oxidizing radicals on the photoactive sites. Xuzhuang et al. [136] have fabricated a new composite Ti/clay by the reaction between $TiOSO_4$ and a synthetic layered clay laponite. The large number of the anatase crystals and better accessibility to the sites by UV light and reactant molecules are the major factors enhancing the photocatalytic activity. The performance of the catalysts is related to their structural features, and it is found that the catalytic activity increased with increasing size of the anatase crystals in the catalysts, specific surface area, and mesopore size. Li et al. [137] have synthesized monodispersed concentric hollow nanospheres with mesoporous silica shell and anatase TiO_2 core by the combination of sol–gel reaction and distillation–precipitation polymerization. The first synthesis step involved the preparation of cross-linked poly(methacrylic acid) (PMAA) core nanospheres via distillation–precipitation polymerization in the presence of ethylene glycol dimethylacrylate. The next step involved the synthesis of PMAA/TiO_2 composite nanospheres via the sol–gel process, using the cross-linked PMAA nanospheres as the template cores. In the subsequent step, the PMAA/TiO_2 composite nanospheres were coated with a uniform PMAA layer via distillation–precipitation polymerization to produce the PMAA/TiO_2@PMAA core–shell particles. Photocatalytic decomposition of methyl orange in the concentric hollow reactors is followed by an apparent first-order rate constant. The observed rate constant for the concentric hollow nanospheres as photocatalysts seems to be lower than those reported for pure TiO_2 and PMAA/TiO_2@PMAA [138]. The lower reaction rate observed in the present work is probably due to the low content of mesoporous anatase TiO_2 in the hollow nanospheres.

Aronson et al. [139] have grafted TiO_2 onto the pore surface of MCM-41 and FSM-16 by reacting $TiCl_4$ in hexanes with the as-synthesized mesostructured silicate. The TiO_2-grafted MCM-41 samples exhibited good photodegradation efficiency of rhodamine-6G and α-terpineol. Alvaro et al. [44] and Maldotti et al. [140] reported the preparation of a series of structured mesoporous silica, starting from colloidal TiO_2 nanoparticles in combination with TEOS using neutral Pluronic or cationic CTAB as templates. Even though the activity of these new mesostructured materials for the degradation of phenol in aqueous solution is lower than those found for P25, the turnover frequency of the photocatalytic activity is much higher for the mesoporous TiO_2. Also, both mesoporous TiO_2 and mixture of 50% TiO_2 and 50% SiO_2 can induce cyclohexane photooxidation to yield cyclohexanone.

Li et al. [45] have prepared a core–shell SiO_2–TiO_2 photocatalyst using a liquid-phase deposition method. The photocatalytic activity of the core–shell SiO_2–TiO_2 catalyst for the decomposition of Orange II in liquid phase was observed to be comparable with that of P25. Mesoporous SiO_2-modified TiO_2 photocatalysts were prepared by sol–hydrothermal processes, followed by posttreatment with F127-modified silica sol [141, 142]. Mesoporous SiO_2-modified TiO_2 samples exhibited much higher photocatalytic activity for degrading rhodamine B than P25, which is explained mainly by the high photoinduced charge carrier separation rate resulting from the high anatase crystallinity and the large surface area related to the small nanocrystallite size and mesoporous SiO_2 as well as still possessing a certain amount of surface OH group. Morishita et al. [59] have employed Ti-containing mesoporous organosilicas (T-OS), synthesized by a surfactant-templating method with an organosilane precursor, as the photocatalyst, and have studied the effects on the olefin conversion and the epoxide selectivity. The T-OS catalysts demonstrate the same high epoxide selectivity as does T–S, but scarcely improve the olefin conversion. Hu et al. [143] have prepared Ti–MCM-41 mesoporous molecular sieves using TEOS and TPOT as the starting materials and CTAB as a structure-directing agent. It was found that an increase in the Ti content caused the structure of the Ti oxides in Ti–MCM-41 to change from an isolated tetrahedral coordination to adjacent Ti-oxide species with Ti^{4+} of tetrahedral coordination [38, 144]. The photocatalytic reactivity of these catalysts for the decomposition of NO into N_2 and O_2 was found to strongly depend on the local structure of the Ti-oxide species

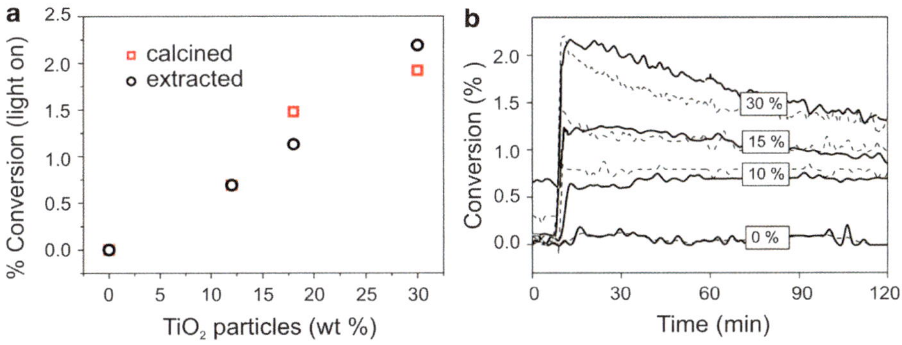

Fig. 12 Photocatalytic activity of extracted (*black*) and calcined (*red*) titania–silica films with different particle contents in the photocatalytic oxidation of NO measured after switching on the light (time 5 min) (**a**) and the corresponding photocatalysis measurement curves of extracted (*solid lines*) and calcined (*dashed lines*) samples (**b**) (adapted from Zhang et al. [144])

including their coordination and distribution, i.e., the charge transfer excited state of the highly dispersed isolated tetrahedrally coordinated Ti oxides acts as the active sites for the photocatalytic decomposition of NO into N_2 and O_2. Ti–MCM-41 showed higher photocatalytic reactivity than Ti–HMS for the decomposition of NO. Rohlfing et al. [61] have fabricated TiO_2–SiO_2 composite films with a high content of crystalline TiO_2 phase and periodic mesoporous structure. While films of pure silica are inactive for photooxidation of NO, the activity of those containing TiO_2 nanocrystals increases almost linearly with the TiO_2 content, approaching the conversion efficiency of 3.9–4% for the films composed solely of TiO_2 particles taken as a reference (Fig. 12). This linearity confirms the homogeneous distribution of the particles and their good accessibility for molecules from the gas phase. Further increases of the photocatalytic activity of those films are expected for thicker films and for films with particle contents over 50 wt%.

TiO_2 colloidal was dispersed within a transparent silica binder to obtain mesoporous structure [145, 146]. Stearic acid was first deposited on the film by spin coating from a solution in tetrahydrofuran. Studies of photodegradation kinetics show that such mesoporous films are at least 15 times more active than films synthesized with a usual microporous silica binder. Moreover, the measured quantum yield efficiency is 1.1%, and the improved photoactivity of the films is obtained as resulting from the closer proximity between the organic molecules and the surface of the TiO_2 crystallites as well as the improved diffusion rate of H_2O and O_2 through the interconnected pore network. Ogawa et al. [46] have prepared transparent self-standing films of titanium-containing (Ti/Si ratio of 1/50) silica-surfactant mesostructured materials from tetramethoxysilane, vinyltrimethoxysilane, TTIP, and octadecyltrimethylammonium chloride. UV irradiation of the titanium-containing nanoporous silica film in the presence of CO_2 and H_2O led to the evolution of CH_4 and CH_3OH, indicating high selectivity for the formation of CH_3OH, showing the characteristic reactivity of the charge transfer excited complexes of the tetrahedrally coordinated titanium oxide species.

3.2.3 Noble Metal/TiO_2 Contacts

To create a space charge separation region (called the Schottky barrier), a semiconductor (TiO_2)–metal (Au, Ag, Pd, and Pt) junction is fabricated. At the interface of the two materials, electrons flow from one material to the other (from the higher to the lower Fermi level) to align the Fermi energy levels [81]. Heterojunction based on the n-type semiconductor and metal was extensively studied, where the ideal case is that the work function of the metal is higher than that of the n-type semiconductor (such as TiO_2), and electrons will flow from the semiconductor into the metal to adjust the Fermi energy levels [1–3, 39, 147–152] (Scheme 4). The formation of the Schottky barrier is a result of the metal having excess negative charges and the semiconductor having excess positive charges [81]. In addition, the Schottky barrier can serve as an efficient electron trap preventing electron–hole recombination in photocatalysis, which often results in an enhanced photocatalytic performance.

Ismail et al. [1–3, 147] have suggested the mechanism of Au, Pt, and Pd/mesoporous TiO_2 for the photooxidation of methanol (Scheme 5). 3D mesoporous TiO_2 network acts as an antenna system transferring the initially generated electrons from the location of light absorption to a suitable interface with the noble metal catalyst and subsequently to the location of the noble metal nanoparticle where the actual electron-transfer reaction will take place. Within this antenna model, it can be envisaged that the overlap of the energy bands of the nanoparticles forming this network will result in unified energy bands for the entire system enabling a quasi-free movement of the photogenerated charge carriers throughout. Consequently, an electron generated by light absorption within one of the nanoparticles forming the network will subsequently be available to promote

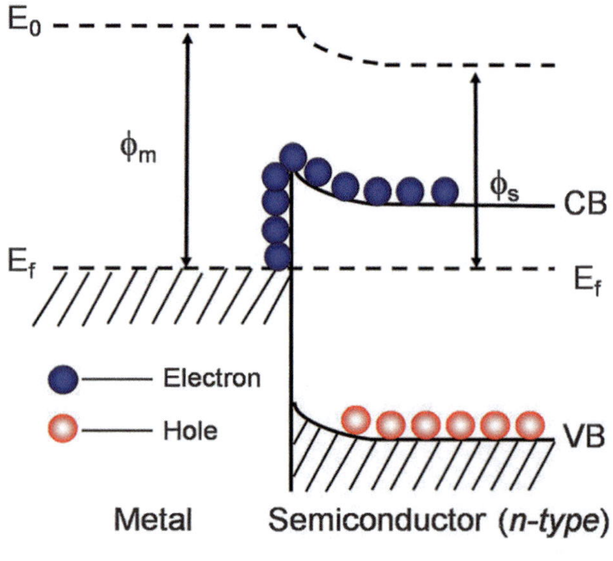

Scheme 4 Schematic of the Schottky barrier (adapted from Wang et al. [81])

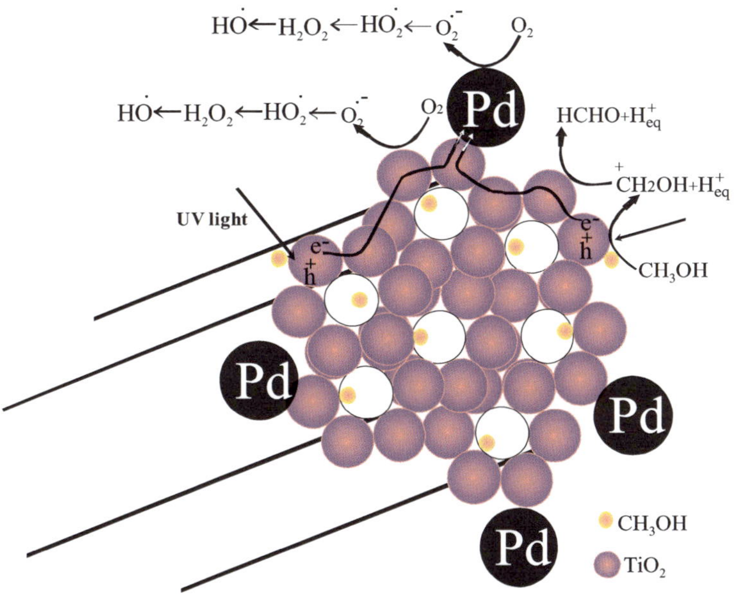

Scheme 5 The proposed antenna and reaction mechanisms for methanol photooxidation to illustrate the enhanced photonic efficiency of mesostructured Pd/TiO_2 photocatalyst; absorption of UV light by the semiconducting nanoparticle promotes an electron from the valence band to the conduction band. The lines in the scheme show cut perpendicular to the c axis of the hexagonal pore system extending infinitely in this direction (adapted from Ismail et al. [147])

redox processes anywhere within the structure. The photonic efficiency of $^{\bullet}OH$ radical formation can, thus, be determined as the ratio of the production rate of HCHO and the incident light intensity.

$$^{\bullet}OH + CH_3OH \rightarrow {}^{\bullet}CH_2OH + H_2O \quad (1)$$

$$^{\bullet}CH_2OH + O_2 \rightarrow HCHO + HO^{\bullet}{}_2 \quad (2)$$

$$TiO_2 - Pd(e^-) + O_2 \rightarrow TiO_2 - Pd + O_2{}^{-\bullet} \quad (3)$$

$$O_2{}^{-\bullet} + 2H^+{}_{eq.} \rightarrow H_2O_2 \quad (4)$$

$$2HO^{\bullet}{}_2 + 2H^+{}_{eq} \rightarrow H_2O_2 + 2^{\bullet}OH \quad (5)$$

$$H_2O_2 + e^- \rightarrow {}^{\bullet}OH + {}^{-}OH \quad (6)$$

Srinvasn et al. [39] have synthesized 3D ordered macroporous (3DOM) TiO_2 by colloidal crystal templating against polystyrene spheres using a metal alkoxide precursor. Macroporous TiO_2 with pore diameter 0.5 μm had the highest first-order rate constant of 0.042 min^{-1} for decomposition of MB, compared to 0.025 min^{-1} for P25 TiO_2. Deposition of Au on the TiO_2 surfaces decreased the reaction rate by covering the surface active sites (Fig. 13). The enhanced activity of the 3DOM structures can be partially attributed to the abundance of anatase

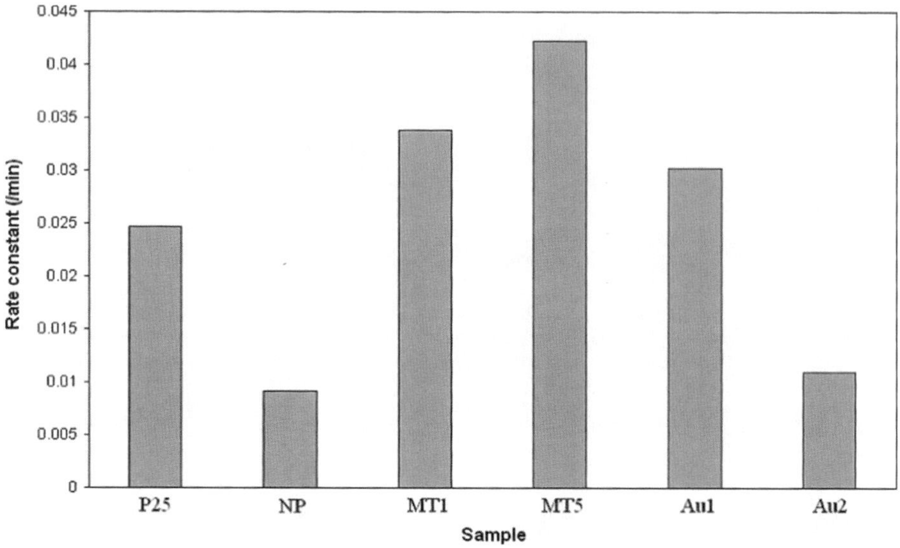

Fig. 13 Reaction rate constants for MB degradation for the photocatalysts; 1 μm 3DOM titania (MT1) and 0.5 μm (MT5) 3DOM titania and 1 mL (Au1) and 5 mL (Au2) of 10^{-2} M chloroauric acid ($HAuCl_4$) (adapted from Srinvasn et al. [39])

(>98%), which is the primary photoactive phase, as compared to P25. Au <0.6 wt% -loaded 3DOM TiO_2 is the optimum Au content. With more complete coverage by Au nanocrystals (as in sample Au_2), the enhancement in photoactivity brought about by the 3DOM structure (MT1) is lost, and the reaction rate (kAu_2) (0.011 min^{-1}) becomes similar to that of powder NP (kNP) (0.009 min^{-1}). In addition, the increased light absorption of Au reduces the light penetration and may contribute to lower the activity.

Bannat et al. [148] have synthesized Au–mesoporous TiO_2 films by the EISA method. For mesoporous TiO_2 films deposited on an ITO layer, the photonic efficiency for NO_x oxidation is higher than for films prepared on glass, because the pore structures are altered. The incorporation of Au results in a significant improvement in the photonic efficiency due to the generation of Schottky barriers, which inhibit the recombination of electron–hole pairs and thereby increase the concentration of photogenerated holes at the film surface reacting with NO. Li et al. [149] have developed a simple method to generate nanoporous organic–inorganic hybrid films and arrays of Au–TiO_2 nanobowls. The photocatalytic activity of a representative hybrid PS-b–PEO–$HAuCl_4$–TiO_2 film in terms of the decomposition of MB is similar to nanostructured TiO_2. On the other hand, Fig. 14 compares the conversion of phenol oxidation and chromium reduction catalyzed by the mesoporous Au–TiO_2 nanocomposites with different Au loadings [23]. When the reactions were conducted under dark condition, no reactions were observed even after 6 h, indicating the really photocatalytic reactions. The conversion of phenol oxidation and chromium reduction continuously increases from 22 to 95%

Fig. 14 Photocatalytic activity of Au–TiO_2 nanocomposites containing 0–5% Au in phenol-oxidation and chromium-reduction reactions (adapted from Li et al. [23])

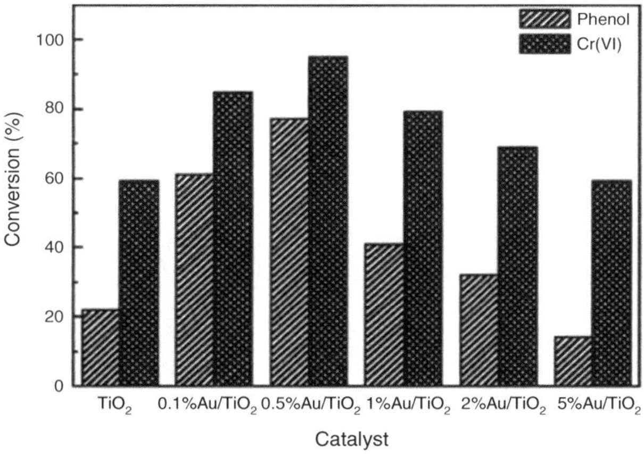

when Au content is increased from 0 to 0.5%. Phenol decomposition is achieved three times when 0.5% of Au was added onto TiO_2, unambiguously suggesting a significantly improved photocatalytic activity. Further increasing Au content, however, decreases catalytic activity. This may be attributed to the presence of a large number of nanoparticles within the pore channels that may also slow down mass transport and reduce the reaction rate. These factors may account for the observed decreasing photocatalytic activity at a higher Au content (e.g., >0.5%).

Ag nanoparticles were introduced into the mesopores using wet impregnation followed by heat treatment [25, 150]. The cubic structured mesoporous TiO_2 had higher stability than the hexagonal structure and could be formed with a high content of nanocrystalline anatase with conservation of the meso-order. It was found that the meso-ordered TiO_2 had a sufficiently high crystallite content to be photoactive for the oxidation of stearic acid and that the reaction mechanism resembled that of a non-meso-ordered TiO_2. It was shown that meso-ordered TiO_2 was photocatalytically active and that the activity was influenced by the presence of Ag nanoparticles. Stathatos et al. [151] reported a transparent mesoporous TiO_2 films which have been deposited on glass slides by a sol–gel method. Ag ion/TiO_2 films, incorporated through the reverse micellar route, are more efficient photocatalysts than pure TiO_2 films and become even more efficient when they are treated with UV radiation. TiO_2 films modified with ruthenium ions are less efficient for photocatalysis, but when they are treated with UV radiation, they also become more efficient photocatalysts than pure TiO_2 films. [151] Wang et al. [26] have prepared highly dispersed Pt nanoparticles embedded in a cubic mesoporous anatase thin film. The diameter of the Pt cluster can be controlled to below 5 nm, and the high dispersion of these clusters gives rise to catalytic activity for the oxidation of CO. The Pt–TiO_2 nanoheterojunctions promote the separation of charge carriers on UV-excited TiO_2, thus significantly improving the photocatalytic activity of porous Pt–TiO_2 composites toward killing bacteria cells of *M. lylae*. Multilayered films of TiO_2 with ordered cubic mesoporosity were grown by via layer-by-layer deposition on a conductive FTO substrate using dip coating and subsequent calcination at 400°C [152]. The photocatalytic gas-phase

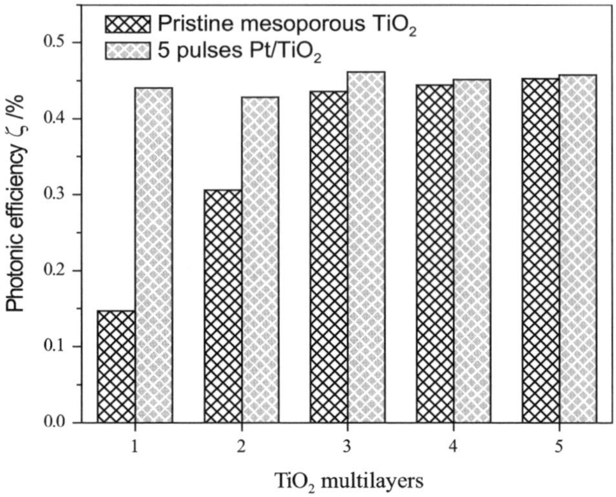

Fig. 15 Comparison of photocatalytic activities of multilayer films pristine TiO_2 and $Pt–TiO_2$ nanocomposite films for the degradation of CH_3CHO in the gas phase (adapted from Ismail et al. [152])

oxidation of acetaldehyde served as a test reaction to characterize the activity in the gas phase of both pristine TiO_2 and $Pt–TiO_2$ single-layer and multilayer films. The ordered mesoporous pristine TiO_2 and $Pt–TiO_2$ nanocomposites exhibited significantly higher photoactivity than commercial Pilkington Activ™ Glass and dense TiO_2 films (Fig. 15). Moreover, for pristine TiO_2 films, those consisting of three layers (about 650 nm in thickness) showed to be sufficient to achieve a maximum photonic efficiency of $\zeta = 0.45\%$. For the $Pt–TiO_2$ system, however, a single-layer film with a total thickness of only about 220 nm shows an almost identical activity. It has been shown that the photocatalytic oxidation of CH_3CHO occurs in the pores of TiO_2 and its surface. Molecular oxygen adsorbed on the TiO_2 surface prevents the recombination of electron–hole pairs by trapping electrons; superoxide ions are thus formed. $^{\bullet}OH$ radicals are formed from holes reacting with either H_2O or OH^- adsorbed on the TiO_2 surface. The radicals $^{\bullet}OH$ and $O_2^{\bullet -}$ are widely accepted as primary oxidants in heterogeneous photocatalysis. These oxidative species can easily oxidize CH_3CHO. The oxidizing power of the $^{\bullet}OH$ radicals is strong enough to break C–C bonds and C–H bonds of CH_3CHO adsorbed on the surface of TiO_2 leading to the formation of CO_2 and H_2O. The overall process of the photocatalytic CH_3CHO decomposition can be described by the following equation:

$$2CH_3CHO + 5O_2 \xrightarrow[\text{UV light}]{\text{Pt/TiO}_2\text{ film}} 4CO_2 + 4H_2O$$

4 Other Mesoporous Photocatalyst Materials

Ag_2S/MCM-41 photocatalysts were prepared by ion exchange method and used for the photocatalytic degradation of MB [153]. A photocatalyst containing 20 wt% Ag_2S has the maximum efficiency on photodegradation of methylene blue. The

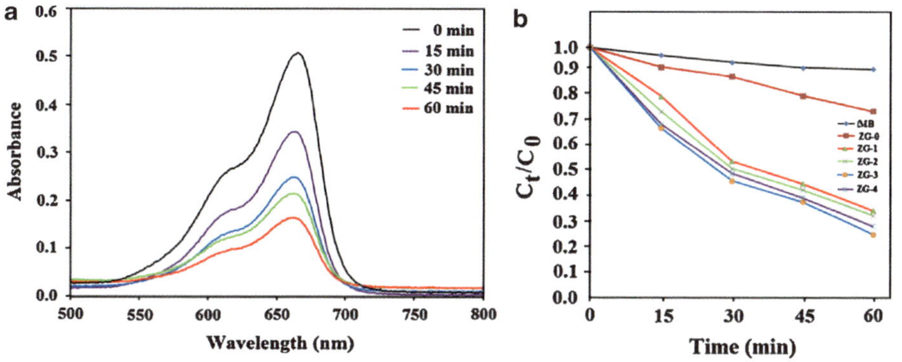

Fig. 16 (**a**) Absorption spectra of MB aqueous solutions in the presence of ZnS–RGO composites with 1.5 wt% GO and (**b**) degradation rate of MB at different intervals with and without catalyst (adapted from Sookhakiana et al. [154])

kinetics of photocatalytic degradation of dye is of the pseudo-first order with $k = 0.0343$. A facile one-pot method for the fabrication of high-quality self-assembled hierarchically ordered macro-mesoporous ZnS microsphere-reduced graphene oxide (RGO) composite without the use of templates or surfactants is described [154]. The incorporation of reduced graphene oxide as an excellent electron-transporting material effectively suppresses the charge recombination. Figure 16a illustrates the optical absorption spectra of MB aqueous solution with 10 mg of the as-prepared ZG-3 composite after exposure to visible-light irradiation for different interval times. Further experiments were performed to compare the effect of graphene oxide on the catalytic activity of the as-prepared ZnS microsphere, and the results are shown in Fig. 16b. It can be seen that the ZnS–RGO composites show a significant improvement and higher efficiency in the photodegradation of MB compared to the pure ZnS. The efficiency for ZnS is 27%, and nearly 73% of the primary dye still remained in the solution for pure ZnS. For the ZnS–RGO composites at the same time interval with the ZnS, the efficiency has increased to 79%.

Mesoporous ZnO/SiO_2 mixed composites were prepared from different amounts of ZnO nanoparticles [155]. All the materials were tested for the photodegradation of rhodamine B with good results, the best catalysts being represented by the composites containing 10 and 20 wt% nanoparticle loading. Both catalysts display higher turnover numbers (TONs) than unstructured ZnO nanoparticles and commercial ZnO nanopowder. Higher TONs were obtained in the case of ZnO-10 and ZnO-20, both displaying an increased activity compared to the commercial nanosized ZnO (TON = 1.6) and unstructured nanoparticles. A novel mesoporous Sb–SnO_2 electrode with high specific surface area and excellent electrocatalytic oxidation performance is fabricated [156]. The mesoporous electrode is served in ketoprofen removal. Compared with conventional Sb–SnO_2 electrode, ketoprofen is completely decomposed on the mesoporous electrode after 3 h, and the corresponding kinetic constant is 0.93 h^{-1}, 2.3 times higher than that of the respective conventional electrode. This may be attributed to the following:

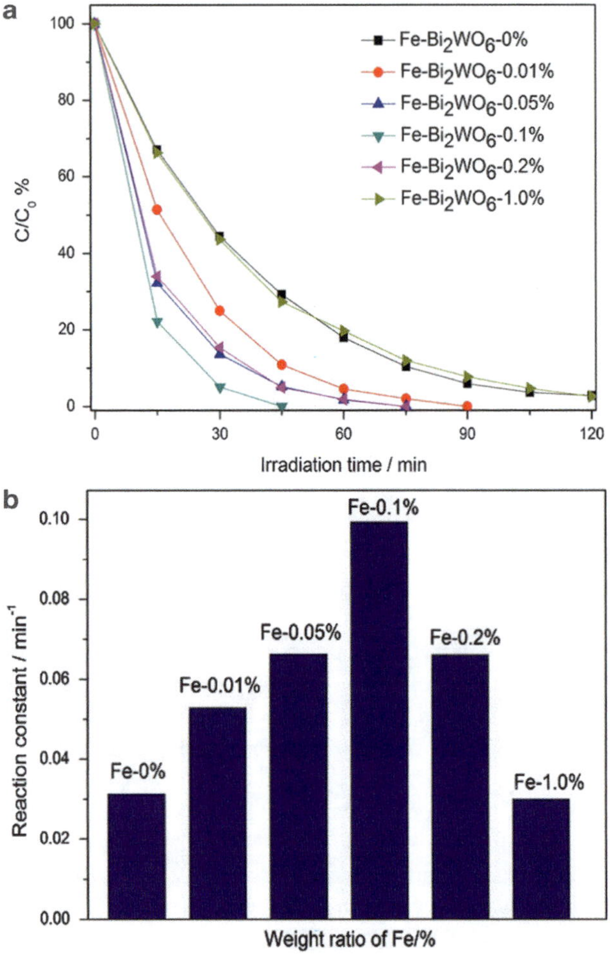

Fig. 17 (**a**) Photocatalytic activities of Fe–Bi_2WO_6-*x* % samples under visible-light irradiation. (**b**) The relationship between k and the weight ratio of Fe ions (adapted from Guo et al. [157])

(1) mesopore structure provides more in situ active sites to accelerate $^{\bullet}OH$ generation and ketoprofen oxidation; (2) mesopore channel effectively promotes the adsorption of organic pollutants and improved contaminant diffusion. The results indicated that the addition of 0.1 wt% Fe ions could evidently improve the activity of Bi_2WO_6 [157]. The apparent reaction rate of Fe–Bi_2WO_6-0.1% (k, 0.099 min^{-1}) was 3.2 times higher than that of Fe–Bi_2WO_6-0% (k, 0.031 min^{-1}). Figure 17a shows the relative removal rate of gaseous toluene against irradiation time for Fe–Bi_2WO_6-x% under visible-light irradiation. Figure 17b illustrates the relationship between apparent reaction rate constant (k, min^{-1}) and the content of Fe ions. With the increase in Fe ion concentration, the apparent reaction rate of gaseous toluene degradation was firstly increase and then decreased. The high photocatalytic activity of Fe–Bi_2WO_6-0.1% was partially due to its relative larger surface area and mesoporous structure, which were usually favorable for accelerating diffusive

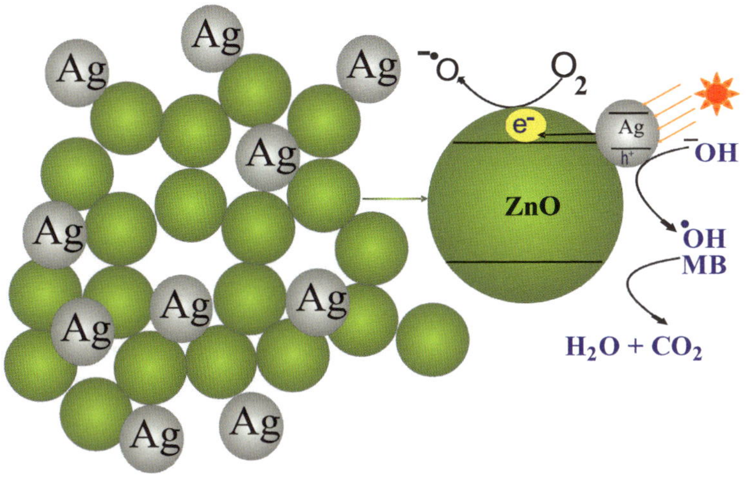

Scheme 6 Illustration of the proposed reaction mechanism of wormlike mesoporous Ag/ZnO nanocrystals for its enhanced photocatalytic performance upon visible-light irradiation for MB photodegradation as a pollutant model. Demonstration of the vital role of Ag to efficiently facilitate photogenerated charge carrier separation and improve the photocatalytic performance of mesoporous ZnO (adapted from Bouzid et al. [158])

transport of photogenerated holes to oxidized species. In a photocatalysis process, a small amount of Fe^{3+} ions act as both electron and hole traps, which enhance the lifetimes of electrons and holes and reduces the e^-/h^+ pair recombination rate.

Mesoporous Ag/ZnO nanocrystals have been successfully synthesized at different Ag contents (0–10 wt%) through a single-step sol–gel method as efficient photocatalyst. [158] The proposed mechanism for improving the photocatalytic performance of mesoporous Ag/ZnO nanocrystals under visible light is the surface conduction electrons of Ag subjected to incident visible light (Scheme 6). The electrons vibrate around the Ag nanoparticles, and electromagnetic wave is reserved and promoted in definite interface between Ag and ZnO nanocrystals, when the oscillation frequency of conduction electrons from Ag corresponds with the frequency of light [159]. Subsequently, plasmonic energy of Ag nanoparticles transfers to ZnO nanocrystals through resonant energy transfer, inducing charge separation in ZnO nanoparticles [160, 161]. The electrons are transferred from Ag to the adsorbed O_2 molecules through the conduction band of ZnO for the formation of superoxide radicals ($O_2^{-\bullet}$). The $O_2^{-\bullet}$ and trapped electrons then combine to produce H_2O, which finally forms $^{\bullet}OH$ radicals [162]. Also, the electrons and holes will react with O_2 and H_2O molecules to form hydroxyl radicals ($^{\bullet}OH$), which are strong oxidants for MB photodegradation which greatly reduce the charge recombination rate. It is reported that the plasmon excited electrons of metal transfer to the conduction band of semiconductor, which is called direct electron transfer instead of resonant energy transfer [163]. However, in Ag/ZnO, the CB level of ZnO is more negative than the work function of Ag (4.26 eV), so the electron transfer from Ag to ZnO is energetically undesirable [163]. The loss of electrons in the Ag

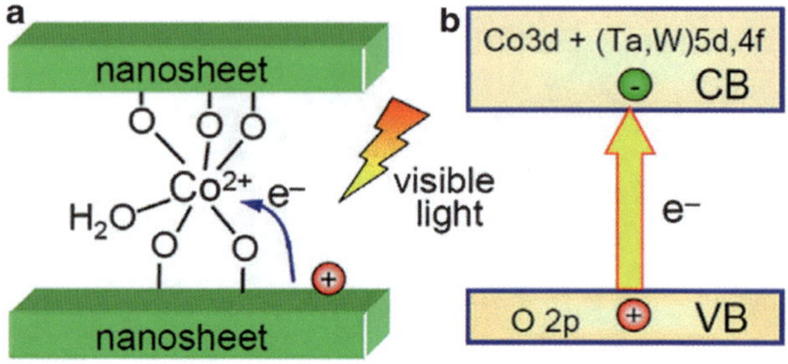

Fig. 18 Photoexcitation in Co-intercalated tantalotungstates (adapted from Lin et al. [164])

nanoparticles is beneficial for generating a large amount of oxidative radicals, such as $^{\bullet}OH$ radicals at the Ag surface and $O_2^{-\bullet}$ at the ZnO surface. The present result is in a good agreement with previously published works [159, 162]. Therefore, the photocatalytic performance of mesoporous Ag/ZnO nanocrystals was completely enhanced with increasing Ag content onto the mesoporous ZnO surface.

Cobalt-intercalated layered either tantalotungstates or tetratitanate nanosheets as high visible-light active photocatalysts have been assembled by tantalotungstatete nanosheets with cobalt ions via an exfoliation–restacking route [164, 165]. Introducing cobalt ions into the interlayered region endowed the nanohybrids with distinct spectral responses in visible-light region, resulted from the hybridization of metal d and f orbits in the conduction band to narrow the bandgap (Fig. 18). The intercalated nanohybrids exhibit high photocatalytic activities in the degradation of methylene blue under visible-light irradiation. High dopant content can be easily achieved by introducing dopant ions into the interlayered regions with a uniform distribution, which will effectively modify the physicochemical properties and the band structure of a layered semiconductor to improve its visible-light photocatalytic activity.

Well-dispersed mesoporous Ta_2O_5 submicrospheres were obtained by tailoring the heating rates (R) of calcining Ta_2O_5 precursor colloidal spheres which were synthesized by the hydrolysis of tantalum glycolate in a mixture of acetone and water [166]. Figure 19 showed the photocatalytic decolorization of RhB and MB by the as-prepared Ta_2O_5 submicrospheres and mesoporous Ta_2O_5, respectively. It can be observed that the decolorization of the MB and RhB solutions in the absence of any photocatalyst occurs very slowly under the simulant light irradiations, whereas the photodecolorization efficiency of the MB and RhB solutions over the as-prepared mesoporous Ta_2O_5 submicrospheres achieves 96.9 and 93.5% after 80 min and 140 min, respectively. It can harvest more lights as well as provide adequate access to the reactant. The unique mesoporosity and thin pore walls can also decrease the transfer distance of the photogenerated charge carriers and effectively restrain their recombination.

$SrTiO_3$ nanocube-dispersed mesoporous silica was prepared [167]. The $SrTiO_3$/SiO_2 nanocomposite exhibited high photocatalytic activity in the decomposition of methylene blue because of the combination of preferential molecular adsorption by

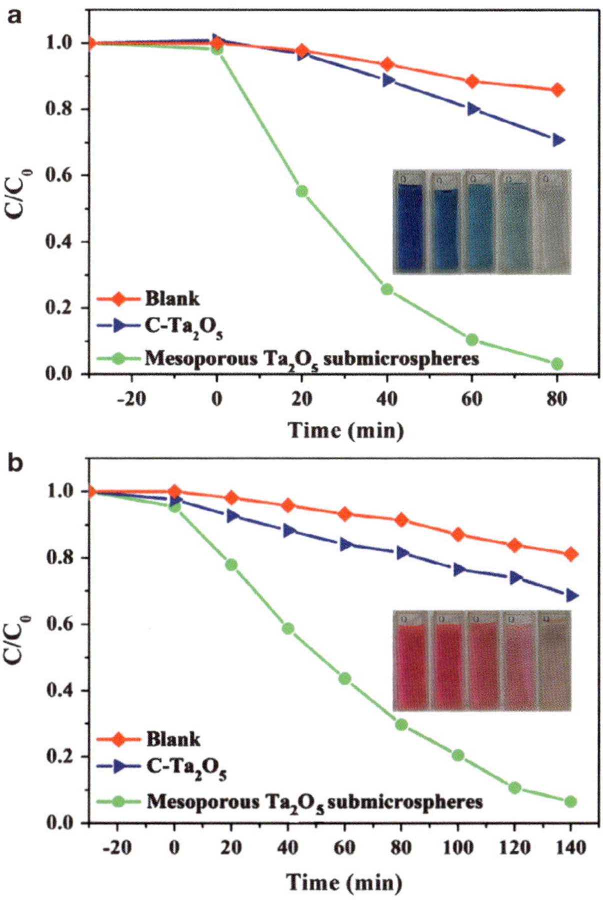

Fig. 19 Photocatalytic decolorization of the MB (**a**) and RhB (**b**) aqueous solutions in the presence/absence of photocatalysts. The *inset pictures* indicate that the color of the aqueous solution containing the mesoporous Ta_2O_5 submicrosphere photocatalysts fades away swiftly with the prolonged irradiation time of the simulant solar light (adapted from Tao et al. [166])

mesoporous silica and photocatalysis by $SrTiO_3$. The methylene blue decomposition rate by the nanocomposite was larger than that of the composite prepared with conventional $SrTiO_3$ and was attributed to its large specific surface area.

Ag nanoparticles are successfully modified onto mesoporous graphitic carbon nitride (mg-C_3N_4) by photo-assisted reduction [168]. Modifying C_3N_4 with Ag increases the conductivity and lowers the energy barrier of the interface reactions. A heterojunction electric field, forms on the interface between the modified Ag and C_3N_4, enhances the separation efficiency of photogenerated electron–hole pairs. Modifying C_3N_4 with Ag significantly improves the adsorption capacity and photocatalytic degradation efficiency. Smaller Ag nanoparticles are much more effective than larger ones in improving photoelectric conversion performance. The 3 wt% Ag/C_3N_4 possessed the best photocatalytic RhB degradation efficiency. Only after 20 min, about 90% RhB have been degraded and all RhBs had been degraded after 25 min under visible light. Modifying Ag on C_3N_4 increased the migration rate of the photogenerated electrons and interfacial electron-transfer ability. A

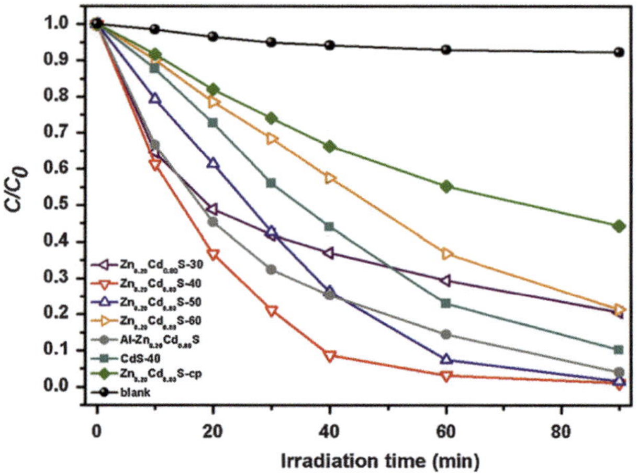

Fig. 20 Photodegradation of MB monitored as the normalized concentration change versus irradiation time (adapted from Xu et al. [169])

heterojunction electric field was formed on the interface between modified Ag and C_3N_4, enhanced the separation efficiency of the photogenerated electron–hole pairs, and prolonged the lifetime of the photogenerated electrons, which eventually made a significant promotion of its photoelectric conversion performance. In the aspect of Ag improving the photoelectric conversion performance, Ag nanoparticles in smaller size are much more effective than those in larger size.

Novel mesoporous $Zn_xCd_{1-x}S$ nanoparticles have been successfully fabricated by two steps [169]. Mesoporous $Zn_xCd_{1-x}S$ exhibits an enhanced photocatalytic performance compared to the pure CdS sample obtained from the same procedure without Zn^{2+}. As shown in Fig. 20, it can be clearly seen that the photocatalytic conversion of MB with $Zn_{0.20}Cd_{0.80}S$ reached as high as 96% after 60 min of irradiation.

5 Conclusions and Outlook

Chapter strategy has been focused on the synthesis of mesoporous semiconductors materials as efficient photocatalysts including synthetic methods, architecture concepts, and fundamental principles that govern the rational design and synthesis. In this chapter, synthesis mechanisms and the corresponding pathways are first demonstrated for the synthesis of mesoporous materials from the surfactant-templating approach. The continuing breakthroughs in the synthesis and modifications of mesoporous metal oxides nanoparticles have brought new properties with improved photocatalysts. The chapter covered the applications of mesoporous semiconductors materials as efficient photocatalysts. This steady progress has demonstrated that mesoporous material nanoparticles are playing and will continue to achieve an important role in the protections of the environment.

References

1. Ismail AA, Bahnemann DW (2011) J Mater Chem 21:11686–11707
2. Ismail AA, Bahnemann DW, Bannat I, Wark M (2009) J Phys Chem C 113:7429–7435
3. Ismail AA, Bahnemann DW (2011) Green Chem 13:428–435
4. Yang P, Zhao D, Margolese DI, Chmelka BF, Stucky GD (1998) Nature 396:152
5. Antonelli DM, Ying JY (1995) Angew Chem Int Ed 34:2014
6. Zhou D, Honma HI (2004) Nat Mater 3:65
7. Lee J, Orilall MC, Warrwn SC, Kamperman M, Disalvo F, Wiesner JU (2008) Nat Mater 7:222
8. Niederberger M, Bartl MH, Stucky GD (2002) Chem Mater 14:4364
9. Ismail AA, Bouzid H (2013) J Colloid Interface Sci 404:127
10. Bettahar MM, Costentin G, Savary L, Lavalley JC (1996) Appl Catal A 145:1–48
11. Grzybowska-Swierkosz B (2000) Top Catal 11/12:23
12. Chappel S, Chen S, Zaban A (2002) Langmuir 18:3336
13. Ito S, Katayama T, Sugiyama T, Matsuda M, Kitamura M, Wada T, Yanagida YS (2003) Chem Mater 15:2824
14. Li B, Wang X, Li M, Yan L (2003) Mater Chem Phys 78:184
15. Nagaveni K, Sivalingam G, Hegde MS, Madras G (2004) Appl Catal B 48:83
16. Huang Y, Ho W, Lee S, Li L, Zhang G, Yu JC (2008) Langmuir 24:3510
17. Choi H, Aantoniou MG, Pelaez M, Delacruz AA, Shoemaker OA, Dionysiou DD (2007) Environ Sci Technol 41:7530
18. Yu JC, Yu JG, Ho WK, Jiang ZT, Zhang LZ (2002) Chem Mater 14:3808
19. Tojo S, Tachikawa T, Fujitsuka M, Majima T (2008) J Phys Chem C 112:14948
20. Sinha AK, Suzuki K (2005) J Phys Chem B 109:1708
21. Schattka JH, Shchukin D, Jia G, Antonietti JM, Caruso RA (2002) Chem Mater 14:5103
22. Tayade RJ, Kulkarni RG, Jasra RV (2006) Ind Eng Chem Res 45:5231
23. Li H, Bian Z, Zhu J, Huo Y, Li H, Lu Y (2007) J Am Chem Soc 129:4538
24. Mills A, Lee S-K (2002) J Photochem Photobiol A Chem 152:233
25. Ollis D, Pelizzetti E, Serpone N (1991) Environ Sci Technol 25:1522
26. Ismail AA, Bahnemann DW (2011) J Phys Chem C 115:57841
27. Chen L, Yao B, Cao Y, Fan K (2007) J Phys Chem C 111:11849
28. Shiraishi Y, Saito N, Hirai T (2005) J Am Chem Soc 127:12820
29. Zhan S, Chen D, Jiao X, Tao C (2006) J Phys Chem B 110:11199
30. Madhugiri S, Sun B, Smirniotis PG, Feraris JP, Balkus KJ Jr (2004) Microporous Mesoporous Mater 69:77
31. Yu JC, Zhang L, Yu J (2002) Chem Mater 14:4647
32. Wang X, Yu JC, Chen Y, Wu L, Fu X (2006) Environ Sci Technol 40:2369
33. Bian Z, Zhu J, Wang S, Cao Y, Qian X, Li H (2008) J Phys Chem C 112:6258
34. Kong L, Chen H, Hua W, Zhang S, Chen J (2008) Chem Commun 4977
35. Yu JC, Zhang L, Zheng Z, Zhao J (2003) Chem Mater 15:2280
36. Yang X, Wang Y, Xu L, Yu X, Guo Y (2008) J Phys Chem C 112:11481
37. Dong W, Sun Y, Lee CW, Hua W, Lu X, Shi Y, Zhang S, Chen J, Zhao D (2007) J Am Chem Soc 129:13894
38. Thangaraj A, Kumar R, Mirajkar SP, Ratnasamy P (1991) J Catal 130:1
39. Srinvasn M, White T (2007) Environ Sci Technol 41:4405
40. Soni SS, Henderson MJ, Bardeau J-F, Gibaud A (2008) Adv Mater 20:1493
41. Li X, Zhuang Z, Li W, Pan H (2012) Appl Catal Gen 429–430:31
42. Yu JC, Li G, Wang X, Hu X, Leung CW, Zhang Z (2006) Chem Commun 2717
43. Pan JH, Lee WI (2006) Chem Mater 18:847
44. Alvaro M, Aprile C, Benitez M, Carbonell E, Garcıa H (2006) J Phys Chem B 110:6661
45. Li G, Bai R, Zhao XS (2008) Ind Eng Chem Res 47:8228
46. Ogawa M, Ikeue K, Anpo M (2001) Chem Mater 13:2900

47. Wang XC, Yu JC, Ho CM, Hou YD, Fu XZ (2005) Langmuir 21:2552
48. Lei Z, Xiao Y, Dang L, Hu G, Zhang J (2007) Chem Mater 19:477–484
49. Choi H, Sofranko AC, Dionysiou DD (2006) Adv Funct Mater 16:1067
50. Beyers E, Cool P, Vansant EF (2005) J Phys Chem B 109:10081
51. Wang H, Miao J-J, Zhu J-M, Ma H-M, Zhu J-J, Chen H-Y (2004) Langmuir 20:11738
52. Lee A-C, Lin R-H, Yang C-Y, Lin M-H, Wang W-Y (2008) Mater Chem Phys 109:275
53. Wark M, Tschirch J, Bartels O, Bahnemann D, Rathousky J (2005) Microporous Mesoporous Mater 84:247
54. Patarin J, Lebeau B, Zana R (2002) Curr Opin Colloid Interface Sci 7:107
55. Shibata H, Ogura T, Mukai T, Ohkubo T, Sakai H, Abe M (2005) J Am Chem Soc 127:16396
56. Xu H, Zhang L (2009) J Phys Chem C 113:1785
57. Tian G, Fu H, Jing L, Xin B, Pan K (2008) J Phys Chem C 112:3083
58. Dai K, Peng T, Chen H, Liu J, Zan L (2009) Environ Sci Technol 43:1540
59. Morishita M, Shiraishi Y, Hirai T (2006) J Phys Chem B 110:17898
60. Hoffmann MR, Martin ST, Choi WY, Bahnemann DW (1995) Chem Rev 95:69
61. Rohlfing DF, Szeifert J, Yu MQ, Kalousek V, Rathousky J, Bein T (2009) Chem Mater 21:2410
62. Kalousek V, Tschirch J, Bahnemann D, Rathousky J (2008) Superlattices Microstruct 44:506
63. Tian B, Yang H, Liu X, Xie S, Yu C, Fan J, Tu B, Zhao D (2002) Chem Commun 1824
64. Liu Y, Li J, Wang M, Li Z, Liu H, He P, Yang X, Li J (2005) Cryst Growth Des 5:1643
65. Lu Y, Hoffmann MR, Yelamanchili S, Terrenoire A, Schrinner M, Drechsler M, Moller MW, Breu J, Ballauff M (2009) Macromol Chem Phys 210:377
66. Yu J, Liu WH, Yu A (2008) Cryst Growth Des 8:930
67. Sugimoto T, Zhou XP (2002) J Colloid Interface Sci 252:347
68. Liu J, An T, Li G, Bao N, Fu G, Sheng J (2009) Microporous Mesoporous Mater 124:197
69. Soler-Illia GJ, de Louis AA, Sanchez AC (2002) Chem Mater 14:750
70. Yu J, Zhang L, Cheng B, Su Y (2007) J Phys Chem C 111:10582
71. Peng T, Zhao D, Dai K, Shi W, Hirao K (2005) J Phys Chem B 109:4947
72. Liu Z, Sun DD, Guo P, Leckie JO (2007) Chem Eur J 13:1851
73. Zhu L, Liu K, Li H, Sun Y, Qiu M (2013) Solid State Sci 20:8–14
74. Wang H-W, Kuo C-H, Lin H-C, Kuo I-T, Cheng C-F (2006) J Am Ceram Soc 89:3388
75. Yu J, Zhou M, Cheng B, Yu H, Zhao X (2005) J Mol Catal A 227:75
76. Wessels K, Minnermann M, Rathousky J, Wark M, Oekermann T (2008) J Phys Chem C 112:15122
77. Matsumoto Y, Ishikawa Y, Nishida M, Ii S (2000) J Phys Chem B 104:4204
78. Wang X, Yu JC, Hou Y, Fu X (2005) Adv Mater 17(1):99
79. Wang K, Yao B, Morris MA, Holmes JD (2005) Chem Mater 17:4825
80. Kim DS, Kwak S-Y (2009) Environ Sci Technol 43:148
81. Wang H, Zhang L, Chen Z, Hu J, Li S, Wang Z, Liu J, Wang X (2014) Chem Soc Rev 43:5234–5244
82. Xiang Q, Yu J, Jaroniec M (2012) Chem Soc Rev 41:782–796
83. Jiang L, Zhou G, Mi J, Wu Z (2012) Catal Commun 24:48–51
84. Asahi R, Morikawa T, Ohwaki T, Aoki K, Taga Y (2001) Science 293:269
85. El-Sheikh SM, Zhang G, El-Hosainy HM, Ismail AA, O'Shea K, Falaras P, Kontos AG, Dionysiou DD (2014) J Hazard Mater 280:723–733
86. Burda C, Lou Y, Chen X, Samia ACS, Stout J, Gole JL (2003) Nano Lett 3:1049
87. Zhou Z, Wang J, Zhou S, Liu X, Meng G (2008) Catal Commun 9:568–571
88. Rengifo-Herrera JA, Pierzcha K, Sienkiewicz A, Forrob L, Kiwi J, Pulgarin C (2009) Appl Catal B 88:398–406
89. Chen X, Burda C (2008) J Am Chem Soc 130:5018–5019
90. Yang X, Cao C, Hohn K, Erickson L, Maghirang R, Hamal D, Klabunde K (2007) J Catal 252:296–302
91. Fang J, Wang F, Qian K, Bao H, Jiang Z, Huang W (2008) J Phys Chem C 112:18150

92. Chi B, Zhao L, Jin T (2007) J Phys Chem C 111:6189
93. Cong Y, Zhang J, Chen F, Anpo M (2007) J Phys Chem C 111:6976
94. Martínez-Ferrero E, Sakatani Y, Boissière C, Grosso D, Fuertes A, Fraxedas J, Sanchez C (2007) Adv Funct Mater 17:3348–3354
95. Grosso D, Cagnol FG, Soler-Illia AA, Crepaldi EL, Amenitsch H, Brunet-Bruneau A, Bourgeois A, Sanchez C (2004) Adv Funct Mater 14:309
96. Amenitsch H, Albouy PA, Sanchez C (2003) Chem Mater 15:4562
97. Kim C-S, Shin J-W, Cho Y-H, Jang H-D, Byund H-S, Kim T-O (2013) Appl Catal A Gen 455:211
98. Thind SS, Wu G, Chen A (2012) Appl Catal B 111–112:38
99. Di Valentin C, Pacchioni G, Selloni A (2005) Chem Mater 17:6656
100. Liu R, Ren Y, Shi Y, Zhang F, Zhang L, Tu B, Zhao D (2008) Chem Mater 20:1140
101. Meng Y, Gu D, Zhang FQ, Shi YF, Yang H, Li F, Yu Z, Tu CZ, Zhao BDY (2005) Angew Chem Int Ed 44:7053
102. Zhang D, Yang D, Zhang H, Lu C, Qi L (2006) Chem Mater 18:3477–3485
103. Pan JH, Zhang X, Du AJ, Sun DD, Leckie JO (2008) J Am Chem Soc 130:11256
104. Zhou JK, Lv L, Yu J, Li HL, Guo P-Z, Sun H, Zhao XS (2008) J Phys Chem C 112:5316
105. Yang HG, Zeng HC (2004) J Phys Chem B 108:3492–3495
106. Yu J, Wang W, Cheng B, Su B-L (2009) J Phys Chem C 113:6743
107. Yu JC, Ho W, Yu J, Hark SK, Iu K (2003) Langmuir 19:3889–3896
108. Vidal H, Kapar J, Pijolat M, Colon G, Bernal S, Cordon A, Perrichon V, Fally F (2001) Appl Catal B 30:75
109. Georgieva J, Armyanov S, Valova E, Poulios I, Sotiropoulos S (2007) Electrochem Commun 9:365
110. Siedl N, Baumann SO, Elser MJ, Diwald O (2012) J Phys Chem C 116:22967–22973
111. Bedja I, Kamat PV (1995) J Phys Chem 99:9182–9188
112. Xuan S, Jiang W, Gong X, Hu Y, Chen Z (2009) J Phys Chem C 113:553
113. Kim TW, Ha H-W, Paek M-J, Hyun S-H, Baek I-H, Choy J-H, Hwang S-J (2008) J Phys Chem C 112:14853
114. Zhu J, Ren J, Huo Y, Bian Z, Li H (2007) J Phys Chem C 111:18965
115. Arnal P, Corriu RJP, Leclercq D, Mutin PH, Vioux A (1996) J Mater Chem 6:1925
116. Mori K, Kondo Y, Morimoto S, Yamashita H (2008) J Phys Chem C 112:397
117. Zhou W, Fu H, Pan K, Tian C, Qu Y, Lu P, Sun C-C (2008) J Phys Chem C 112:19584
118. Zhang X, Zhang L, Xie T, Wang D (2009) J Phys Chem C 113:7371
119. Fan X, Chen X, Zhu S, Li Z, Yu T, Ye J, Zou Z (2008) J Mol Catal A 284:155
120. Rolison DR (2003) Science 299:1698
121. Bell AT (2003) Science 299:1688
122. Yu JC, Wang XC, Fu XZ (2004) Chem Mater 16:1523
123. Rodrigues S, Ranjit KT, Uma S, Martyanov IN, Klabunde KJ (2005) Adv Mater 17:2467
124. Liu S, Yu J, Mann S (2009) J Phys Chem C 113:10712
125. Nakamura R, Okamoto A, Osawa H, Irie H, Hashimoto K (2007) J Am Chem Soc 129:9596
126. Livingston JV (2005) Trends in water pollution research. Nova, New York
127. Kim TW, Hwang S-J, Park Y, Choi W, Choy J-H (2007) J Phys Chem C 111:1658
128. Stone VF, Davis RJ (1998) Chem Mater 10:1468
129. Peng TY, Zhao D, Song HB, Yan CH (2005) J Mol Catal A 238:119
130. Chu S-Z, Inoue S, Li K, Wada D, Haneda H, Awatsu S (2003) J Phys Chem B 107:6586
131. Chen Y-C, Pu Y-C, Hsu Y-J (2012) J Phys Chem C 116:2967–2975
132. Mao A, Park NG, Han GY, Park JH (2011) Nanotechnology 22:175703–175709
133. Wang Z, Zhang F, Xue Y, Cui B, Guan JN (2007) Chem Mater 19:3286
134. Li G, Zhao XS (2006) Ind Eng Chem Res 45:3569
135. Li Y, Kim S-J (2005) J Phys Chem B 109:12309
136. Xuzhuang Y, Yang D, Huaiyong Z, Jiangwen L, Martins WN, Frost R, Daniel L, Yuenian S (2009) J Phys Chem C 113:8243

137. Li G, Kang ET, Neoh KG, Yang X (2009) Langmuir 25:4361
138. Wang XH, Li JG, Kamiyama H, Moriyoshi Y, Ishigaki TJ (2006) J Phys Chem B 110:6804
139. Aronson BJ, Blanford CF, Stein A (1997) Chem Mater 9:2842
140. Maldotti A, Molinari A, Amadelli R, Carbonell E, Garcia H (2008) Photochem Photobiol Sci 7:819
141. Kang C, Jing L, Guo T, Cui H, Zhou J, Fu H (2009) J Phys Chem C 113:1006
142. Shi KY, Chi YJ, Yu HT, Fu HG (2005) J Phys Chem B 109:2546
143. Hu Y, Martra G, Zhang J, Higashimoto S, Coluccia S, Anpo M (2006) J Phys Chem B 110:1680
144. Zhang W, Froba M, Wang J, Tanev PT, Wong J, Pinnavaia TJ (1996) J Am Chem Soc 118:9164
145. Allain E, Besson S, Durand C, Moreau M, Gacoin T, Boilot J-P (2007) Adv Funct Mater 17:549
146. Besson S, Ricolleau C, Gacoin T, Jacquiod C, Boilot J-P (2003) Microporous Mesoporous Mater 60:43
147. Ismail AA, Bahnemann DW, Robben L, Wark M (2010) Chem Mater 22:108
148. Bannat I, Wessels K, Oekermann T, Rathousky J, Bahnemann D, Wark M (2009) Chem Mater 21:1645
149. Li X, Peng J, Kang J-H, Choy J-H, Steinhart M, Knoll W, Kim DH (2008) Soft Matter 4:515
150. Alberius PCA, Frindell KL, Hayward RC, Kramer EJ, Stucky GD, Chmelka BF (2002) Chem Mater 14:3284
151. Stathatos E, Petrova T, Lianos P (2001) Langmuir 17:5025
152. Ismail AA, Bahnemann DW, Rathouský J, Yarovyi V, Wark M (2011) J Mater Chem 21:7802
153. Pourahmad A (2012) Superlattices Microstruct 52:276–287
154. Sookhakiana M, Amina YM, Basirun WJ (2013) Appl Surf Sci 283:668
155. Collard X, El Hajj M, Su B-L, Aprile C (2014) Microporous Mesoporous Mater 184:90
156. Fan J, Zhao G, Zhao H, Chai S, Cao T (2013) Electrochim Acta 94:21–29
157. Guo S, Li X, Wang H, Dong F, Wu Z (2012) J Colloid Interface Sci 369:373
158. Bouzid H, Faisal M, Harraz FA, Al-Sayari SA, Ismail AA (2014) Catal Today, http://dx.doi.org/10.1016/j.cattod.2014.10.011
159. Gao P, Ng K, Sun DD, Hazard J (2013) Mater 262:826–835
160. Cushing SK, Li J, Meng F, Senty TR, Suri S, Zhi M, Li M, Bristow AD, Wu N (2012) J Am Chem Soc 134:15033–15041
161. Furube A, Du L, Hara K, Katoh R, Tachiya M (2007) J Am Chem Soc 129:14852–14853
162. Yu H, Ming H, Zhang H, Li H, Pan K, Liu Y, Wang F, Gong J, Kang Z (2012) Mater Chem Phys 137:113
163. Lin Y-G, Hsu Y-K, Chen Y-C, Wang S-B, Miller JT, Chen L-C, Chen K-H (2012) Energy Environ Sci 5:8917–8922
164. Lin B, Xu B, He L, Fan X, Qu H (2013) Microporous Mesoporous Mater 172:105
165. Liu H, Lin B, He L, Qu H, Sun P, Gao B, Chen Y (2013) Chem Eng J 215–216:396
166. Tao C, Xu L, Guan J (2013) Chem Eng J 229:371–377
167. Katagiri K, Miyoshi Y, Inumaru K (2013) J Colloid Interface Sci 407:282
168. Bua Y, Chen Z, Li W (2014) Appl Catal B 144:622
169. Xu X, Lu R, Zhao X, Zhu Y, Xu S, Zhang F (2012) Appl Catal B 125:11

Spectroscopic Methods for Investigating Reaction Pathways

Russell F. Howe

Abstract This chapter reviews the use of infrared spectroscopy and electron paramagnetic resonance (EPR) spectroscopy to investigate reaction pathways in photocatalysis. In the case of infrared spectroscopy, examples are given from four different experimental methods for obtaining spectra of photocatalysts and adsorbed species: transmission, diffuse reflectance IR Fourier transform (DRIFT), attenuated total reflectance (ATR) and reflection–absorption infrared spectroscopy (RAIRS),which is applicable to single-crystal surfaces. EPR spectroscopy has been employed to observe trapped charge species (electrons and holes) and radical intermediates produced by reaction of electrons or holes with adsorbed species. Examples of both are given.

Keywords ATR, DRIFT, EPR spectroscopy, Radical intermediates, RAIRS, Transmission FTIR, Trapped electrons, Trapped holes

Contents

R.F. Howe (✉)
Chemistry Department, University of Aberdeen, Aberdeen AB24 3UE, Scotland
e-mail: r.howe@abdn.ac.uk

D.W. Bahnemann and P.K.J. Robertson (eds.), *Environmental Photochemistry Part III*, Hdb Env Chem (2015) 35: 267–300, DOI 10.1007/698_2014_255,

1 Introduction

In all forms of heterogeneous catalysis, spectroscopic techniques play an important role in developing an understanding of reaction mechanisms. A reaction *mechanism* ultimately involves transition states which cannot be directly observed spectroscopically. A reaction *pathway*, on the other hand, involves reactants, active sites, reaction intermediates and products which exist long enough to be identified as distinct chemical species on the catalyst surface. Reaction mechanisms cannot be convincingly deduced however until reaction pathways are known, and this is where spectroscopy comes into its own.

In conventional heterogeneous catalysis, the techniques most commonly employed to identify adsorbed reactants, intermediates and products are Fourier transform infrared spectroscopy (FTIR), Raman, NMR and UV–visible spectroscopy. Numerous reviews are available illustrating the application of these techniques to study reaction pathways in conventional heterogeneous catalysis [1–3]. Note the distinction between studying reaction pathways and characterising catalysts. In the latter case, all of the techniques of modern materials chemistry and surface science are applicable.

In photocatalysis, the problem of identifying adsorbed species is not dissimilar to that in conventional heterogeneous catalysis, whether the reaction is performed in the gas phase or in the presence of a solvent. Photocatalysis offers an experimental advantage, in principle, of occurring at ambient temperatures. On the other hand, the need to couple in photoexcitation may add an experimental complexity. The initiating reactants in photocatalysis may be regarded as valence band holes and conduction band electrons in the photocatalyst. These are short-lived at room temperature, but can be studied by time-resolved optical spectroscopies which do not have a counterpart in conventional catalysis.

Ultimately, the questions which need to be answered to establish a reaction pathway for a heterogeneously photocatalysed reaction are:

- What happens to conduction band electrons and valence band holes once they are produced by band-gap irradiation of the semiconductor catalyst?
- How are reactants adsorbed on the surface of the catalyst?
- How do conduction band electrons and valence band holes react with adsorbed reactants?
- What intermediate species are subsequently formed on the catalyst surface or in the solution phase?
- How do these intermediates form the final reaction products, and does this happen on the catalyst surface or in the solution phase?

In this article, the ways in which infrared spectroscopy and electron paramagnetic resonance (EPR) spectroscopy can address these questions will be reviewed by describing examples from the recent literature. In a subject as complex as photocatalysis, no one spectroscopic technique can provide all of the answers, and the challenge in practice is seeing how to link together the information obtained from different techniques to build up a satisfactory picture of the reaction pathway (and ultimately inferring a reaction mechanism).

2 Infrared Spectroscopy

The traditional experimental approach to measuring infrared spectra from heterogeneous catalysts, at least for gas-phase reactions, is by transmission. The infrared beam passes through a pressed disk of powdered catalyst mounted in a reaction cell which may be a high vacuum static cell allowing dosing of adsorbed molecules from the gas phase (and in the case of photocatalysis injection of light) or alternatively a flow cell in which reactants are injected into a flowing gas stream. Provided the surface area of the powdered catalyst is sufficiently high, the measured FTIR spectra will contain measureable contributions from surface species, as well of course as from the bulk catalyst. Preparing a pressed disk which is sufficiently thin to transmit infrared radiation while being sturdy enough to survive treatment can be a problem. Some authors have overcome this problem by pressing the powder into a metal mesh which provides mechanical stability. Alternatively, a thin film of the catalyst may be deposited onto an infrared transparent substrate. Examples of both of these approaches are described below.

An alternative experimental approach is diffuse reflectance IR Fourier transform (DRIFT) spectroscopy. In this case, absorption spectra of surface species are obtained by analysing infrared radiation diffusely reflected from the surface of the powdered catalyst. Commercial in situ cells are available, and it is straightforward to couple UV–visible radiation into the cell via a focussed lamp or a fibre optic.

Neither transmission FTIR nor DRIFT is an appropriate method for use in photocatalytic reactions in the liquid phase, since the intense absorption of infrared radiation by the solvent (particularly water) prevents useful data being obtained. However, the alternative method of attenuated total reflectance (ATR) spectroscopy has been shown by several groups to be useful for photocatalysis in aqueous systems. In this case the infrared beam samples the catalyst via a high refractive index internal reflection element, and the path length through the aqueous phase is negligible.

Surface science studies on model single-crystal surfaces are proving enormously helpful in understanding active sites on heterogeneous catalysts. The well-defined nature of such surfaces and the comparative ease of applying computational modelling to them are both attractive features. In the case of titania photocatalysts, the external reflectance technique (sometimes referred to as reflection–absorption infrared spectroscopy (RAIRS)) has recently been used to obtain infrared spectra from single-crystal titania surfaces for the first time.

In the following sections, examples of recent work from the literature utilising these different experimental approaches will be described.

2.1 Transmission FTIR

The utility of this technique is illustrated here with several published studies of surface sites on titania photocatalysts and photocatalysed oxidation of alcohols over titania.

The hydroxyl groups on titania surfaces play an important role in the photocatalytic reactivity. In particular, surface-hydroxylated centres have been proposed to act as trapping sites for valence band holes which migrate to the surface:

$$Ti^{4+}(OH)^{-} + h^{+} \rightarrow Ti^{4+}(OH)^{\cdot}$$

and for conduction band electrons

$$Ti^{4+}(OH)^{-} + e^{-} \rightarrow Ti^{3+}(OH)^{-}$$

although EPR studies described in Sect. 3.1 indicate that the reality is more complicated than that depicted by these simple schemes. Nevertheless, understanding the nature of the surface hydroxyl groups (and other surface sites) is an important part of understanding reaction pathways.

Delana et al. [4] have recently applied the transmission FTIR technique to characterise the hydroxyl groups on P25 titania. (P25 is a commercially available titania comprising 80% anatase and 20% rutile.) They mounted self-supporting pressed disks of P25 in a high vacuum cell allowing in situ treatment. Figure 1 below shows a series of spectra obtained after outgassing P25 at successively higher temperatures from 323 to 773 K.

The complex pattern of bands in the 3,800–2,600 cm^{-1} region after outgassing at low temperatures contains at least nine components. Some of these are due to OH-stretching modes of adsorbed water molecules, as indicated by the presence of an accompanying H_2O-bending mode at 1,620 cm^{-1}. Seventy-five percent of the adsorbed water was removed by outgassing at 473 K, and more than 90% at 573 K, according to the decrease in intensity of the 1,620 cm^{-1} band. The bands remaining after high-temperature outgassing are due to OH-stretching modes of surface hydroxyl groups, of which there appear to be at least six different types. The authors attribute bands above 3,680 cm^{-1} to linear (Ti-OH) groups and those below 3,680 cm^{-1} to bridging Ti(OH)Ti groups. At outgassing temperatures above 773 K (not shown), hydroxyl groups were removed from the surface due to condensation reactions between terminal and bridging hydroxyl groups.

The heterogeneity of surface sites on P25 reflected in three different types of terminal hydroxyl group and three different types of bridging hydroxyl group was consistent with high-resolution TEM images of P25 which showed the presence of a variety of local terminations of anatase and rutile crystal planes resulting in corner, edge and step sites as well as terraces.

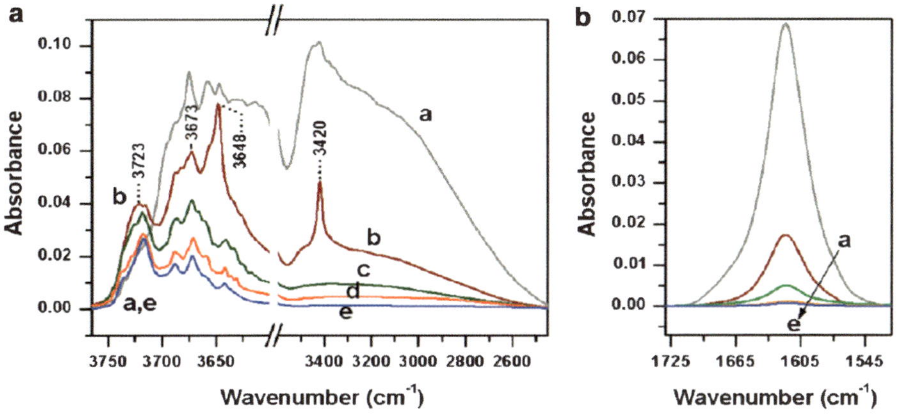

Fig. 1 (**A, B**) FTIR spectra of P25 titania after outgassing in vacuo at (*a*) 323 K; (*b*) 473 K; (*c*) 573 K; (*d*) 673 K; (*e*) 773 K. Reproduced with permission from [4]

In the same work the authors also measured low-temperature infrared spectra of carbon monoxide physically adsorbed on the P25 surfaces dehydrated to different extents. CO interacts only with coordinatively unsaturated Ti^{4+} sites which may be exposed by desorption of water molecules or by dehydroxylation at higher temperatures. The multiplicity of CO-stretching bands seen further illustrates the heterogeneity of the surface sites. At lower outgassing temperatures, spectra of adsorbed CO were dominated by bands assigned to CO adsorbed on pentacoordinated Ti^{4+} sites, suggesting that these sites are exposed during desorption of molecularly adsorbed H_2O. At higher degassing temperatures, the loss of surface hydroxyl groups was accompanied by increasing contributions from bands due to CO adsorbed on four-coordinate Ti^{4+} sites, suggesting that the surface hydroxyl groups (both terminal and bridging) are present on such lower coordination sites.

The heterogeneity of surface sites on P25 revealed by these infrared studies should also be reflected in corresponding studies of photoreactivity. In an earlier study, Wu et al. [5] have examined by transmission FTIR the photooxidation of alcohols over P25. In their work, catalyst samples mounted as powders pressed into a tungsten grid were outgassed at 723 K prior to exposure to reactant molecules. As described above, such pretreatment should remove all molecularly adsorbed water, exposing pentacoordinated Ti^{4+} plus terminal and bridging hydroxyl groups. After exposure to methanol at room temperature, bands were observed due to molecularly adsorbed methanol and to surface methoxy groups. The molecular species was largely desorbed at 473 K, whereas much higher temperatures were needed to remove the methoxy species. Two different types of methoxy group were identified: a monodentate (terminal) species, $TiOCH_3$, and a bidentate (bridging) species, $Ti(OCH_3)Ti$. On heating in vacuo both of these species decayed at the same rate; there was no interconversion between them, suggesting that they were formed on different types of surface site.

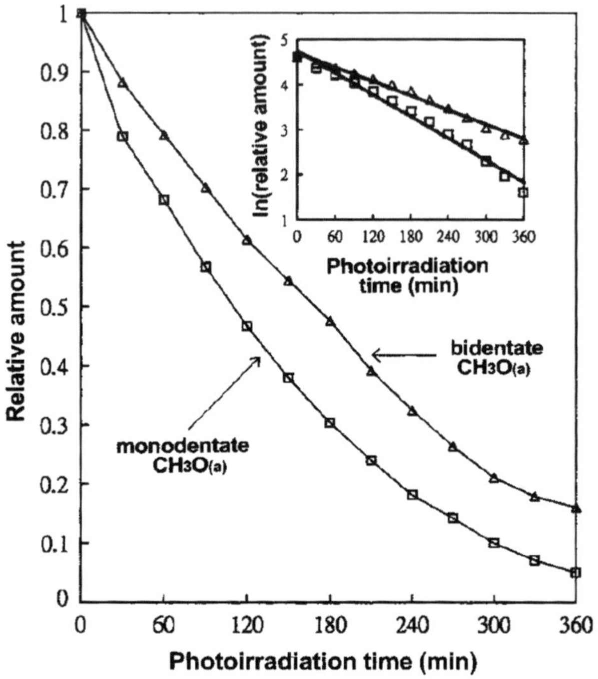

Fig. 2 Relative amounts of monodentate and bidentate methoxy groups on P25 during irradiation in oxygen at room temperature (Reproduced with permission from [5])

The photoreactivity of the two different types of methoxy group in oxygen was however found to be different. Figure 2 shows a plot of the relative intensities of infrared bands due to the two different methoxy species versus time of UV irradiation in oxygen at room temperature. It can be seen that the monodentate species reacts at 1.5 times the rate of the bidentate species. Other changes occurring in the spectra at the same time were an increase in the baseline absorption (not commented on by the authors, but probably due to accumulation of conduction band electrons, as discussed further below) and the appearance of new bands due to adsorbed formate (HCOO(a)) and gas-phase CO_2.

From their observations the authors suggest a reaction pathway in which the initiating step is capture of a valence band hole by an adsorbed $TiOCH_3$ species to form $TiOCH_2^{\cdot}$ radicals. These radicals were considered to react with O_2 to form $OCH_2O_2^{\cdot}$ peroxy radicals. At the same time, O_2 was considered to trap conduction band electrons to form superoxide ions O_2^-. Reaction of the peroxy radicals with superoxide ions followed to form an unstable tetra-oxide species, which dissociated to formate ions (observed by infrared) and adsorbed water.

Similar experiments were reported with adsorbed ethanol. In this case, the monodentate ethoxide species reacted with oxygen under UV irradiation at 1.7 times the rate of the bidentate species, and the adsorbed reaction product detected by infrared spectroscopy was adsorbed acetate.

Although this infrared study could directly detect only the initial and final adsorbed species, the clear distinction in photoreactivity between the monodentate and bidentate alkoxy species is an important result. It relates to the crucial step of valence band hole trapping by adsorbed reactants and indicates that this initial electron transfer step occurs more readily with the monodentate species. The possible role of various radical intermediates in the subsequent reaction pathway is discussed again below under the heading of EPR spectroscopy.

The importance of methoxy species in trapping of valence band holes is supported by recent theoretical studies of methanol on single-crystal anatase and rutile surfaces which show that the highest occupied molecular orbitals of adsorbed methoxy species lie just below the valence band maximum in titania. According to these theoretical studies, the methoxy species provide a more favourable hole-trapping site than adsorbed methanol, as seen in the infrared studies [6, 7].

The role of surface methoxy groups is also highlighted by the very recent study of Panayatov et al. [8] on methanol photooxidation over rutile nanoparticles. This group also used the experimental approach of pressing their rutile powder into a tungsten grid in order to measure transmission spectra. On rutile, as on P25, infrared spectra of methanol adsorbed at room temperature showed bands due to molecularly adsorbed methanol, monodentate and bidentate methoxy groups. The molecular species was shown to be hydrogen bonded either to surface hydroxyl groups or to bridging oxide ions, whereas the methoxy species were considered to result from dissociation of methanol at 5-coordinate Ti^{4+} sites and adjacent oxide ions:

$$Ti^{4+} + CH_3OH + O^{2-} \rightarrow Ti^{4+} - OCH_{3^-} + OH^-$$

Striking changes occurred in the infrared spectra of methanol on rutile when the sample was subsequently irradiated in vacuo. A large increase in absorbance across the entire spectrum was seen, with a broad maximum at around 1,800 cm^{-1}. This infrared absorption is attributed to two different optical processes: the direct infrared excitation of shallow trapped electrons and the acoustic phonon-mediated infrared excitation of free conduction band electrons. The infrared detection of excited electrons in anatase has been reported earlier [9, 10], but recombination of conduction band electrons and valence band holes usually precludes their observation above 200 K. In this case, however, effective trapping of valence band holes by adsorbed methoxy species allows the infrared signature of conduction band electrons to persist at room temperature.

Irradiation in vacuo also caused some reduction in intensity of the infrared bands of the methoxy species, but this was almost completely reversed when irradiation was stopped. The authors attributed these changes to electric field effects in the rutile particles (the presence of high concentrations of conduction band electrons) rather to any irreversible chemical reaction. A small fraction (~3%) of the methoxy groups were converted however to formate species under these conditions. The authors consider that initial hole trapping by adsorbed methoxy species can be followed by further electron transfer from the resulting radical to the conduction band of the rutile in what is overall a 2-electron oxidation process:

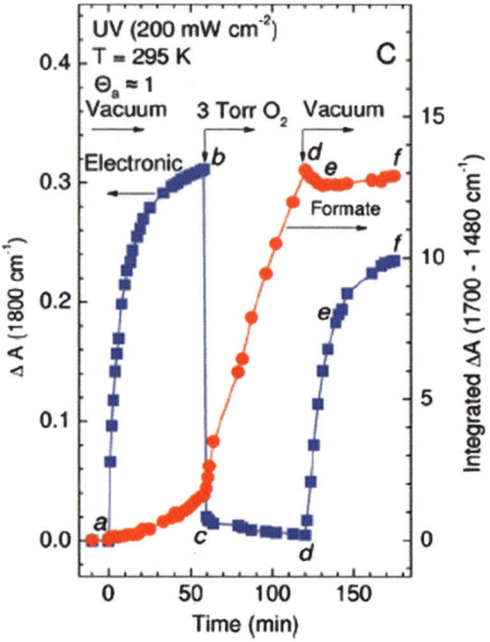

Fig. 3 Time evolution of infrared signatures of conduction band electrons and adsorbed formate groups during irradiation of methanol adsorbed on nanorutile in vacuo, on addition of oxygen and on subsequent evacuation. Reproduced with permission from [8]

$$Ti^{4+}OCH_{3^-} \rightarrow Ti^{4+} - OCH_2 + H^+ + 2e$$

although no formaldehyde species could be detected in the infrared spectra.

Addition of oxygen as an electron scavenger following irradiation in vacuo caused rapid quenching of the infrared signature of conduction band electrons and an order of magnitude increase in the intensity of the infrared bands due to adsorbed formate. Figure 3 illustrates these effects.

The fast quenching of the conduction band electrons on addition of oxygen is attributed to electron scavenging by oxygen, forming the superoxide ion (this chemistry is directly detected in EPR experiments). Removal of the conduction band electrons in this way enhances the trapping of valence band holes by adsorbed methoxy groups, which are converted to formate. When oxygen is removed, conduction band electrons are again detected.

The direct interconversion of methoxy groups to adsorbed formate was shown by monitoring the intensities of the respective infrared bands as a function of time during irradiation in oxygen. As shown in Fig. 4, the presence of isosbestic points confirms that no intermediate species detectable by infrared spectroscopy are involved.

The authors present a reaction pathway for this conversion involving formyl intermediates which are not detected directly because of their low concentrations and short lifetimes. The most significant finding from this infrared study however is that the main role of oxygen in the photocatalytic oxidation of methanol over rutile is as an electron scavenger; it plays no direct role as an oxidant.

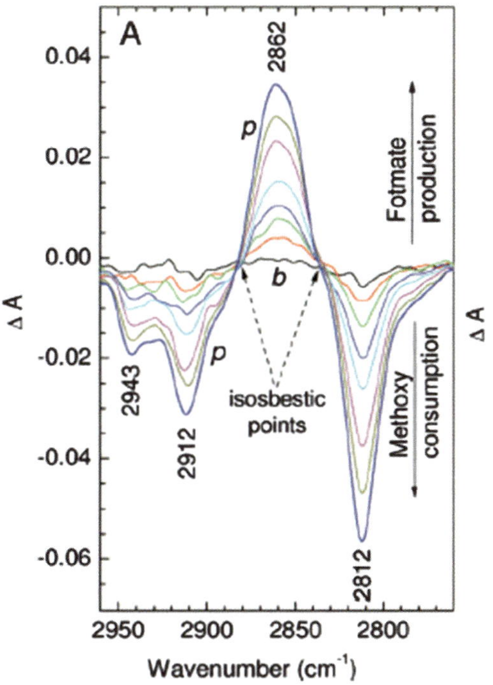

Fig. 4 Conversion of methoxy groups to adsorbed formate when nanorutile is irradiated in the presence of oxygen. Reproduced with permission from [8]

2.2 *DRIFT*

The diffuse reflectance technique offers an advantage in principle over transmission FTIR in that contributions from the bulk catalyst should contribute less to the spectrum, allowing a greater surface sensitivity. This advantage was first exploited by Szczepankiewicz et al. [9] in their studies of surface species on P25 and anatase titania under UV irradiation. This group used a commercially available environmental chamber mounted in a diffuse reflectance accessory into which light from a 1 kW xenon arc lamp could be focussed.

Irradiation of the titania samples in vacuo caused a dramatic decrease in surface reflectivity which decayed over a period of minutes when the light was switched off. This is the same phenomenon noted later in transmission measurements (described in Sect. 2.2) as an increase in infrared absorbance attributed to conduction band electrons. The baseline relaxed much more quickly to its initial position if oxygen was added or if the temperature was raised. At the same time, clear changes were noted in the ν(OH) region of the spectrum, as shown below.

The outgassed photocatalyst showed at least five different bands in this region, four of which lie below 3,680 cm^{-1} and therefore (following reference [4]) must be due to bridging hydroxyl groups. The 3,417 cm^{-1} band was not seen in pure anatase and was therefore attributed to the rutile component of P25. On UV irradiation, the 3,417 and 3,647 cm^{-1} bands were reduced in intensity, and a band at 3,716 cm^{-1}

Fig. 5 DRIFT spectra of P25 titania before (*a*) and after (*b*) UV irradiation in vacuo (*c*) are the difference spectrum (*b-a*). Reproduced with permission from [9]

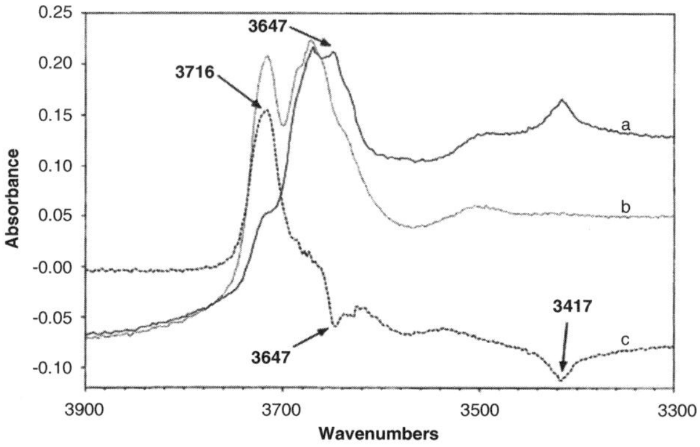

showed a marked increase in intensity. The authors attribute this band to $Ti^{3+}OH^{-}$ species, i.e. to hydroxyl groups bound to surface-trapped electrons. A similar loss of the 3,647 cm^{-1} band and a growth of a 3,716 cm^{-1} band were seen with pure anatase, although these changes occurred much more slowly than in the case of P25.

The 3,716 cm^{-1} band was also seen to grow on irradiation in oxygen, but in this case accompanied by a more intense band at 3,683 cm^{-1} (which can possibly be seen as a weak shoulder in Fig. 5 above). This new band was attributed to surface-bound hydroxyl radicals $Ti^{4+}OH$ resulting from oxidation of $Ti^{4+}OH^{-}$ by valence band holes. The authors argue that this becomes possible because of scavenging of conduction band electrons by oxygen, although they suggest that the hole-trapping reaction competes with the experimentally observed hole-induced lattice oxygen desorption.

A similar experimental set-up for DRIFT measurements has been used by Yu and Chuang [11] to examine in situ the photooxidation of ethanol over P25. As in the transmission spectra described in Sect. 2.1 above, the DRIFT spectra showed clearly the presence of both molecularly adsorbed and dissociated ethanol (ethoxide species). The relative coverages of C_2H_5OH(ads), CH_3CH_2O and H_2O (ads) profoundly influenced the subsequent photooxidation pathways detected by DRIFT.

At low coverages of adsorbed ethanol, the major intermediate species detected by FTIR during photooxidation was adsorbed formate (HCOO(ads)), and the final reaction products were CO_2(g) and adsorbed H_2O. The authors suggest that this reaction pathway involves attack of hydroxyl radicals (formed by reaction of valence band holes with adsorbed water) on adsorbed ethoxide species. Hydrogen abstraction from the β-carbon of ethoxide will form CH_2CH_2OH radicals which then undergo C–C bond scission.

At high coverages of adsorbed ethanol, the spectra show that adsorbed water has been displaced. Formate species are no longer detected, and the intermediate species seen are adsorbed acetaldehyde, acetic acid and acetate. In this case, valence band holes were considered to react directly with adsorbed ethanol, abstracting hydrogen from the α-carbon of ethoxide groups to produce acetate species. Under these conditions, the increased infrared absorbance due to conduction band electrons is also detected, since no adsorbed oxygen is present to intercept these. Only ethoxide species adsorbed directly at hole-generating sites can be oxidised, and the initial rate of oxidation is slow. Only after a significant decrease in the adsorbed ethoxide concentration and corresponding increase in adsorbed water does the reaction rate increase as the hydroxyl radical pathway begins to operate.

The higher surface sensitivity of the DRIFT experiment compared with transmission has allowed in this case a much more detailed (and complex) picture to be built up of the reaction pathways in ethanol oxidation over P25. The distinction between direct hole attack on adsorbed ethoxide and indirect hydroxyl radical attack (via adsorbed water) is particularly important.

The same group has also applied the DRIFT technique to study a reaction much loved by the photocatalytic community, the degradation of methylene blue (MB) over P25 [12]. Changes occurring in the FTIR spectrum of adsorbed MB during UV irradiation are extremely complex, and the reader is referred to a full discussion of the assignments of the many bands observed. The key finding in this paper is that the different functional groups in MB are lost at different rates. This is illustrated in Fig. 6 below, which plots the intensities of selected infrared bands as a function of time. The decomposition of MB is considered to proceed via (1) demethylation; (2) breaking of the central aromatic ring and then the side aromatic rings; (3) conversion of the fragments produced from the first two steps to smaller intermediate species and (4) final conversion of these to CO_2, H_2O, NH_4^+ and SO_4^2.

The authors consider that the initiating step in demethylation of adsorbed MB is hydrogen abstraction from CH bonds. This occurs indirectly through hydroxyl radicals formed by valence band hole attack on adsorbed water. Direct attack of valence band holes on adsorbed MB was ruled out on the grounds that positively charged MB could not approach the catalyst surface closely enough in the presence of adsorbed water. This proposal was supported by the observation that OD groups from adsorbed D_2O were lost when MB was irradiated on a deuterated titania surface. Furthermore, the rate of the initial demethylation step was significantly slowed in the presence of adsorbed ethanol. This was attributed to the competing reaction of hydroxyl radicals with adsorbed ethanol. The authors could not rule out however from their infrared observations a possible contribution to the reaction pathway from photosensitisation of the titania by MB, that is, absorption of the visible light component of the incident radiation by MB to produce an excited state which then injects an electron into the conduction band of the titania, forming an MB^+ radical cation which subsequently demethylates.

All of the DRIFT studies described above used broadband radiation to initiate the photochemistry, which (as seen in the case of MB) does not allow a clear-cut distinction to be made between band-gap excitation of the titania photocatalyst and possible photosensitisation by adsorbed reactants. Ramakrishnan et al. [13] have recently described a modified DRIFT experiment in which a UV diode coupled to a fibre-optic light guide was used to irradiate a gold–titania photocatalyst for the

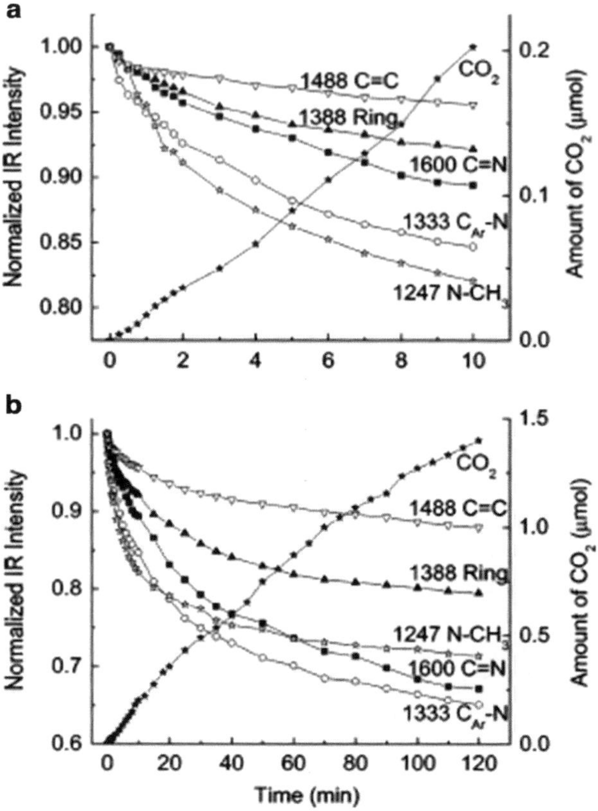

Fig. 6 Time dependence of infrared bands observed during the decomposition of MB over P25 under UV irradiation (measured by DRIFT). (**a**) during first 10 minutes. (**b**) during 2 hours. Reproduced with permission from [12]

oxidation of NO_2. In their modified DRIFT cell, 365 nm light was directed onto the surface of the photocatalyst powder via a small prism, while the diffusely scattered infrared radiation was collected with standard DRIFT optics, and the entire cell was operated as a flow reactor connected to a gas-phase NOx analyser.

Figure 7 shows DRIFT spectra of a 0.5 wt% Au–TiO_2 catalyst exposed to NO_2 before and after the start of 365 nm irradiation. The new bands appearing are due to adsorbed nitrite, nitrite and nitrosyl species. Their intensities were higher on Au–TiO_2 than on TiO_2 alone, consistent with the higher catalytic activity of the gold-containing catalyst, and some additional bands were present, which were not found on TiO_2 alone. The authors did not attempt a detailed kinetic analysis of the growth and decay of various species, but this paper demonstrates well the potential of using selected wavelength DRIFT studies of reaction pathways in gas-phase photocatalysis.

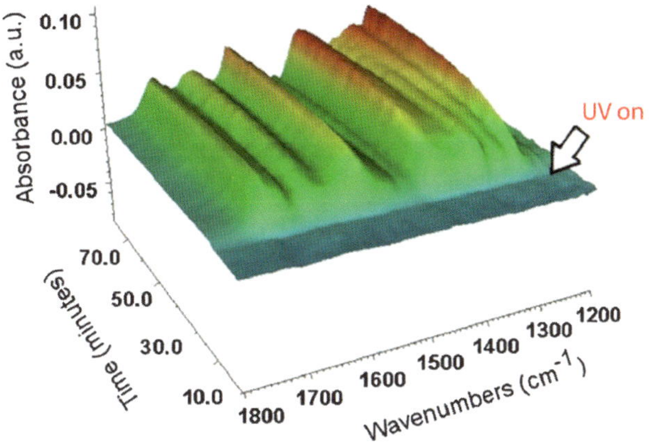

Fig. 7 DRIFT spectra of Au–TiO_2 catalyst following 365 nm irradiation in the presence of NO_2. Reproduced with permission from [13]

2.3 ATR

Photocatalysis in the liquid phase represents a large fraction of all published reports. Transmission and DRIFT infrared spectroscopy cannot however generally be used for such systems because of prohibitively high infrared absorption by the solvent (despite attempts that have been made to overcome this with very short path length transmission cells [14]). The ATR technique reduces the solvent contribution by confining the infrared beam within a crystal of high refractive index material such as ZnSe, which is then coated externally with a thin layer of the photocatalyst of interest. The infrared beam undergoing multiple internal reflections within the crystal samples a few microns of the photocatalyst coating and any adsorbed species, with relatively little contribution from the surrounding solvent.

Mendive et al. [15] have used ATR to monitor changes in the infrared spectrum of titania nanoparticles when subjected to UV-A irradiation in water. Figure 8 shows spectra obtained. These are plotted as difference spectra relative to the spectrum of the titania recorded in the dark. Despite the short infrared path length through the solvent in the ATR experiment, the ν(OH) bands of water absorb too strongly to be fully subtracted, and no useful information is obtained in this region of the spectrum. Likewise, the strong absorption by ν(Ti-O) vibrations of the titania below 1,000 cm^{-1} blanks out this region of the spectrum. Two important observations can be made however. There is an increase in baseline absorbance with time of irradiation. This is similar to that seen in transmission and DRIFT experiments when TiO_2 is irradiated in the presence of valence band hole scavengers and is attributed to photoexcitation of electrons into the conduction band. The presence of this effect on irradiation in water implies that water in some manner is acting as a hole scavenger, i.e. being oxidised. Secondly, there is a gradual appearance of a new band at 1,640 cm^{-1} due to the bending mode of adsorbed water. In their paper Mendive et al. [15] argue that this results from de-aggregation of the titania powder due to thermal energy from hole/electron recombination, causing an increase in available surface area of the catalyst.

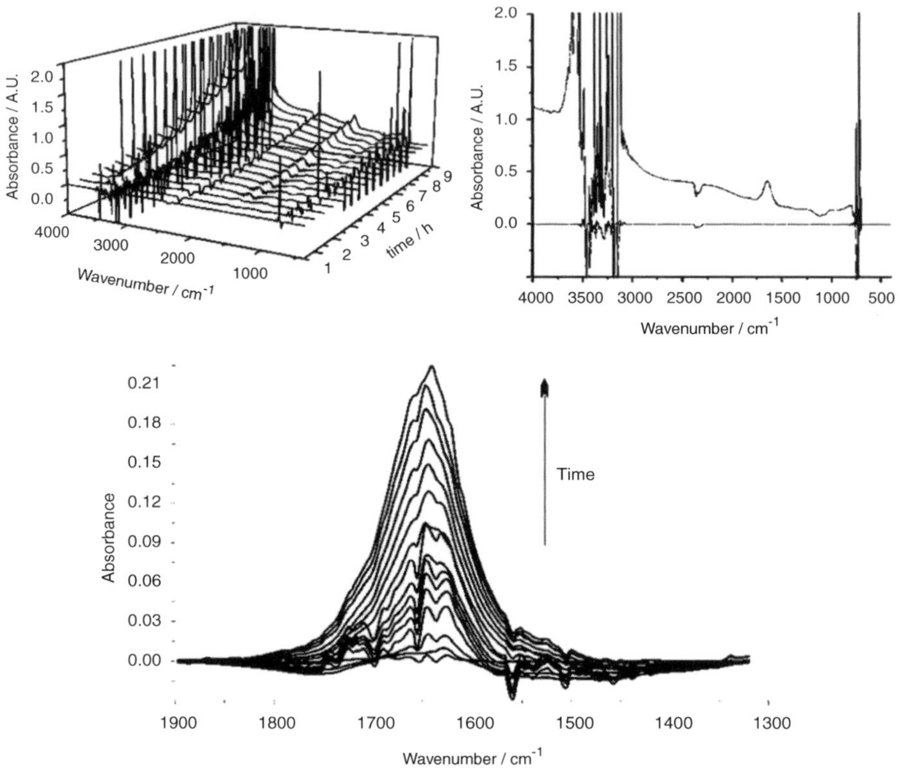

Fig. 8 ATR difference spectra of nanocrystalline anatase during exposure to UV-A radiation in the presence of liquid water. Reproduced with permission from [15]

Infrared studies by transmission and DRIFT of the photooxidation of ethanol over titania in the gas phase were described above. Gong et al. [16] have used the ATR technique to investigate the same reaction in the liquid phase over titania and platinised titania photocatalysts. Exposure of P25 deposited on the ATR element to aqueous ethanol solutions caused displacement of adsorbed water from the surface (negative changes in the ν(OH) and HOH-bending regions) and appearance of new bands due to adsorbed ethanol. Very little difference was seen with platinised titania.

Photooxidation of aqueous ethanol over pristine P25 with broadband UV–visible radiation caused an initial rapid upward shift in the baseline, accompanied by growth of negative features indicating loss of adsorbed water and adsorbed ethanol. The baseline shift is clearly due to the production of conduction band electrons, notwithstanding the fact that the experiments were not conducted under oxygen-free conditions. Some conversion of adsorbed ethanol does occur under these conditions, but no clear evidence for reaction intermediates or products was obtained.

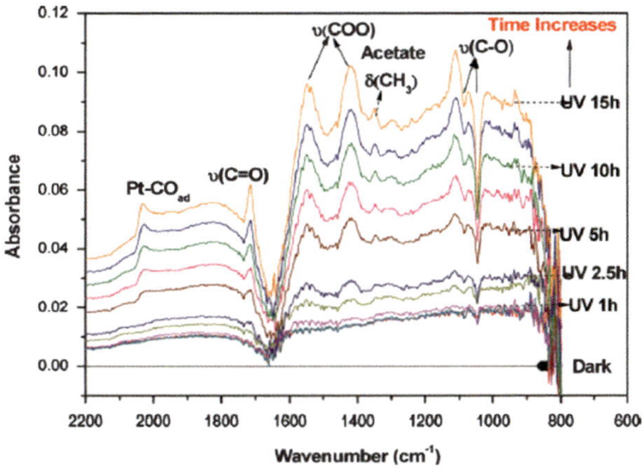

Fig. 9 ATR difference spectra recorded during the photooxidation of aqueous ethanol over platinised P25. Reproduced with permission from [16]

Over platinised titania, on the other hand, clear infrared evidence for formation of intermediate species was obtained. The other striking difference from pristine P25 was the almost complete absence of shifts in the baseline absorbance during irradiation of platinised titania. Figure 9 shows difference spectra recorded in this experiment. Bands due to acetaldehyde and acetate are seen to appear at an early state of illumination and to grow further with time. The identity of these intermediates was confirmed by adsorbing them separately onto fresh catalysts. Extended illumination gave rise to an additional band at ~2,050 cm^{-1} which is characteristic of CO adsorbed on platinum.

Key findings from this study were that:

- In oxygenated aqueous ethanol, photooxidation over pristine P25 is slow because reductive activation of di-oxygen does not occur fast enough, resulting in the accumulation of electrons in the conduction band.
- Platinisation of the P25 results in a large increase in activity, supporting the view that platinum acts as a sink for conduction band electrons, facilitating reduction of di-oxygen.
- Photooxidation of ethanol over platinised P25 occurs via acetaldehyde and acetic acid intermediates, but complete mineralisation of acetic acid is slow and constitutes the rate-limiting step.
- The observation of CO adsorbed on platinum as a potential catalyst poison raises concerns about the efficacy of sustained photooxidation over submerged catalysts.

The authors [16] comment further on several limitations of the ATR technique which warrant further study. In particular, they mention the difficulty of tracking gaseous reaction products with this technique, the separation of spectral responses from adsorbed species versus liquid-phase species, and the problem of making the technique quantitative (also an issue with the DRIFTS technique).

Other applications of the ATR method include the studies by Burgi et al. of the photo-assisted mineralisation of dicarboxylic acids [17–19]. In the most recent of these, titania-supported gold catalysts are compared with pristine titania.

2.4 RAIRS

Reflection–absorption infrared spectroscopy is a technique frequently employed in surface science to observe species adsorbed on single-crystal surfaces, usually under UHV conditions. The infrared beam is reflected externally from the surface of the sample crystal at grazing incidence. The enhanced sensitivity achieved at grazing incidence is further amplified, at least in the case of metal surfaces, by an image dipole effect, and the so-called surface selection rule means that only vibrations involving a dipole moment change perpendicular to the crystal surface are detected [20]. The image dipole effect is not expected to be present on oxide single-crystal surfaces, which makes the task of observing RAIRS spectra from titania single crystals particularly challenging.

Takahashi and Yui [21] used the RAIRS technique with polarisation modulation to observe hydroxyl groups on single-crystal rutile surfaces. In these experiments, polarisation of the incident infrared beam is rapidly modulated between parallel and perpendicular to the direction of incidence, and the difference between them in the reflected beam measured in real time. These processes effectively cancel out signals from isotropic components such as water vapour and molecularly adsorbed water. The concept is illustrated in Fig. 10 below.

Spectra measured with polarised light without modulation are dominated by bands due to water vapour and possibly adsorbed water, but show in addition a band at 3,279 cm^{-1} which is more intense with parallel polarisation. Application of polarisation modulation totally suppresses the contributions from water vapour and adsorbed water leaving only the 3,279 cm^{-1} band (the base line curvature is an artefact of the polarisation modulator). Since no bands were detected in the modulated spectrum in the HOH-bending region (~1,640 cm^{-1}), the authors attribute the 3,279 cm^{-1} band to the OH-stretching mode of bridging hydroxyl groups on the (110) rutile surface. Support for this assignment came from the presence of a similar band on the (100) surface but its complete absence from the (001) surface, which contains no bridging oxide ions.

A frequency of 3,279 cm^{-1} is much lower than that seen for OH-stretching modes on dehydrated surfaces. The authors suggest that the ~400 cm^{-1} shift is due to hydrogen bonding of the hydroxyl group to adsorbed water molecules, which would weaken the OH bond. It is nevertheless surprising that such a low-frequency band has not been seen in earlier studies, and hydrogen bonding to surface hydroxyl groups normally results in a substantial broadening of the OH-stretching band, as well as a shift to lower frequencies.

Interestingly, UV irradiation of the rutile surfaces caused no change in the intensity of the 3,279 cm^{-1} band, ruling out an increase in surface hydroxyl groups

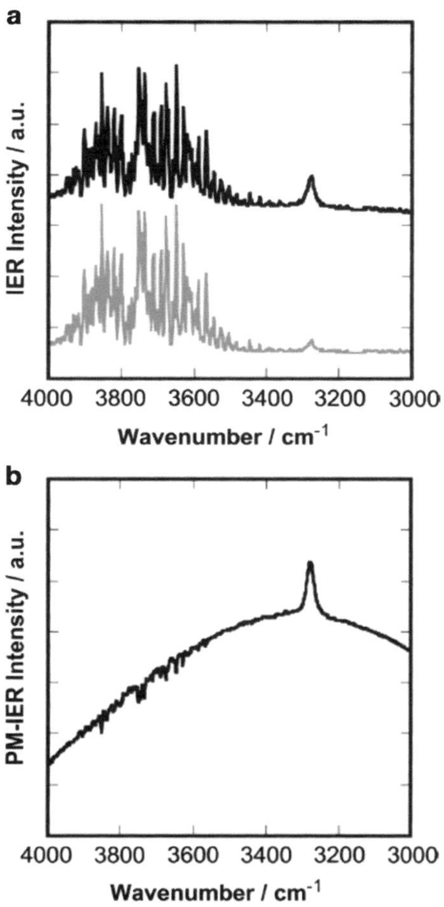

Fig. 10 (**a**) RAIRS spectra of a hydrated rutile (110) surface measured with unmodulated parallel (*dark*) and perpendicular (*light*) polarised light. (**b**) The corresponding polarisation-modulated spectrum. Reproduced with permission from [21]

as an explanation for the observed UV-enhanced wettability of the rutile surfaces. A downward shift of 2 cm^{-1} in the 3,279 cm^{-1} band under UV irradiation was reported, which was reversed when the light was switched off, but more work would be required to establish an explanation for this very small effect.

Xu et al. [22] have recently applied the RAIRS technique (without polarisation modulation) to observe the photooxidation of CO on single-crystal (110) rutile and (101) anatase surfaces under UHV conditions. CO adsorption gives a band at 2,180 cm^{-1} attributed to CO bound to fivefold Ti^{4+} sites exposed on both surfaces. UV irradiation in the presence of oxygen causes a gradual reduction in intensity of this band, accompanied by the appearance of a new band at 2,340 cm^{-1} attributed to physisorbed CO_2. As shown in Fig. 11, the new band due to CO_2 has the opposite sign to that due to CO; this is said by the authors to be a consequence of the different surface orientation of the CO_2 in the RAIRS experiment. The other striking feature of these results is the order of magnitude faster rate of CO oxidation on the anatase (101) surface than on rutile (110). The authors of this work report also some

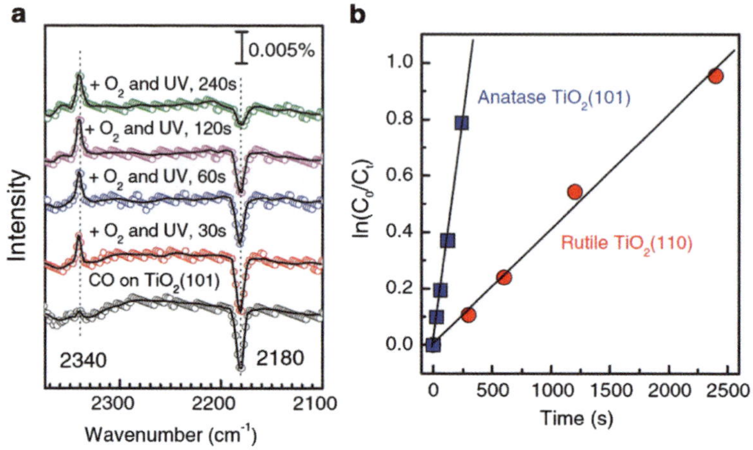

Fig. 11 (**a**) RAIRS spectra of CO photooxidation over anatase (101); (**b**) rates of disappearance of the 2,180 cm^{-1} band over anatase and rutile. Reproduced with permission from [22]

microwave photoconductivity experiments which show a much longer lifetime of conduction band electrons in the anatase. They suggest that the lower rate of photooxidation of CO over rutile (110) is due to the faster hole/electron recombination rate on this surface and speculate that the rutile may be behaving as a direct band-gap semiconductor, in contrast to anatase.

3 EPR Spectroscopy

The technique of electron paramagnetic resonance (EPR, also known as electron spin resonance, ESR) is capable of observing paramagnetic species at levels down to 10^{12} spins. Application of the technique to heterogeneous catalysis, particularly to the characterisation of transition metal ions and of adsorbed paramagnetic probe molecules, has been recently reviewed [23]. In the context of photocatalysis, where most elementary steps in proposed reaction mechanisms involve one-electron transfers, EPR spectroscopy should in principle be capable of observing most of the intermediate species postulated. In this section, examples are given of how EPR spectroscopy can be very successful in some cases, but not in others.

The background theory of EPR spectroscopy is described elsewhere [24–26]. The method can be applied to both the solid–gas and solid–liquid interfaces. It is essential in studying photocatalysts to exclude atmospheric oxygen (unless of course it is a reactant in the study) since molecular O_2 is paramagnetic and can broaden signals from other paramagnetic species on catalyst surfaces. Oxygen is also a very efficient scavenger of conduction band electrons, which can interfere with attempts to observe other electron transfer processes. The preferred sampling method is to contain the photocatalyst in a quartz cell which can be degassed on a vacuum line and/or exposed to reactants before sealing and placing in the EPR sample cavity. In situ irradiation is then possible; light can be focussed onto the sample through a port in the side of the cavity and spectra recorded before, during or after irradiation. One limitation of the method is that spectra cannot usually be recorded at room temperature. Spin–lattice relaxation means that many signals (such as Ti^{3+} resulting from electron trapping in TiO_2) become broadened beyond detection at room temperature and must be recorded at cryogenic temperatures. Short-lived species such as radical intermediates in a photocatalytic reaction may have lifetimes too short to be detected at room temperature, since the conventional EPR experiment involves sweeping the magnetic field over a timescale of seconds or more. On the other hand, spectra can be recorded at temperatures down to 4 K, which may stabilise such short-lived intermediates and allow them to be characterised in detail.

This review considers application of EPR spectroscopy to two aspects of photocatalysis: observation of charge trapping processes following band-gap irradiation and observation of radical intermediates in photocatalytic reactions. EPR has also been extensively used to characterise dopant and defect states in titania (see, e.g. [27]) but that lies outside the scope of this review.

3.1 EPR Observations of Charge Trapping

The early EPR experiments of Howe and Graetzel [28, 29] first showed evidence for trapping of conduction band electrons at titanium sites (to form paramagnetic Ti^{3+}) and of valence band holes at oxide ions (to form paramagnetic O^-) when hydrated anatase samples were UV irradiated under vacuum at low temperatures. Ti^{3+} contains a single d-electron, and the g-tensor components of the signal observed ($2 > g_{perpendicular} > g_{parallel}$) are consistent with Ti^{3+} in octahedral or axially distorted octahedral symmetry. The trapped electron site was identified in [28, 29] as an interstitial Ti^{3+} cation, although some authors have preferred to assign the signal to Ti^{3+} substituted in the anatase lattice. Very recently, Chiesa et al. have re-examined the Ti^{3+} species in reduced anatase using the pulsed EPR technique of HYSCORE, which detects the very small hyperfine couplings between Ti^{3+} and surrounding oxide ions which have been enriched with ^{17}O (nuclear spin = 5/2) [30]. The signal previously attributed to interstitial Ti^{3+} shows a hyperfine coupling to ^{17}O of less than 2 MHz, too small to be associated with a single interstitial Ti^{3+} cation. The authors propose that the so-called interstitial Ti^{3+} is in fact an unpaired electron delocalised over several adjacent Ti^{3+} lattice sites, and they present several other arguments in favour of the trapped electron being associated with lattice rather than interstitial titanium ions.

A more detailed and quantitative study of the light-induced formation of trapped holes and electrons in anatase has been reported more recently by Berger et al. [10]. These authors observed the same trapped hole signal seen by Howe and Graetzel, plus two different trapped electron signals. One was the so-called

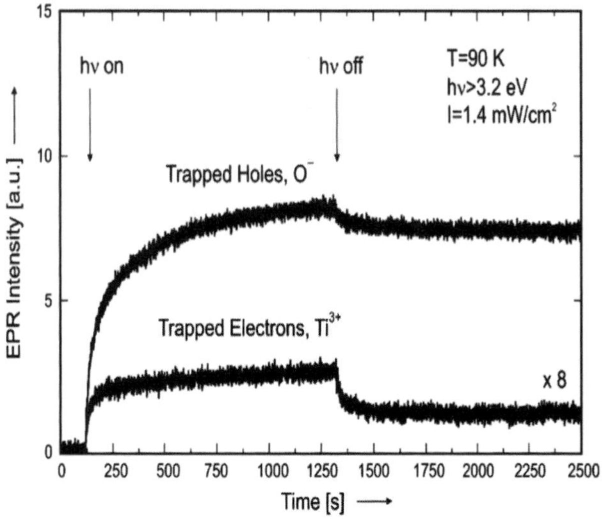

Fig. 12 Time dependence of trapped hole and trapped electron signals on UV irradiation of anatase in vacuo at 90 K. Reproduced with permission from [10]

interstitial Ti^{3+} described above (and probably in fact a bulk lattice species), and a second with different g-tensor components attributed to unspecified Ti^{3+} sites. The new features of this work were estimation of the relative concentrations of the trapped holes and electrons and measurement of the kinetics of their formation and decay. Figure 12 shows the concentrations of trapped holes and trapped electrons as a function of UV exposure time at 90 K. The most striking feature of these data is the tenfold difference in intensity between the O^- (trapped hole) and Ti^{3+} (trapped electron) signals. The authors attribute this discrepancy to the presence of EPR-silent electrons in the conduction band or in delocalised very shallow trap sites. Support for this conclusion came from parallel observations of in situ infrared spectra measured under similar (although not identical) conditions. These showed on UV irradiation in vacuo an increase in the baseline absorption due to conduction band or shallow trap state electrons, as discussed in the previous section of this chapter. Importantly, the kinetics of growth and decay of the infrared baseline absorption matched quite well those of the growth and decay of the trapped electron signals in the EPR experiments. Both the trapped electron and trapped hole signals grew in intensity with half-lives of seconds when UV irradiation was started, rather than the nano- to pico-second timescales determined in laser flash photolysis studies. One difference is that the EPR experiments were performed at 90 K rather than room temperature; however it must be concluded that the trapped species detected by EPR are not necessarily the same as those whose optical absorption signatures are monitored in the flash photolysis experiments. In the absence of oxygen (which quickly quenches the trapped electron signals through formation of superoxide ions):

$$Ti^{3+} + O_2 \rightarrow Ti^{4+} + O_2^{-}$$

the trapped hole and trapped electron signals were found to be stable over a period of hours at 90 K, although recombination occurred immediately on warming to room temperature.

Further studies of the kinetics of hole and electron trapping in anatase nanoparticles have been reported by Ke et al. [31]. These authors measured the rates of growth and decay of the trapped electron and trapped hole signals at different temperatures and fitted the data to a kinetic model. They distinguish between shallow trap sites (not detected by EPR) and deep trap sites (responsible for the observed EPR signals). The lifetimes of electrons in shallow trap sites were considered to be too short for them to be detected by EPR spectroscopy ($<10^{-9}$ s). The observed time dependences of appearance of the EPR signals were fitted with double exponentials. The fast initial rise in both signals was attributed to direct carrier trapping in deep trap sites, followed by a slower process of carrier hopping between shallow trap sites until a vacant deep trap is encountered. The initial rate of deep trapping of electrons (22×10^{-3} s^{-1}) was found to be more than twice that of holes (9.8×10^{-3} s^{-1}), consistent with the larger effective mass of the hole in TiO_2, whereas the rates of the slower process were quite comparable for holes and electrons (~10^{-3} s^{-1}).

Subsequent decay of the trapped hole and trapped electron signals when the light is turned off was attributed to recombination. The trapped hole decay was fitted to a single exponential with a rate constant of 0.76×10^{-3} s^{-1} at 10 K. The trapped electron signal showed an initial much faster decay than this (13×10^{-3} s^{-1}), which the authors attribute to recombination of trapped electrons with valence band holes, followed by a slower process with the same rate constant as the hole decay, which was attributed to recombination of trapped holes with trapped electrons.

A further piece of information obtained from these studies was the activation energy for diffusion of electrons out of trap sites when the temperature is raised. The authors distinguish between surface and inner trapped electrons and report activation energies of 1.4 and 3.3 meV, respectively. They make the important point that EPR spectroscopy is able to identify and evaluate the energetics and kinetics of trap states occurring on timescales not readily accessible by ultrafast absorption spectroscopies.

EPR spectroscopy has been used more recently by Macdonald et al. [32] to investigate electron trapping in nanocrystalline rutile. In this case, the number of trapped electrons detected following irradiation depends strongly on the level of light intensity used. Figure 13 shows results of an experiment in which the nanorutile was irradiated in vacuo at 80 K using light of selected wavelengths (band-pass filters). A Ti^{3+} signal of trapped electrons is seen only when the photon energy approaches or exceeds the rutile band gap. A trapped hole signal also appears at this energy (partially obscured by a background signal from the EPR cryostat). The incident light intensity through the filters was measured to be ~30 mW cm^{-2}.

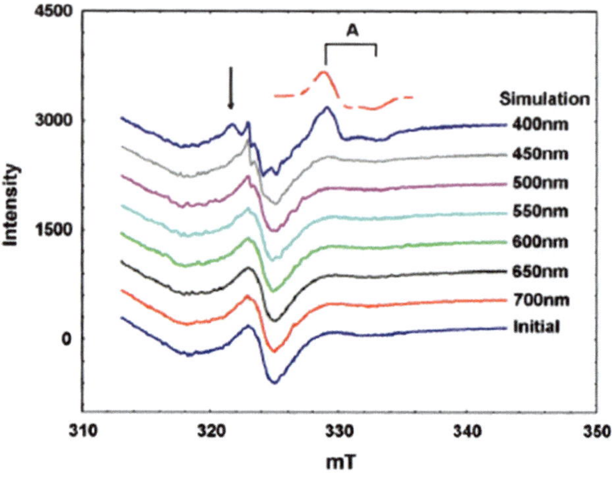

Fig. 13 EPR spectra of nanocrystalline rutile irradiated in vacuo at 77 K with light of successively shorter wavelengths. Reproduced with permission from [32]

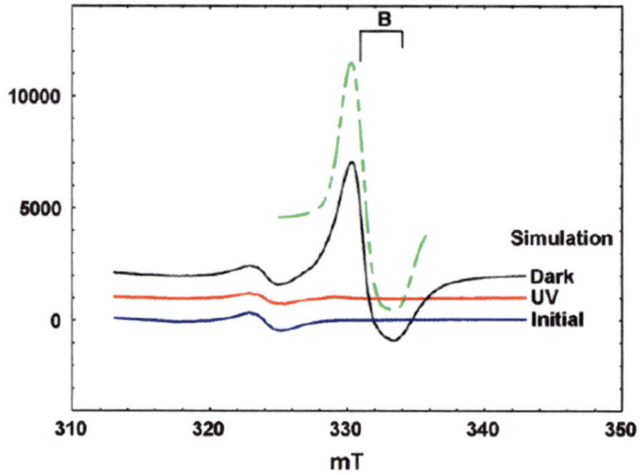

Fig. 14 EPR spectra recorded during and following irradiation of nanocrystalline rutile in vacuo at 4 K with broadband light (320–900 nm^{-1}, 300 mW cm^{-2}). Reproduced with permission from [32]

When the same nanorutile was irradiated in vacuo with full intensity broadband radiation from a xenon arc lamp (320–900 nm, 300 mW cm^{-2}) almost no trapped electrons were seen, but an intense Ti^{3+} signal appeared when the irradiation was stopped (Fig. 14). This process was completely reversible. The Ti^{3+} signal completely disappeared when the light was turned on again and was fully restored in the dark. The spectra in Fig. 14 were recorded at 4 K, but similar effects were seen at 80 K.

The Ti^{3+} signal obtained following broadband excitation is an order of magnitude more intense than that seen with 400 nm excitation and occurs at higher field ($g\perp = 1.972$ compared with $g\perp = 1.977$), indicating the presence of two different electron trap states in the nanorutile. The trapped electrons in nanorutile are remarkably stable. The Ti^{3+} signals cannot be measured at room temperature because of spin–lattice relaxation, but provided oxygen was completely excluded, the trapped electrons could still be observed at 4 K or 80 K after brief warming to room temperature.

The presence of two different electron trap states in nanocrystalline rutile was confirmed by performing the same experiment with a low-surface area large-crystal rutile sample. In this case only the lower field Ti^{3+} signal ($g\perp = 1.977$) was observed following broadband high-power irradiation in vacuo at 80 K. This signal was assigned to trap states in the bulk of the rutile and the higher field ($g\perp = 1.972$) signal which dominates the spectra of high surface area nanocrystalline rutile to surface trap states.

Recent pulsed EPR (HYSCORE) measurements on a thermally reduced rutile sample by Livraghi et al. [33] observed the $g\perp = 1.977$ signal and from analysis of ^{17}O hyperfine coupling deduced that it was due to interstitial Ti^{3+} ions in the sample.

The appearance of intense trapped electron signals when irradiation is stopped and the subsequent bleaching of these signals when irradiation is resumed can be explained by a simple model of trap states within the band gap of the rutile. Under high-power irradiation, most photoexcited electrons are located in the conduction band and invisible to EPR spectroscopy. Electrons trapped under these conditions are immediately excited back into the conduction band. When irradiation is stopped, electrons are trapped in bulk and/or surface trap states. When irradiation is resumed, the trapped electrons are returned to the conduction band, and the EPR signals are bleached. Under irradiation at lower powers, the rate of de-trapping is correspondingly lower, and a higher concentration of trapped electrons is seen.

The kinetics of the electron trapping in nanorutile have been studied in detail by measuring EPR signal intensities as a function of time (Macdonald and Howe, manuscript in preparation). The data could be fitted with a double exponential similar to that used by Ke et al. for anatase [31], although the rate constants obtained are noticeably smaller (e.g. $3 \times 10^{-3}\ s^{-1}$ at 77 K for the fast initial step and $0.4 \times 10^{-3}\ s^{-1}$ for the slower step). The rapid initial step was found to be almost temperature independent (activation energy ~4 meV), whereas the slower step showed a significant activation energy (~46 eV). The kinetic measurements could not differentiate between the surface trap states and the bulk trap states, but since the spectra are dominated by the surface traps in the case of the nanocrystalline rutile, the kinetic parameters were assumed to apply to the surface trap states. The model proposed by Ke et al. to explain the trapping kinetics in anatase is appropriate also for rutile: the rapid process is due to direct trapping of conduction band electrons, while the slow step is due to hopping of electrons between shallow trap states before becoming trapped. Both steps occur more slowly in rutile at 77 K than in anatase at 10 K.

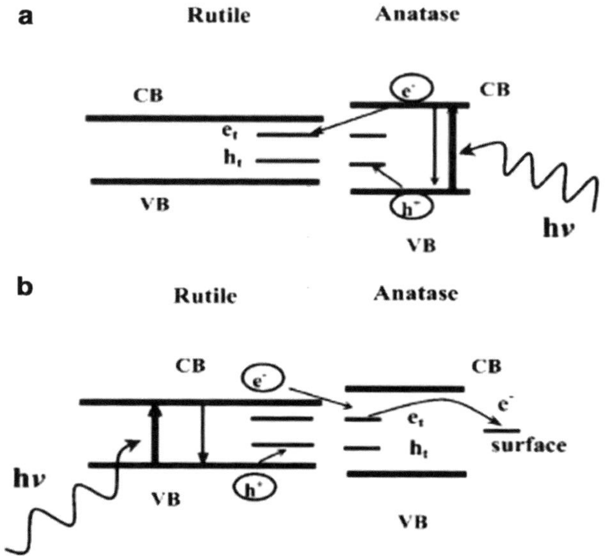

Fig. 15 Two possible models for electron transfer and trapping in mixed phase P25. (**a**) excitation of anatase, electron transfer to trap sites in rutile. (**b**) excitation of rutile, electron transfer to trap sites in anatase. Reproduced with permission from [34]

One additional measurement possible in such in situ studies is an estimate of the depth of the trap states below the conduction band. This was done [34] by measuring the wavelength dependence of the de-trapping (bleaching) process. The trapped electron signal was produced by initial broadband high-power irradiation (followed by a dark period to allow full trapping). The sample was then irradiated with single-wavelength light (band pass of 10 nm) from a monochromator and the extent of signal bleaching measured, before restoring the full intensity with broadband light. The threshold wavelength of light needed to initiate de-trapping was found to be ~900 nm, indicating that the trap states lie about 1.3 eV below the conduction band edge.

EPR spectroscopy has also been used to investigate electron trapping in Degussa (now Evonik) P25 titania, which as noted above comprises ~80% anatase and ~20% rutile. The origin of the enhanced photocatalytic activity of P25, which in many cases exceeds that of both anatase and rutile alone, has continued to be widely debated in the photocatalysis community. Two different models have been discussed in the literature: either the rutile phase acts as a sink for conduction band electrons produced in the majority anatase phase, thereby reducing electron/hole recombination in the anatase phase (model A), or conduction band electrons produced in the rutile phase are transferred to trap states in the anatase phase, again reducing electron/hole recombination (model B, see Fig. 15).

Hurum et al. carried out low-temperature in situ EPR experiments with P25 which they argue support model B. Figure 16 shows EPR spectra presented in that work [34]. Irradiation of an anatase sample at 10 K in vacuo with UV light (<400 nm) gave the signal expected for trapped electrons (Ti^{3+}) in anatase. This signal was not observed at all when the same sample was irradiated with visible light (>400 nm). In contrast, a signal due to electrons trapped in the rutile phase

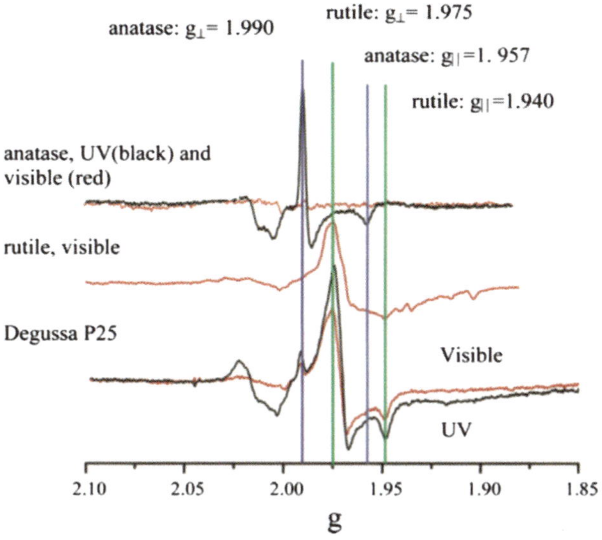

Fig. 16 EPR spectra obtained on irradiation of aqueous suspensions of anatase, rutile and P25 in vacuo at 4 K with visible and/or UV light. Reproduced with permission from [34]

was observed when a rutile sample was irradiated with visible light. In the case of P25, traces of the anatase Ti^{3+} signal were detected when visible light excitation was employed.

The authors attribute these observations as support for model B above. Only the rutile phase in P25 can absorb light of wavelength >400 nm. The observation of trapped electrons in the anatase phase under these conditions suggests that under these conditions some electron transfer is occurring from the rutile to the anatase.

It is noteworthy however that the number of anatase-trapped electrons detected in P25 on UV irradiation is much fewer than those detected in pure anatase, notwithstanding the fact that anatase is the majority phase in P25. This suggests strongly that the reverse process (model A) may also be occurring. Nevertheless model B may suggest a reason for the enhanced photocatalytic activity of P25 if, as suggested in [34], the rutile phase is acting as an antenna for visible light photons (>400 nm). In a later paper, the same authors [35] showed, by separating the different particle size fractions in P25, that the number of rutile-trapped electrons increased in the larger size fractions, suggesting (not surprisingly) that electron transfer between the phases is particle size dependent.

EPR evidence favouring model A, on the other hand, has been presented by Komaguchi et al. [36]. These authors irradiated pre-reduced samples of anatase, rutile and P25. For the pure phases, irradiation with white light (including UV) in vacuo at 77 K caused complete bleaching of the Ti^{3+} EPR signals, which were restored to their original intensity when irradiation was stopped. This behaviour, as described above for rutile, is attributed to promotion of trapped electrons into the conduction band under irradiation. The process is reversed when irradiation is stopped. For P25, on the other hand, the number of electrons trapped in the rutile phase was enhanced following irradiation. This was attributed to transfer of electrons from the conduction band of anatase into trap states in the rutile phase (model A). The same experiments carried out with physical mixtures of anatase and rutile did not show this effect, confirming that the intimacy of the anatase/rutile interface is critical to electron transfer in either direction.

3.2 EPR Observation of Radical Intermediates

Given the acknowledged importance of radical intermediates in photocatalytic reactions, it is surprising that EPR spectroscopy has so far not been widely used to study the one-electron oxidation or reduction products of hole or electron transfer to reactants adsorbed on photocatalyst surfaces. The most comprehensive study of radical formation on titania surfaces to date is that by Shkrob et al. [37], who irradiated aqueous suspensions of titania, hematite and goethite in the presence of very many different organic substrates. The primary purpose of this work was to explore possibilities for light-induced chemistry on extraterrestrial planetary surfaces. However, many of the conclusions reached are relevant to photocatalytic reactions of organic compounds studied in terrestrial laboratories.

For example, Fig. 17 shows spectra obtained following irradiation at 77 K of aqueous dispersions of nanoparticulate anatase in the presence of sodium acetate or acetic acid. The spectrum in Fig. 17a is dominated by a 4-line pattern characteristic of the methyl radical. Also present is a Ti^{3+} signal due to trapped electrons. Formation of methyl radicals from acetic acid over platinised titania catalysts had been previously reported to occur in the context of the photocatalytic Kolbe reaction [38]:

$$CH_3COOH \rightarrow CH_4 + CO_2$$

and can be envisaged to involve attack on adsorbed acetate or acetic acid by photo-produced valence band holes

$$CH_3COO^- + h^+ \rightarrow [CH_3COO]^{\cdot} \rightarrow CH_3 + CO_2$$

The removal of valence band holes in this manner then allows conduction band electrons to be trapped, i.e. the acetic acid can be regarded as a hole scavenger.

On raising the temperature, these workers found that a second radical signal appeared which was assigned to the carboxymethyl radical $^{\cdot}CH_2CO_2^-$ (Fig. 17b). The identification of this radical was based on comparison with the product of irradiation of disodium malonate on hematite nanoparticles. In this case the spectrum was dominated by the carboxymethyl radical, presumably formed by oxidative cleavage of the C–C bond.

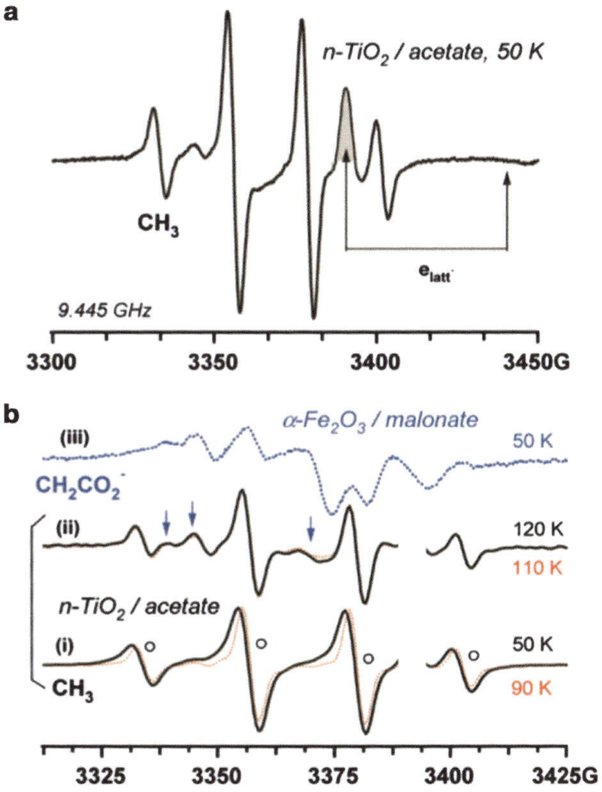

Fig. 17 EPR spectra obtained from irradiation of aqueous dispersion of anatase in sodium acetate at 77 K. (**a**) Spectrum measured at 50 K. (**b**) Spectra measured at higher temperatures. Also shown for comparison is the spectrum obtained from irradiating an aqueous dispersion of nanoparticulate hematite (Fe_2O_3) in the presence of sodium malonate. Reproduced with permission from [37]

The explanation for the changes in spectra as the temperature is raised is that when adsorbed methyl radicals become mobile, they can extract hydrogen from the parent molecule:

$$CH_3^{\cdot} + CH_3COO^- \rightarrow CH_4 + {}^{\cdot}CH_2COO^-$$

The dissociative electron transfer from the adsorbed acetate is strongly exergonic, since even when the hydrogens in the methyl group of acetate are replaced with electronegative groups such as CN or F, the corresponding $^{\cdot}CH_2CN$ and $^{\cdot}CF_3$ radicals were detected.

Similar reaction pathways were inferred for longer chain carboxylic acids and aminocarboxylic acids, where the alternative process of electron trapping by protonated amino groups resulting in deamination does not occur. Figure 18 shows for example spectra obtained by irradiation of aqueous dispersions of anatase in the presence of glycine. At low temperatures the spectrum observed shows trapped electrons (Ti^{3+}) plus a 5-line signal which was attributed to the protonated aminomethyl radical. This assignment was confirmed by simulation of the spectrum

Fig. 18 EPR spectra obtained from irradiation of an aqueous dispersion of anatase in the presence of glycine at 77 K. Reproduced with permission from [37]

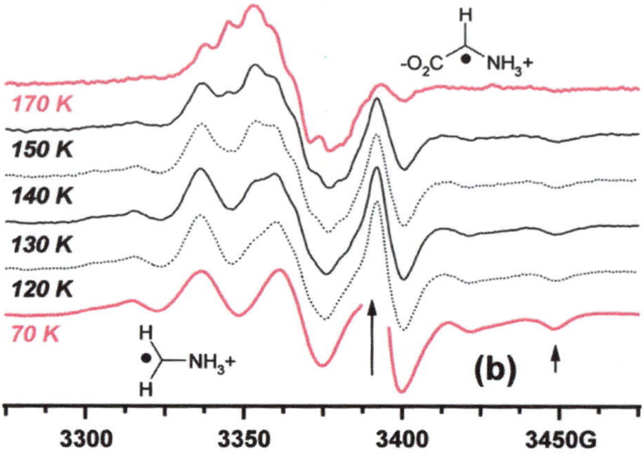

from DFT calculations. On raising the temperature, the spectrum converts to that of the protonated carboxyaminomethyl radical, presumably formed by a similar hydrogen abstraction to that seen with acetic acid.

Aromatic carboxylates showed variable behaviour. Benzoic acid, 4,4'-dibenzoic acid and naphthalene-2,3-dicarboxylic acid failed to yield free radicals when irradiated in the presence of aqueous dispersions of anatase, whereas for picolinic, phthalic, mellitic and 1,2,4,5-benzene tetracarboxylic acids the radicals expected for oxidative (single) decarboxylation were observed.

The authors suggest that the main criterion for radical formation is the stability of the initially formed trapped hole species and the energy gain on decarboxylation. They estimated these energetics using DFT calculations and showed that where the trapped hole species is more stable, fragmentation does not occur; the trapped hole species recombines with conduction band electrons, and no radical products are detected.

The use of carboxylic acids to generate free radicals when irradiated in the presence of TiO_2 has been exploited recently by Manley et al. [39] to carry out free radical coupling and addition reactions on a preparative scale. Figure 19 below shows examples of the chemistry occurring; EPR spectroscopy was used in this work to demonstrate the presence of free radical intermediates generated by valence band hole attack on the carboxylic acid precursors. Figure 20a shows for example a spectrum obtained by in situ irradiation of a dispersion of P25 in acetonitrile in the presence of *t*-butyl-carboxylic acid at 80 K. This shows the 10-line signal expected for the *t*-butyl radical, superimposed on higher field signals of trapped electrons (Ti^{3+}). The central 4 lines of the *t*-butyl radical signal were also detected during room temperature irradiation in benzene (Fig. 20b) (the higher dielectric constant of acetonitrile at room temperature precludes its use at this temperature). The trapped electron signals are not detected at room temperature because of rapid spin–lattice recombination. The *t*-butyl radical lines are much narrower at room temperature than at 80 K, which may suggest that the radical has desorbed from the titania

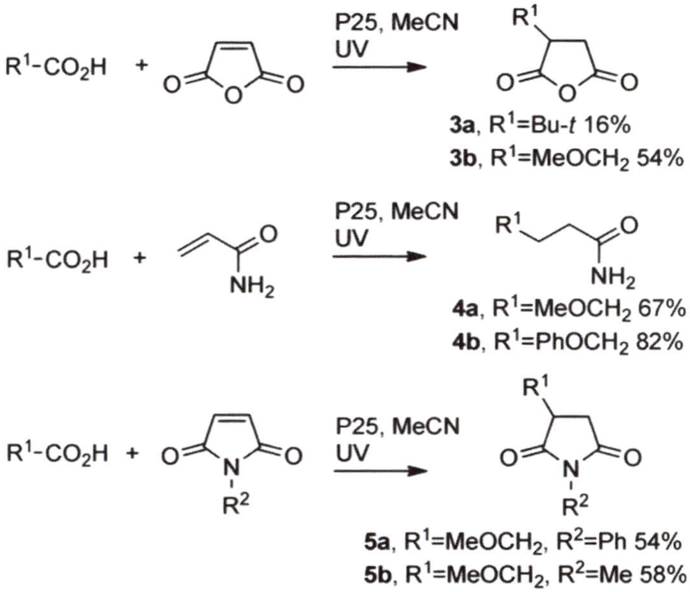

Fig. 19 Examples of free radical addition reactions initiated by carboxylic acid precursors over P25 photocatalysts. Reproduced with permission from [39]

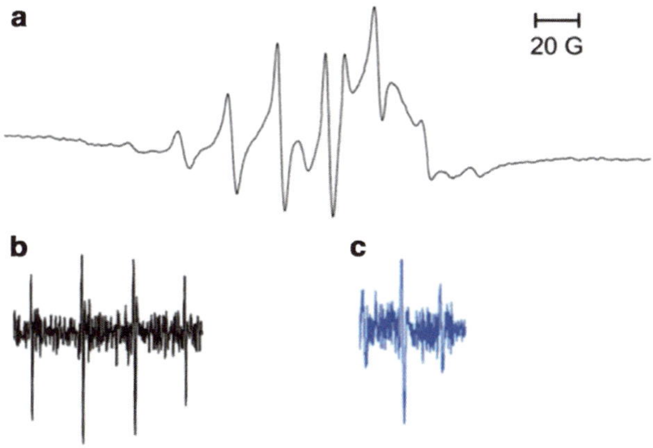

Fig. 20 EPR spectra of (**a**) P25 titania irradiated in situ at 80 K in the presence of *t*-butyl-carboxylic acid in acetonitrile; (**b**) P25 irradiated in situ at 298 K in the presence of *t*-butyl-carboxylic acid in benzene; (**c**) P25 irradiated in situ at 298 K in the presence of phenoxyacetic acid in benzene. Reproduced with permission from [39]

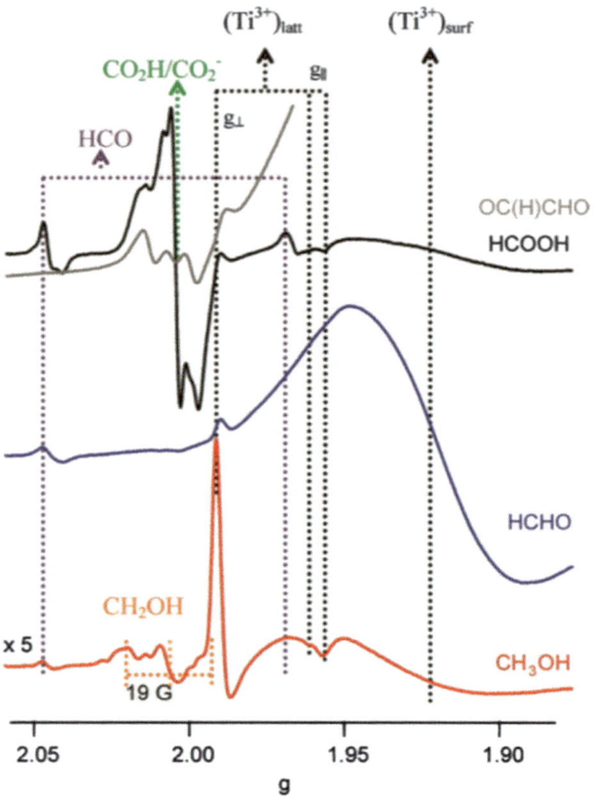

Fig. 21 EPR spectra recorded at 5 K after illumination of TiO_2 in the presence of 2 M formic acid, formaldehyde and methanol, respectively. Reproduced with permission from [40]

surface into the solvent. The narrowing could however also be caused by increased mobility of the adsorbed radical on the titania surface. When phenoxyacetic acid was used as the precursor, the room temperature spectrum showed clearly the triplet signal expected for the phenoxymethyl radical ($PhOCH_2^{\cdot}$) (Fig. 20c). At 80 K, the spectrum was dominated by a complex multiplet due to the aromatic radical cation, although traces of the triplet signal could also be detected. As seen also by Shkrob et al. [37] the situation with aromatic carboxylic acids is more complex and evidently involves in this case a balance between decarboxylation and electron transfer from the aromatic ring to the valence band of the semiconductor, which dominates at lower temperatures.

In a later study from the Argonne group, Dmitrijevic et al. studied by EPR and transient absorption spectroscopy the interfacial charge transfer to formic acid, formaldehyde and methanol on the surface of titania nanoparticles [40]. Figure 21 shows spectra recorded at 5 K following irradiation of frozen aqueous dispersions at 77 K.

In the case of formic acid, the spectra show two overlapping radical signals. The CO_2^- radical anion (probably present in these aqueous suspensions as the protonated form CO_2H) is formed by valence band hole attack on adsorbed formate:

$$HCOO^- + h^+ \rightarrow CO_2^- + H^+$$

The second signal with a large anisotropic proton coupling (120, 124 and 135 gauss components) is due to the formyl radical $HCO^{\cdot}$. This is formed by reaction of formic acid with photogenerated electrons:

$$HCOOH + e \rightarrow HCO^{\cdot} + OH^-$$

The formic acid is thus acting as both an electron scavenger and a hole scavenger, and very few trapped electrons (Ti^{3+} signals) are seen.

With formaldehyde, on the other hand, a very high concentration of trapped electrons is detected, indicating that this molecule is a very inefficient electron acceptor, while formyl radicals are formed through scavenging of valence band holes:

$$HCHO + h^+ \rightarrow HCO^{\cdot} + H^+$$

The high yield of trapped electrons also indicates that the formyl radicals cannot accept further electrons. Note that most of the Ti^{3+} is the broad higher field signal usually assigned to surface trap sites, possibly stabilised in this case by adsorbed formaldehyde.

According to these measurements, methanol is a much less efficient hole scavenger than formaldehyde. The EPR signals are 5× less intense under the same conditions, and the largest Ti^{3+} signal is that due to bulk trap states. The authors attribute this difference to the weaker binding of methanol to the titania surface. Hole trapping by methanol produces a weak signal of the hydroxymethyl radical:

$$CH_3OH + h^+ \rightarrow {}^{\cdot}CH_2OH + H^+$$

The hydroxymethyl radicals have a large negative potential on the titania surface and can inject electrons into the titania to form formaldehyde (which may trap a further hole to give formyl radicals, traces of which can be seen in the spectra).

The spectra reported in [37] show that radical yields in the case of ethanol are even lower for methanol, indicating that multiple electron transfer is more favourable for ethanol.

Work in the author's laboratory has found that titania valence band holes can also attack carbon–silicon bonds in substituted silane compounds. It is known [41] that titania will photocatalyse the addition of aromatic benzyl silanes to electron-deficient double bonds. The mechanism postulated in [41] for this reaction involves oxidative cleavage of the carbon–silicon bond by valence band holes to create the benzyl radical. EPR spectroscopic studies [42] involving in situ irradiation of reactants adsorbed on titania at 77 K showed that in this temperature all 4 of the

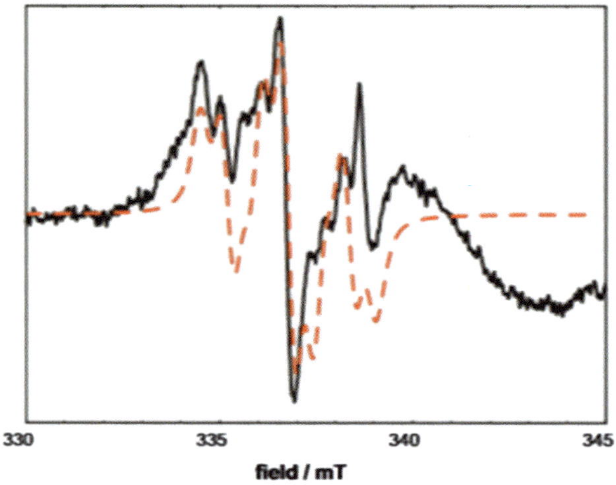

Fig. 22 EPR spectrum of the *o*-methoxybenzyl radical formed by irradiation of a dispersion of anatase in acetonitrile in the presence of *o*-methoxybenzyltrimethylsilane. Reproduced with permission from [42]

carbon–silicon bonds in *o*-methoxybenzyltrimethylsilane were cleaved, generating methyl radicals and the *o*-methoxybenzyl radical. The methyl radical signals could be removed by raising the temperature, leaving a clear signal of the benzyl radical. This is compared with a computer simulation in Fig. 22. The benzyl radical signal is superimposed on Ti^{3+} signals due to trapped electrons, confirming that the radical is generated from valence band holes.

4 Concluding Remarks

This review has aimed to show with selected examples how the spectroscopic techniques of FTIR and EPR can provide a level of detailed information about reaction pathways in photocatalysis which was hitherto unavailable. An overwhelming conclusion from these studies is that events occurring at the semiconductor surface are crucial to reaction mechanisms. In situ spectroscopic measurements will continue to play a vital role in mechanistic studies.

A second conclusion is that almost all of the spectroscopic studies to date have been focussed on titania photocatalysts. There is a clear need to extend studies of the type described here to other semiconductor systems, particularly in the context of the ongoing search for visible light-active catalysts.

References

1. Niemantsverdriet JW (2007) Spectroscopy in catalysis. 3rd edition, Wiley VCH, Berlin Elsevier. (ISBN 978-3-527-31651-9)
2. Weckuysen BM (ed) (2012) In-situ spectroscopy of catalysts. American Scientific Publishers, Valencia, California (ISBN 1-58883-026-8)
3. Haw JF (ed) (2002) In-situ spectroscopy in heterogeneous catalysis. Wiley VCH, Berlin
4. Delana C, Fois E, Coluccia S, Martra G (2010) J Phys Chem C 114:21531
5. Wu WC, Chuang CC, Lin JL (2000) J Phys Chem B 104:8719
6. Tiloca A, Selloni A (2004) J Phys Chem B 108:19314
7. Zhao J, Yang J, Petek H (2009) Phys Rev B 80:235416
8. Panayotov D, Burrows SP, Morris JR (2012) J Phys Chem C 116:6673
9. Szczepankiewicz SH, Colussi AJ, Hoffmann MR (2000) J Phys Chem B 104:9842
10. Berger T, Sterrer M, Diwald O, Knozinger E, Panayatov D, Thompson TL, Yates JT (2005) J Phys Chem B 109:6061
11. Yu Z, Chuang SSC (2007) J Catal 246:118
12. Yu Z, Chuang SSC (2007) J Phys Chem C 111:13813
13. Ramakrishnan G, Zhao S, Han W, Orlov A (2011) Chem Eng J 170:445
14. Sa J, Anderson JA (2008) Appl Catal B 77:409
15. Mendive C, Hansmann D, Bredow T, Bahnemann D (2011) J Phys Chem C 115:19676
16. Gong D, Subramaniam VP, Highfield JG, Tang Y, Lai Y, Chen Z (2011) ACS Catal 1:864
17. Dolamic I, Burgi T (2006) J Phys Chem B 110:14898
18. Dolamic I, Burgi T (2007) J Catal 248:268
19. Hu X, Burgi T (2012) Appl Catal A. doi:10.1016/j.apcata.2012.09.017
20. Friedbacker G, Bubert H (eds) (2011) Surface and thin film analysis. 2nd edition, Wiley-VCH, Berlin
21. Takahashi K, Yui H (2009) J Phys Chem C 113:20322
22. Xu M, Gao Y, Moreno EM, Kunst M, Muhler M, Wang Y, Idriss H, Woll C (2011) Phys Rev Lett 106:138302
23. Bruckner A (2010) Chem Soc Rev 39:4673
24. Weil JA, Bolton JR (2007) Electron paramagnetic resonance. Wiley, Hoboken, New Jersey
25. Drescher M, Jeschke G (eds) (2012) EPR spectroscopy: applications in chemistry and biology. Springer-Verlag, Berlin
26. Brustolon M (2009) EPR spectroscopy: a practitioners toolkit. Wiley, Hoboken, New Jersey
27. Chiesa M, Giamello E, Che M (2010) Chem Rev 110:1320
28. Howe RF, Graetzel M (1985) J Phys Chem 89:4495
29. Howe RF, Graetzel M (1987) J Phys Chem 91:3906
30. Chiesa M, Paganini MC, Giamello E (2011) J Phys Chem 115:25413
31. Ke SC, Wang TC, Wong MS, Gopal NO (2006) J Phys Chem B 110:11628
32. Macdonald IR, Howe RF, Zhang X, Zhou W (2010) J Photochem Photobiol A 216:238
33. Livraghi S, Maurelli S, Paganini MC, Chiesa M, Giamello E (2011) Angew Chem Int Ed 50:8038
34. Hurum DC, Agrios AG, Gray KA, Rajh T, Thurnauer MC (2003) J Phys Chem B 107:4545
35. Hurum DC, Agrios AG, Crist SE, Gray KA, Rajh T, Thurnauer MC (2006) J Electron Spectors 150:155
36. Komaguchi K, Nakano H, Araki A, Harima Y (2006) Chem Phys Lett 428:338
37. Shkrob I, Cheremisov S (2009) J Phys Chem C 113:17138
38. Kraeutler B, Bard AJ (1978) J Am Chem Soc 100:5985
39. Manley DW, McBurney RT, Miller P, Howe RF, Rhydderch S, Walton JC (2012) J Am Chem Soc 134:13580
40. Dimitrijevic N, Shkrob I, Gosztola DJ, Rajh T (2012) J Phys Chem C 116:878
41. Cermenati L, Richter C, Albini A (1998) Chem Commun 805
42. Macdonald IR, Rhydderch S, Holt E, Grant N, Storey JMD, Howe RF (2012) Catal Today 182:39

Fundamentals and Applications of the Photo-Fenton Process to Water Treatment

Fernando S. García Einschlag, André M. Braun, and Esther Oliveros

Abstract Among Advanced Oxidation Processes (AOPs), the Fenton process and the photochemically enhanced or assisted Fenton process, commonly called photo-Fenton, are considered to be among the most efficient for the oxidative degradation of a large variety of organic contaminants in aqueous systems. These processes, based on the generation of highly oxidizing species (hydroxyl radicals and possibly others) from hydrogen peroxide and Fe ions, may be counted among the few methods that are actually applied on a technical scale for an abiotic (pre-)treatment of wastewaters. With close to 5,000 articles published on this topic during the last decade, covering both fundamental aspects and applications, this chapter is restricted to a selective overview of the photo-Fenton process applied to water treatment. It briefly recalls the fundamentals of the Fenton reaction, describes the main lines of research for process enhancement and economic feasibility, summarizes the essentials determining the primary process parameters, and discusses the present state of technical development and its priorities.

Keywords Degradation of organic pollutants, Fenton and photo-Fenton processes, Photochemical water treatment, Technical development

---

F.S. García Einschlag
Instituto de Investigaciones Fisicoquímicas Teóricas y Aplicadas (INIFTA), Departamento de Química, Facultad de Ciencias Exactas, Universidad Nacional de La Plata, CCT La Plata-CONICET, Casilla de Correo 16, Sucursal 4, 1900 La Plata, Argentina

A.M. Braun
Engler-Bunte-Institut, Karlsruher Institut für Technologie (KIT), 76131 Karlsruhe, Germany

E. Oliveros (✉)
Laboratoire des Interactions Moléculaires et Réactivité Chimique et Photochimique (IMRCP), UMR 5623-CNRS/UPS, Université Toulouse III (Paul Sabatier), 118, route de Narbonne, 31062 Toulouse Cédex 9, France
e-mail: oliveros@chimie.ups-tlse.fr

D.W. Bahnemann and P.K.J. Robertson (eds.), *Environmental Photochemistry Part III*, Hdb Env Chem (2015) 35: 301–342, DOI 10.1007/698_2013_247,

Contents

1 Introduction

Early signs of the deterioration of the quality of surface waters due to the rapid industrialization and urbanization after World War II appeared already in the 1970s.

Increasing contamination of the aquatic environment as a consequence of human activities (agricultural, industrial, household) has led to more and more stringent regulations concerning water resources worldwide (e.g., European Water Framework Directive: http://ec.europa.eu/environment/water; United States Environmental Protection Agency: http://water.epa.gov). Following progress in analytical techniques, legislative frameworks have been implemented to limit the introduction of xenobiotic substances (such as pesticides, fungicides, herbicides, detergents, pharmaceuticals) into water bodies and diminish continuous contamination and environmental risks. Nevertheless, lack of adequate water treatment facilities and the absence of safe drinking water supply still remain tremendous problems in large parts of the planet. In countries where the infrastructures are well developed, sewage and wastewater treatment plants (WWTP) rely mainly on activated sludge biological treatment [74], a low-cost technology largely applied in urban areas for purifying and recycling wastewaters into the environment and for preparing drinking water from natural water resources. However, biological treatment cannot cope with nonbiodegradable or biocidal substances that remain most often difficult and expensive to eliminate from wastewaters by complementary physical (adsorption on activated carbon, filtration) and chemical (coagulation–flocculation–sedimentation, precipitation, chlorination, ozonization) conventional treatments. Elimination of nonbiodegradable and toxic organic pollutants (agrochemical and pharmaceutical residues in particular) still represents one of the most important challenges in water treatment.

During the last decades, alternative abiotic oxidation technologies, classified as Advanced Oxidation Processes (AOPs), have been extensively investigated in a search for simple, low-cost, on-site technologies: (i) as a pretreatment to condition pollutants by partial oxidative degradation for a faster abatement in biological treatment stations, (ii) as post-treatment to eliminate biocidal residues at the exit of biological treatment processes, (iii) as a method to recycle water for industrial use, (iv) as a means to decrease the dissolved organic carbon (DOC) or chemical oxygen demand (COD) content before release into the environment, and (v) as the method of choice to detoxify landfill leachates. Most AOPs rely on the catalytic, electrochemical, and/or photochemical generation of *hydroxyl radicals* ($HO^•$), strong oxidizing species known to initiate the oxidative degradation of a variety of organic contaminants in aqueous solutions (e.g., [71, 101, 167, 182]). Indeed, the hydroxyl radical has a very high standard electrode potential ($E_0 = 2.73$ V vs. NHE) [31, 184], an argument frequently used to explain its high reactivity as an oxidant. In fact, $HO^•$ reacts with high rate constants (10^6–10^9 L mol^{-1} s^{-1}) with a large variety of organic compounds [31, 66]: reaction occurs by hydrogen abstraction (Reaction 1), electrophilic addition to π-systems (e.g., Reaction 2), or electron transfer (Reaction 3) [101, 175]. Depending on the organic substrate, electron transfer reactions to produce organic radical cations and hydroxide could be in line with the electrochemical argument, and in such cases, oxidation by hydrogen transfer (Reaction 1) could be explained by a succession of electron and proton transfers (Reactions 3 and 4). However, electrophilic additions to aromatic systems are common and the relation between the rate of such additions and the oxidation potential of $HO^•$ is still unknown.

$$RH + HO^• \rightarrow R^• + H_2O \tag{1}$$

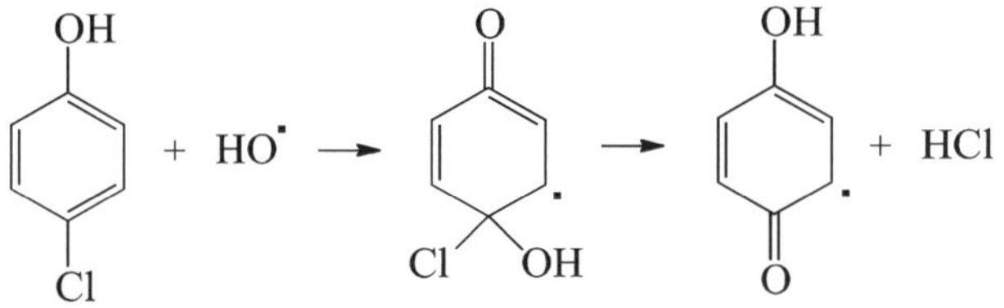

(2)

$$RH + HO^• \rightarrow HR^{•+} + HO^- \tag{3}$$

$$HR^{•+} \rightarrow R^• + H^+ \tag{4}$$

The primary reactions 1–4 initiate the oxidative degradation of most organic substrates, provided molecular oxygen (O_2) is present to trap efficiently the generated alkyl, alkenyl, or cyclohexadienyl radicals yielding peroxyl radicals (Reaction 5) that generate subsequently thermally instable (hydro)peroxides. The peroxyl radicals might therefore be considered as the key intermediates for the thermal chain reaction of oxidative degradation (Reaction 6) that decreases the amount of DOC, eventually until complete mineralization to CO_2, H_2O, and inorganic acids (if heteroatoms are present in the organic pollutant structure).

$$R^{\bullet} + O_2 \rightarrow RO_2^{\bullet} \rightarrow\rightarrow ROOH, ROOR' \quad (5)$$

$$\text{Oxidized intermediates} \overset{HO^{\bullet},}{\rightarrow} \overset{O_2}{\rightarrow} H_2O + CO_2(+\text{inorganic acids}) \quad (6)$$

Some limitations of the abiotic water treatment based on the reactivity of $HO^{\bullet}$ must be mentioned: (i) poly-halogenated alkanes are inert or show very slow reactivity towards hydrogen abstraction by $HO^{\bullet}$ and (ii) polyfluorinated and polychlorinated alkenes and benzene derivatives exhibit too low electron densities of the π-system. Another limitation concerns the formation of highly stabilized radicals or their precursors in the course of the sequence of reactions leading to the mineralization of the organic substrate. Examples to be mentioned are phenoxyl radicals (e.g., [150]) and cyanuric acid formed by oxidation of triazine herbicides (e.g., [38, 63, 134]).

A large variety of AOPs based on the production of $HO^{\bullet}$ have been developed for water treatment. Among these AOPs, photochemical methods represent an important group of water treatment technologies [56]. Early reviews appeared in 1993 [101, 128]. Photochemical AOPs may be classified into a few main types: (i) processes involving the photolysis of oxidants, such as hydrogen peroxide and ozone, using ultraviolet (UV) radiation (UV/H_2O_2 and UV/O_3 processes, e.g., [59, 101]); (ii) heterogeneous photocatalysis based on the irradiation of semiconductors, predominantly titanium dioxide (TiO_2) (e.g., [60, 71, 76, 128]); (iii) vacuum-UV photolysis of water using high-energy UV radiation (e.g., [64]); and (iv) the *photochemically enhanced* or *assisted Fenton process*, commonly called *photo-Fenton* (e.g., [14, 59, 60, 110, 140, 154, 170]).

Among AOPs, the Fenton and photo-Fenton processes are considered to be the most efficient for the remediation of wastewaters from chemical, pharmaceutical, and dye industries [140]. More than a century ago, Fenton [49] published the first report on the oxidation of an organic compound (tartaric acid) in aqueous solutions containing hydrogen peroxide (H_2O_2) and ferrous salts. Later on, Haber and Weiss [67] came to the conclusion that the reactive oxidant generated by the Fenton's reagent was the hydroxyl radical (Reaction 7, coordinated water molecules were excluded for simplification), and Barb et al. proposed an oxidation mechanism involving a sequence of radical reactions initiated by this species [8–10].

$$Fe^{2+} + H_2O_2 \rightarrow Fe^{3+} + HO^- + HO^{\bullet} \quad (7)$$

More than 2 decades later, the extensive work by Walling and coworkers provided further evidence of the involvement of $HO^{\bullet}$ in the Fenton oxidation of various organic compounds [177–179]. Although experimental evidence of the formation of higher valent iron at +IV (ferryl) or +V (Fe(V)=o species) oxidation states has been reported in some cases [25, 139], the radical mechanism has been largely accepted for most reactions in acidic systems [140].

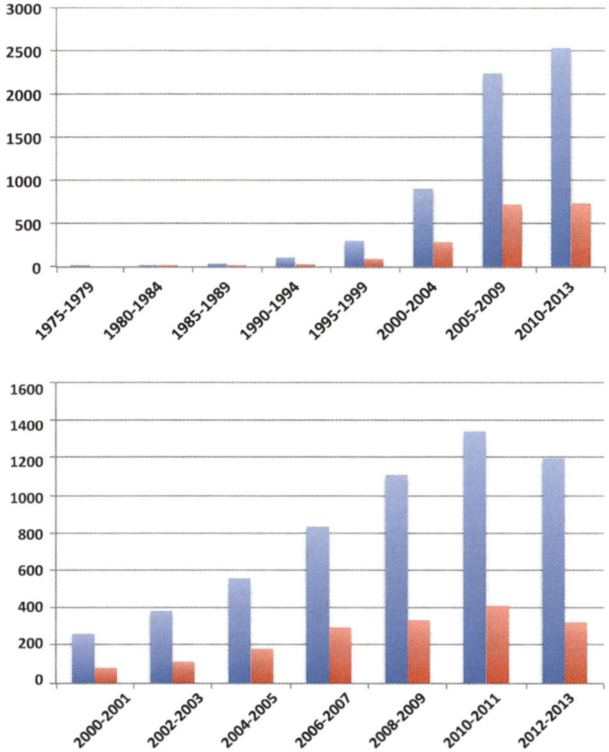

Fig. 1 Results of a literature search (SciFinder 09/09/2013) using the combined keywords "Fenton and water treatment" (in *blue*) and "photo-Fenton and water treatment" (in *red*) since the year 1975 (*upper graph*) and since 2000 (*lower graph*)

A few publications already reported in the 1960s potential applications of the Fenton reaction to wastewater treatment [21, 43]. However, research at the laboratory scale and development work on this topic began in the early 1990s and increased spectacularly during the last decade. A literature search using the combined keywords "Fenton and water treatment" resulted in over 6,000 references, among them 4,770 published between 2005 and 2013, and more than 2,500 from 2010 up to now (Fig. 1). Since the year 2000, more than 300 reviews and 700 patents dealing, at least in part, with different aspects of the Fenton and the photo-Fenton processes have been released. *Several factors may contribute to explain this enthusiastic interest.*

The Fenton and related reactions make use of environmentally benign, convenient, and rather economic reagents. Indeed, in comparison with other oxidants, hydrogen peroxide is relatively cheap and easy to handle; it is miscible with water in all proportions (in contrast to ozone) and decomposes to water and oxygen at ambient temperature and therefore does not represent any threat for the environment. Iron is one of the most abundant nontoxic elements in the earth crust where it is present as Fe(II) and Fe(III). It is found in the aquatic environment such as groundwaters, lakes, and oceans, as well as in the atmosphere in aerosols, fogs, rain drops, and urban clouds [190, 191]. Therefore, the Fenton process seemed to be most promising, as only H_2O_2 and nonhazardous Fe(II) salts are needed to generate the reactive species ($HO^•$) in a well-stirred reactor, and, as an additional advantage, the process operates at ambient temperature and atmospheric pressure. Moreover, albeit with limited efficiency, the Fe(III) species produced in Reaction 7 are recycled to Fe(II) by reaction with H_2O_2 leading to a catalytic cycle for $HO^•$ production, thus decreasing the amount of Fe(II) salt to be used (e.g., [52], Sect. 2.1). In the middle of the 1990s, it was clearly recognized that a large variety of organic substances (acids, alcohols, aldehydes, ketones, ethers, amines, aromatics, dyes,...) could be efficiently oxidized by the Fenton's reagent and potential applications to the treatment of industrial wastewaters of various origins were reported (e.g., [20]).

It was also established in the 1990s that the Fenton process could be accelerated and higher degradation efficiencies achieved, *when the reaction system was irradiated by ultraviolet (UV) and/or visible light (photo-Fenton process)* (e.g., [7, 14, 25, 75, 153, 154, 166, 185]). Irradiation of the reaction system countermands some of the problems of the technical development of a Fenton process: (i) the relatively slow rate of overall oxidative degradation, i.e., of the DOC decrease in the treated water, due to the slow thermal recycling of Fe(II) and (ii) the partial degree of mineralization (40–60%), as Fe(III) is complexed by intermediates of the oxidative degradation of the target substrate (mostly carboxylic acids). Indeed, the photochemical reduction of Fe(III), irrespective of the complexing ligands, leads to an enhanced recycling of Fe(II) (Sect. 3.1) that can further react with H_2O_2 to produce $HO^•$ (Reaction 7). Therefore, the photo-Fenton process is generally more effective for the abiotic treatment of wastewater than the corresponding thermal process. As a consequence, investigations using both artificial radiation sources and *solar light* were undertaken, and the number of publications concerning the photo-Fenton process followed the same growing trend as the literature on the Fenton process (Fig. 1). However, the former process being more complex to develop due to the necessary implementation of adequate light sources and/or solar collectors, the related literature represents less than one-third of that on the Fenton process (Fig. 1). Nevertheless, a literature search using the combined keywords "Photo-Fenton and water treatment" resulted in about 1,750 publications since the year 2000, among them 370 concerning the solar process. Specific reviews on the photo-Fenton process are scarce [14, 154, 165]. The most comprehensive review on the Fenton process (including photo-Fenton and electro-Fenton) was published in 2006 by Pignatello et al. [140]. Other reviews most often present the photo-Fenton process among other AOPs (e.g., [59, 60, 84, 182]) or other photochemical AOPs (e.g., [104]) or focus on some specific aspects, such as types of contaminants or wastewater of particular origin (e.g., [55]). The use of solar irradiation has opened a high application potential for large-scale solar photochemical installations and specific reviews on this topic have been published [108–110].

Although the photo-Fenton process has proven its potential for the treatment of wastewater, only few reports on developments at an industrial or large pilot scale may be found in the literature (e.g., [11, 126, 149]). Besides the specific problems related to the design of photochemical reactors, some inherent drawbacks of the Fenton process still persist: pH range limited to mildly acidic conditions; production of sludge due to precipitation of iron hydroxides/oxides after neutralization of the reaction system and problem of its disposal or its recycling; and decreased efficiency and additional costs of immobilized/supported iron catalysts that might be used at neutral pH and recycled.

This chapter is restricted to a selective overview of the photo-Fenton process applied to water treatment. It briefly recalls the fundamentals of the Fenton reaction, describes the main lines of research for process enhancement and economic feasibility, summarizes the essentials determining the primary process parameters, and discusses the present state of technical development and its priorities.

2 Fenton and Fenton-Like Processes

2.1 Catalysis and Fe(II) Recycling

The Fenton reaction efficiently produces $HO^{\bullet}$ from H_2O_2 and a Fe(II) salt in a pH range from 2.5 to 3.5 (Reaction 7), $[Fe(H_2O)_6]^{2+}$ being the dominant aqua complexes under these conditions (Reactions 7′) [140, 168, 178]. The bimolecular rate constant of Reaction 7 lies between 40 and 80 $M^{-1}\,s^{-1}$ in acidic solution.

$$\left[Fe(H_2O)_6\right]^{2+} + H_2O_2 \rightarrow \left[Fe(H_2O)_6\right]^{3+} + HO^- + HO^{\bullet} \qquad (7')$$

The limitation of the pH range is primarily related to the speciation of Fe(III): the hexaaquo complex $[Fe(H_2O)_6]^{3+}$ exists in strongly acidic conditions in the absence of complexing counterions and undergoes hydrolysis as the pH increases (to $[Fe(H_2O)_5(OH)]^{2+}$, $[Fe(H_2O)_4(OH)_4]^{+}$, binuclear species, …) and poorly reactive ferric oxyhydroxides ($Fe_2O_3.nH_2O$) start precipitating at relatively low pH [169]. Concomitantly to the production of $HO^{\bullet}$, Reaction 7 leads to the oxidation of Fe(II) to Fe(III). An important feature of the Fenton reagent is that Fe(III) is reduced to Fe (II) by reaction with H_2O_2 (Reaction 8, with simplified notation for the aqua complexes), leading to a catalytic cycle for Fe(II) and $HO^{\bullet}$ production as long as H_2O_2 is present (e.g., [52]).

$$Fe^{3+} + H_2O_2 \rightarrow Fe^{2+} + HO_2^{\bullet} + H^+ \quad (8)$$

$$Fe^{3+} + H_2O_2 \leftrightarrows [Fe - OOH]^{2+} + H^+ \quad (8a)$$

$$[Fe - OOH]^{2+} \rightarrow Fe^{2+} + HO_2^{\bullet} \quad (8b)$$

Therefore, the Fenton reaction is a Fe(II)-mediated reductive homolysis of H_2O_2 [140] and is often called Fe(II)-catalyzed decomposition of H_2O_2. Recycling of Fe (II) through Reaction 8 has two noteworthy consequences: (i) Fe(III) may be added instead of Fe(II) for initiating the Fenton process (most often called "Fenton-like" process) and (ii) iron may be used at concentrations below the 1:1 stoichiometric $[H_2O_2]$:[Fe(II)] ratio required by Reaction 7 itself.

The use of iron salts at relatively low amounts is of advantage for technical applications, as it reduces ferric oxyhydroxides' production upon pH increase (neutralization) and therefore sludge formation at the end of the process line. The use of ferric salts is economically advantageous and avoids handling problems with air sensitive Fe(II). Experiments with sludge containing Fe(III) yielded qualitatively positive results [96], but there is no indication on how such a result could be transferred to a large-scale process. However, it is important to note that Fe(III) ions react much slower than Fe(II) ions with H_2O_2: the apparent rate constant for the overall Reaction 8, i.e., (8a) + (8b), varies between 10^{-2} and 10^{-3} M^{-1} s^{-1} [110, 120, 179], values several orders of magnitude smaller than those of Reaction 7. If complexation of Fe (III) by H_2O_2 (equilibrium 8a) is inhibited, the Fenton process is slowed down and may completely stop. Hence, the regeneration of Fe(II) is the rate-limiting step in the catalytic iron cycle of most Fenton or Fenton-like systems, and the overall process efficiencies are strongly dependent on reduction pathways of Fe(III) species other than the reaction between Fe(III) and H_2O_2 (Sect. 2.2). Nevertheless, two other reactions with higher rate constants contribute to the reduction of Fe(III): its reactions with the hydroperoxyl radical ($HO_2^{\bullet}$, Reaction 9) and with its conjugated base, the superoxide anion ($O_2^{\bullet-}$, Reaction 10), with reported rate constants of approx. 10^3 M^{-1} s^{-1} and 5×10^7 M^{-1} s^{-1}, respectively [86, 100, 110, 140].

$$Fe^{3+} + HO_2^{\bullet} \rightarrow Fe^{2+} + O_2 + H^+ \quad (9)$$

$$Fe^{3+} + O_2^{\bullet-} \rightarrow Fe^{2+} + O_2 \quad (10)$$

Yet, for reaction conditions typically used for Fenton and Fenton-like processes (i. e., pH $\leq$3.5 and $[H_2O_2]$ from 10^{-1} to 10^{-2} M), and in the absence of organic compounds or intermediates capable of reducing Fe(III), the overall Reaction 8 mainly governs the reduction of Fe(III) [120, 138]. It should be noted that superoxide ($O_2^{\bullet-}$)-mediated Fe(III) reduction (Reaction 10) may become important in natural environments, where higher pH values and much lower H_2O_2 concentrations are frequent.

Hydroxyl radicals produced by the Fenton reaction oxidize organic pollutants, but are also scavenged by both components of the Fenton's reagent (Reactions 11 and 12). The rate constants of these reactions are high [31] and the hydroperoxyl radical (Reaction 12) is poorly reactive toward most organic substrates compared to $HO^{\bullet}$ [18]. Therefore optimal concentrations of iron and H_2O_2 depend largely on the reactivity of the organic pollutant with $HO^{\bullet}$ and on its concentration.

$$Fe^{2+} + HO^{\bullet} \rightarrow Fe^{3+} + HO^{-} \tag{11}$$

$$H_2O_2 + HO^{\bullet} \rightarrow H_2O + HO_2^{\bullet} \tag{12}$$

One of the principal obstacles for a technical application of a Fenton or a Fenton-like process at its optimal efficiency remains the slow rate of Fe(III) reduction and, hence, of Fe(II) recycling, even under conditions of optimized concentrations of H_2O_2 [125]. It should be noted that organic intermediates of the oxidative degradation as well as some inorganic ions may affect Fe(II) recycling considerably in a positive or in a negative way (Sects. 2.2 and 3.1, respectively).

2.2 *Organic Intermediates, Fe(II)/(III) Redox Reactions, and Fe(III) Complexation*

Given the ease of changing the oxidation state between +2 and +3 [34], the nature of the ligands of iron plays a central role within the manifold of reactions involved in the production of $HO^{\bullet}$ in Fenton and Fenton-like processes. Ligands that coordinate with oxygen atoms tend to stabilize the +3-oxidation state and decrease the reduction potential of the couple Fe(III)/Fe(II), whereas ligands that coordinate with nitrogen or sulfur atoms tend to stabilize the +2-oxidation state and increase the reduction potential of the couple Fe(III)/Fe(II). Thus, the pronounced dependence of redox potentials on complex formation determines the reactivity of the catalyst material [114].

As mentioned in Sect. 2.1, the rate of regeneration of Fe(II) through Reaction 8 being very slow, the overall process efficiency is in most cases strongly dependent on the reduction of Fe(III) by other pathways. In fact, the recycling of the catalyst is sensitive to changes in the coordination sphere of Fe(III) and to the ambient redox potential of the reaction mixture [190]. These changes are primarily due to reaction intermediates formed in the course of the oxidative degradation of organic substrates. Intermediates may impact Fe speciation, not only through redox processes (due to the formation of oxidizing/reducing species), but also through complexation reactions (e.g., by the formation of strong complexes with Fe cations). Therefore, either enhancement or inhibition of the catalytic process may be observed depending on the chemical structures of organic pollutants and the in situ formed intermediates.

Short-lived intermediates are mainly C-centered radicals, resulting from hydrogen abstraction by $HO^{\bullet}$ from C–H bonds and $HO^{\bullet}$ addition to C=C bonds or aromatic systems (Reactions 1 and 4) [31, 175]. Although less common, radical cations may be formed by electron transfer (Reaction 3). Further reactions of these species with O_2 and/or water molecules lead to the formation of alcohols (hydrolysis of radical cations), hydroxylated aromatic compounds, and carbonyl and carboxyl derivatives.

C-centered radicals and semiquinone-like intermediates are known to be strong reductants that may participate in the recycling of Fe(II) (Reaction 13), in particular under conditions of inefficient O_2 trapping. In addition, most aromatic compounds are hydroxylated during their initial oxidation steps, yielding hydroquinone-like intermediates (*ortho*- or *para*-substituted dihydroxybenzenes) that are able to reduce Fe(III) species (Reaction 14).

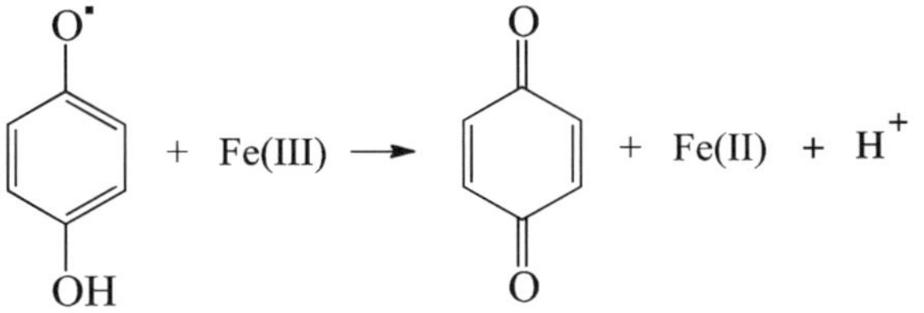

(13)

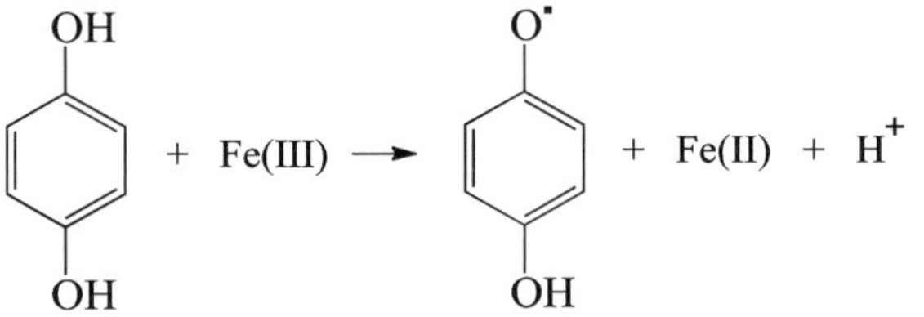

(14)

Moreover, even in the presence of O_2, benzoquinone- and semiquinone-like intermediates are readily reduced by $HO_2^{\bullet}$, as well as by C-centered radicals, hence enhancing the catalytic cycle [110]. Reaction manifolds involving Fe(III)-reducing intermediates usually lead to an autocatalytic behavior (Fig. 2) during the oxidative degradation of aromatic substrates in Fenton-like systems [42, 110, 120, 121].

Polyhydroxylated aromatics (e.g., *ortho*-hydroquinone-like intermediates) and carbonyl and carboxyl compounds might also be strong complexing agents. Multiple-step oxidation of aromatic substrates leads to the production of carboxylic and di-carboxylic acids. *Ortho*-quinone-like intermediates react to yield *cis*-muconic acid derivatives, whereas *para*-quinone-like intermediates are oxidized to succinic and acetic acid analogs [82]. These and other carboxylic compounds form strong complexes with Fe(III) that are stable in the dark and substantially decrease the availability of Fe(III) for thermal reduction [45, 110, 158, 172, 190]. Benzoic acid derivatives with a hydroxyl group in *ortho* position with respect to the carboxyl group form stable complexes with Fe(III), both functions participating in the complexation [121]. In fact, complexation of Fe(III) by intermediates of the oxidative degradation of organic compounds may prevent its complexation with H_2O_2 and therefore the reduction of Fe(III) (Reactions 8a and 8b). Consequently, the overall process is inhibited because of a strong deceleration of Fe(II) recycling due to the sequestration of Fe(III) in more or less inert Fe(III) complexes [189, 190]. Formation of stable complexes between Fe(III) and oxidation by-products, especially carboxylic acids of

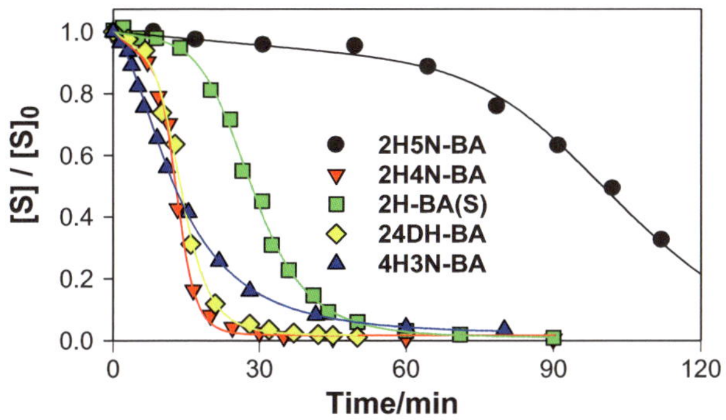

Fig. 2 Normalized concentration profiles of benzoic acid derivatives (S) during the Fenton-like process; reaction conditions: $[S]_0 = 1.0$ mM, $[H_2O_2]_0 = 5.2$ mM, $[Fe(III)]_0 = 0.1$ mM, pH 3.0, 25°C, Hg medium-pressure lamp (HPK 125) with Pyrex sleeve (adapted from [121]); salicylic (2H-BA), 2,4-dihydroxy-benzoic (24DH-BA), 2-hydroxy-5-nitrobenzoic (2H5N-BA), 4-hydroxy-3-nitrobenzoic (4H3N-BA), 2-hydroxy-4-nitrobenzoic (2H4N-BA) acids

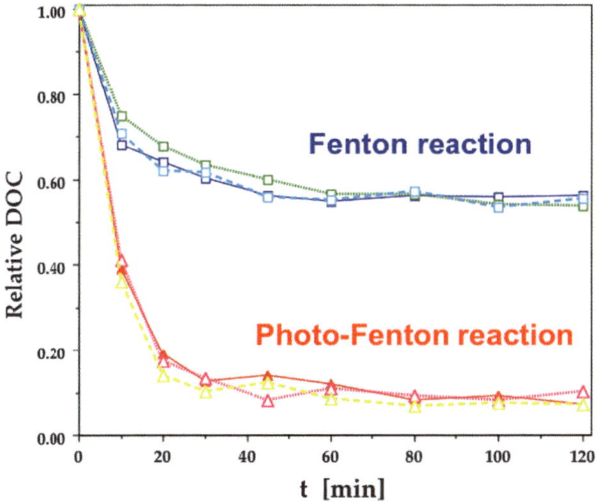

Fig. 3 Mineralization of polyvinylalcohol (PVA) of different molecular weights (15,000, 49,000, 100,000) by Fenton and photo-Fenton processes. Relative DOC values vs. time of reaction. [PVA]: 8.33×10^{-3} M of (C_2H_4O) units or 200 mg C L^{-1}, $[H_2O_2] = 3.33 \times 10^{-2}$ M, $[Fe^{2+}] = 4.16 \times 10^{-4}$ M, reaction system purged with compressed air, Hg medium-pressure lamp (TQ 150) in Pyrex sleeve, reaction system: 250 mL in a semi-batch equipment (adapted from [24])

low molecular weight (e.g., ferrioxalate $Fe(C_2O_4)_3^{3-}$), is one of the primary reasons why substrate oxidation arrives at a standstill and only partial mineralization (40–60%) of most organic pollutants is achieved by Fenton and Fenton-like processes in the dark (e.g., [24], Fig. 3).

2.3 High-Valent Oxoiron Intermediates

Besides the classical mechanism involving the formation of $HO^•$ (Reaction 7), other species have been proposed as reactive intermediates in the Fenton reaction (e.g., [25, 90, 139]). Although the corresponding mechanistic schemes have been often controversial, evidence in favor of the involvement of a high-valent oxoiron moiety with the iron in the +IV or +V oxidation state has been presented. Such a "ferryl" species (denoted as $Fe = O^{2+}$) may be formed by the reaction of Fe(II) and Fe(III) chelated by polycarboxylate or pyridyl-type ligands with H_2O_2 or organic peroxides. A Fe(IV) ferryl complex has also been formed in water by the reaction of Fe^{2+} with ozone [85] but seems to be a much weaker oxidant than $HO^•$ [112]. Ferryl-generating conditions may also lead to a different distribution of intermediate products of oxidation. For example, hydroxylated anilines were formed by reaction of $HO^•$ produced by H_2O_2 photolysis with 2,4-dimethylaniline, whereas 2,4-dimethyl phenol was the most important intermediate under Fenton conditions. A Fe(IV) oxoiron moiety was proposed to result from an inner-sphere two-electron transfer reaction within a hydrated Fe(II)–H_2O_2 complex [25]. The overall rate of substrate oxidation or oxidative degradation might therefore vary depending on the manifold of primary reactions. It was shown that the apparent rate constant of consumption of 2,4-dimethylaniline largely depends on the oxidation potential of the Fe (II) complex used and varies in the opposite manner for the pathway involving addition of $HO^•$ and for the electron transfer pathway (Fe(IV) intermediate) [26].

3 Photochemically Enhanced Fenton Processes

3.1 Photochemistry of Fe(III) Complexes: Fe(II) Recycling

Transition metal complexes usually have several accessible electronically excited states under irradiation with UV and/or visible light. These excited states may react through various pathways, such as electron transfer (due to the population of diverse charge transfer states [CT]), dissociation, substitution, rearrangement (from excited ligand field states [LF]), and ligand-centered reactions (due to the population of intra-ligand excited states [IL]) [73]. In addition, relaxation processes to the ground state, including radiationless and radiative (fluorescence, phosphorescence) deactivation, may compete strongly with the pathways of chemical reaction. Consequently, a wide range of efficiencies and quantum yields (Φ) can be observed [73, 160].

Fe(III) cations are mostly present as hexa-coordinated, high-spin, labile complexes which absorb in the UV–visible spectral region and readily undergo photochemical reduction of Fe(III) to Fe(II) [160]. In their electronic absorption spectra, the spin-forbidden LF bands are generally very weak and hidden in the tails of intense spin-allowed ligand to metal charge transfer (LMCT) and/or IL bands. The lifetimes of Fe(III) complexes' excited states are usually very short, and most of them do not show luminescence [159, 160]. The decay to the ground state occurs mainly through non-radiative physical deactivation or involves a primary step of electron transfer. This photoinduced electron transfer (PET) may be an inner- or an outer-sphere process [34]. Depending on the ligands coordinated to Fe(III), an inner-sphere electron transfer may generate $HO^{\bullet}$, $HO_2^{\bullet}$, and/or organic C-centered radicals (e.g., Reactions 15–18, with one negatively charged ligand L, HO^-, or HOO^-). Uncharged ligands (such as H_2O) will lose a positively charged entity (e.g., proton) subsequent to the electron transfer (Reaction 19).

$$Fe(OH)(H_2O)_5{}^{2+} + H_2O + hv \rightarrow Fe(H_2O)_6{}^{2+} + HO^{\bullet} \quad (15)$$

$$Fe(OOH)(H_2O)_5{}^{2+} + H_2O + hv \rightarrow Fe(H_2O)_6{}^{2+} + HO_2{}^{\bullet} \quad (16)$$

$$Fe(L)(H_2O)_5{}^{2+} + H_2O + hv \rightarrow Fe(H_2O)_6{}^{2+} + L^{\bullet} \quad (17)$$

$$Fe(L)_n(H_2O)_{6-n}{}^{(3+)-(n-)} + hv \rightarrow Fe(L)_{n-1}(H_2O)_{6-(n+1)}{}^{(2+)-((n-1)-)} + L^{\bullet} \quad (18)$$

$$Fe(H_2O)_6{}^{3+} + H_2O + hv \rightarrow Fe(H_2O)_6{}^{2+} + HO^{\bullet} + H^+ \quad (19)$$

The photolysis of Fe(III)-coordination compounds may therefore contribute efficiently to the Fe(II) recycling (Reactions 15–19) and, consequently, to the production of $HO^{\bullet}$ by Reaction 7 and the degradation of organic pollutants present in the reaction system.

The quantum yields of Fe(II) production by photolysis of Fe(III) complexes ($\Phi_{Fe(II)}$, Sect. 4.3) are usually wavelength dependent [160] and typically decrease with increasing wavelength of the incident radiation. In fact, the probability of the separation of the products resulting from the electron transfer reaction increases with the difference between the vibrational-rotational energy of a given LMCT state populated and the threshold energy of the primary photoproducts formation [159]. An interaction between IL- and photoredox-reactive LMCT states can also affect the wavelength dependence of $\Phi_{Fe(II)}$ and, in general, increase the yield of the Fe(III) reduction.

3.2 Quantum Yields and Quantum Efficiencies

The quantum yield (Φ_λ) represents a measure of the efficiency of a photochemical reaction occurring under monochromatic irradiation at a given wavelength λ. It is defined as the ratio of the rate of consumption of a reactant (or formation of a product) and the photon flux absorbed by the reactive species (Eq. 20, [28, 29]).

$$\Phi_{M,\lambda} = \frac{d[M]/dt}{P_{p,a,\lambda}} \quad (20)$$

where $d[M]/dt$ is the rate of transformation of M (mol L^{-1} s^{-1}) and $P_{p,a,\lambda}$ is the photon flux absorbed by M at the wavelength of irradiation λ (einstein L^{-1} s^{-1}); $P_{p,a,\lambda}$ is related to the incident photon flux ($P_{p,0,\lambda}$) by the Lambert–Beer law:

$$P_{\mathrm{p,a},\lambda} = P_{\mathrm{p,0},\lambda}\left(1 - 10^{-A_\lambda}\right) = P_{\mathrm{p,0},\lambda}\alpha_\lambda \tag{21}$$

with A_λ: absorbance of M at the wavelength λ ($A_\lambda = \varepsilon_\lambda l[M]$, where ε_λ: molar absorption coefficient of M (L mol^{-1} cm^{-1}) and l: optical path length (cm)) and α_λ: absorption factor.

The overall efficiency of the photochemical Fe(III) reduction is one of the key parameters of the photo-Fenton process. The quantum yields of the photochemical Fe(II) production ($\Phi_{\mathrm{Fe(II)}}$) are known to be wavelength dependent. Besides, the photo-Fenton process is generally carried out under polychromatic irradiation, where solar light, medium- or high-pressure mercury arcs, medium-pressure sodium arcs, black light, or fluorescent tubes are used [79, 84, 140].

Assuming a single photochemically reactive Fe(III) substrate absorbing the incident radiation, a polychromatic quantum efficiency for Fe(II) production ($\eta_{\mathrm{Fe(II)}}$) within the wavelength domain λ_1 to λ_2 may be experimentally evaluated using Eq. (22):

$$\eta_{\mathrm{Fe(II)}} = \frac{\mathrm{d[Fe(II)]}/\mathrm{d}t}{\sum_{\lambda_1}^{\lambda_2} P_{\mathrm{p,a},\lambda}} \tag{22}$$

where d[Fe(II)]/dt is the rate of Fe(II) formation (mol L^{-1} s^{-1}) and $\sum_\lambda P_{\mathrm{p,a},\lambda}$ is the summation of absorbed photon fluxes in small finite wavelength intervals within the range λ_1 to λ_2.

In general, lamp providers give the spectral distribution of the radiant power emitted by the lamp ($P_{e,\lambda}$, W nm^{-1}) or the distribution in small wavelength intervals. The spectral photon flux ($P_{p,\lambda}$, photon s^{-1} nm^{-1}) is related to the spectral radiant power ($P_{e,\lambda}$, W nm^{-1}) according to

$$P_{\mathrm{p},\lambda} = P_{\mathrm{e},\lambda}/E_{\mathrm{p},\lambda} \tag{23}$$

where $E_{\mathrm{p},\lambda}$ ($= hc/\lambda$) is the energy of a photon of wavelength λ (m), with h: Planck constant (J s photon^{-1}) and c: speed of light (m s^{-1}). To express the spectral photon flux in units of einstein L^{-1} s^{-1}, $P_{\mathrm{p},\lambda}$ (photon s^{-1}) should be divided by the Avogadro number (N_A) and corrected for the volume of solution irradiated (V_{irr}, L).

Therefore, the total photon flux absorbed in the range λ_1 to λ_2 ($P_{\mathrm{p,a}}$, einstein L^{-1} s^{-1}) may be calculated as

$$P_{\mathrm{p,a}} = \sum_{\lambda_1}^{\lambda_2} P_{\mathrm{p,a},\lambda} = (1/N_\mathrm{A}V_{\mathrm{irr}})\,(1/hc)\sum_{\lambda_1}^{\lambda_2} \lambda P_{\mathrm{e},\lambda}\left(1 - 10^{-A_\lambda}\right) \tag{24}$$

If only the relative spectral distribution of the radiant power emitted by the lamp ($S_{\mathrm{e},\lambda} = P_{\mathrm{e},\lambda}/P_\mathrm{e}$) is provided and the total radiant power (P_e) is unknown, the latter may be determined by chemical actinometry [29, 92]. The actinometer should have spectrophotometric characteristics related to those of the photochemical reaction investigated and similar solubility properties. In the case of photochemically enhanced Fenton- or Fenton-like processes, ferrioxalate may be used for that purpose (known quantum yield of Fe(II) production, $\Phi_{ac,\lambda}$, under irradiation in a wide wavelength range) (Eq. 25, [23, 92, 140]).

$$P_e = \frac{\left(dn_{Fe(II)}/dt\right)}{\sum_{\lambda_1}^{\lambda_2}\Phi_{ac,\lambda}\left(S_{e,\lambda}/E_{p,\lambda}\right)\left(1-10^{-A_{ac,\lambda}}\right)} \tag{25}$$

where the subscript "ac" stands for actinometer and $dn_{Fe(II)}/dt$ is the number of Fe(II) species formed during the irradiation time (may be determined by complexation of Fe(II) by 1,10-phenanthroline).

The overall rate of the photochemical Fe(II) production depends on the contribution of the various Fe(III) species present in solution. These species may exhibit different absorption properties and quantum yields, both of them being wavelength dependent. Consequently, the concentration of soluble iron, the presence and concentration of different Fe(III) coordinating compounds, and the inner filter effects of other radiation absorbing species in the system may play a crucial role for the efficiency of the photo-Fenton reaction. Moreover, the pH value of the reaction system is strongly affecting the nature of the complexes formed [110, 190].

In the presence of different photochemically reactive species contributing to the production of Fe(II), the total photon flux absorbed at a given wavelength ($P_{a,\lambda}$) is expressed as (Lambert–Beer law):

$$P_{a,\lambda} = P_{0,\lambda}\left(1-10^{-A_j}\right) = P_{0,\lambda}\left(1-10^{-l\sum_j \varepsilon_{j,\lambda}c_j}\right) \tag{26}$$

where c_j is the concentration of compound j (mol L^{-1}); see Eq. (21) for the definition of the other terms in Eq. (26).

The fraction of photons absorbed by the jth species at wavelength λ ($f_{j,\lambda}$) is given by

$$f_{j,\lambda} = \frac{\varepsilon_{j,\lambda}\ c_j}{\sum_j \varepsilon_{j,\lambda}\ c_j} \tag{27}$$

and the corresponding photon flux absorbed within the spectral domain of λ_1 to λ_2 is

$$P_{a,j} = \sum_{\lambda_1}^{\lambda_2} f_{j,\lambda} P_{a,\lambda}\ d\lambda \tag{28}$$

The incident photon flux available in a given photochemical reactor may be determined for a defined spectral domain by either actinometry or radiometric measurements. Combined with the data of spectrophotometric measurements, the absorbed photon flux of complex reaction systems ($\sum_j P_{a,j}$) might be evaluated taking into account that it will change in the course of the process of oxidative degradation of one or a mixture of organic substrates. Moreover, some of the compounds absorbing in the range between λ_1 and λ_2 might not contribute to the production of Fe(II) (inner filters), and the evaluated absorbed photon flux would be in most cases higher than the actual value.

Solar photochemical reactors are usually equipped with radiometers measuring permanently the direct and/or the diffuse incident radiation. The data measured (most often in W m^{-2}) may be used to evaluate and control the performance of solar reactor geometries [108]. Values of the solar radiant power transmitted through a photoreactor tube ($P_{e,S}$, W) exposed to sunlight may be also determined by ferrioxalate actinometry. Knowing the spectral distribution of solar light ($S_{e,S,\lambda}$) and the transmittance (T_λ) of the optical material used for the construction of the reactor, $P_{e,S}$ is obtained using Eq. (29) (similar to Eq. 25) [152]. Values of $P_{e,S}$ under different weather conditions may be correlated with the responses from the solar sensors (radiometers).

$$P_{e,S} = \frac{\left(\mathrm{d}n_{\mathrm{Fe(II)}}/\mathrm{d}t\right)}{\sum_{\lambda_1}^{\lambda_2} \Phi_{ac,\lambda} T_\lambda \left(S_{e,S,\lambda}/E_{p,\lambda}\right)\left(1 - 10^{-A_{ac,\lambda}}\right)} \tag{29}$$

Chemical actinometry is very valuable for the evaluation and optimization of existing photochemical reactors as well as for the technical development of photochemical processes that imply the design of new photochemical reactors or the up-scaling of existing ones. Such investigations are essential for the development of hybrid systems combining solar irradiation and artificial light sources.

3.3 *Photochemical Reactivity of Inorganic Fe(III) Complexes*

3.3.1 Photochemical Reactivity of Fe(III) Aqua Complexes

Among the inorganic complexes of Fe(III), aqua complexes undoubtedly play a central role in Fenton and Fenton-like systems. Iron speciation calculations show that, in the absence of other ligands, $Fe(H_2O)_6{}^{3+}$, $Fe(OH)(H_2O)_5{}^{2+}$, $Fe(OH)_2(H_2O)_4{}^{+}$, and $Fe_2(\mu\text{-}OH)_2(H_2O)_8{}^{4+}$ are the only relevant soluble ferric species below pH 4. Under pH conditions typically used in photo-Fenton processes (i.e., pH 2–4) [110, 140], the most important species are $Fe(H_2O)_6{}^{3+}$ and $Fe(OH)(H_2O)_5{}^{2+}$ (also referred to in the literature as Fe^{3+} or $Fe_{aq}{}^{3+}$ and $Fe(OH)^{2+}$, respectively).

Quantum mechanical calculations have shown that the photochemical reactivity of these Fe(III) complexes is due to LMCT transitions involving nonbonding *p*-orbitals of the coordinating O-atoms and empty *d*-orbitals of the metal center. Consequently, excitation into the LMCT absorption band leads to an inner-sphere electron transfer reducing Fe(III) and oxidizing coordinated H_2O or hydroxide (HO^-) (Reactions 19 and 15, Sect. 3.1). Therefore, the photolysis of Fe(III) aqua complexes contributes to both Fe(II) recycling and $HO^•$ generation.

At pH values of approx. 3, $Fe(OH)(H_2O)_5^{2+}$ was found to be the dominant hydroxo-complex providing optimal conditions for a photochemical Fe(II) recycling [110, 140]. $Fe(OH)(H_2O)_5^{2+}$ exhibits a stronger LMCT absorption at longer wavelengths ($\lambda_{max} : \sim 300$ nm, $\varepsilon_{300} : \sim 2 \times 10^3$ M^{-1} cm^{-1}) than that of $Fe(H_2O)_6^{3+}$ ($\lambda_{max} : \sim 240$ nm, $\varepsilon_{240} : \sim 4 \times 10^3$ M^{-1} cm^{-1} and $\varepsilon_{300} : \sim 1 \times 10^2$ M^{-1} cm^{-1}) [15, 34]. The bathochromic shift of the LMCT transition of $Fe(OH)(H_2O)_5^{2+}$ was attributed to a higher HOMO-orbital energy of the HO^- ligands than that calculated for the HOMO orbital of coordinated H_2O [106]. The quantum yield of Fe(II) production ($\Phi_{Fe(II)}$) was determined to be ≈ 0.2 for $Fe(OH)(H_2O)_5^{2+}$ at $\lambda_{max} = 300$ nm, about six times higher than $\Phi_{Fe(II)}$ for $Fe(H_2O)_6^{3+}$ (≈ 0.05 at $\lambda < 280$ nm) [15].

The higher $\Phi_{Fe(II)}$ of the former might be explained by postulating an intermediate excited state $[(Fe–^•OH)(H_2O)_5^{2+}]^*$ or an intermediate ground state $[(Fe–^•OH)(H_2O)_5^{2+}]$ from which a dissociation of $HO^•$ would require less structural changes than the release of $HO^•$ and H^+ from $[(Fe–OH_2)(H_2O)_5^{2+}$ [106]. For practical purposes, the polychromatic quantum efficiency of Fe(II) production ($\eta_{Fe^{2+}}$) was determined with aqueous $Fe_2(SO_4)_3$ to be 0.14 (pH 3, Hg medium-pressure lamp, Pyrex filter) [25].

3.3.2 Photochemical Reactivity of Fe(III)–Peroxo Complexes

The photolysis of Fe(III)–peroxo complexes (e.g., Reaction 16), formed in aqueous solutions of Fe(III) and H_2O_2, also contributes to Fe(II) recycling and enhances the rate of H_2O_2 consumption compared to the corresponding reaction in the dark (Reaction 8b). In contrast to the photolysis of Fe(III) hydroxocomplexes (Reactions 15 and 19), the thermal and photochemical reactions of Fe(III)–peroxo complexes do not generate $HO^•$, but the much less reactive perhydroxyl radical ($HO_2^•$) and/or its conjugated base, the superoxide anion ($O_2^{•-}$), depending on the pH value of the reaction system [19]. For Reaction 16, the polychromatic efficiency of Fe^{2+} formation ($\eta_{Fe^{2+}}$) was determined to be 0.33 from mixtures of $Fe_2(SO_4)_3$ and H_2O_2 (pH 3, Hg medium-pressure lamp, Pyrex) [25]. The contribution of this process to the overall rate of Fe(II) recycling is expected to be significant at relatively high H_2O_2 concentrations. In fact, speciation calculations showed that, at pH 3 and with 500 mM of H_2O_2, the concentration of $Fe(OOH)(H_2O)_5^{2+}$ accounts for about 50% of the concentration of $Fe(OH)(H_2O)_5^{2+}$. In addition, with ε-values of ~ 500-1000 M^{-1} cm^{-1}, the Fe(III)–peroxo complex exhibits a significant absorption in the spectral range between 350 and 500 nm [52].

3.3.3 Effects of Inorganic Anions on the Photo-Fenton Process

As already mentioned in Sect. 2.2, the type of iron salt used as catalyst [51, 129] or the presence of various inorganic ions [40, 98, 107, 140] may also affect the (photo-)Fenton processes and the oxidation of the organic pollutants.

Although the rate of Reaction 7 is practically independent of the inorganic ferrous salt used [40], inorganic ions may alter the kinetics [48, 98] as well as the reaction mechanism [51, 129] of the photo-Fenton process. While some anions may form stable ferric complexes, others may react with $HO^{•}$ to yield secondary radicals. Scavenging of $HO^{•}$ by inorganic ions usually yields less reactive species that may slow down the process, lead to different intermediates, or even be unreactive toward the substrate and intermediate oxidation products [51, 129].

At pH 3, the dominant photoactive species $Fe(OH)(H_2O)_5^{2+}$ [48] is not altered by bicarbonate (HCO_3^-), and the effect on the efficiency of the photo-Fenton process would be due to scavenging of $HO^{•}$ (e.g., Reaction 30). Therefore, this effect will depend on the HCO_3^- concentration [33] and on the rate constant of its reaction with $HO^{•}$ [31].

$$HCO_3^- + HO^{\bullet} \rightarrow HCO_3^{\bullet} + HO^- \tag{30}$$

The rate constant of the reaction of nitrate (NO_3^-) with $HO^{•}$ is rather small [54], but due to the known photochemical reactivity of NO_3^-, its effect on the course of the photo-Fenton process can only be disregarded for irradiation wavelengths above 320 nm. Given the wavelength dependence of both the absorption coefficient and the quantum yield of NO_3^- consumption, the photochemical generation of $HO^{•}$ upon electronic excitation of NO_3^- in the UV-C (below 280 nm) and UV-B (280–315 nm) spectral ranges may be important.

Perchlorate ions (ClO_4^-) are practically inert, but phosphate ions form insoluble complexes with Fe(III) at mildly acidic pH. Sulfate (SO_4^{2-}) ions reduce the reactivity of Fe(III) ions through coordination but only at relatively high concentrations, and the use of ferrous/ferric sulfate at millimolar concentrations does not affect the performance of the Fenton reaction. Iron(III) ions and F^- form strong complexes that are catalytically inactive in the Fenton reaction. The complexation constant of Fe(III) by Cl^- is smaller than those by PO_4^{3-} and SO_4^{2-}, and comparatively high Cl^- concentrations would be needed to substantially modify the distribution of Fe(III) complexes. The Fe(III) complexes of $H_2PO_4^-$, SO_4^{2-}, or Cl^- (X^-) might contribute to the photochemical recycling of Fe(II) (Reaction 31), but they exhibit much smaller $\Phi_{Fe(II)}$ than the aqua complex $Fe(OH)(H_2O)_5^{2+}$ [15, 37, 48, 98]. For instance, $FeSO_4^+$ is about 40–50 times less photochemically reactive than $Fe(OH)^{2+}$.

$$Fe(X)(H_2O)_5{}^{2+} + H_2O + h\nu \rightarrow Fe(H_2O)_6{}^{2+} + X^{\bullet} \quad (31)$$

Chloride and bromide ions are efficient $HO^{\bullet}$ scavengers at relatively low concentrations (mM range [39]) and form halogen radicals ($X^{\bullet}$) and radical ions ($X_2^{\bullet -}$) that yield organohalogen compounds by addition to organic substrates (e.g., [163]). Hence, the competition for $HO^{\bullet}$ scavenging leads to a decrease of the overall rate of mineralization and must be taken into account when high concentration of, e.g., NaCl are present, for example, in seawater from off-shore crude oil production [107, 117].

3.4 Photochemical Reactivity of Organic Fe(III) Complexes

A number of organic compounds, especially those acting as bi- and polydentate ligands, form complexes with Fe(III). The absorption properties and the photochemical reactivity of Fe(III) chelates are strongly dependent on the ligand structure. Typically, these complexes have higher molar absorption coefficients in the near-UV and visible spectral regions than the aqua complexes [34, 110, 140]. Excitation of their LMCT state leads in most cases to the reduction of Fe(III) to Fe (II) and to one-electron oxidation of the chelating ligand (Reactions 17 and 18). An example is given below in the case of a bidentate ligand with a single negative charge (LL^-, Reaction 32).

$$Fe(III)(LL)_3 + 2H_2O + h\nu \rightarrow Fe(II)(LL)_2(H_2O)_2 + LL^{\bullet} \quad (32)$$

The wavelength-dependent quantum yields of Fe(II) formation ($\Phi_{Fe(II)}$) are largely influenced by the ligand structure and may vary from $\sim 10^{-4}$ to ~ 1 in the UV/visible spectral range.

3.4.1 Fe(III) Complexes with Aliphatic Carboxylates

The carboxylate group ($RCOO^-$) is one of the most common functional groups of dissolved organic compounds present in natural waters [116]. Besides, aliphatic carboxylic acids are intermediate products of the oxidative degradation of organic compounds and may accumulate in aquatic ecosystems due to their relative high stability. However, their coordination to transition metal ions may lead to more or less photochemically reactive complexes. In this way, carboxylic acids, such as oxalic, tartaric, maleic, citric, isocitric, succinic, malonic, pyruvic, glyoxylic, formic, and acetic acids, may accelerate the photochemical redox cycling of Fe (II/III) in aqueous solutions [1].

Besides dissociation of the organic ligand, the LMCT-excited states of Fe(III) carboxylate complexes deactivate through an inner-sphere electron transfer process yielding Fe(II) and a carboxylate radical $RCOO^{\bullet}$ (Reaction 33) [57, 142, 160, 180]. The generation of a long-lived radical complex ($[Fe{-}^{\bullet}OOC{-}R]^{2+}$) as intermediate was also postulated. The carboxylate radical $RCOO^{\bullet}$ subsequently undergoes decarboxylation (Reaction 34) [73, 160, 180], thus significantly contributing to the mineralization of organic pollutants.

$$Fe(OOCR)(H_2O)_5{}^{2+} + H_2O + h\nu \rightarrow Fe(H_2O)_6{}^{2+} + RCOO^{\bullet} \quad (33)$$

$$RCOO^{\bullet} \rightarrow R^{\bullet} + CO_2 \quad (34)$$

The impact of Reaction 33 on the overall efficiency of Fe(II) recycling depends on the Fe(III) speciation and, therefore, varies with the nature of the carboxylate ligand, the initial iron concentration, the ratio of ligand to metal ion concentrations and the pH value of the reaction system [1, 45, 57, 160, 180]. Various studies comparing the relative rates of photochemical reduction of different iron (III)–carboxylate complexes have been published [1, 57, 81, 93, 183]. The measured quantum yields of Fe(II) production ($\Phi_{Fe(II)}$) vary in a wide range, usually from 0.1 to ~1.0.

The pH effect has been investigated for a series of aliphatic carboxylate complexes [1]. At pH values lower than 3 and at high ratios of ligand to metal ion concentrations, the order of reactivity was found to be: oxalate > tartrate > malate > citrate > isocitrate > succinate > formate, whereas for lower ratios, the order was oxalate > tartrate > citrate > malate > isocitrate. An increase of pH to 4.0 resulted in higher quantum yields for all listed carboxylates except oxalate.

In addition, the overall efficiency of Fe(II) production may be affected by the concentration of dissolved O_2 due to competitive reactions of O_2 and Fe(III) with the C-centered radicals derived from decarboxylation (Reactions 34–36 [45, 160, 189]). For example, the experimentally determined $\Phi_{Fe(II)}$ were about ten times lower in air-equilibrated solutions than in Ar-saturated solutions for the Fe (III)–oxalate system.

$$R^{\bullet} + O_2 \rightarrow RO_2{}^{\bullet} \quad (35)$$

$$R^{\bullet} + Fe(III) \rightarrow R^{+} + Fe(II) \quad (36)$$

In the presence of different substrates, the production of Fe(II) may also be affected by various secondary reactions that may alter the nature and the distribution of the intermediate and final reaction products [57, 180, 190]. Typically, the quantum of Fe(II) formation may be larger than the primary quantum yield of Fe (III) reduction due to these secondary reactions.

3.4.2 Ferrioxalate

Similarly to Fe(III)–carboxylate complexes, the photochemical redox reactions of Fe(III)–polycarboxylate complexes usually yield decarboxylated products. They are known to be photochemically reactive species ($\Phi_{Fe(II)} > 0.4$), and their photolysis contributes significantly to the mineralization of the organic content of the reaction system [140].

Potassium ferrioxalate ($K_3Fe(C_2O_4)_3$) is one of the first and most probably best investigated Fe(III) chelate. Parker [133] studied for the first time in detail the photochemical reduction of Fe(III) and concomitant oxidation of oxalate ($^-OOCCOO^-$) and subsequently introduced potassium ferrioxalate as a standard chemical actinometer for photon flux measurements [72]. The photolysis is efficient up to 450 nm and, depending on the experimental conditions, the apparent $\Phi_{Fe(II)}$ may reach values higher than 1 [29, 92]. This is the result of redox reactions between the oxidized ligand formed by photolysis ($C_2O_4^{\bullet -}$, Reaction 37) and the Fe(III) species present in the reaction system (e.g., Reaction 38). In the process, the oxalate ligand is oxidized to CO_2.

$$Fe(C_2O_4)_3^{3-} + h\nu \rightarrow Fe(C_2O_4)_2^{2-} + C_2O_4^{\bullet -} \tag{37}$$

$$Fe(C_2O_4)_3^{3-} + C_2O_4^{\bullet -} \rightarrow Fe(C_2O_4)_3^{4-} + 2CO_2 \tag{38}$$

3.4.3 Fe(III) Complexes with Aromatic Ligands

The quantum efficiency for electron transfer from a ligand to Fe(III) is much smaller for aromatic ligands such as phenolate ($C_6H_5O^-$) or benzoate ($C_6H_5COO^-$) than for succinate ($^-OOCCH_2CH_2COO^-$) and oxalate ($^-OOCCOO^-$) [12, 36, 94, 143, 144]. The poor efficiency of photochemical reduction of Fe(III)–benzoate complexes may be explained by the high energy of the intermediate phenyl radical. Fe(III) chelates with polyphenolate ligands exhibit strong LMCT absorption bands in the visible spectral range between 400 and 600 nm. But in contrast to the aliphatic polycarboxylate complexes, $\Phi_{Fe(II)}$ rarely exceeds 10^{-4} [36, 94]. In such cases, irradiation can barely enhance the rate of oxidative degradation of the organic pollutants compared to the Fenton process in the dark [158].

The Fe(III) complex of sulfosalicylic acid (2-hydroxy-5-sulfobenzoic acid) is photochemically inert, a result that has been attributed to an ultrafast back-electron transfer in the Frank–Condon excited state [143, 144].

4 Photo-Fenton Process: Current State of Development

4.1 Context

Fenton and Fenton-like processes operated under irradiation usually exhibit faster substrate transformation and much faster and higher DOC removals and may demand lower catalyst concentrations than corresponding thermal processes [110]. Fe(III) complexes are transformed to Fe(II) species by photochemical redox reactions based on the oxidation of the coordinated ligands (Sect. 3) or other sacrificial donors [34]. The photochemical reactivity of a large number of dissolved Fe(III) species is beneficial to the overall AOP, because the recycled Fe (II) complexes react with H_2O_2 to produce $HO^{•}$ (Reaction 7) initiating the oxidative degradation of target pollutants, while the oxidation of coordinated ligands enhances the rate and the level of DOC consumption [140].

From an oversimplifying point of view, the photochemical part of the photo-Fenton process might be considered as more or less independent of the (thermal) Fenton process and as serving solely, or at least predominantly, to enhance the latter. Nevertheless, the nature of the radiation sources and the radiant power of the photochemical installation, the design and dimensions of the photochemical reactors, and the residence time of the treated water are closely linked to the primary parameters of the thermal process. Moreover, the efficiencies of the individual steps involved in the complex reaction manifold of the process, particularly under irradiation, are most often difficult to evaluate. Therefore, up-scaling of laboratory results to pilot and industrial installations is especially challenging. In this context, besides mechanistic and kinetic studies, empirical statistical modeling (mainly based on the experimental design and artificial neural network methodologies) has been proven effective for the optimization of operating conditions during the technical development of the overall process (e.g., [3, 5, 11, 58, 115, 122, 125, 126, 137, 161]). Some of the important parameters affecting the photo-Fenton process are discussed in the following sections.

4.2 Radiation Sources

4.2.1 Artificial Radiation Sources

Fe(III) complexes exhibit different absorption spectra depending on their ligand structure and absorb light more or less efficiently in the UV and visible spectral regions. Radiation sources with dominant emission in these wavelength domains are known and used on industrial levels since many decades [30]. *Hg medium-pressure (MP) arcs* with quartz protection tubes exhibit emission lines at 254, 313, 367, 405, 436, 546, and 578 nm and are suitable to be used for photo-Fenton processes. The UV-B (280–320 nm) and UV-A emission of Hg MP arcs is increased by Fe or Ga doping [30]. These arcs are inserted into cooling wells for avoiding excessive heating. If borosilicate glass wells are employed, irradiation is limited to the UV-A (320–400 nm) and visible spectral regions. For mechanistic investigations, it may be of interest to restrict the spectral region of irradiation. However, for a technical development, quartz wells, albeit more expensive, may be of advantage as they are transparent to the relatively intense radiation emitted at 313 nm. With the use of quartz wells, the UV-C emission (254 nm) of these arcs is also at disposal and might contribute to the overall efficiency of the process, as a consequence of H_2O_2 photolysis (Reaction 39) or of the photolysis of organic compounds present in the reaction system.

$$H_2O_2 + hv \rightarrow 2\ HO^{\bullet} \tag{39}$$

The importance of Reaction 39 depends on the H_2O_2 concentration. Despite a rather high quantum yield of photolysis, H_2O_2 exhibits a very low molar absorption coefficient at 254 nm [22]. Consequently, competition for absorption (inner filter effects) by the dissolved organic compounds and different ferric species would require sufficiently high H_2O_2 concentrations, so that H_2O_2 photolysis might contribute significantly to the enhancement of the oxidative degradation.

The continuous solar spectrum at the earth surface is best reproduced by *high-pressure Xe arcs*. However, these arcs are point sources [29] that are very difficult to install in large-scale photochemical reactors dedicated to AOPs. On a laboratory scale, solar simulators for indoor testing are commercially available. Recently, small Xe arcs came onto the market for the use in laboratory scale tubular reactors, but so far, it is not known if a technical development for large-scale applications is planned.

Light-emitting diodes (LEDs) are commercially available for use in the UV-A and visible spectral regions. Single LEDs and small LED arrays may be fitted to micro- and mini-photoreactors that are rapidly gaining interest mostly for research purposes.

4.2.2 Solar Irradiation

Sunlight (UV-B, UV-A, and visible spectral regions) is highly convenient for the excitation of a large number of inorganic and organic Fe(III) complexes and Fe(III) chelates [34, 180, 189] and might be an economical asset for the technical development of photo-Fenton processes.

The technical development of a solar light enhanced Fenton process should take into account the inherent characteristics of solar radiation that vary with geographical location, season, time of the day, and atmospheric conditions. Three major constraints have to be considered: (i) the day and night cycle, (ii) the high variability of the radiant power during day time mainly due to the attenuation by clouds, and (iii) the strong IR radiation that is absorbed by water and increases the reaction temperature. Solar overheating may be prevented by cooling devices. If continuous operation over long periods of time would be needed, hybrid installations would be of advantage. In this case, solar light and artificial radiation sources should be combined in such a way that the spectral distribution of the incident photons and the radiant power should not vary in a significant way in order to ensure stable operating conditions [30]. Different types of photochemical reactors for solar irradiation are used and are briefly described in the following section.

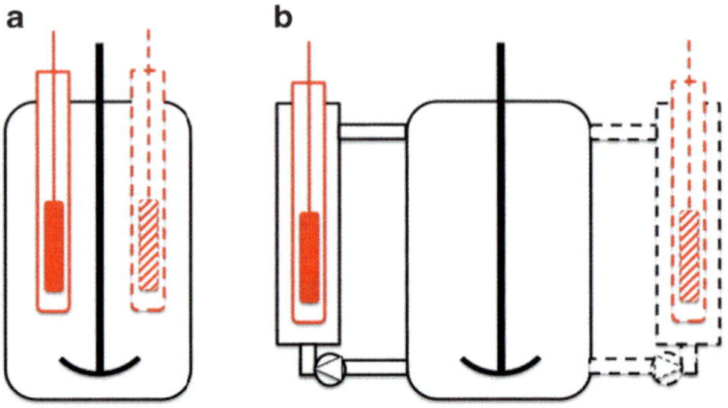

Fig. 4 Schematic representation of (**a**) an immersion-type photochemical reactor (batch regime) and (**b**) a circulating batch installation, equipped with one or several lamps or photochemical reactors

4.3 Equipment Overview and Process Regimes

4.3.1 Photo-Fenton Process with Artificial Radiation Sources

The impact of the photochemical and photochemically initiated reactions on the overall oxidative process is decisive for large-scale applications. Fenton processes are mostly conducted in stirred reactors tanks that may be equipped for efficient air supply. Extensive foaming has to be taken into account. Combination of such processes with a photochemical unit may be achieved in two ways: (i) by immersing one or several lamp wells into the stirred tanks (batch immersion-type photoreactor, Fig. 4a) or (ii) by circulating the reaction system between the tank and one or several photochemical reactors (circulating batch installation, Fig. 4b).

Immersion type reactors (Fig. 4a) are frequently used for laboratory scale experiments, but large stirred tank reactors are usually avoided, because stirring will cause torsional forces on and resonant vibrations of the immersed lamp wells increasing the risk of damage. Circulating batch installations (Fig. 4b) are mostly chosen for the experimental work during technical development and for large-scale applications. Their early use for laboratory scale experiments allows to test various reactor designs and to evaluate optimal flux conditions for a given photochemical installation. Circulating batch regimes are advantageous for large-scale AOPs requiring multistep oxidative degradation and low final DOC values. A circulating batch installation (Fig. 4b) allows to adapt the process to the varying conditions in terms of volume per unit of time, nature and concentration of the pollutants. Indeed, the number of photochemical reactors might be adjusted depending on the electrical power needed.

Circulating batch installations may be readily transformed for continuous regimes (Fig. 5) that might be used for simple substrate oxidations (e.g., discoloration), for treatment of waters of low initial DOC, or for disinfection. The process equipment is advantageously mounted after a reservoir acting as a buffer tank in case of variable feeds or pollutant concentrations of the wastewater to be treated.

The geometry and dimensions of photochemical reactor(s) depend mainly on the type of light source chosen, on cooling requirements, on the absorbance in the

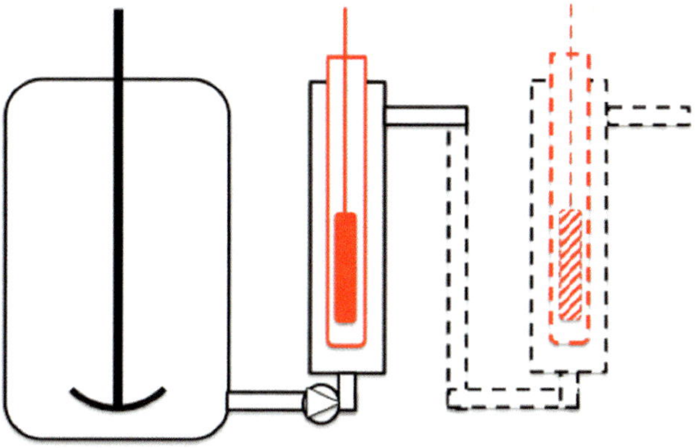

Fig. 5 Schematic representation of an installation for a continuous process regime with one or several photochemical reactors

actinic spectral region, and on the flux of the wastewater to be treated. Among artificial light sources, Hg medium- and high-pressure arcs must be operated with a cooling system, consisting in most cases of a refrigerated closed circuit of tap water. The installation of a closed cooling circuit allows to control the use of the lamp(s) independently of the flux of the reaction system and represents therefore an important factor of operational security. LED arrays are usually thermoregulated by air or Peltier cooling modules.

Large-scale photochemical installations for circulating batch or continuous regimes using artificial light sources are in general *cylindrical immersion type reactors* [29]. The outer diameter of the immersed well that contains the source of radiation determines the irradiated surface and therefore the exitance (photon flux density [28]). The latter represents an important process parameter for optimization and up-scaling tasks. Means for air introduction, for the addition of iron salts or complexes and H_2O_2, as well as for the control of foaming are best installed at the reservoir. Depending on the antifoaming device and the importance of air bubbling, stirring the liquid located in the reservoir is optional.

4.3.2 Solar Photo-Fenton Process

The fundamentals and the development and the different types of solar photochemical reactors are well described in several reviews by Malato et al. [108–110].

Flatbed reactors are exposed to the direct solar irradiation and might be equipped with a tracking system. These installations are mounted without solar concentrators but collect advantageously diffuse radiation. They are used for small pilot scale experiments or applications due to the limited natural incident photon flux density. Flatbed reactors might be conceived as open falling film or as flow-through reactors, the latter disposing of channel arrangements or grids to enhance turbulence and prevent inhomogeneous flow conditions. Flow-through reactors are also chosen to prevent evaporation under conditions of intense solar heating.

Installations with concentrators reflect and focus the incident solar radiation on a reactor tube made of borosilicate or quartz, hence amplifying the photon flux density on the surface of the reactor tube by a factor depending on the reflector geometry. Tracking parabolic trough concentrators (PTC or CPTC) were originally developed for solar thermal applications and, according to model calculations, may reach concentration ratios of 10 on receiver tubes of 3 cm of diameter [164]. Their design limits their use to direct solar irradiation. Plant installations for photochemical purposes operate with two-axis tracking PTC with lower solar concentration factors [55]. In contrast to thermal applications, solar heating must be limited by high water fluxes through the solar photochemical reactor and a thermoregulation at the reservoir of the circulating batch process. Small reactor (receiver) tubes and high fluxes of the reaction system also ensure good turbulence conditions. Compound parabolic concentrators (CPC) are mostly used when installations without tracking system but increased incident photon densities are in demand. The particular geometry of the collector permits the use of direct and diffuse solar radiation. Although model calculations predict solar concentration factors up to 10 [151], concentration factors at the pilot installations at, e.g., the Plataforma Solar de Almeria (PSA, Spain) do not exceed 1.5 [55]. The advantages of the relatively small size of the concentrators and the use of a receiver tube, which is in fact a tubular photochemical reactor with positive irradiation geometry [29], allow easy up-scaling of the installation by increasing the number of collectors.

Transparent holographic concentrators with factors up to 1.4 [162] might stimulate research in the field of solar photochemistry and favor the use of flatbed reactors for limited size applications.

The PSA is one of the most important European centers of research and development of solar photochemical processes where different types of collectors of technical scale are available. Besides the treatment of waters spiked with model pollutants and of wastewaters at different levels of technical development [108–110], a large-scale CPC installation was mounted and operated successfully for the treatment of wastewaters in combination with a biological treatment station [127]. A related economic feasibility study of pesticide elimination from industrial wastewater by a combination of solar Fenton process and membrane bioreactors yielded treatment costs of 122 euro m^{-3} [155]. For a treatment of landfill leachates by combining a granulate biofilter, ozonization, and solar Fenton process, costs were calculated to be in the range between 3 and 6 euro m^{-3} depending on the sequence of processes used and on initial COD [32]. Solar photo-Fenton processes have also been brought on pilot scale level for decontamination (e.g., [11, 55, 117, 122, 158]) and disinfection of water (e.g., [130, 141]) and the degradation of emerging (micro) pollutants at low concentrations (e.g., [17, 89, 148]). The use of heterogeneous catalysts was investigated on solar pilot scale reactors using zeolites loaded with Fe(III) [65].

4.4 *Process Parameters: pH, Iron(II)/(III), and Hydrogen Peroxide Concentrations*

The concentrations of Fe(II)/Fe(III) and H_2O_2 are primary process parameters affecting the rate of substrate oxidation and oxidative degradation. The hydroxyl radical may be generated stoichiometrically by reacting an Fe(II) salt with H_2O_2 (Reaction 7). However, if Fe(II) may be used in catalytic amounts owing to recycling thermal and photochemical reactions (Sects. 2.1 and 3.1), the amount of H_2O_2 introduced should be high enough to ensure oxidation of the target pollutant and of the intermediate by-products, if a significant DOC abatement or even complete mineralization is required. As expected, the H_2O_2 to iron molar ratios depend largely on the type and concentration of organic pollutant and published values vary in a wide range (typically between 100 and 1000, e.g., [91, 140]).

If Fe(II)/Fe(III) salts and H_2O_2 are added at once at the start of the process, high H_2O_2 concentrations might diminish the rate of oxidation due to competitive trapping of $HO^{\bullet}$ by H_2O_2 (Reaction 12). But the major disadvantage of such a procedure is the disregard of the possibility to vary and optimize the H_2O_2 concentration as one of the major process parameters, both at the start and during the experiment. Some authors prefer therefore procedures, where a predetermined (optimized) amount of Fe(II)/Fe(III) salt is added to the reaction system and an H_2O_2 solution of predefined concentration is added periodically or continuously during reaction (irradiation) time (e.g., [126, 145, 147, 186]). Laboratory experiments showed that the yield of oxidative degradation of 4-chlorophenol, expressed as the decrease in COD, could be linearly correlated with the concentration of H_2O_2 introduced and that the periodic addition of H_2O_2 yielded a constant efficiency of mineralization [4]. Interestingly, the yield of oxidative degradation did not depend on the concentration of Fe(II) (in the range of 5 to 20 mg L^{-1}). In a large pilot scale experiment [126], 500 L of industrial wastewater containing 3,4-xylidine (2700 mg C L^{-1}, initial pH = 3) were irradiated with a 10 kW Hg medium-pressure lamp in a Pyrex cooling well under continuous addition of H_2O_2 (35% w/w in water). Under the optimal conditions, xylidine was eliminated in 30 min and more than 90% of the initial DOC was removed after 2 h of irradiation, using about 4 moles of H_2O_2 and 0.4 mole of Fe^{2+} (ferrous sulfate) per mole of xylidine degraded. Biological tests showed that as soon as xylidine was completely transformed the dissolved organic matter remaining was biodegradable and not toxic toward the sludge of the communal biological treatment station.

The importance of the O_2 concentration in the oxidative degradation process should be stressed (Reactions 5 and 6, Sect. 1). Indeed, a lack of O_2 may increase the probability of combination and disproportionation of the intermediate radicals. Nevertheless, oxygenation in aerated tanks, or letting the treated water fall into open reservoirs, is generally sufficient to ensure high enough O_2 concentrations, but antifoaming measures or installations could be required. Some studies suggest that O_2 may, at least in part, substitute H_2O_2 [68, 173]. Channeling reaction pathways toward O_2 rather than H_2O_2 consumption is of particular interest for the technical development of the photo-Fenton process. However, this phenomenon should be more generally investigated for various types of pollutants.

Another limitation for large-scale applications of Fenton and photo-Fenton processes is the range of pH within which these methods of wastewater treatment can be efficiently applied [140] (as mentioned in Sect. 2.1). In order to comply with the optimal pH value between 2.5 and 4, addition of inorganic acid (or base) may be required to adjust the pH of wastewaters originating from various sources. However, the pH conditions required for an optimal performance of the process are different from those required for releasing the treated wastewater into the environment or into a biological treatment station. Pre- or post-treatment pH adjustments result in an increase of the salinity of the aqueous reaction system that might affect the efficiency of the AOP, e.g., by anion interaction (Sect. 3.4.3). The post-treatment to increase the pH from strongly acidic to neutral values required for discharge leads to the precipitation of amorphous ferric oxyhydroxides. Their separation by flocculation and filtration or sedimentation is part of the techniques used for deferrization (e.g., [176]). The loss of iron in general and the controlled elimination of the iron oxide sludge in particular may represent non-negligible cost factors that must be taken into account when evaluating the economical feasibility of a (photo-)Fenton process for a given wastewater. It should be noted, however, that iron may be used in relatively low concentrations, even lower in the photo-Fenton process due to a more effective Fe(II) recycling (Sect. 3.1). Besides, some economical studies have shown that, even taking into account the costs of sludge disposal, the Fenton, photo-Fenton, and solar Fenton processes were more efficient and less expensive than other AOPs, such as UV/H_2O_2 or UV/ozone or TiO_2 photocatalysis (e.g., [14, 91, 136, 154]).

Mohr's salt ($((NH_4)_2Fe(SO_4)_2{\cdot}6H_2O)$) has been used successfully as catalyst in Fenton and photo-Fenton processes for the oxidative degradation of chloral hydrate and 1,1,1-trichoroethane at pH 7. Phosgene as an intermediate of the degradation had to be hydrolyzed to enhance mineralization, and the Fenton process was developed to plant scale [149].

According to a report of the European Medecine Agency [44], there is so far no regulatory assessment as far as the oral intake of Fe(II)/Fe(III) is concerned, but the agency proposes a permitted exposure (PDE) of 260 $\mu g\ kg^{-1}\ day^{-1}$ for a 50 kg patient. There are, however, directives of the Council of the European Union [35] for drinking water with limiting values lower than 200 $\mu g\ L^{-1}$, i.e., 7.6×10^{-6} mol L^{-1} of iron. Large-scale applications of the (photo-)Fenton process might therefore call for the use of deferrization equipment to avoid local or regional accumulation of Fe(III) in surface waters as well as in drinking water.

4.5 *Modified Photo-Fenton Processes*

Quite a number of research groups work on modifications of the (photo-)Fenton process to diminish the technical and economical impacts of its pH dependence and to enhance catalyst recycling. Strategies to minimize the problems related to pH limitations and sludge formation and disposal include the development of efficient heterogeneous (photo-)Fenton catalysts and of complexing agents that maintain iron ions in solution within a larger pH range.

4.5.1 Heterogeneous Photo-Fenton Systems

Preventing precipitation and loss of Fe(III) by immobilizing the ions on a solid support may be accomplished using various support materials and different types of fixation. While an easy separation and the possibility of working without pH adjustments are advantages of this approach, the drawback is generally diminished reaction rates compared to the homogeneous photo-Fenton process [110]. A *heterogeneous photo-Fenton process* requires solid catalysts (i) in which the electronically excited Fe(III) ions are in interaction with a reducing agent and (ii) that do not leach under irradiation due to Fe(II) formation or oxidative degradation of the support.

Several types of solid photo-Fenton catalysts have been investigated. Solid iron compounds consisting of various iron oxides might be used as powders. For instance, ground goethite (FeO(OH)) was used for the oxidation of tetrachloroethene (TCE) at pH 7 with a rate of conversion of 22% relative to that of a homogeneous Fenton process catalyzed with $Fe_2(SO_4)_3$ [171]. It is interesting to note that the addition of $HO^{\bullet}$ scavengers did not alter the result leading to the assumption that the reaction manifold could involve another highly reactive oxidant such as Fe(IV). Investigations with α-Fe_2O_3 lead to the conclusion that this catalyst (as well as other iron oxide catalysts) may work as n-type semiconductors (bandgap = 2.2 eV [174]) able to oxidize water as well as adsorbed pollutants at its surface. Dimension and morphology, which can be modified by varying the method of preparation, affect the absorption of radiation and interaction with potential reductants. Nano-size hematite particles are known to be the most stable iron oxide catalysts [181] and fulfill well the requirements set for efficient photocatalysts. Iron or composite metal/iron particles may be immobilized on inorganic supports, such as silicas (e.g., [78, 118]), zeolites (e.g., [113, 152]), alumina [80], clays [47], or soil particles [135]. Organic supports for nano-size iron oxide particles comprise among others collagen [105] and activated carbon [119]. Iron(II/III) might be complexed by functionalized polymers such as carboxylate-modified PTFE [41, 50], Nafion® (polymer perfluorosulfonic acid [111]), chitosan [99], or ion exchange resins [46, 123]. However, it was demonstrated that sulfonated ion exchange resins bearing coordinated Fe(II) underwent oxidative degradation and dissolution under photo-Fenton conditions [187]. Efficiency of the heterogeneous catalysis is a controversial issue, as the contribution of the homogeneous catalysis by leached iron must be discriminated from the true heterogeneous catalysis. Iron leaching is in many cases only qualitatively assessed and depends on the morphology of iron oxides, on the nature of the support, and on the pH of the bulk aqueous phase [69]. Moreover, electronic excitation of chromophores contained in solid surfaces is less efficient compared to that in a homogenous reaction system and is very difficult to quantify [27].

Zero-valent iron is also used as a heterogeneous Fenton catalyst [16, 149]. Similarly to iron oxides, nano-size particles of zero-valent iron might be loaded on inorganic or organic supports (e.g., [62, 70]) or scrap iron might be used [2, 53]. Little is known about the mechanism of the zero-valent catalysis. In fact, given the acidic pH of the Fenton process, Fe(II) ions might be slowly released from zero-valent iron providing a simple and relatively cheap way to introduce a catalytic concentration of Fe(II) into wastewaters to be treated. Alternatively, it may be assumed that the decomposition of H_2O_2 occurs at the surface of the zero-valent iron particles. Zero-valent iron might also be used for a reductive pretreatment of nitro-compounds (e.g., TNT [124], *p*-chloronitrobenzene [97]) enhancing the subsequent oxidative degradation by (photo-)Fenton processes.

The electro-Fenton process is used to generate Fe(II) and/or H_2O_2 [131]. In the first case, the method is an alternative way to enhance the Fe(II) recycling of the Fenton process by the cathodic reduction of added Fe(III) (Reaction 40); in the second case, the electrochemical production of H_2O_2 (Reaction 41) replaces its continuous addition as mentioned earlier (Sect. 4.4). Using a sacrificial iron or steel anode, Fe(II) may also be produced electrochemically (Reaction 42) providing a controlled concentration of the catalyst.

$$Fe(H_2O)_6{}^{3+} + e^- \rightarrow Fe(H_2O)_6{}^{2+} \tag{40}$$

$$O_2 + 2\ H^+ + 2\ e^- \rightarrow H_2O_2 \tag{41}$$

$$Fe^0 + 6\ H_2O \rightarrow Fe(H_2O)_6{}^{2+} + 2\ e^- \tag{42}$$

The photoelectron-Fenton process provides an additional enhancement of the electrochemical recycling of Fe(II) (Reaction 40). The latter can only take place at the surface of the cathode. The same holds for the recycling of Fe(II) from Fe(III) oxalates [61].

4.5.2 Chelate-Assisted Photo-Fenton Systems

The addition of chelating agents was investigated as another means to diminish pH limitations and enhance the efficiency of the photo-Fenton process [140, 154, 166]. In fact, chelating ligands compete favorably with hydroxide for coordination and corresponding Fe chelates are soluble over an extended pH range. Fe(III) chelates exhibit usually higher absorption coefficients, a larger overlap with the solar spectrum, and higher quantum efficiencies of Fe(III) reduction than inorganic Fe(III) complexes typically present in photo-Fenton systems (e.g., [83, 84, 88, 154]). As organic chelating agents are also reactive toward $HO^\bullet$, they must be added in excess or continuously supplied and may therefore significantly increase the DOC, as well as the toxicity of the wastewater if they are not completely depleted at the end of the treatment. Chelation may also interfere with the thermal Fe(III)-reduction pathways (Sect. 2.1, Reactions 8a and 8b). Thus, from an economical point of view, the addition of a chelating agent should be carefully evaluated taking into account the nature and the organic load of the wastewater. Modified Fenton systems operated in the presence of a chelating agent at near-neutral pH values may be suitable for the treatment of wastewaters containing relatively low pollutant concentrations or for in situ environmental remediation [166]. In contrast, for heavily loaded effluents, the classical photo-Fenton process at pH 3 is more appropriate.

Oxalate and *citrate* have been frequently investigated for *chelate-assisted photo-Fenton processes* (e.g., [6, 87, 158, 188]), because their Fe(II) complexes react efficiently with H_2O_2 in the pH range of 3 to 8, and photochemical redox reactions of the corresponding Fe(III) complexes exhibit moderate to high quantum yields of Fe(II) formation ($\Phi_{Fe(II)}$) [45].

The absorption spectrum of Fe(III) complexes with oxalate as bidentate ligand extends to the visible spectral region and the photolysis of *ferrioxalate complexes* may be carried out using doped Hg medium-pressure lamps and under solar irradiation. Moreover, $\Phi_{Fe(II)}$ of ferrioxalate may reach values higher than 1 (Sect. 3.4.2). The increased efficiency of the ferrioxalate-assisted photo-Fenton process was reported by several authors [6, 77, 83, 95, 98, 146, 154, 190]. The working pH range may be extended up to 6 [86], which is advantageous since the treated effluent may be disposed of without pH adjustment or utilized in hybrid chemical–biological systems [13]. The overall rate of oxidative degradation depends on the pH and tends to decrease with increasing pH in the presence of H_2O_2. This complexity may be related to the coexistence in solution of complexes with one, two, and three oxalate ligands in the coordination sphere of Fe(III), their relative fractions depending on the pH and on the concentrations of Fe(III) and oxalate ions [86, 95, 142, 188]. The pH-dependent efficiency of the process reveals to be even more complex as secondary thermal reactions with Fe(II/III) species (e.g., superoxide) are also pH dependent.

It should be recalled that the formation of Fe(III)-oxalato complexes inhibits the thermal decomposition of H_2O_2, and, subsequently, the use of irradiation is mandatory for an efficient redox cycling of the Fe(II) catalyst [146, 190].

Citric acid (cit) is a tri-carboxylic acid that forms Fe(III) complexes with 1:1 or 2:2 stoichiometry. Fe(cit) is the dominant species at low pH, whereas at pH > 4, $Fe(OH)(cit)^{-1}$ and $Fe_2(OH)_2(cit)_2{}^{-2}$ are formed [33]. In the pH range of 4.6 to 8.0, more than 95% of the dissolved Fe(III) species was calculated to be $Fe(OH)(cit)^{-1}$ [188]. Citric acid used as a chelating ligand may be seen as an acidifying agent [156], but its chelating characteristics inhibit the precipitation of Fe(III) hydroxides and oxides as the pH value is increased [6, 33]. However, at neutral pH, Fe(II) and Fe(III) ions are almost quantitatively complexed by citrate, and the rates of generation of $HO^{\bullet}$, and therefore of substrate oxidation, may be smaller than those known for non-modified Fenton systems [102]. Quantum yields of Fe(II) formation ($\Phi_{Fe(II)}$) in the pH range 3 to 7 and at different wavelengths of irradiation (366 and 436 nm) have been reported to vary from 0.4 to 0.2 [132, 157]. The decrease of Φ_{Fe}

$_{(II)}$ values with increasing pH indicates that $Fe(OH)(cit)^{-1}$ exhibits a lower photochemical reactivity than Fe(cit) [188] and might explain the decreasing efficiency of oxidative degradation with increasing pH [114]. In spite of a lower $\Phi_{Fe(II)}$ for citrate Fe(III) complexes than that for ferrioxalate, citrate is readily available, is biodegradable, and can be used at higher pH values than oxalate (up to pH 9) [114].

In addition to oxalate and citrate, other ligands, such as tartrate, pyruvate, and malonate, have been investigated as photoactive Fe(III)–carboxylate complexes [45, 180, 183]. Aminopolycarboxylic acids can form stable water-soluble complexes in a wide pH range, with metal ions and more particularly with iron ions [81]. Polydentate ligands such as ethylenediaminotetraacetic acid (EDTA) and ethylenediamine-*N*,*N'*-disuccinic acid (EDDS) lead to an efficient stabilization and solubilization of iron at neutral pH values [81, 103]. However, EDTA is toxic and difficult to degrade [158], whereas EDDS is naturally present in soil, yields benign photodegradation products, and is markedly faster degraded than EDTA [81]. Recent studies show that the degradation of various target pollutants upon irradiation of the Fe(III)–EDDS complex under different conditions can be effective both in homogeneous and heterogeneous systems. Therefore, EDDS might be considered as a promising chelating agent for larger scale investigations [81, 88, 103].

5 Conclusion

The results of fundamental investigations, technical evaluations, and optimizations in the domain of photo-Fenton processes applied to the treatment of wastewaters containing a large variety of pollutants are recorded in more than 1,800 publications. There are, however, only relatively few reports on up-scaling efforts and large-scale projects. They illustrate in principle the feasibility and the benefit of the photo-Fenton process to reclaim polluted waters: (i) as a pre- or post-treatment in combination with biological treatment, (ii) as a method to recycle water for industrial use, (iii) as a means to decrease the DOC or COD content before release into the environment, and (iv) as the method of choice to detoxify landfill leachates. Among the different AOPs, only ozonization, eventually in combination with H_2O_2 and UV radiation, the photolysis of H_2O_2 and H_2O, and the photo-Fenton process are presently used or exhibit the potential to be used on an industrial level. Among the technical details that need to be improved are (i) catalyst stability and process efficiency at neutral pH and (ii) the separation or recycling of the spent catalyst. Recent findings increase the prospects to find technically and economically improved solutions for the first, and several proposals for the latter were investigated at pilot scale.

As it is the case for all photochemical processes, limits of electrical power restrict the emitted photon flux per radiation source and thereby the capacity of the treatment process in terms of pollutant concentration and volume per unit of time. In this respect, the photo-Fenton process exhibits the highest capacity among all photochemical processes. Results obtained with plant scale reactors comprising artificial light sources or arrays of solar reactors of comparable absorbed photon fluxes exhibit equivalent efficiencies. Despite higher costs of investment, solar radiation would render considerable savings in electrical energy consumption. Yet, the day/night cycle and the incident photon flux depending on weather conditions remain a severe handicap as far as process control and capacity are concerned. Hybrid installations, where variations of the incident solar photon flux would be compensated by artificial light sources, might be an advantageous technical solution.

References

1. Abrahamson H, Rezvani A, Brushmiller J (1994) Photochemical and spectroscopic studies of complexes, of iron(III) with citric acid and other carboxylic acids. Inorg Chim Acta 226:117–127
2. Ali MEM, Gad-Allah TA, Badawy MI (2013) Heterogeneous Fenton process using steel industry wastes for Methyl Orange degradation. Appl Water Sci 3:263–270
3. Arslan-Alaton I, Ayten N, Olmez-Hanci T (2010) Photo-Fenton-like treatment of the commercially important H-acid: process optimization by factorial design and effects of photocatalytic treatment on activated sludge inhibition. Appl Catal B Environ 96(1–2): 208–217
4. Bacardit J, Oller I, Maklonado MI, Chamrro E, Malato S, Esplugas S (2007) Simple models for the control of photo-Fenton by monitoring H_2O_2. J Adv Oxid Technol 10:219–228
5. Balanosky E, Herrera F, Lopez A, Kiwi J (2000) Oxidative degradation of textile waste water. Modeling reactor performance. Water Res 34:582–596
6. Balmer ME, Sulzberger B (1999) Atrazine degradation in irradiated iron/oxalate systems: effects of pH and oxalate. Environ Sci Technol 33:2418–2424
7. Bandara J, Morrison C, Kiwi J, Pulgarin C, Peringer PJ (1996) Degradation/decoloration of concentrated solutions of orange II. Kinetics and quantum yield for sunlight induced reactions via Fenton type reagents. Photochem Photobiol A Chem 99:57–66
8. Barb WG, Baxendale JH, George P, Hargrave KR (1949) Reactions of ferrous and ferric ions with hydrogen peroxide. Nature 163:692–694
9. Barb WG, Baxendale JH, George P, Hargrave KR (1951) Reactions of ferrous and ferric ions with hydrogen peroxide. Part I. The ferrous ion reaction. Trans Faraday Soc 47:462–500
10. Barb WG, Baxendale JH, George P, Hargrave KR (1951) Reactions of ferrous and ferric ions with hydrogen peroxide. Part II. The ferric ion reaction. Trans Faraday Soc 47:591–616
11. Bassam A, Salgado-Transito I, Oller I, Santoyo E, Jimenez AE, Hernandez JA, Zapata A, Malato S (2012) Optimal performance assessment for a photo-Fenton degradation pilot plant driven by solar energy using artificial neural networks. Int J Energy Res 36(14):1314–1324
12. Bates H, Uri N (1953) Oxidation of aromatic compounds in aqueous solution by free radicals produced by photo-excited electron transfer in iron complexes. J Am Chem Soc 75: 2754–2759
13. Batista A, Nogueira Pupo R (2012) Parameters affecting sulfonamide photo-Fenton degradation – iron complexation and substituent group. J Photochem Photobiol A Chem 232:8–13

14. Bauer R, Fallmann H (1997) The photo-Fenton oxidation – a cheap and efficient wastewater treatment method. Res Chem Intermediat 23(4):341–354
15. Benkelberg H, Warneck P (1995) Photodecomposition of iron(III) hydroxo and sulfato complexes in aqueous solution: wavelength dependence of OH and $SO_4^{\bullet -}$ quantum yields. J Phys Chem 99:5214–5221
16. Bergendahl JA, Thies TP (2004) Fenton's oxidation of MTBE with zero-valent iron. Water Res 38:327–334
17. Bernabeu A, Palacios S, Vicente R, Vercher RF, Malato S, Arques A, Amat AM (2012) Solar photo-Fenton at mild conditions to treat a mixture of six emerging pollutants. Chem Eng J 198–199:65–72
18. Bielski BHJ, Cabelli DE (1991) Highlights of current research involving superoxide and perhydroxyl radicals in aqueous solutions. Int J Radiat Biol 59(2):291–319
19. Bielski BHJ, Cabelli DE, Arudi RL, Ross AB (1985) Reactivity of perhydroxyl/superoxide radicals in aqueous solution. J Phys Chem Ref Data 14:1041–1100
20. Bigda RJ (1995) Consider Fenton's chemistry for wastewater treatment. Chem Eng Prog 91:62–66
21. Bishop DF, Stern G, Fleischman M, Marshall LS (1968) Hydrogen peroxide catalytic oxidation of refractory municipal waste waters. Ind Eng Chem Process Des Dev 7:110–117
22. Bolton JR, Cater SR (1994) Homogeneous photodegradation of pollutants in contaminated waters. In: Helz GR, Zepp RG, Crosby DG (eds) Aquatic and surface photochemistry. Lewis Publishers, Boca Raton
23. Bossmann SH, Oliveros E, Goeb S, Kantor M, Goeppert A, Braun AM, Lei L, Yue PL (2001) Oxidative degradation of polyvinyl alcohol by the photochemically enhanced Fenton reaction: evidence for the formation of supermacromolecules. Prog React Kinet Mec 26:113–137
24. Bossmann SH, Oliveros E, Goeb S, Kantor M, Goeppert A, Lei L, Yue PL, Braun AM (2001) Degradation of polyvinylalcohol (PVA) by homogeneous and heterogeneous catalysis applied to the photochemically enhanced Fenton reaction. Water Sci Technol 44:257–262
25. Bossmann SH, Oliveros E, Goeb S, Siegwart S, Dahlen EP, Pavayan L Jr, Straub M, Wörner M, Braun AM (1998) New evidence against hydroxyl radicals as reactive intermediates in the thermal and photochemically enhanced Fenton reaction. J Phys Chem A 102: 5542–5550
26. Bossmann SH, Oliveros E, Kantor M, Niebler S, Bonfill A, Shahin N, Wörner M, Braun AM (2004) New insights into the mechanisms of the thermal Fenton reactions occurring using different iron(II)-complexes. Water Sci Technol 49:75–80
27. Brandi RJ, Citroni MA, Alfano OM, Cassano AE (2003) Absolute quantum yields in photocatalytic slurry reactors. Chem Eng Sci 58:979–985
28. Braslavsky SE et al (2007) Glossary of terms used in photochemistry. Pure Appl Chem 79:293–465
29. Braun AM, Maurette MT, Oliveros E (1991) Photochemical technology (trans: Ollis DF, Serpone N). Wiley, Chichester
30. Braun AM, Peschl GH, Oliveros E (2012) Industrial photochemistry. In: Griesbeck A, Oelgemoeller M, Ghetti F (eds) CRC handbook of organic photochemistry and photobiology, vol 1, 3rd edn. CRC, Boca Raton
31. Buxton G, Greenstock CL, Helman WP, Ross AB (1988) Critical review of rate constants for reactions of hydrated electrons, hydrogen atoms and hydroxyl radicals ($^{\bullet}OH/^{\bullet}O-$) in aqueous solutions. J Phys Chem Ref Data 17:513–886
32. Cassano D, Zapata A, Brunetti G, Del Moro G, Di Iaconi C, Oller I, Malato S, Mascolo G (2012) Comparison of several combined/integrated biological-AOPs setups for the treatment of municipal landfill leachate: minimization of operating costs and effluent toxicity. Chem Eng J 172(1):250–257
33. Chen Y, Liu Z, Wang Z, Xue M, Zhu X, Tao T (2011) Photodegradation of propanolol by Fe(III)-citrate complexes: kinetics, mechanism and effect of environmental media. J Hazard Mater 194:202–208

34. Ciesla P, Kocot P, Mytych P, Stasicka Z (2004) Homogeneous photocatalysis by transition metal complexes in the environment. J Mol Catal A Chem 224:17–33
35. Council of the European Union (1998) Directive 98/83/C. Off J Eur Commun L330/32–L330/40
36. Cunningham K, Goldberg M, Weiner E (1988) Mechanism for aqueous photolysis of adsorbed benzoate, oxalate, and succinate on iron oxyhydroxide (goethite) surfaces. Environ Sci Technol 22:1090–1097
37. David F, David P (1976) Photoredox chemistry of iron(III) chloride and iron(III) perchlorate in aqueous media. A comparative study. J Phys Chem 80:579–583
38. De Laat J, Doré M, Suty H (1995) Oxydation de S-triazines par les procédés d'oxydation radicalaire. Sous-produits de réaction et constantes cinétiques de reaction. Rev Sci Eau/J Water Sci 8:23–42
39. De Laat J, Truong Le G (2006) Effects of chloride ions on the iron(III)-catalyzed decomposition of hydrogen peroxide and on the efficiency of the Fenton-like oxidation process. Appl Catal B Environ 66:137–146
40. De Laat J, Truong Le G, Legube B (2004) A comparative study of the effects of chloride, sulfate and nitrate ions on the rates of decomposition of H_2O_2 and organic compounds by Fe(II)/H_2O_2 and Fe(III)/H_2O_2. Chemosphere 55:715–723
41. Ding Z, Dong Y, Li B (2012) Preparation of a modified PTFE fibrous photo-Fenton catalyst and its optimization towards the degradation of organic dye. Int J Photoenergy. Article ID 121239. http://dx.doi.org/10.1155/2012/121239
42. Du Y, Zhou M, Lei L (2006) Role of the intermediates in the degradation of phenolic compounds by Fenton-like process. J Hazard Mater 136:859–865
43. Eisenhauer HR (1964) Oxidation of phenolic wastes. J WPCF 36:1116–1128
44. European Medicines Agency (2008) http://www.ema.europa.eu/docs/en_GB/document_library/Scientific_guideline/2009/09/WC500003586.pdf. Accessed 30 April 2013
45. Faust B, Zepp R (1993) Photochemistry of aqueous iron(III)-polycarboxylate complexes: roles in the chemistry of atmospheric and surface waters. Environ Sci Technol 27:2517–2522
46. Feng J, Hu X, Yue PL (2004) Degradation of salicylic acid by photo-assisted Fenton reaction using Fe ions supported on strongly acidic ion exchange resin as catalyst. Chem Eng J 100:159–165
47. Feng J, Hu XJ, Yue PL, Zhu HJ, Lu GQ (2003) A novel laponite clay-based Fe nanocomposite and its photocatalytic activity in photo-assisted degradation of Orange II. Chem Eng Sci 58:679–685
48. Feng W, Nansheng D (2000) Photochemistry of hydrolytic iron (III) species and photo induced degradation of organic compounds. A minireview. Chemosphere 41:1137–1147
49. Fenton HJH (1894) Oxidation of tartaric acid in presence of iron. J Chem Soc Trans 65: 899–911
50. Ferney Gonzales-Bahamon L, Mazille F, Benitez L, Pulgarín C (2011) Photo-Fenton degradation of resorcinol mediated by catálysts based on iron species supported on polymers. J Photochem Photobiol A Chem 217:201–206
51. Franch M, Ayllón J, Peral J, Domènech X (2004) Fe(III) photocatalyzed degradation of low chain carboxylic acids implications of the iron salt. Appl Catal B Environ 50:89–99
52. Gallard H, De Laat J, Legube B (1999) Spectrophotometric study of the formation of iron (III)-hydroperoxy complexes in homogeneous aqueous solutions. Water Res 33:2929–2936
53. Ganesan R, Thanasekaran K (2011) Decolourisation of textile dyeing wastewater by modified solar photo-Fenton oxidation. Int J Environ Sci 1:1168–1176
54. Garcia Einschlag F, Felice J, Triszcz J (2009) Kinetics of nitrobenzene and 4-nitrophenol degradation by UV irradiation in the presence of nitrate and nitrite ions. Photochem Photobiol Sci 8:953–960
55. Gernjak W, Malato Rodriguez S, Maldonato Rubio MI, Fuerhacker M (2006) Solar photo-Fenton treatment of EU priority substances – process parameters and control strategies. Editorial CIEMAT, Madrid

56. Glaze W, Kang J-W, Chapin DH (1987) The chemistry of water treatment processes involving ozone, hydrogen peroxide and ultraviolet radiation. Ozone Sci Eng 9:335–352
57. Glebov E, Pozdnyakov I, Grivin V, Plyusnin V, Zhang X, Wu F, Deng N (2011) Intermediates in photochemistry of Fe(III) complexes with carboxylic acids in aqueous solutions. Photochem Photobiol Sci 10:425–430
58. Göb S, Oliveros E, Bossmann SH, Braun AM, Nascimento CAO, Guardani R (2001) Optimal experimental design and artificial neural networks applied to the photochemically enhanced Fenton reaction. Water Sci Technol 44:339–345
59. Golgate PR, Pandit AB (2004) A review of imperative technologies for waste water treatment II: hybrid methods. Adv Environ Res 8:553–597
60. Golgate PR, Pandit AB (2004) A review of imperative technologies for waste water treatment I: oxidation technologies at ambient conditions. Adv Environ Res 8:501–551
61. Gomathi H (2000) Chemistry and electrochemistry of iron complexes. Bull Electrochem 16(10):459–465
62. Gomathi Devi L, Girish Kumar S, Mohan Reddy K, Munikrishnappa C (2009) Photo degradation of Methyl Orange an azo dye by advanced Fenton process using zero valent metallic iron: influence of various reaction parameters and its degradation mechanism. J Hazard Mater 164:459–467
63. Gonzalez MC, Braun AM, Bianco Prevot A, Pelizzetti E (1994) Vacuum-ultraviolet (VUV) photolysis of water: mineralization of atrazine. Chemosphere 28:2121–2127
64. Gonzalez MC, Oliveros E, Wörner M, Braun AM (2004) Vacuum-ultraviolet photolysis of aqueous reaction systems. J Photochem Photobiol C Photochem Rev 5:225–246
65. Gonzalez-Olmos R, Martin MJ, Georgi A, Kopinke F-D, Oller I, Malato S (2012) Fe-zeolites as heterogeneous catalysts in solar Fenton-like reactions at neutral pH. Appl Catal B Environ 25:51–58
66. Haag WR, Yao CD (1992) Rate constants for the reaction of hydroxyl radicals with several drinking water contaminants. Environ Sci Technol 26:1005–1013
67. Haber F, Weiss J (1934) The catalytic decomposition of hydrogen peroxide by iron salts. Proc R Soc A 134:332–351
68. Haddou M, Benoit-Marquié F, Maurette M-T, Oliveros E (2010) Oxidative degradation of 2,4-dihydroxybenzoic acid by the Fenton and photo-Fenton processes: kinetics, mechanisms and evidence for the substitution of H_2O_2 by O_2. Helv Chim Acta 93:1067–1080
69. Han Y-F, Phonthammachai N, Ramesh K, Zhong Z, White T (2008) Removing organic compounds from aqueous medium via wet peroxidation by gold catalysts. Environ Sci Technol 42:908–912
70. Hansson H, Kaczala F, Marques M, Hogland W (2012) Photo-Fenton and Fenton oxidation of recalcitrant industrial wastewater using nanoscale zero-valent iron. Int J Photoenergy. Article ID 531076. http://dx.doi.org/10.1155/2012/531076
71. Hashimoto K, Irie H, Fujishima A (2005) TiO_2 photocatalysis: a historical overview and future prospects. Jpn J Appl Phys 1(44):8269–8285
72. Hatchard CG, Parker CA (1956) A new sensitive chemical actinometer. II. Potassium ferrioxalate as a standard chemical actinometer. Proc R Soc Lond A 235:518–536
73. Hennig H (1999) Homogeneous photocatalysis by transition metal complexes. Coord Chem Rev 182:101–123
74. Henze M, Harremoës P, La Cour Jansen J, Arvin E (2000) Wastewater treatment: biological and chemical processes, 3rd edn. Springer, Berlin
75. Herrera F, Kiwi J, Lopez A, Nadtochenko V (1999) Photochemical decoloration of Remazol Brilliant Blue and Uniblue A in the presence of Fe^{3+} and H_2O_2. Environ Sci Technol 33: 3145–3151
76. Herrmann JM (1999) Heterogeneous photocatalysis: fundamentals and applications to the removal of various types of aqueous pollutants. Catal Today 53:115–129

77. Hislop K, Bolton J (1999) The photochemical generation of hydroxyl radicals in the UV-vis/ferrioxalate/H_2O_2 system. Environ Sci Technol 33:3119–3126
78. Ho CH, Huang YJ, Huang YH (2010) Degradation of azo dye Reactive Black B using an immobilized iron oxide in a batch photo-fluidized bed reactor. Environ Eng Sci 27: 1043–1048
79. Hong J, Lu S, Zhang C, Qi S, Wang Y (2011) Removal of Rhodamine B under visible irradiation in the presence of Fe^0, H_2O_2, citrate and aeration at circumneutral pH. Chemosphere 84:1542–1547
80. Hsueh CL, Huan YH, Chen CY (2006) Novel activated alumina-supported iron oxide-composite as a heterogeneous catalyst for photooxidative degradation of Reactive Black 5. J Hazard Mater 129:1–3
81. Huang W, Brigante M, Wu F, Hanna K, Mailhot G (2012) Development of a new homogeneous photo-Fenton process using Fe(III)-EDDS complexes. J Photochem Photobiol A Chem 239:17–23
82. Huang Y-H, Huang Y-J, Tsai H-C, Chen H-T (2010) Degradation of phenol using low concentration of ferric ions by the photo-Fenton process. J Taiwan Inst Chem Eng 41(6): 699–704
83. Huang Y, Tsai S, Huang Y, Chen C (2007) Degradation of commercial azo dye Reactive Black B in photo/ferrioxalate system. J Hazard Mater 140:382–388
84. Ikehata K, El-Din MG (2006) Aqueous pesticide degradation by hydrogen peroxide/ultraviolet irradiation and Fenton-type advanced oxidation processes: a review. J Environ Eng Sci 5:81–135
85. Jacobsen F, Holcman J, Sehested K (1998) Reactions of the ferryl ion with some compounds found in cloud water. Int J Chem Kinet 30:215–221
86. Jeong J, Yoon J (2005) pH effect on OH radical production in photo/ferrioxalate system. Water Res 39:2893–2900
87. Kim M-K, Kong S-H (2006) Modified photo-Fenton reaction. Methyl tert-butyl ether (MTBE). J Soil Groundwater Environ 11(6):69–75
88. Klamerth N, Malato S, Agüera A, Fernandez Alba A, Mailhot G (2012) Treatment of municipal wastewater treatment plant effluents with modified photo-Fenton as a tertiary treatment for the degradation of micro pullutants and disinfection. Environ Sci Technol 46:2885–2992
89. Klamerth N, Malato S, Agüera A, Fernandez-Alba A (2013) Photo-Fenton and modified photo-Fenton at neutral pH for the treatment of emerging contaminants in wastewater treatment plant effluents: a comparison. Water Res 47(2):833–840
90. Kremer ML (1999) Mechanism of the Fenton reaction. Evidence for a new intermediate. Phys Chem Chem Phys 1:3595–3605
91. Krichevskaya M, Klauson D, Portjanskaja E, Preis P (2011) The cost evaluation of advanced oxidation processes in laboratory and pilot-scale experiments. Ozone Sci Eng 33:211–223
92. Kuhn HJ, Braslavsky SE, Schmidt R (2004) Chemical actinometry. Pure Appl Chem 76: 2105–2146
93. Kuma K, Nakabayashi S, Matsunaga K (1995) Photoreduction of Fe(III) by hydroxycarboxylic acids in seawater. Water Res 29:1559–1569
94. Kunkley H, Vogler A (2003) Photoredox reactivity of iron(III) phenolates in aqueous solution induced by ligands-to-metal charge transfer excitation. Inorg Chem Commun 6:1335–1337
95. Kusic H, Koprivanac N, Bozic N (2011) Treatment of chlorophenols in water matrix by UV/ferrioxalate system: Part I. Key process parameter evaluation by response surface methodology. Desalination 279:258–268
96. Laine D, Cheng F (2007) The destruction of organic pollutants under mild reaction conditions: a review. Microchem J 85:183–193
97. Le C, Liang J, Wu J, Li P, Wang X, Zhu N, Wu P, Yang B (2011) Effective degradation of para-chloronitrobenzene through a sequential treatment using zero-valent iron reduction and Fenton oxidation. Water Sci Technol 64:2126–2131

98. Lee Y, Jeong J, Lee C, Kim S, Yoon J (2003) Influence of various reaction parameters on 2.4-D removal in photo/ferrioxalate/H_2O_2 process. Chemosphere 51:901–912
99. Lee Y, Lee W (2010) Degradation of trichloroethylene by Fe(II) chelated with cross-linked chitosan in a modified Fenton reaction. J Hazard Mater 178(1–3):187–193
100. Lee C, Yoon J (2004) Determination of quantum yields for the photolysis of Fe(III)-hydroxo complexes in aqueous solution using a novel kinetic method. Chemosphere 57:1449–1458
101. Legrini O, Oliveros E, Braun AM (1993) Photochemical processes for water treatment. Chem Rev 93:671–698
102. Lewis S, Lynch A, Bachas L, Hamson S, Ormsbee L, Bhattacharyya D (2009) Chelate-modified Fenton reaction for the degradation of trichloroethylene in aqueous and two-phase systems. Environ Eng Sci 26(4):849–859
103. Li J (2012) 17β-estradiol degradation photoinduced by iron complex, clay and iron oxide minerals: effect of the iron complexing agent ethylenediamine-n,n′-disuccinic acid. PhD thesis. Universite Blaise Pascal, U.F.R. Sciences et Technologies. http://tel.archives-ouvertes.fr/docs/00/71/92/47/PDF/2010CLF22030.pdf
104. Litter MI (2005) Introduction to photochemical advanced oxidation processes for water treatment. In: Boule P, Bahnemann D, Robertson P (eds) The handbook of environmental chemistry, Part 2. Environmental photochemistry. Springer, Berlin
105. Liu X, Tang R, He Q, Liao X (2010) Fe(III)-loaded collagen fiber as a heterogeneous catalyst for the photo-assisted decomposition of Malachite Green. J Hazard Mater 174:687–693
106. Lopes L, de Laat J, Legube B (2002) Charge transfer of iron(III) monomeric and oligomeric aqua hydroxo complexes: semiempirical investigation into photoactivity. Inorg Chem 41: 2505–2517
107. Luna A, Chiavone-Filho O, Machulek A Jr, de Moraes J, Nascimento C (2012) Photo-Fenton oxidation of phenol and organochlorides (2,4-DCP and 2,4-D) in aqueous alkaline medium with high chloride concentration. J Environ Manage 111:10–17
108. Malato S, Blanco J, Vidal A, Alarcon D, Maldonado MI, Caceres J, Gernjak W (2003) Applied studies in solar photocatalytic detoxification: an overview. Sol Energy 75:329–336
109. Malato S, Blanco J, Vidal A, Richter C (2002) Photocatalysis with solar energy at a pilot plant scale: an overview. Appl Catal B Environ 37:1–15
110. Malato S, Fernandez-Ibañez P, Maldonado M, Blanco J, Gernjak W (2009) Decontamination and disinfection of water by solar photocatalysis: recent overview and trends. Catal Today 147:1–59
111. Maletzky P, Bauer R, Lahnsteiner J, Pouresmael B (1999) Immobilisation of iron ions on nafion® and its applicability to the photo-Fenton method. Chemosphere 38:2315–2325
112. Martire DO, Caregnato P, Furlong J, Allegreti P, Gonzalez MC (2002) Kinetic study of the reactions of oxoiron(IV) with aromatic substrates in aqueous solutions. Int J Chem Kinet 34:488–494
113. Melian-Cabrera I, Kapteijn F, Moulijn JA (2006) Tooling up heterofeneous catalysis through Fenton's chemistry. Detemplation and functionalization of micro- and mesoporous materials. In: Gaigneaux EM et al (eds) Scientific bases for the preparation of heterogeneous catalysts. Elsevier, Philadelphia
114. Miller D, Buettner R, Aust S (1990) Transition metals as catalysts of "auto-oxidation" reactions. Free Radic Biol Med 8:95–108
115. Mosteo R, Ormad P, Mozas E, Sarasa J, Ovelleiro JL (2006) Factorial experimental design of winery wastewaters treatment by heterogeneous photo-Fenton process. Water Res 40(8): 1561–1568
116. Nansheng D, Feng W, Fan L, Mei X (1998) Ferric citrate-induced photodegradation of dyes in aqueous solutions. Chemosphere 36:3101–3112
117. Nascimento CAO, Teixeira ACSC, Guardani R, Quina FH, Chiavone-Filho O, Braun AM (2007) Industrial wastewater treatment by photochemical processes based on solar energy. J Sol Energy Eng 129:45–52

118. Navalon S, Alvaro M, Garcia H (2010) Heterogeneous Fenton catalysts based on clays, silicas and zeolites. Appl Catal B Environ 99:1–26
119. Navalon S, Dhakshinamoorthy A, Alvaro M, Garcia H (2011) Heterogeneous Fenton catalysts based on activated carbon and related materials. ChemSusChem 4(12):1712–1730
120. Nichela D, Carlos L, García Einschlag F (2008) Autocatalytic oxidation of nitrobenzene using hydrogen peroxide and Fe(III). Appl Catal B Environ 82:11–18
121. Nichela D, Haddou M, Benoit-Marquié F, Maurette MT, Oliveros E, García Einschlag SF (2010) Degradation kinetics of hydroxy- and hydroxynitro derivatives of benzoic acid by Fenton-like and photo-Fenton techniques: a comparative study. Appl Catal B Environ 98:171–179
122. Nogueira KRB, Nascimento CAO, Guardani R, Teixeira ACSC (2012) Feasibility study of a solar reactor for phenol treatment by the photo-Fenton process in aqueous solution. Chem Eng Technol 35(12):2125–2132
123. Noorjahan M, Durga Kumari V, Subrahmanyam M, Panda L (2005) Immobilized Fe(III)-HY: an efficient and stable photo-Fenton catalyst. Appl Catal B Environ 57:291–298
124. Oh SY, Chiu PC, Kim BJ, Cha DK (2003) Enhancing Fenton oxidation of TNT and RDX through pretreatment with zero-valent iron. Water Res 37(17):4275–4283
125. Oliveros E, Goeb S, Bossmann SH, Braun AM, Nascimento CAO, Guardani R (2000) Waste water treatment by the photochemical enhanced Fenton reaction: modeling and optimization using experimental design and artificial neural networks. In: Hu X, Yue PL (eds) Sustainable energy and environmental technology. Proceedings of 3rd Asia Pacific conference. World Scientific, Singapore
126. Oliveros E, Legrini O, Hohl M, Müller T, Braun AM (1997) Industrial waste water treatment: large scale development of a light-enhanced Fenton reaction. Chem Eng Proc 36:397–405
127. Oller I, Malato S, Sanchez-Perez JA, Gernjak W, Maldonado MI, Perez-Estrada LA, Pulgarin C (2007) A combined solar photocatalytic-biological field system for the mineralization of an industrial pollutant at pilot scale. Catal Today 122:150–159
128. Ollis DF, Al-Ekabi H (eds) (1993) Photocatalytic purification and treatment of water and air. Elsevier, Amsterdam
129. Orozco S, Bandala E, Arancibia-Bulnes C, Serrano B, Suarez-Parra R, Hernandez-Perez I (2008) Effect of iron salt on the color removal of water containing the azo-dye Reactive Blue 69 using photo-assisted Fe(II)/H_2O_2 and Fe(III)/H_2O_2 systems. J Photochem Photobiol A Chem 198:144–149
130. Ortega-Gomez E, Esteban García B, Ballesteros Martín MM, Fernández Ibáñez P, Sánchez Pérez JA (2013) Inactivation of Enterococcus faecalis in simulated wastewater treatment plant effluent by solar photo-Fenton at initial neutral pH. Catal Today 209:195–200
131. Oturan N, Sirés I, Oturan MA, Brillas E (2009) Degradation of pesticides in aqueous medium by electro-Fenton and related methods. A review. J Environ Eng Manage 19:235–255
132. Ou X, Quan X, Chen S, Zhang F, Zhao Y (2008) Photocatalytic reaction by Fe(III)-citrate complex and its effect on the photodegradation of atrazine in aqueous solution. J Photochem Photobiol A Chem 197:382–388
133. Parker CA (1954) Induced autoxidation of oxalate in relation to the photolysis of potassium ferrioxalate. Trans Faraday Soc 50:1213–1221
134. Pelizzetti E, Carlin V, Minero C, Pramauro RM, Vincenti M (1992) Degradation pathways of atrazine under solar light and in the presence of TiO_2 colloïdal particles. Sci Total Environ 123(124):161–169
135. Pereira MC, Oliveira LCA, Murad E (2012) Iron oxide catalysts: Fenton and Fenton-like reactions – a review. Clay Miner 47(3):285–302
136. Perez M, Torrades F, Garcia-Hortal JA, Domenech X, Perel J (2002) Removal of organic contaminants in paper pulp treatment effluents under Fenton and photo-Fenton conditions. Appl Catal B Environ 36:63–74

137. Perez-Moya M, Graells M, Buenestado P, Mansilla HD (2008) A comparative study on the empirical modeling of photo-treatment process performance. Appl Cat B Environ 84(1–2): 313–323
138. Pignatello J (1992) Dark and photoassisted Fe^{3+}-catalyzed degradation of chlorophenoxy herbicides by hydrogen peroxide. Environ Sci Technol 26:944–951
139. Pignatello JJ, Liu D, Huston P (1999) Evidence for an additional oxidant in the photoassisted Fenton reaction. Environ Sci Technol 33:1832–1839
140. Pignatello J, Oliveros E, Mac Kay A (2006) Advanced oxidation processes for organic contaminant destruction based on the Fenton reaction and related chemistry. Crit Rev Environ Sci Technol 36:1–84; (2007) Erratum 37:273–275
141. Polo-Lopez MI, Oller I, Fernández-Ibáñez P (2013) Benefits of photo-Fenton at low concentrations for solar disinfection of distilled water. A case study: Phytophthora capsici. Catal Today 209:181–187
142. Pozdnyakov I, Kel O, Plyusnin V, Grivin V, Bazhin N (2008) New insight into photochemistry of ferrioxalate. J Phys Chem A 112:8316–8322
143. Pozdnyakov I, Plyusnin V, Grivin V, Vorobyev D, Bazhin N, Pages S, Vauthey E (2006) Photochemistry of Fe(III) and sulfosalicylic acid aqueous solutions. J Photochem Photobiol A Chem 182:75–81
144. Pozdnyakov I, Plyusnin V, Tkachenko N, Lemmetyinen H (2007) Photophysics of Fe(III)–sulfosalicylic acid complexes in aqueous solutions. Chem Phys Lett 445:203–207
145. Prato-Garcia D, Bultrón G (2012) Evaluation of three reagent dosing strategies in a photo-Fenton process for the decolorization of azo dye mixtures. J Hazard Mater 217–218:293–300
146. Prato-Garcia D, Vazquez Medrano R, Hernandez Exparza M (2009) Solar photoassisted advanced oxidation of synthetic phenolic wastewaters using ferrioxalate complexes. Sol Energy 83:306–315
147. Prieto-Rodriguez L, Oller I, Zapata A, Agueera A, Malato S (2011) Hydrogen peroxide automatic dosing based on dissolved oxygen concentration during solar photo-Fenton. Catal Today 161(1):247–254
148. Prieto-Rodríguez L, Spasiano D, Oller I, Fernández-Calderero I, Agüera A, Malato S (2013) Solar photo-Fenton optimization for the treatment of MWTP effluents containing emerging contaminants. Catal Today 209:188–194
149. Prousek J, Palackova E, Priesolova S, Markova L, Alevova A (2007) Fenton- and Fenton-like AOPs for wastewater treatment: from laboratory-to-plant-scale application. Sep Sci Technol 42(7):1505–1520
150. Quici N, Litter MI, Braun AM, Oliveros E (2008) Vacuum-UV photolysis of aqueous solutions of citric and gallic acid. J Photochem Photobiol A Chem 197:306–312
151. Rabl A (1976) Comparison of solar concentrators. Sol Energy 18:93–111
152. Rios-Enriquez M, Shahin N, Durán-de-Bazúa NC, Lang J, Oliveros E, Bossmann SH, Braun AM (2004) Optimization of the heterogeneous Fenton-oxidation of the model pollutant 2,4-xylidine using the optimal experimental design methodology. Sol Energy 77:491–501
153. Ruppert G, Bauer R, Heisler G (1993) The photo-Fenton reaction – an effective photochemical wastewater treatment process. J Photochem Photobiol A Chem 73:75–78
154. Safarzadeh-Amiri A, Bolton JR, Cater SR (1996) The use of iron in advanced oxidation technologies. J Adv Oxid Technol 1:18–26
155. Sanchez Perez JA, Roman Sanchez IM, Carra I, Cabrera Reina A, Casas Lopez JL, Malato S (2013) Economic evaluation of a combined photo-Fenton/MBR process using pesticides as model pollutant. Factors affecting costs. J Hazard Mater 244–245:195–203
156. Seol Y, Javandel I (2008) Citric acid-modified Fenton's reaction for the oxidation of chlorinated ethylenes in soil solution systems. Chemosphere 72(4):537–542, http://www.osti.gov/bridge/servlets/purl/937444-TriKoa/937444.pdf
157. Silva M, Trovo A, Nogueira R (2007) Degradation of the herbicide tebuthiuron using solar photo-Fenton process and ferric citrate complex at circumneutral pH. J Photochem Photobiol A Chem 191:187–192

158. Silva M, Vilegas W, Zanoni M, Pupo Nogueira R (2010) Photo-Fenton degradation of the herbicide tebuthiuron under solar irradiation: Iron complexation and initial intermediates. Water Res 44:3745–3753
159. Sima J (2001) Mechanism of photoredox reactions of iron(III) complexes containing Salen-type ligands. Croat Chem Acta 74:593–600
160. Sima J, Makanova J (1997) Photochemistry of iron(III) complexes. Coord Chem Rev 160: 161–189
161. Simunovic M, Kusic H, Koprivanac N, Bozic AL (2011) Treatment of simulated industrial wastewater by photo-Fenton process: Part II. The development of mechanistic model. Chem Eng J 173(2):280–289
162. Singh SN, Saw P, Kumar R (2012) Holography: new breakthrough in solar power conversion technology. Might be used for small flatbed reactors. Int J Eng Sci Technol 4:2485–2492
163. Sirtori C, Zapata A, Malato S, Agueera A (2012) Formation of chlorinated by-products during photo-Fenton degradation of pyrimethanil under saline conditions. Influence on toxicity and biodegradability. J Hazard Mater 217–218:217–223
164. Sulaiman F, Abdullah N, Singh BSM (2012) Comparison of solar concentrators. World Acad Sci Eng Technol 72:99–103
165. Sun C, Chen C, Ma W, Zhao J (2011) Photodegradation of organic pollutants catalyzed by iron species under visible light irradiation. Phys Chem Chem Phys 13(6):1957–1969
166. Sun Y, Pignatello JJ (1993) Photochemical reactions involved in the total mineralization of 2,4-D by iron(3+)/hydrogen peroxide/UV. Environ Sci Technol 27:304–310
167. Suty H, De Traversay C, Cost M (2004) Applications of advanced oxidation processes: present and future. Water Sci Technol 49:227–233
168. Sychev AY, Isaak VG (1995) Iron compounds and mechanisms of the homogeneous catalysis of the activation of O_2 and H_2O_2 and of the oxidation of organic substrates. Russ Chem Rev 64(12):1105–1129 (translated from Uspekhi Khimii 64(12):1183–1208)
169. Sylva RN (1972) The hydrolysis of iron(III). Pure Appl Chem 22:115–130
170. Tarr AM (2003) Chemical degradation methods for wastes and pollutants. Environ Sci Pollut Control Ser 26:165–200
171. Teel AL, Warberg CR, Atkinson DA, Watts RJ (2001) Comparison of mineral and soluble iron Fenton's catalysts for the treatment of trichloroethylene. Water Res 35:977–984
172. Tokumura M, Morito R, Hatayama R, Kawase Y (2011) Iron redox cycling in hydroxyl radical generation during the photo-Fenton oxidative degradation. Dynamic change of hydroxyl radical concentration. Appl Catal B Environ 106:565–576
173. Utset B, Garcia J, Casado J, Domenech X, Peral J (2000) Replacement of H_2O_2 by O_2 in Fenton and photo-Fenton reactions. Chemosphere 41:1187–1189
174. Vergara-Sanchez J, Perez-Orozco JP, Suárez-Parra R, Hernández-Pérez I (2012) Degradation of Reactive Red 120 azo dye in aqueous solution using homogeneous/heterogeneous iron systems. Rev Mex Ing Quim 11:121–131
175. Von Sonntag C, Schuchmann HP (1997) Peroxyl radicals in aqueous solutions. In: Alfassi ZB (ed) Peroxyl radicals. Wiley, New York
176. Waite TD (2002) Challenges and opportunities in the use of iron in water and wastewater treatment. Rev Environ Sci Biotechnol 1:9–15
177. Walling C (1975) Fenton's reagent revisited. Acc Chem Res 8:125–131
178. Walling C (1998) Intermediates in the reactions of Fenton type reagents. Acc Chem Res 31:155–157
179. Walling C, Goosen A (1973) Mechanism of the ferric ion catalyzed decomposition of hydrogen peroxide. Effect of organic substrates. J Am Chem Soc 95:2987–2991
180. Wang Z, Chen X, Ji H, Ma W, Chen C, Zhao J (2010) Photochemical cycling of iron mediated by dicarboxylates: special effect of malonate. Environ Sci Technol 44:263–268
181. Wang C, Liu H, Sun Z (2012) Heterogeneous photo-Fenton reaction catalyzed by nonsized iron oxides for water treatment. Int J Photoenergy. Article ID 801694 http://dx.doi.org/10.1155/2012/801694. Accessed 30 April 2013

182. Wang JL, Xu LJ (2012) Advanced oxidation processes for wastewater treatment: formation of hydroxyl radical and application. Crit Rev Environ Technol 42:251–325
183. Wang L, Zhang C, Mestankova H, Wu F, Deng N, Pan G, Bolte M, Mailhot G (2009) Photoinduced degradation of 2,4-dichlorophenol in water: influence of various Fe(III) carboxylates. Photochem Photobiol Sci 8:1059–1065
184. Wardman P (1989) Reduction potentials of one-electron couples involving free radicals in aqueous solution. J Phys Chem Ref Data 18:1637–1755
185. Wu K, Xie Y, Zhao J, Hidaka H (1999) Photo-Fenton degradation of a dye under visible light irradiation. J Mol Catal A Chem 144:77–84
186. Yamal-Turbay E, Graells M, Pérez-Moya M (2012) Systematic assessment of the influence of hydrogen peroxide dosage on caffeine degradation by the photo-Fenton process. Ind Eng Chem Res 51(13):4770–4778
187. Zahorodna M, Oliveros E, Wörner M, Bogoczek R, Braun AM (2008) Dissolution and mineralization of ion exchange resins: differentiation between heterogeneous and homogeneous (photo-)Fenton processes. Photochem Photobiol Sci 7:1480–1492
188. Zepp R (1992) Hydroxyl radical formation in aqueous reactions (ph 3-8) on iron(II) with hydrogen peroxide: the photo-Fenton reaction. Environ Sci Technol 26:313–319
189. Zuo Y (1995) Kinetics of photochemical/chemical cycling of iron coupled with organic substances in cloud and fog droplets. Geochim Cosmochim Acta 59:3123–3130
190. Zuo Y, Hoigné J (1992) Formation of hydrogen peroxide and depletion of oxalic acid in atmospheric water by photolysis of iron(III)-oxalato complexes. Environ Sci Technol 26: 1014–1022
191. Zuo Y, Hoigné J (1994) Photochemical decomposition of oxalic, glyoxalic and pyruvic acid catalyzed by iron in atmospheric waters. Atmos Environ 28:1231–1239

Index

D.W. Bahnemann and P.K.J. Robertson (eds.), *Environmental Photochemistry Part III*, Hdb Env Chem (2015) 35: 343–346, DOI 10.1007/698_2015,

Printed by Printforce, the Netherlands